EARTHQUAKE ENGINEERING AND SOIL DYNAMICS II—

RECENT ADVANCES IN GROUND-MOTION EVALUATION

Proceedings of the specialty conference
sponsored by the Geotechnical Engineering Division
of the American Society of Civil Engineers

in cooperation with the
Association of Engineering Geologists
Earthquake Engineering Research Institute
Seismological Society of America
Utah Section of ASCE
Structural Engineers Association of Utah

June 27-30, 1988
Park City, Utah

Edited by J. Lawrence Von Thun

Geotechnical Special
Publication No. 20

Published by the
American Society of Civil Engineers
345 East 47th Street
New York, New York 10017-2398

ABSTRACT

Ground-motion determination for analysis and design of engineering structures is a shared, interdisciplinary task involving geologists, seismologists, geophysicists, geotechnical engineers, and earthquake engineers. Considerations in developing ground motions representing all of these disciplines are presented beginning with a current assessment of geologic and seismic hazard on an overall or regional scale and continuing with discussions on: local site effects measurement of strong ground motion, seismologic characterization of ground motion; determination of laboratory and in-situ properties to enhance analysis of ground-motion transmission; practical development of ground motion for structural analysis; and ending with an analysis of the effects of strong ground motion and large deformation.

Library of Congress Cataloging-in-Publication Data

Earthquake engineering and soil dynamics II: recent advances in ground-motion evaluation: proceedings of the specialty conference/sponsored by the Geotechnical Engineering Division of the American Society of Civil Engineers, in cooperation with the Association of Engineering Geologists . . . [et al.] June 27-30, 1988, Park City, Utah; edited by J. Lawrence Von Thun.

    p. cm.
Includes index.
ISBN 0-87262-664-4

    1. Earthquake engineering—Congresses. 2. Soil dynamics—Congresses. I. Von Thun. J. Lawrence. II. American Society of Civil Engineers. Geotechnical Engineering Division. III. Association of Engineering Geologists.
TA654.6.E365 1988
624.1'762—dc19   88-18236
                                            CIP

PREFACE

This publication contains the papers prepared for presentation, review, and discussion at the ASCE Geotechnical Division Specialty Conference held in Park City, Utah, from June 27 through June 30, 1988. The conference, titled Earthquake Engineering and Soil Dynamics II — Recent Advances in Ground Motion Evaluation, focused on all aspects of earthquake ground-motion specification for design and analysis of engineering structures. The topics spanned from regional geologic consideration, through transmission of motions from bed rock to the ground surface, to idealization of ground motions for design.

The publication contains five invited state-of-the-art papers covering five different aspects of ground-motion determination and 23 papers submitted by engineers, geophysicists, and seismologists through a call for papers.

It is the practice of the Geotechnical Engineering Division that each paper published in a Conference Proceedings undergo a peer review before being accepted. The standards for the peer review are essentially the same as those for papers being reviewed for possible publication in the ASCE *Journal of Geotechnical Engineering*. Each paper must receive two positive reviews to be accepted, and must be revised to conform to the mandatory revisions of the reviewers. But, because there is a tight schedule from receipt of papers through review and on to publication, there is not time for more than one cycle of editing and revising for Conference papers. All papers published in this volume are eligible for discussion in the ASCE Journal of Geotechnical Engineering and for ASCE awards.

One of the primary goals of the conference and of this publication was to bring together the various considerations of seismologists, geophysicists, and engineers in portraying ground motions. Hopefully, through this effort, engineers will be able to improve their understanding of earthquake source and ground-motion transmission processes, and seismologists and geophysicists will be able to better understand the engineer's methods and needs in developing representative ground motions for design and analysis.

J. Lawrence Von Thun
Editor

ACKNOWLEDGMENTS

ORGANIZING COMMITTEE

Loren Anderson
Roger Borcherdt
Ricardo Dobry
Liam Finn
Paul Jennings
Richard Woods
Les Youd - *Chairman*

LOCAL ARRANGEMENTS COMMITTEE

Loren Anderson - *Chairman*
Brent Bingham
Jerald Bishop
James Higbee
Walter Jones
Jeffrey Keaton - *Field Trip*
Bill Leeflang
James Nordquist
Stanley Plaisier
Kyle Rollins

CONTENTS

PART I – GEOLOGIC AND SEISMOLOGIC CONSIDERATIONS IN GROUND-MOTION EVALUATION

PART II – CHARACTERIZATION OF SOIL PROPERTIES AND SITE CHARACTERISTICS FOR TRANSMISSION OF GROUND MOTIONS

---

*State-of-the-Art Report

viii

GEOLOGIC CHARACTERIZATION OF SEISMIC SOURCES: MOVING INTO THE 1990s

DAVID P. SCHWARTZ[*]

INTRODUCTION

As a geologist who spends much of his field time in 3- to 4-m-deep trenches excavated across major active faults like the San Andreas and Wasatch, questions of when?, where?, and how large? are never far from mind. The ability to answer these, whether estimating the time of the next earthquake, a maximum earthquake magnitude, the amount of potential surface displacement on an active fault, or the probability of exceeding a particular level of ground motion, rests on our ability to correctly characterize a seismic source. Seismic source characterization is the quantification of the size(s) of earthquakes that a fault can produce and the distribution of these earthquakes in space and time. As such, source characterization provides the basis for evaluating the long-term seismic potential at particular sites of interest.

In the late 1960s and early 1970s, largely in response to expansion of nuclear power plant siting and the issuance of a code of federal regulations by the Nuclear Regulatory Commission referred to as Appendix A, 10CFR100, the need to characterize the earthquake potential of individual faults for seismic design took on greater importance. Appendix A established deterministic procedures for assessing the seismic hazard at nuclear power plant sites. Bonilla and Buchanan (1970), using data from historical surface-faulting earthquakes, developed a set of statistical correlations relating earthquake magnitude to surface rupture length and to surface displacement. These relationships, which have been refined and updated (Slemmons, 1977; Bonilla et al., 1984), along with the relationship between fault area and magnitude (Wyss, 1979), and seismic moment and moment magnitude (Hanks and Kanamori, 1979), have served as the basis for selecting maximum earthquakes in a wide variety of design situations (Schwartz et al., 1984). A related concept that developed at about the same time and that has also seen widespread use is the idea that a seismic source can produce two types of earthquakes, a "maximum credible" event or simply a "maximum" earthquake, which is the largest conceivable, and a "maximum probable" event, which is smaller and more frequent.

It is clear that the correlations between earthquake magnitude and fault parameters can provide reasonable estimates of the magnitude or surface displacement associated with future earthquakes on a fault when appropriate input values are used. However, in applying these correlations to actual siting situations, there is often much uncertainty, and there has frequently been great controversy. Perhaps no better example can be found than the diversity of conclusions regarding the seismic design parameters for the proposed Auburn Dam on the American River east of Sacramento, California. Reports on these were issued by the U.S. Bureau of Reclamation, the U.S. Geological Survey, Woodward-Clyde Consultants, and five additional independent consultants to the Bureau of Reclamation. Estimates of the magnitude of the maximum earthquake on a fault in the vicinity of the dam ranged from 6.0 to 7.0; the closest approach of the source of the maximum earthquake ranged from less than 0.8 km to 8 km; estimates of the focal depth of the maximum event varied from 5 km to 10 km; the amount of the surface displacement expected during the maximum event varied from 25 cm to 3 m; and estimates of the recurrence interval of the maximum earthquake ranged from 10,000 to 85,000 years. Characteristics of expectable faulting within the dam foundation similarly had a wide range of estimated values: the maximum earthquake was 5.0 to 7.0; displacement per event was less than 2.5 cm to 1 m; and the recurrence interval of an event in the foundation was 260,000 to about 1 million years. This variability reflects, to a large degree, the differences in perception among the various consultants or groups regarding both the physical basis for quantifying

[*]U.S. Geological Survey, 345 Middlefield Road MS/977, Menlo Park, CA 94025

1

a particular fault parameter and the general understanding of fault behavior.

During the past ten years the integration of geological, seismological, and geophysical information has led to a much better, though still far from complete, understanding of the relationships between faults and earthquakes in space and time. Geological studies, especially trenching and geomorphic analysis, mapping of coseismic surface faulting and secondary deformation from historical earthquakes, and investigations of fault zone structure in both unconsolidated sediments and bedrock have led to some of the most exciting and important contributions to the understanding of earthquake behavior (Hanks, 1985; Allen, 1986; Schwartz, 1987; Crone, 1987). Such investigations are now referred to as paleoseismology (Wallace, 1981), seismic geology, and earthquake geology. They have demonstrated that individual past large-magnitude earthquakes can be recognized in the geologic record and that the timing between events can be measured. Additionally, they have yielded information on fault slip rate, the amount of displacement during individual events, and the elapsed time since the most recent event. These data can be used in a number of different ways and have led to the development of new approaches to quantifying seismic hazard.

The objective of the present paper is to discuss leading-edge directions in paleoseismology and seismic geology, particularly as they relate to characterizing seismic sources. The paper builds on earlier articles that discuss some of these trends (Schwartz and Coppersmith, 1986; Schwartz, 1987). There are several areas that appear to be especially important as we move into the 1990s. These are: fault segmentation, which provides a physical framework for evaluating both the size and potential location of future earthquakes on a fault zone; earthquake recurrence models, which provide information on the frequency of different size earthquakes on a fault; and long-term earthquake potential, an area in which significant advances have been made through development of earthquake hazard models that use probabilistic methodology to incorporate the uncertainties in seismic source characterization and the evolving understanding of the earthquake process.

THE GEOLOGIC DATA BASE

Figure 1 is a schematic diagram showing the types of geologic data that can be obtained for individual faults, and the applications of each to the evaluation of seismic hazard.

Slip Rate

Slip rate is the net tectonic displacement on a fault during a measurable period of time (Figure 2). In recent years a great deal of emphasis has been placed on obtaining slip-rate data, and published rates are available for many faults. For example, Clark et al. (1984) have produced a major compilation of 160 slip rates for 81 faults in California. An important aspect of the compilation is the rating of individual slip rates based on reliability of both the measured displacement and the age estimate of the displaced datum. Slip rates are an expression of the long-term, or average, activity of a fault. In a general way, they can be used as an index to compare the relative activity of faults. Slip rates are not necessarily a direct expression of earthquake potential. While faults with high slip rates generally generate large-magnitude earthquakes, those with low rates may do the same, but with longer periods of time between events. Slip rates reflect the rate of strain energy release on a fault, which can be expressed as seismic moment. Because of this they are now being used to estimate earthquake recurrence on individual faults, especially in probabilistic seismic hazard analyses.

Recurrence Intervals

A recurrence interval is the time period between successive geologically recognizable earthquakes. The excavation of trenches across faults has proven to be a tremendously successful technique for exposing stratigraphic and structural evidence of past individual earthquakes in the geologic record (Figure 3). The recognition of geomorphic features such as tectonic terraces and individual

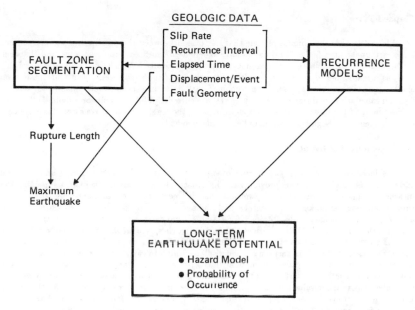

GEOLOGIC DATA

Slip Rate
Recurrence Interval
Elapsed Time
Displacement/Event
Fault Geometry

FAULT ZONE
SEGMENTATION

RECURRENCE
MODELS

Rupture Length

Maximum
Earthquake

LONG-TERM
EARTHQUAKE POTENTIAL
● Hazard Model
● Probability of
Occurrence

FIGURE 1. Relationship between geologic data and aspects of fault behavior and seismic hazard evaluation.

stream offsets, morphometric analysis of fault scarps, and evidence of past liquefaction also provide direct information on the number of past events for many faults. Where datable material is found, the actual intervals between successive events can be determined, although in many cases only average recurrence intervals can be estimated. Data on recurrence intervals can be combined with information on displacement during each event to develop fault-specific recurrence models.

L

DISPLACED PROVO TERRACE

13,500 yr B.P.

13,500 yr B.P. L'

CNVTD
11.5-13.5 m

SH
28.5 m

40

METERS

0

FIGURE 2. Example of slip rate calculation for a normal fault. The displaced datum is a 13,500-year-old terrace surface displaced by the Wasatch fault. The stippled band is the projection of the surface across the fault zone. The fault scarp at this location is 28.5 m high but the cumulative net vertical tectonic displacement (CNVTD), which is the displacement used for the slip rate, is 11.5-13.5 m. This yields a slip rate of 0.85-1.0 mm/yr. for the past 13,500 years. Similar types of measurement can be made for strike-slip and reverse faults. Vertical scale = horizontal scale. (Modified from Swan *et al.*, 1980).

Elapsed Time

Elapsed time is the amount of time that has passed between the present and the most recent large earthquake on a fault. Many faults have experienced repeated late Pleistocene and Holocene surface-faulting earthquakes but have not ruptured historically. With trenching and geomorphic analysis it is possible to identify and estimate the timing of the most recent event. Information on elapsed time is desirable because, when combined with data on recurrence intervals, it provides the basis for calculating conditional probabilities of the occurrence of future events on a fault. Differences in the timing of the most recent surface rupture along the length of a fault zone are also extremely useful in identifying segments that may behave independently.

Displacement Per Event

Displacement per event is the amount of slip that occurs at the surface during an individual earthquake. Geologic studies are providing this information for past earthquakes and assume that the measured displacement occurred coseismically; that is, most occurred simultaneously with seismic rupture, although an unknown percentage could be associated with post-seismic adjustment called afterslip. Displacements may be obtained, for example, from measurements of displaced stratigraphic horizons, the thickness of colluvial wedges observed in trenches, stream offsets, the heights of tectonic terraces on the upthrown side of faults, and inflections in fault scarp profiles. Displacement reflects the energy associated with an earthquake and displacement data can be used as input for calculating maximum earthquakes. Because the amount of coseismic slip generally varies in some systematic way along the length of a surface rupture, care must be taken to evaluate the degree to which a particular displacement value reflects a minimum, maximum, or average displacement for that event. Displacement per event data for repeated earthquakes at a point on a fault, coupled with the timing of the events, provide a basis for formulating recurrence models.

Fault Geometry

The geometry of a fault is defined by its surface orientation, its dip, and its down-dip extent. For many faults, and particularly dip-slip faults, changes in the strike of the fault at the surface, especially when coupled with major changes in lithology, may aid in assessing the location of fault segment boundaries. For strike-slip faults dips are generally vertical or very high angle, but for dip-slip faults the dip at depth may vary considerably from the surface dip. Some normal faults may decrease in dip with depth (become listric), whereas seismogenic thrust or reverse faults often steepen with depth. Seismic reflection data and seismicity data such as focal mechanisms can provide constraints on dip. The thickness of the seismogenic or brittle crust in a region determined from the depth distribution of seismicity also places constraints on the down-dip extent of the part of a fault that exhibits brittle behavior. Fault dip and down-dip seismogenic extent define fault width. Fault length and fault width are the key parameters for quantifying the fault area that is used to estimate magnitude and seismic moment.

Dating Past Earthquakes

Though individual past earthquakes can be recognized in the geologic record, the actual dating of an event can be difficult. This is especially true throughout large parts of the western United States, especially in the Great Basin, where charcoal and other datable organic material are not commonly found in faulted deposits. To overcome this, important advances have been made in the dating of Quaternary deposits (Pierce, 1986). One very recent technique is morphologic dating of fault scarps by modeling fault-scarp degradation with diffusion-equation model mathematics (Hanks et al., 1984; Nash, 1980, 1984; Hanks and Wallace, 1985). At present, this approach appears reliable for distinguishing between early, middle, and late Holocene single-event scarps. Factors such as scarp height, microclimate, vegetation, material properties, and scarp orientation may control rates of scarp degradation (Mayer, 1984; Pierce and Colman, 1986). As they are more systematically

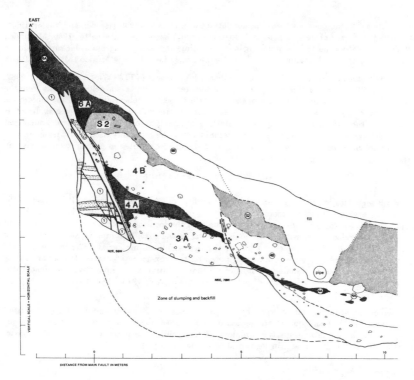

FIGURE 3. Detail of trench log across the Wasatch fault near Kaysville, Utah. Colluvial units 3A, 4A/4B/S2, and 6A are three distinct deposits, each of which represents erosion of the fault scarp following a surface-faulting earthquake. This was the first trench across the Wasatch fault excavated specifically to define earthquake recurrence. (From Swan *et al.*, 1980.)

studied and incorporated into modeling, it may be possible to distinguish between events that are more closely spaced (1000–2000 years or less) in time.

An important advance in radiocarbon dating has been the development of accelerator mass spectrometry (Donahue *et al.*, 1983; Kutschera, 1983). This technique can provide dates on samples as small as 2 to 5 mg and should lead to the dating of many deposits that contain charcoal in amounts too small for conventional β-counting. New and revised dendrologically-corrected calibration curves for converting radiocarbon ages to calendar years have been published for periods back to about 10,000 years (Stuiver and Pearson, 1986; Pearson and Stuiver, 1986; Kromer *et al.*, 1986). These dendro corrections must be made when using radiocarbon dates to calculate recurrence intervals and slip rates. Also, very precise radiocarbon dates with a precision of less than 20 years, as compared to the standard 50 to 100 years, can now be obtained using large amounts of sample, long counting times, and special counters.

Estimates of the age of faulted surfaces are also benefiting from increased understanding of soil development and soil chemistry. Soils tend to become better developed with increasing age and area-specific soil chronosequences are being widely recognized (Birkeland, 1985). Harden (1982) described a soil profile development index that quantifies relative strength of soil profile development using changes in soil chemistry, texture, and soil-horizon thickness. These provide a basis for estimating soil ages within a chronosequence where some absolute age control is available (Harden and Taylor, 1983). Similarly, the degree of development of pedogenic calcium carbonate (Machette, 1985; McFadden and Tinsley, 1985) seems to be a promising index for estimating the age of geomorphic surfaces; it has been used to estimate the timing of events on the La Jencia fault in New Mexico (Machette, 1986).

Another dating technique with promise is thermoluminescence (TL) (Wintle and Huntley, 1982; Forman, 1987). Thermoluminescent properties of minerals like quartz and feldspar result from the effects of ionizing radiation. TL can be measured when these minerals are heated in the laboratory. Exposure to sunlight has the same effect as heating, and zeroes the thermoluminescence. It is this zeroing and the subsequent buildup of TL starting at the time of burial of mineral grains that provide the basis for dating sediments. This technique may be especially useful for dating deposits that contain sparse organic material such as loess that can be deposited in graben or on alluvial fan surfaces, or possibly scarp-derived colluvial deposits.

FAULT ZONE SEGMENTATION

Fault segmentation is emerging as an important field of earthquake research. It is based on the common observation that fault zones, especially long ones, do not rupture along their entire length during a single earthquake. Studies are suggesting that the location of rupture is not random, that there are physical controls in the fault zone that define the extent of rupture and divide a fault into segments, and that segments can persist through many seismic cycles. The recognition and identification of rupture segments have the potential to provide new insights into characterizing seismic sources and understanding controls of rupture initiation and termination. Inherent in the concept of segmentation is the idea of persistent barriers (Aki, 1979, 1984) that control rupture propagation.

Geological, seismological, and geophysical observations from recent and ongoing paleoseismic studies and historical earthquakes are discussed below. These examples present the basic concepts of segmentation and also discuss the complexities, variability, and uncertainties in fault behavior that must be taken into account in segmentation modeling.

Examples of Segmentation

Wasatch Fault Zone, Utah. The Wasatch is a 370-km-long normal-slip fault that has not had a historical surface-faulting earthquake but has produced large-magnitude scarp-forming earthquakes throughout the Holocene. Based on historical surface ruptures on normal faults in the Great Basin, which have ranged in length from about 25 to 65 km, only a part of the Wasatch fault zone will be expected to rupture in a future earthquake, and with a length comparable to the Great Basin historical events. A segmentation model for the Wasatch fault zone (Schwartz and Coppersmith, 1984; Bruhn et al., 1987; Machette et al., 1987; Wheeler and Krystinik, 1987) is shown on Figure 4. Each Wasatch segment has been defined using a variety of observations including surface fault geometry and structure, differences in slip rate, timing of the most recent earthquake and prior events, gravity data, and geodetic data. From north to south, the length and orientation of the segments are: (1) Collinston segment, 25 km, N20°W; (2) Brigham City segment, 40 km, N10°W; (3) Weber segment, 50 km, N10°W; (4) Salt Lake City segment, 35 km, convex east N20°E to N30°W; (5) American Fork segment, 22 km, N25°W; (6) Provo segment, 30 km, N25°W to N25°E; (7) Nephi segment, 35 km, N11°E; and (8) Levan segment, 35 km, convex west.

The proposed segment boundaries may represent structurally complex transition zones ranging from a few to more than ten kilometers across. To varying degrees, boundaries selected on the basis of paleoseismic and geomorphic observations are coincident with changes in the surface trend of the fault zone; major salients in the range front; intersecting east-west or northeast structural trends observed in the bedrock geology of the Wasatch Range; gaps in Holocene surface faulting; cross faults and transverse structural trends interpreted from gravity data (Zoback, 1983); and geodetic changes (Snay et al., 1984). Smith and Bruhn (1984) showed a strong spatial correlation between segment boundaries and the margins of major thrust faults of Late Jurassic to Early Tertiary age.

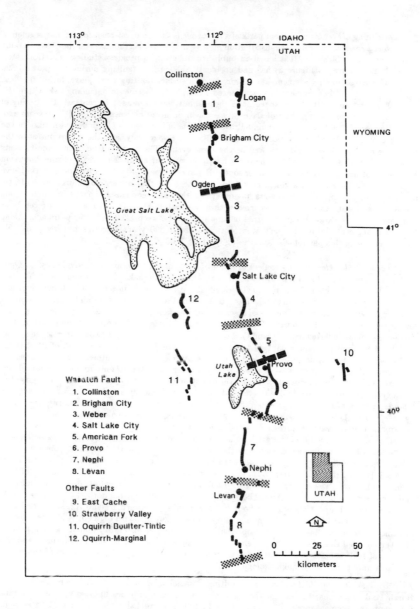

FIGURE 4. Segmentation model for the Wasatch fault zone, Utah. Stippled bands define segment boundaries identified by Schwartz and Coppersmith (1984); dashed bands are additional boundaries interpreted by Machette *et al.* (1987).

The timing of events on adjacent parts of a fault zone is the strongest basis for defining segments. Figure 5 is a space-time plot of the distribution of large magnitude earthquakes on the Wasatch fault during the past 6000 years. It is based on published data and in-progress studies. The Collinston segment (not shown on Figure 5) has had no identifiable surface faulting during the past 13,500 years. The Brigham City segment has not produced a surface-faulting earthquake during the past 3400 years. The Weber segment has experienced multiple displacements, including two within the past 1200 years and with the most recent of these about 500 years ago. In contrast, the timing of the most recent event along the Salt Lake City segment occurred between 1100 and 1900 years ago. The behavior of the American Fork and Provo segments is less clear. The preferred timing of the most recent event is extremely close, 500 and 600 years, respectively, and the uncertainties in the age dates overlap; these could very well be the same event. The same is true of the oldest events about 5300 to 5500 years ago. On the other hand, an event at about 2400 years on the American Fork segment has no correlatives north or south. Based on these observations American Fork–Provo could behave as a) a master segment containing a subsegment that occasionally fails independently or b) two truly independent segments having events that are very closely spaced in time. Along the Nephi segment one event has occurred within the past 1100 years and possibly as recently as 300 years ago; two earlier events occurred on this segment between 3800 and 5200 years ago. Along the Levan segment the most recent event occurred about 800 years ago; this has been the only earthquake along this part of the fault during the past 7500 years.

Oued Fodda Fault, Algeria. An excellent example of fault zone segmentation is provided by the Oued Fodda fault, which produced the M_s 7.3 El Asnam, Algeria earthquake of 10 October, 1980. Yielding *et al.* (1981) and King and Yielding (1984) described this earthquake in terms of fault geometry and rupture propagation and termination. Basic features of the surface rupture and segmentation are shown on Figure 6. Thirty kilometers of coseismic surface faulting occurred on a northeast trending thrust fault with secondary normal faulting on the upper plate. This rupture is composed of three distinct segments, referred as A, B, and C. The southern segment contains two smaller segments, A1 and A2. Local and teleseismic data showed that the earthquake occurred at a depth of 10 to 15 km and was a complex rupture event. The main shock nucleated at the southwest end of segment A and propagated 12 km northeast where a second rupture of equal seismic moment occurred and ruptured 12 km further northeast; a smaller third rupture occurred and propagated along segment C. Geologically, coseismic surface displacement during the 1980 earthquake decreases at each segment boundary, the strikes of the segments differ, and there is a gap in the main thrust rupture and an *en echelon* step between the southern and central segments. There are also differences in long-term deformation along each as expressed by the degree of development of folds on the hanging wall of the thrust. A well-developed anticline with an amplitude of more than 200 m occurs along segment B, amplitude of the anticline decreases to less than 100 m along A2, and the amplitude along A1 is less than 30 m before the anticline dies out toward the south end of the segment. The slip distribution from the 1980 earthquake corresponds closely with the observed differences in the amount of long-term deformation. The average net slip in 1980 was greatest on segment B, decreased along A2, and decreased again along A1. Aftershocks show that strike-slip faulting normal to the trend of the surface rupture occurs at the segment boundaries, specifically between A1 and A2, and A and B. In addition, aftershocks indicate differences in dip between segments, with segment A having a steeper dip than segment B. Based on these observations, Yielding *et al.* (1981) and King and Yielding (1984) concluded that the 1980 displacement pattern was similar to past surface ruptures, and that features of fault geometry and barriers that control the nucleation and propagation of rupture on this fault have persisted through geologic time.

Lost River Fault Zone, Idaho. Surface faulting associated with the 28 October, 1983, M_s7.3 Borah Peak, Idaho earthquake on the Lost River fault zone provides another excellent example for examining segmentation. The Lost River fault is a normal-slip fault zone that extends for approximately 140 km from Arco to Challis. In 1983 it ruptured along 36 km of its length (Crone and Machette, 1984; Crone *et al.*, 1987). Scott *et al.* (1985) suggested that the zone may be composed of five or six segments characterized by different geomorphic expression, structural relief, and timing of most recent displacement. The segmentation model for the fault zone is shown on Figure 7. The

WASATCH FAULT ZONE RECURRENCE

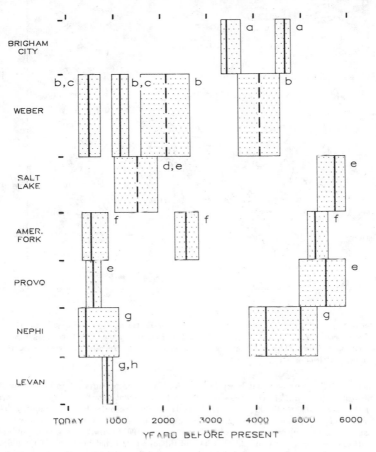

Sources of data: a. S. Personius in Machette et al., in prep.
b. A. Nelson in Machette et al., in prep.
c. Swan et al. (1980)
d. Lund and Schwartz (1986)
e. Schwartz et al. (1988)
f. Machette et al. (1987)
g. Schwartz and Coppersmith (1984)
h. M. Jackson, M.A. Thesis, U Colorado, in prep.

FIGURE 5. Space-time plot of large magnitude, scarp-forming earthquakes on the Wasatch fault zone during the past 6000 years. Solid line is best estimate of timing of event; dashed line is less well-constrained age; stippled box is uncertainty in date.

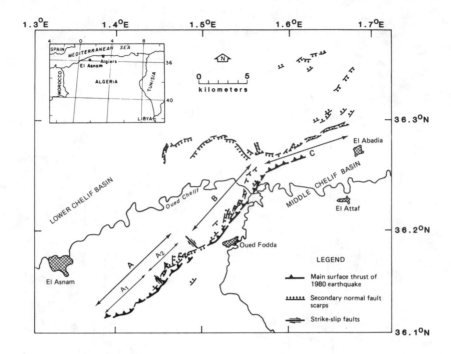

FIGURE 6. Map showing coseismic surface rupture from the 1980 El Asnam, Algeria, earthquake and the segmentation model for the Oued Fodda fault (modified from King and Yielding, 1984). Fault segments A, B, and C are defined by differences in geomorphic expression, seismicity, coseismic slip, geometry, and long-term rates of deformation.

Arco segment has high scarps of Quaternary age but there has been no surface faulting during the past 30,000 years (Malde, 1987). The Pass Creek segment has discontinuous, poorly defined scarps in deposits that are 15,000 to 30,000 years old, but no latest Pleistocene or Holocene faulting. The Mackay segment did not rupture in 1983; however, it has continuous scarps of late Pleistocene and Holocene age. Trenching investigations (Schwartz and Crone, 1988) show this segment produced only one earthquake during the past 12,000 years and this event occurred between 4300 and 6800 years ago. The Thousand Springs segment was the source of the 1983 earthquake. Scarp profiles and trenching show that slip distribution in 1983 faithfully reproduced the slip distribution of the pre-1983 event. Reconstruction and morphometric analysis of the pre-1983 scarp suggests the event occurred about 6000 to 8000 years ago (Hanks and Schwartz, 1987). The Warm Spring segment is north of the Willow Creek Hills. In 1983 it experienced discontinuous surface rupture with displacements averaging 10–20 cm. Radiocarbon dating of inplace burns in and near the base of scarp-derived colluvial wedges observed in trenches, and the degree of soil carbonate development, suggest only one event prior to 1983 occurred on this segment during the past 12,000 years; this paleoearthquake appears to have occurred shortly before 5500–6200 years ago. The North segment has no late Quaternary fault scarps.

    In 1983 surface rupture initiated at the south end of the Thousand Springs segment and propagated northwest. At the south end a 25-cm-high scarp that formed in 1983 is coincident with a fault scarp of approximately the same height that defines the pre-1983 event at this location. South

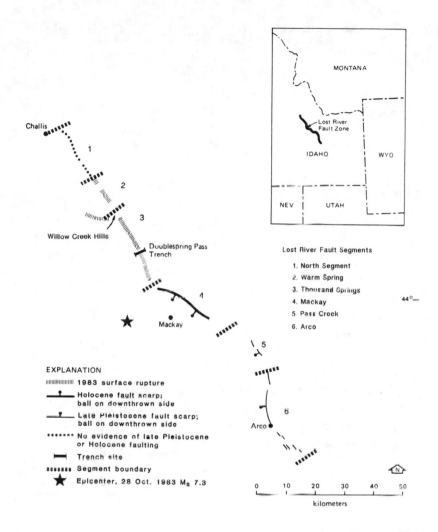

FIGURE 7. Segmentation model for the Lost River fault zone, Idaho (modified from Scott *et al.*, 1985).

of this point the strike of the range front changes sharply, transverse faults occur in the bedrock of the range, and a gap of 4 km in late Pleistocene-Holocene faulting occurs until the scarps of the Mackay segment are reached. Thirty kilometers to the northwest, the 1983 rupture was essentially arrested at a complex zone of bedrock structure transverse to the rangefront called the Willow Creek Hills. Crone *et al.* (1987) interpret this structure as a barrier to rupture propagation and a segment boundary.

There is little doubt, in hindsight, that if the Lost River fault zone had been studied with

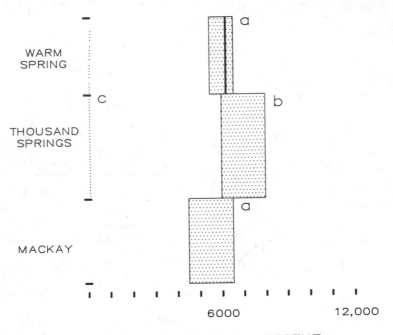

LOST RIVER FAULT ZONE RECURRENCE

WARM
SPRING

c

THOUSAND
SPRINGS

MACKAY

6000   12,000

YEARS BEFORE PRESENT

Sources of data: a.   Schwartz and Crone (1988)
                    b.   Hanks and Schwartz (1987)
                    c.   Crone et al. (1987)

FIGURE 8. Space-time plot of large magnitude, scarp-forming earthquakes on the Lost River fault zone during the past 12,000 years. Solid line is best estimate of timing of event; stippled box is uncertainty in date.

segmentation in mind prior to 1983 a similar model would have been developed based on geometric, structural, and geomorphic changes, and the Thousand Springs segment would have been identified as an independent seismic source. In detail, however, there is some ambiguity concerning Warm Spring–Thousand Springs segmentation. The pre-1983 earthquake could have been somewhat larger and completely ruptured both as a master segment during a single event. This is within the constraints of present timing estimates (Figure 8). Alternatively, both segments are consistently independent seismogenic sources. In this case, stress at the Willow Creek Hills barrier triggered slip on the Warm Spring segment, and the segment will produce a separate event at some future time.

San Andreas Fault Zone, California. The San Andreas fault zone also provides an example for evaluating segmentation, primarily because most of the fault has ruptured in historical time and the amount and extent of slip are known reasonably well. Allen (1968) recognized differences in the historical behavior of various parts of the San Andreas fault zone and identified four segments. Wallace (1970) further subdivided the fault into seven segments based on historical earthquakes and fault creep. A generalized segmentation model is shown on Figure 9. From north to south the segments are: 1) North Coast, which ruptured in 1906; 2) San Francisco Peninsula, which had low slip values in 1906 and was likely the location of an $M7(?)$ event in 1838; 3) Central Creeping, which has a historical creep rate of about 30 mm/yr; 4) Parkfield, representing a transition from the Central Creeping segment to the locked Cholame segment and producing $M6$ events about every 22 years; 5) Cholame, which ruptured in 1857 and is a transition to the high slip Carrizo segment; 6) Carrizo, which ruptured in 1857 and produced slip of 9 to 12 m per event during each of the past three earthquakes; 7) Mojave, a section of the fault along which slip averaged about 4 m in 1857 and during prior events; 8) San Bernadino Mountains, which has had no certain historical rupture or creep and is geometrically complex; and 9) Coachella Valley, which has historical low-level creep and last had a major surface faulting earthquake around 1680 (Sieh, 1986).

Recently developed historical and paleoseismicity data, particularly recurrence from trenching studies and information on displacement per event gathered at different points along the zone (Sieh, 1978, 1984; Hall, 1984; Bakun and McEvilly, 1984; Sieh and Jahns, 1984; Weldon and Sieh, 1985), indicate that long term differences in the behavior of individual segments do occur. The question remains as to how well the historically defined segments have persisted as distinct units through at least the past several seismic cycles and whether they are accurate guides to future behavior.

New dating of events raises intriguing questions about the nature of segmentation along the southern San Andreas, particularly the section commonly referred to as the Mojave segment (Cajon Pass to Tejon Pass). The 1857 event extended from the vicinity of Parkfield to just north of Cajon Creek (CC); it nucleated in the Parkfield–Carrizo Plain area (Figure 10). At Pallett Creek (PC) the event prior to 1857 (event X of Sieh, 1978) was initially dated at 1720 ± 50 (Sieh, 1978; 1984) and now has a revised radiocarbon age of 1785 ± 32 (Sieh et al., 1988). Dendrochronological studies on trees along the fault at Wrightwood (W) suggest an event in 1812 (Jacoby et al., 1987) and Sieh et al. (1988) correlate this with event X. The extent of the 1812 event is unknown. If pre-1857 single-event stream offsets of several meters observed in the Littlerock area north of Pallett Creek are associated with 1812, they indicate a rupture length of many tens of kilometers. Figure 10 shows two segmentation models permitted by the observations. In 10a the 1812 event nucleates outside of the Mojave segment, possibly in the San Gorgonio Pass area (star) and extends an unknown distance north of Pallett Creek. The 1857 rupture overlaps it (stippled area). In this model the Mojave is not an independent rupture segment but rather a broad overlap zone between long rupture segments, as suggested by Schwartz and Coppersmith (1984). In 10b the 1812 event extends from Cajon Creek to Three Points (TP), defining an independent segment that occurs along the low slip section of events such as the 1857 rupture; in this case the section from Cajon Creek to San Gorgonio Pass is a separate segment with an elapsed dating back past 1812. A sequence of late Holocene slip rates along the length of the Mojave segment and better constraints on slip per event would go a long way toward resolving alternative models.

Segmentation and Seismic Hazard Assessment

Faults are geometrically and mechanically segmented at a variety of scales. Segments may represent the repeated coseismic rupture during a single event on a long fault and be many tens to hundreds of kilometers in length, they may represent a part of the rupture associated with an individual faulting event and be only a few kilometers long, or they may represent local inhomogeneities along a fault plane and be only a few tens or hundreds of meters in length. It is the rupture of fault lengths of tens to hundreds of kilometers that represents the greatest hazard and results in sufficient surface deformation to allow fault behavior to be tracked through time.

As the examples discussed above show, the identification of independent rupture segments can be difficult and the methodology for segmentation modeling is in the early stages of development. Ideally, the repeated occurrence of moderate to large earthquakes along a fault could show the degree to which ruptures repeat spatially, slip distribution remains similar or varies during successive

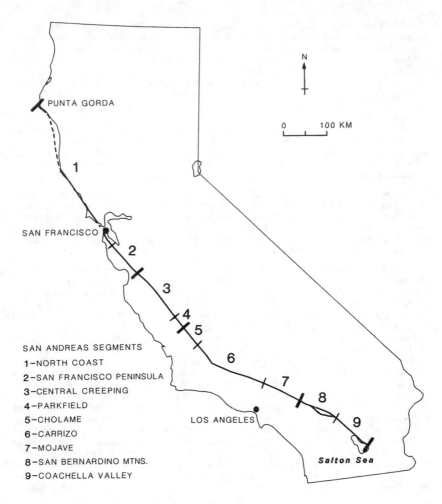

SAN ANDREAS SEGMENTS
1–NORTH COAST
2–SAN FRANCISCO PENINSULA
3–CENTRAL CREEPING
4–PARKFIELD
5–CHOLAME
6–CARRIZO
7–MOJAVE
8–SAN BERNARDINO MTNS.
9–COACHELLA VALLEY

FIGURE 9. Segmentation model for San Andreas fault zone, California. Heavy lines define major segments; short lines define subsegments and transition zones that may be independent segments.

events, adjacent segments rupture together, ruptures are arrested at or bypass apparent structural or geometric barriers, and subsegments rupture to produce events smaller than the expected maximum or characteristic earthquake. However, there are few data of this type for shallow crustal faults. Therefore, paleoseismic data, particularly the timing of past events, slip per event and its distribution along the length of a fault, and slip rate are critical for defining segments and for modeling earthquake recurrence.

Figures 11 and 12 show conceptual segmentation models for strike-slip and normal faults, respectively. They reflect the variability in fault behavior suggested by historical and paleoseismic observations. Each case has its particular geologic expression and effect on hazard evaluation.

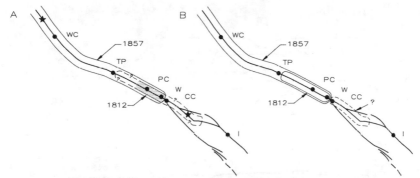

FIGURE 10. San Andreas segmentation scenarios. A) 1857 and proposed 1812 earthquakes nucleate at stars and overlap in Mojave segment; Mojave does not rupture as an independent segment; B) Mojave is an independent subsegment; San Bernadino (dashed line) also ruptures as an independent segment.

Figure 11 is inspired, in part, from observations that a) for many long surface ruptures, such as the 1857 and 1906 San Andreas events and 1940 Imperial earthquake, coseismic slip varies along long lengths of fault and b) the amount of point-specific slip per event repeats. In 11a, there are no persistent segments, and both point-specific slip and earthquake magnitude are variable. In 11b, slip along a master segment (*i.e.*, 1857) repeats periodically. Regions of low slip in the large event define sub or "piggy-back" segments that fill in with smaller events. Segment boundaries would be locations where the amount of slip changes significantly. This type of segmentation and the recurrence associated with it would be expressed paleoseismically by variability in the number and timing of events at points along the fault and by constant slip rate along its length. In 11c, the slip distribution during a large event repeats itself in successive events. Single-event displacements at any point along the fault are essentially constant. Geologically, events correlate temporally but slip rate would vary systematically along the fault. Lengths of fault with low slip do not define independent slip-deficient segments. This is explained by off-fault deformation. The volume of crust in which the fault is located responds to a uniform deformation rate; however, some of this slip is partitioned into folds and other faults. Consequently, the rate is maintained across the volume but slip variability accumulates on individual structural elements. In this model, overlap occurs toward the end of the master segment. Figure 11d expands on 11c. Here, the slip distribution of two large events nucleating at A and B, and a zone of overlap, are shown. This is expressed paleoseismically by a greater number of events in the overlap zone, some of which could be closely spaced in time, reflecting the close timing of A and B. In this case, the overlap zone is not an independent source of earthquakes, but is driven by the recurrence at nucleation sites A and B.

Figure 12 shows several possible segmentation models for normal faults. Normal faults can be tracked through time more easily than strike-slip faults, largely because scarp-derived colluvial wedges from individual earthquakes are generally well-preserved, easily recognizable, and provide, along with scarp profiles, information on the amount of coseismic slip. Typically, locations of low slip, especially near segment boundaries, do not fill in during subsequent events; long-term slip rates vary systematically and are frequently lowest near segment boundaries (often coincident with the ends of ranges). In 12a, the repeated difference in timing is a strong basis for independent rupture segments. In 12b, a subsegment may occasionally fail as an independent event. In 12c, an event on one segment may extend into or trigger slip on an adjacent segment, which itself fails independently at a later time. In 12d, segments are independent but events are closely spaced in time, making distinction of separate events difficult within resolution of age-dating techniques (the occurrence of which is probably all too frequent for the paleoseismologist!)

As more segmented faults are observed and studied, the physical characteristics and controls of

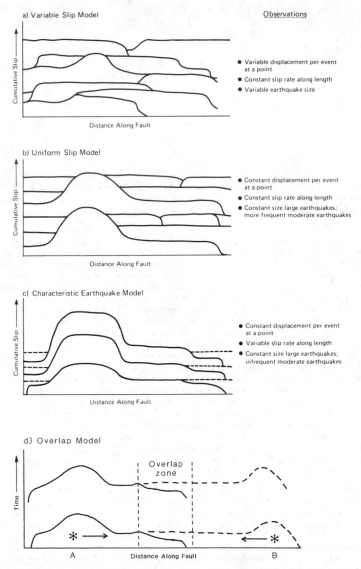

a) Variable Slip Model

Observations

- Variable displacement per event at a point
- Constant slip rate along length
- Variable earthquake size

b) Uniform Slip Model

- Constant displacement per event at a point
- Constant slip rate along length
- Constant size large earthquakes; more frequent moderate earthquakes

c) Characteristic Earthquake Model

- Constant displacement per event at a point
- Variable slip rate along length
- Constant size large earthquakes; infrequent moderate earthquakes

d) Overlap Model

FIGURE 11. Conceptual segmentation/slip/recurrence models for strike-slip faults. A–C from Schwartz and Coppersmith (1984). D is expansion of C.

segmentation will become more readily identifiable and better understood. The best types of data that provide information on segmentation are those that quantify differences in behavior along the length of a fault during its most recent seismic cycle. The most definitive is the difference in the

NORMAL FAULT SEGMENTATION MODELS

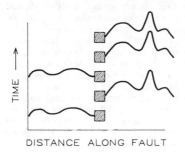

DISTANCE ALONG FAULT

a. Independent

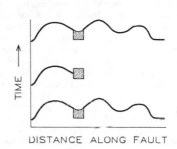

DISTANCE ALONG FAULT

b. Master segment/
   subsegment

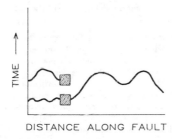

DISTANCE ALONG FAULT

c. Overlap/triggered
   slip

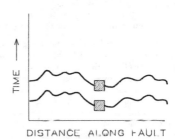

DISTANCE ALONG FAULT

d. Independent behavior
   but adjacent segments
   rupture within uncertain—
   ties of age dates

FIGURE 12. Segmentation models for normal faults. Curvilinear lines are hypothetical slip distributions during a single event; stippled box is segment boundary/barrier.

timing of the most recent event, followed by differences in the timing of older events as indicated by paleoseismological recurrence data. Differences in representative slip rates are also important. For many faults it appears that surface geology and changes in fault geometry have a one-to-one correlation with, and are an expression of, rupture processes occurring at seismogenic depths. Major changes in the strike of the fault, the occurrence of significant lithologic changes, and the presence of transverse geologic structures may supply additional information that can be used to recognize fault segments.

   Implicit in segmentation modeling is the concept that segments can persist as generally discrete units through significant periods of time and, therefore, that each segment ruptures separately. There are no geologic data that presently preclude the possibility that some ruptures may cross segment boundaries or that generally independent adjacent segments may rupture together during the same event. However, in instances where the amount of surface slip during a historical earthquake can be compared to the size of prior events at the same location it is often observed that displacement during successive events has been essentially the same. This indicates that the slip distribution along the fault, and by inference the rupture length, has remained relatively constant. This is the physical basis for the characteristic earthquake. When the amount and rate of short-term (*i.e.*, Holocene) deformation on a fault segment can be compared to the amount of long-term deformation, there is often a good correspondence. For example, Schwartz and Coppersmith (1984) showed that Wasatch fault segments defined on the basis of paleoseismicity data are also reflected by systematic changes in the elevation of the Wasatch range. The elevation of the range is highest where late Quaternary slip rates are fastest and recurrence intervals are shortest, the elevation of the range decreases at segment boundaries and where Holocene scarps die out, and it is lowest at each end of the Wasatch range where paleoseismicity data reveal the lowest Holocene slip rates and the longest recurrence intervals along the fault zone. Wallace (1987) notes similar relationships for the Lost River fault zone. Along the south-central segment of the San Andreas fault, the parts of the segment that had the largest amounts of coseismic slip in 1857 also have the higher long-term slip rates. Aki (1979, 1984) suggested that strong, stable barriers to rupture propagation persist through many repeated earthquakes and the observations noted above are consistent with this. For dip-slip faults these barriers appear to be mainly transverse geologic structures inherited from previous stress regimes that permit decoupling of adjacent segments. Sibson (1985, 1986) has shown that for several historical strike-slip surface ruptures *en echelon* steps in the fault, which are commonly referred to as dilational jogs, appear to be structural features that arrest rupture propagation. King and Nabelek (1985) compared changes in surface fault geometry with epicentral locations and showed that for a number of strike-slip faults bends appear to be critical locations of earthquake initiation.

The independent behavior of fault segments has important implications for seismic hazard evaluation. Segmentation models, when well constrained, provide a physical basis for the selection of rupture lengths used in the calculation of maximum earthquakes. Also, if a fault is segmented, the potential hazard posed by each segment may be different. For example, variability in segment length will mean variability in the size of the maximum earthquake. Moreover, recognition of segments and of the differences in the behavior of each will be extremely important for long-range earthquake forecasting. Information on the difference in the elapsed time and on the recurrence interval for each segment can be used to assess where along a fault zone the next major event will most likely occur and to calculate the probability of that event. This provides a basis for selecting parts of a fault zone for more intensive investigation such as deployment of instrument arrays and other aspects of short-term earthquake prediction.

EARTHQUAKE RECURRENCE

Earthquake recurrence models describes the rate or frequency of occurrence of earthquakes of various magnitudes, up to the maximum, on a fault or in a region. The evaluation of recurrence geologically rests on the ability to recognize past events, evaluate the size of each event, and date the interval between events.

Characteristic Earthquake Model

Statistical studies of the historical seismicity of large regions have shown that the number of earthquakes is exponentially distributed with earthquake magnitude. The general form of this recurrence model is the familiar Gutenberg–Richter exponential frequency-magnitude relationship.

$$\log N(m) = a - bm$$

where $N(m)$ is the cumulative number of earthquakes of magnitude m or greater than m, and a and b are constants. This is often termed a "constant b-value" model. In the general absence of fault-specific seismicity data, it has commonly been assumed that the exponential recurrence model is as appropriate to individual faults as it is to regions. However, recent geologic studies of late

Quaternary faults strongly suggest that the exponential recurrence model is not appropriate for expressing earthquake recurrence on individual faults or fault segments.

The major new concept in earthquake recurrence is that of the characteristic earthquake. The characteristic earthquake model states that many individual faults and fault segments tend to generate essentially the same size or characteristic earthquakes having a relatively narrow range of magnitudes at or near the maximum that can be produced by the geometry, mechanical properties, and state of stress of that fault or segment. Schwartz and Coppersmith (1984) reached this conclusion using displacement-per-event data from trenching and geomorphic studies along the Wasatch and south central San Andreas faults. These data showed that at a point the amount of displacement during successive surface-faulting earthquakes remained essentially constant. A major implication of their study is that earthquake recurrence on an individual fault does not conform to an exponential (constant b-value) model. Wesnousky et al. (1983, 1984) obtained similar results through a comparison between recurrence based on geologic slip rates of Quaternary faults and the 400-year historical seismicity record of Japan. They concluded that the seismicity of individual faults is in poor agreement with the Gutenberg–Richter relation. They explained this observation with the maximum moment model, a variation of the characteristic earthquake in which recurrence is expressed as the repeat of only the maximum event.

Other observations support the concept of characteristic earthquakes. The Parkfield segment of the San Andreas fault (Figure 9) appears to produce a characteristic event (Bakun and McEvilly, 1984). The 1934 and 1966 events had identical epicenters, magnitudes, fault-plane solutions, unilateral southeastward ruptures, foreshocks, and lateral extent of aftershocks, the location and extent of the 1934 and 1966 surface displacements were extremely similar; intensity patterns for the 1901, 1922, 1934, and 1966 events are similar; and seismic moments based on seismograms for the 1922, 1934, and 1966 events are very nearly equal. The 1983 Borah Peak, Idaho earthquake also appears to have been a characteristic earthquake. Mapping of the surface rupture (Crone and Machette, 1985; Crone et al., 1987) and trenching (Schwartz and Crone, 1985) show that slip distribution of the 1983 event faithfully repeated, both in location and amount, the slip distribution of the one pre-1983 earthquake that had occurred on this segment of the Lost River fault zone during the past 12,000–15,000 years. Similar behavior is observed along the south central segment of the San Andreas fault where location-specific slip during the 1857 earthquake appears to repeat the amount of displacement of at least the two prior events. This has been shown to be the case at Wallace Creek, where surface offset during the 1857 earthquake was $9.5 + 0.5$ m and offset during the two prior events was 12.3 ± 1.2 m and 11.4 ± 2.5 m (Sieh and Jahns, 1984), as well as at other sites along that fault segment.

The 1983 Borah Peak, Idaho, Earthquake—A Characteristic Event. The 1983 Borah Peak, Idaho earthquake is an excellent example of a characteristic event. In 1976 a trench was excavated across a fault scarp of the Lost River fault zone that was developed in a Pinedale-age outwash fan (approximately 12,000 to 15,000 years old) at Doublespring Pass Road. Relations in this trench suggested that only one major surface-faulting event had occurred since formation of the fan surface (Hait and Scott, 1978). This event has now been shown to be about 6000–8000 years old. As part of the evaluation of the 1983 earthquake, a parallel trench was excavated to re-expose the pre-1983 earthquake relationships and observe the changes that occurred in 1983. A generalized log of the 1984 trench is shown on Figure 13. Within this trench, correlative stratigraphic marker horizons occur on both sides of the main fault and can be traced across the graben. Because of this, the complete post-fan faulting history is exposed, and measurement of pre-1983 displacements can be made and compared with 1983 displacements. Mapping and analysis of the stratigraphic and sructural relationships in the trench (Schwartz and Crone, 1985) indicate the following sequence of events:

1. *Pre-1983 surface faulting.* The fan surface was displaced and a series of graben and a horst were produced across a 40-m-wide zone west of the main scarp.

2. *Deposition of scarp-derived colluvium.* This occurred at and west of the main fault (meters

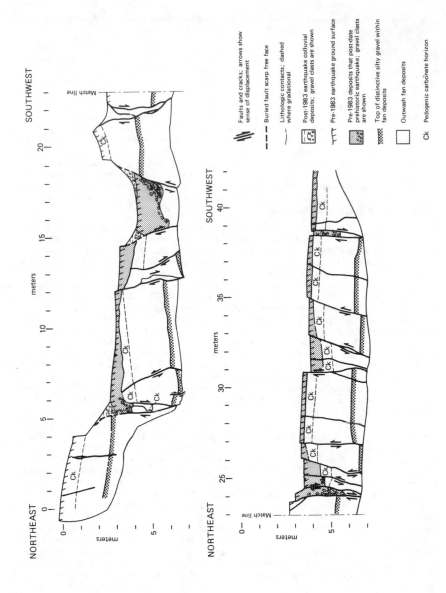

FIGURE 13. Generalized log of trench across the Lost River fault and surface rupture from the 1983, Borah Peak, Idaho, earthquake at Doublespring Pass Road (from Schwartz and Crone, 1985). The trench location is shown on Figure 7.

5 through 10) and in graben (meters 15 through 19; 24 through 27). Fissure infills also developed (meters 24 and 39). These deposits are shown by the grey stippled pattern.

3. *Continued colluviation and the development of an organic A-horizon (slanted pattern) at the pre-1983 ground surface.*

4. *1983 surface faulting.* All pre-1983 faults were reactivated, one new trace developed, and the existing colluvial wedge at the main fault was backtilted to the east.

5. *Deposition of scarp-derived colluvium.* Post-faulting colluvial deposits (dashed units) buried fault scarp free faces and are prominently developed at the main fault (meter 5) and in a graben (meters 15 and 19).

Important conclusions can be drawn from the new Doublespring Pass trench regarding the size of past events. Surface displacement that occurred in 1983 mimicked displacement from the prior event in both style and amount. All individual pre-1983 faults in the trench were reactivated, including small graben and the well-defined horst. Displacement across the main fault was similar for both events as was displacement on many of the synthetic and antithetic faults.

Crone et al. (1987) mapped the distribution of displacement along the length of the 1983 surface rupture. In-progress mapping suggests that scarp heights from 1983 were extremely similar to heights developed during the one pre-1983 event along the entire length of the fault; small 1983 scarps are associated with small pre-1983 scarps and large 1983 scarps are associated with large pre 1983 scarps. This is shown graphically on Figure 14 where 1983 scarps of 0.25 m, 1.8 m, and 1.0 m are associated with pre-1983 scarps of the same height. The pattern of faulting is also remarkably consistent. This is observed not only in the trench, but also at other locations along the fault where other graben and existing en echelon scarps were all reactivated in 1983 It appears that point-specific slip distribution along the length of the fault, and therefore the size of the earthquake, was essentially the same for the past two events.

Characteristic Earthquakes and Seismic Hazard. By combining the recurrence intervals for large-magnitude earthquakes developed from geologic data with the recurrence for smaller-magnitude events developed from seismicity data, a characteristic earthquake recurrence model is derived that has the general form shown on Figure 15. Notice that the geologic data represented by the box on Figure 15 include the uncertainty in both the recurrence intervals and the magnitude of the paleoseismic events. The model has a distinctive non-constant b-value that changes from values of about 1.0 in the small-magnitude range to lower values of about 0.2–0.4 in the moderate-to-large magnitude range. The low b-value reflects a recurrence curve anchored at the large-magnitude events and having relatively fewer moderate-magnitude earthquakes than would be expected by b of about 1.0 (Schwartz and Coppersmith, 1984). The implications of this are that for an individual fault, estimates of the frequency of occurrence of large earthquakes based on extrapolation of the frequency of occurrence of small earthquakes, may be subject to considerable error. Likewise, the concept of a "probable" earthquake that is somewhat more likely to occur than the maximum event, and is therefore usually assumed to be somewhat smaller, is probably erroneous.

Although it may be argued that the non-constant slope of the recurrence curve on Figure 15 results from too short a sampling period of historical seismicity data, the distinctive non-constant recurrence relationship suggested by the characteristic earthquake model has also been observed from historical seismicity data along fault zones that have had historically repeated characteristic events. Examples are the Alaskan subduction zone (Utsu, 1971), the Mexican subduction zone (Singh et al., 1983), Greece (Båth, 1983, Japan (Wesnousky et al., 1983), and Turkey (Båth, 1981). These observations strongly suggest that the non-constant slope is real and not merely the result of an inadequate data base.

There are several points regarding the characteristic earthquake that must be kept in mind. First, characteristic refers only to the successive repeat of similar size events. It does not imply that

FIGURE 14. Photographs showing similarity in amount of surface faulting during 1983 Borah Peak
earthquake and the prior event. A) Scarplet of about 25 cm is coincident with a minor break in slope
defining pre-1983 displacement of same amount; B) 1.8-m-high scarp (in shadow) at Doublespring Pass
Road with beveled scarp of about same height from pre-1983 event; C) 1.0-m-high scarp near Arentson
Gulch with similar-sized pre-1983 scarp on top (covered by snow).

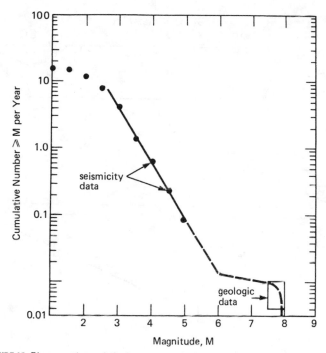

FIGURE 15. Diagrammatic cumulative frequency-magnitude recurrence relationship for an individual fault or fault segment. A low b-value is required to reconcile the small-magnitude recurrence with geologic recurrence, which is represented by the box (from Schwartz and Coppersmith, 1984).

the interval between events is uniform. Second, there is little doubt that characteristic earthquakes are real and likely express the behavior of a large number of faults. However, it is also clear that a spectrum of behavior exists and the characteristic earthquake does not typify, and should not be forced on, all faults or segments.

Uniform vs. Variable Recurrence Intervals

An important aspect of recurrence with regard to understanding and quantifying seismic hazard is the degree to which actual intervals between successive earthquakes are uniform or variable. There are few faults for which there is a long enough record of well-constrained ages of past earthquakes. The data that are available suggest a range of behavior, from the generally uniform to the highly irregular (Figure 16).

Generally uniform recurrence appears to characterize the Parkfield segment of the San Andreas fault, which has had repeated similar size events in 1881(?), 1901, 1922, 1934, and 1966. It is on the basis of these, and of the slip estimated for each, that the Parkfield earthquake prediction experiment was established and 1988 (±5 years at the 95 per cent confidence interval) designated as the center of a time window for the occurrence of the next Parkfield earthquake (Bakun and Lindh, 1985). The Mojave segment of the San Andreas fault has been considered to have an average recurrence interval of 140–200 years based on dating of the past twelve earthquakes at Pallett Creek (Sieh, 1978; 1984). However, Sieh et al. (1988) have re-dated the Pallett Creek sequence using very precise radiocarbon

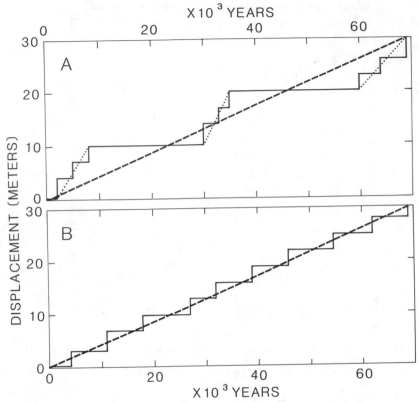

FIGURE 16. Diagrammatic graphs of two types of fault-displacement time histories. A) Temporal clustering of events separated by periods of quiescence (solid line). For a few thousand years during clusters of events the average slip rate (dotted lines) may be sevral times that of the long-term rate averaged over tens of thousands of years (dashed line). B) More regular recurrence of slip events. Dashed line shows an average slip rate, which is the same in both examples. (From Wallace, 1987.)

dating. The results are shown on Figure 17. Although an average recurrence of 131 years can be calculated, these past earthquakes appear to fall into clusters of two and three events separated by intervals of 200 to 330 years rather than by relatively uniform intervals. As discussed, it is likely that Pallett Creek represents an overlap of adjacent segments. These observations, if correct, suggest that recurrence on the San Andreas reflects temporal clustering controlled, in large part, by the longer recurrence of strong parts of the fault (sections with large surface offset and recurrence intervals of several hundred years such as the Carrizo Plain) that fail more or less sequentially during a short period of time, for example 1812, 1857, and 1906. Superimposed on this broad framework could be localized variability at ends of long ruptures, at slip steps, or at zones of structural complexity. In this regard, the apparent variability at Pallett Creek could, depending on the segmentation model, represent generally uniform recurrence for the zone.

It is likely that along major transform boundaries like the San Andreas where the rate of application and source of stress are relatively constant and the rate of strain accumulation is high,

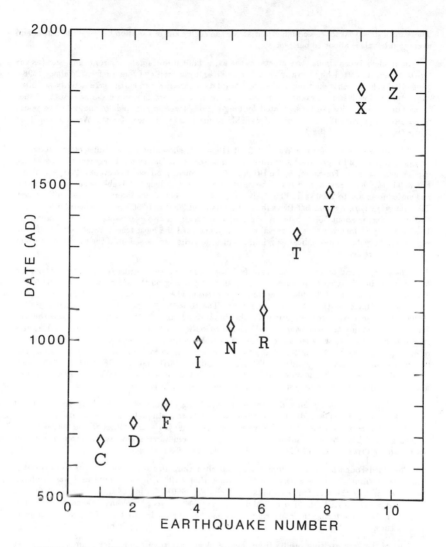

FIGURE 17. Dates of historical and paleoearthquakes at Pallett Creek on the Mojave segment of the San Andreas fault. Error bars (diamonds and lines) indicate 95 per cent confidence interval. (From Sieh et al., 1988.)

major faults and fault segments may tend toward a generally uniform, though not precisely time-predictable, behavior. Nishenko and Buland (1987) have looked at historical and geologic recurrence data from fifteen plate boundary faults and segments, twelve of which are subduction zones. While subduction zones differ in many respects from the shallow crustal faults discussed here, the rates of strain accumulation are generally high enough that repeated events have been observed historically.

These data show that for these high strain areas, the coefficient of variation between the actual and average intervals is about 20 per cent.

Intervals between similar-size events on the same fault in intraplate environments, such as the Great Basin, are often highly variable and can differ by a factor of four or five. Wallace (1984, 1985, 1987) showed that on a regional scale intraplate recurrence and earthquake activity are often characterized by clusters of events (a cycle) that are concentrated in small areas or zones and that these bursts of activity can be separated by long periods (tens to hundreds of thousands of years) of quiescence (Figure 16). Examples of variable recurrence can be seen for the Wasatch and Lost River fault zones.

The space-time plot for the Wasatch fault (Figure 5) shows the repeat times of characteristic events along each of the proposed segments. It can be seen that for any given segment the recurrence can be quite variable. For example, the Brigham City segment produced two events 1200 years apart but 3400 years have now passed since the most recent event. Along the Weber segment recurrence has varied from 600 to 2000 years. The Salt Lake segment has had one interval of about 4000 years. The American Fork segment has two subequal intervals of 2000 and 2500 years. The interval on the Provo segment is about 5000 years. Along the Nephi segment two events occurred approximately 1000 years apart followed by an interval of 3000 years until the most recent event. On the Levan segment radiocarbon dates indicate a minimum of 6700 years had passed until the most recent event about 800 years ago.

There are several other ways to quantify Wasatch recurrence. During the past 6000 years the fault has produced fifteen to seventeen earthquakes (depending on the segmentation model) for an average recurrence of 350 to 400 years for an event somewhere along the zone. Figure 5 suggests some partitioning of earthquake activity in time. The entire fault zone south of the Brigham City segment failed in a sequence of large earthquakes between 300 and 1500 years ago. Excluding the Salt Lake segment events are even more tightly clustered between 300 and 1000 years ago. Within this recent active sequence, the average recurrence interval has been 200–250 years. A similar sequence of earthquakes, primarily from Salt Lake south, occurred between 5000 and 6000 years ago. The occurrence of a sequence of moderate to large magnitude earthquakes in 1872 1903, 1915, 1932, 1934, and 1954, along what is now referred to as the Nevada seismic zone (Wallace, 1981), may be a historical analog to the youngest cluster along the Wasatch fault.

The Lost River fault zone shows temporal clustering (Figure 8). Trenching and dating of charcoal and volcanic ashes indicate that there were no large earthquakes from at least 12,000 years ago to middle Holocene time. Then, between 5500 and 8000 years ago, although the actual timing of events could have been considerably closer, the northern three segments produced a series of scarp-forming earthquakes. The fault then held until the 1983 earthquake.

The Oued Fodda fault also exhibits temporal clustering. Based on observations from trenches across the 1980 surface rupture Swan (1987) showed that a 100,000-year-old paleosol was displaced 10 m vertically by three or four earthquakes (including 1980). Dating of these by radiocarbon in faulted deposits indicated that all occurred during the past 1500 years, yielding an average recurrence interval of 450 years during the clustering. This had been preceeded by tens of thousands of years of quiescence.

One of the very exciting results from these geologic investigations is that temporal clustering along adjacent, and often independent, segments occurs and can be deciphered in the geologic record. This has important implications for the understanding of recurrence and for seismic hazards evaluation. It raises the possibility that segments adjacent to historical ruptures may have a higher potential to produce the next earthquakes, particularly if they can be shown to have similar elapsed times. Alternatively, the hazard might be smaller if the fault zone has recently released most of its accumulated strain. Focused studies to collect data on timing of past earthquakes for both the historical and adjacent segments could yield information on the degree to which failure of adjacent segments has, or has not, been closely coupled in time.

Slip Rate and Earthquake Recurrence

As discussed, geologic studies have been successful at identifying prehistoric earthquakes in the geologic record and at estimating recurrence intervals between surface-faulting earthquakes. Unfortunately, these types of paleoseismicity data are not presently available for most faults. The late Quaternary geologic slip rate, however, can frequently be obtained and is being used to constrain fault-specific earthquake recurrence relationships for seismic hazard analysis.

Fault slip rates offer the advantage over historical seismicity data of spanning several seismic cycles of large-magnitude earthquakes on a fault, and they can be used to estimate average earthquake frequency. Their use, however, requires a number of assumptions and each of these must be carefully considered when using a slip rate to calculate recurrence on a specific fault. These assumptions are: (1) all slip measured across the fault is seismic slip, unless fault creep has been recognized; (2) surface measurements of slip rate are representative of slip at seismogenic depths; (3) it is appropriate to use the slip rate, which is an average and which does not allow for short-term fluctuations in rate to be recognized, (4) the slip rate measured at a point is representative of the fault; and (5) the slip rate is applicable to the future time period of interest.

Two basic approaches have been developed for using geologic slip rates. The first, proposed by Wallace (1970), allows average earthquake recurrence intervals to be calculated by dividing the slip rate into the displacement per event. Slemmons (1977) developed this further and arrived at relationships between recurrence intervals, magnitude, and slip rate. This general approach assumes that only one size earthquake, usually the maximum, occurs and the displacement per event used represents this event. However, because earthquakes with magnitudes less than the maximum also occur on the fault, less of the total slip rate is available for the maximum event. Therefore the maximum event may have a somewhat longer recurrence interval than would be calculated assuming no other slip events occur.

The second approach is based on the assumption that the slip rate reflects the rate at which strain energy (seismic moment) accumulates along the fault and is available for release. Seismic moment, M_0, is the most physically meaningful way to describe the size of an earthquake in terms of static fault parameters:

$$M_0 = \mu AD$$

where μ is the rigidity or shear modulus (usually taken to be about 3×10^{11} dyne/cm^2), A is the area of fault plane undergoing slip during the earthquake, and D is the average displacement over the slip surface (Aki, 1966). The seismic moment rate $\dot{M}_0$, which is the rate of energy release along a fault, is estimated by (Brune, 1968):

$$\dot{M}_0 = \mu AS$$

where S is the average slip rate along the fault (cm/yr). The seismic moment rate provides an important link between geologic and seismicity data. For example, seismic moment rates determined from fault slip rates in a region may be directly compared with seismic moment rates based on seismicity data (Doser and Smith, 1982).

Once a seismic moment rate has been calculated for a fault, it must be partitioned into various magnitude earthquakes according to an assumed recurrence model. Most commonly, an exponential magnitude distribution is used. Several authors (Smith, 1976; Campbell, 1977; Anderson, 1979; Molnar, 1979; Papastamatiou, 1980) have developed relationships between earthquake recurrence and fault or crustal deformation rates, assuming an exponential magnitude distribution.

As discussed, there is increasing evidence that, at least for some faults, a recurrence model based on the characteristic earthquake may be more appropriate than the exponential model for individual faults and fault segments. Youngs and Coppersmith (1985) developed a generalized recurrence

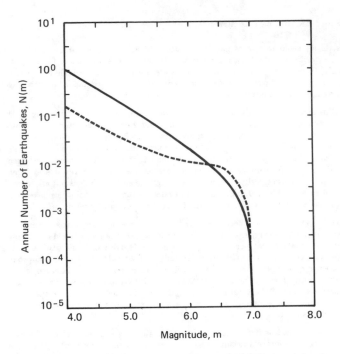

FIGURE 18. Comparison of recurrence relationships based on an exponential magnitude distribution (solid curve) and a characteristic earthquake distribution (dashed curve). Both relationships assume the same maximum magnitude, b-value, and fault slip rate (from Youngs and Coppersmith, 1985).

density function for this model that can be used when fault slip rate data are available. The choice of either the exponential model or the characteristic earthquake model can have a significant impact on the resulting recurrence relationship. Figure 18 compares the earthquake recurrence relationship for a single fault developed using an exponential magnitude distribution (solid curve) with that developed using the characteristic magnitude distribution (dashed curve). Both relationships were developed using the same maximum magnitude, b-value (for the exponential distribution magnitude range), and fault slip rate. As shown on Figure 18, for the same slip rate, use of the characteristic earthquake model rather than a constant b-value model results in a significant reduction in the rate of occurrence of moderate-magnitude earthquakes and a modest increase in the rate of the largest events. This difference can affect the assessment of seismic hazard at a site, depending on whether the moderate-magnitude events or the large events contribute most to the hazard. Wesnousky (1986) has made extensive use of slip rates, and anchored them to a characteristic earthquake model, to estimate recurrence intervals for use in calculations of expected ground acceleration in California during the next 50 years.

One final consideration that is important in assessing earthquake recurrence from fault slip rate (moment rate) is sensitivity to the choice of maximum magnitude used in the analysis. As shown on Figure 19, for the same slip rate (constant moment rate), increasing the maximum magnitude from 6 to 8 results in a dramatic decrease in the recurrence rate for smaller events. This is because the largest earthquakes account for the major part of the total seismic moment rate and adding a single large earthquake requires the subtraction of many smaller events to maintain the same moment rate.

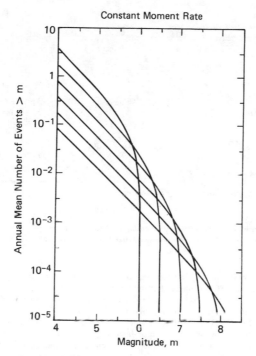

FIGURE 19. Effect of variations in maximum magnitude on the recurrence relationship for a fault when fault slip rate (moment rate) is held constant (from Youngs and Coppersmith, 1985).

Recurrence on "Hidden" Sources

Almost all of the discussion has focused on faults that have clearly defined surface expression. This is because the distinctive surface geologic signature of these faults provides the best opportunity to really understand fault behavior. However, there are regions where earthquake sources are difficult to identify and their behavior even harder to quantify. In these regions, secondary deformation associated with faulting or ground shaking such as liquefaction, folding, and subsidence have the potential to provide important data on earthquake recurrence.

Earthquake Recurrence in the Eastern and Central United States. Although damaging earthquakes have occurred historically in the eastern United States with the 1811–1812 New Madrid, MO and 1886 Charleston, S.C. events being the largest, little progress has been made in identifying individual seismogenic sources. There is uncertainty regarding the association of earthquakes and geologic structure, and seismic sources are defined by areal zones based on the distribution of seismicity rather than by specific faults. This difficulty is due, in large part, to low rates of deformation coupled with rates of erosion and deposition that are high enough to obscure geomorphic evidence of surface faulting. To overcome this, nontraditional approaches need to be used or developed. A major breakthrough has been the recognition and dating of past liquefaction events and secondary surface deformation at sites of historical liquefaction. In the New Madrid seismic zone Russ (1979, 1982) used observations of liquefaction, folding, and faulting in trenches across the Reelfoot scarp to interpret the occurrence of two events with intensities similar to the 1811–1812 earthquake sequence

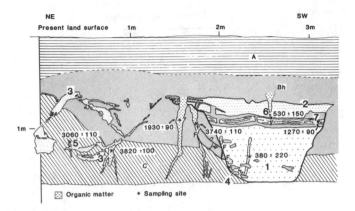

FIGURE 20. Trench log at site of liquefaction during the 1886 Charleston, S.C. earthquake. 1 is a filled crater from a pre-1886 earthquake, which occurred after 3700 and before 1300 years ago. (From Talwani and Cox, 1985.)

during the past 2000 years. He suggested an average recurrence interval of 600 years. Similarly, sites of liquefaction from the 1886 earthquake have yielded information on recurrence in the Charleston, S.C. area (Obermeir et al., 1985; Talwani and Cox, 1985). These studies show that one and possibly two liquefaction events occurred prior to 1886 during the past 3000 to 3700 years (Figure 20). Similarly, Thorson et al. (1986) have attributed sand-filled fissures in late Quaternary glacial deposits in eastern Connecticut to liquefaction from a prehistoric earthquake. It should be kept in mind that the occurrence of liquefaction in the geologic record indicates only that a specific site has experienced a level of ground motion sufficient to cause liquefaction. It does not uniquely define the location of the seismogenic source. It is clear however, that this is an extremely promising field of research for evaluating aspects of eastern United States paleoseismicity.

Another potentially promising approach to evaluating the extent and rate of Holocene deformation in the eastern United States is river response to slow active uplift and subsidence. Ouchi (1985) and Schumm (1986) have discussed the ways in which valley floor morphology and alluvial channel patterns reflect active tectonism, with examples from the Mississippi River where it crosses the Lake County uplift (Missouri), the Mississippi and other streams that cross the Monroe uplift (northeast Louisiana, southeast Arkansas, western Missouri), and streams that cross the Wiggins uplift (south-central Missouri). This could be used to identify and focus studies in areas where active tectonism is occurring but where rates of slip on faults are too low to be recognized with traditional methods.

Folds and Earthquakes. The occurrence of earthquakes in areas where large seismogenic faults are expressed at the surface as folds has clearly raised important questions about seismic hazards and our ability to evaluate them in folded terrains (Stein and King, 1984; Yeats, 1986). The 2 May, 1983, M_s6.7 Coalinga, California earthquake focused attention on active folds and the relationship between folds, faults, and earthquakes. The main shock occurred on a shallow, southwest-dipping reverse fault (Eaton, 1985). The surface expression of the fault is Anticline Ridge, which is part of a series of Pliocene and younger folds that lie along the west side of the San Joaquin Valley. There was no coseismic surface faulting on the west-dipping fault during the main event (Clark et al., 1983). However, coseismic surface displacement did occur on the Nuñez fault, an east-dipping reverse fault west of Anticline Ridge in association with an M_L 5.2 event about one month after the main shock. Maximum slip on the Nuñez fault was 64 cm and cumulative length of the discontinuous surface rupture was 3.3 km (Rymer et al., 1985). Folds similar to Coalinga occur along the west side of the

San Joaquin Valley, in the Transverse Ranges, and in the Los Angeles basin. The 1 October, 1987, M_L 5.5 Whittier Narrows, California earthquake in the Los Angeles basin also occurred on a reverse fault that was expressed at the surface primarily as a fold.

We are just beginning to appreciate the relationship between seismogenic faulting at depth and surface folding. The geometry of surface folds changes along strike and may provide a basis for segmenting fold belts. In addition rates of shortening and uplift can be calculated. It is unclear, at least at the present, how earthquake recurrence intervals can be well-constrained in folded deposits or how we can distinguish between folds that are surface expressions of throughgoing seismogenic faults and those that are not. This is certainly an area for future research.

Pacific Northwest. The interface between the Juan de Fuca and North American plates, which is referred to as the Cascadia subduction zone, has become the subject of great debate with regard to its potential to produce large magnitude earthquakes below the Pacific Northwest. One of the most outstanding features of this subduction zone is the almost complete absence of instrumentally recorded thrust earthquakes along the plate interface and of moderate to large events during at least the past 100 years.

Geologic observations are suggesting that large subduction zone events have occurred during the recent past. On the basis of stratigraphy in coastal lowlands, including liquefaction features and buried soils, Atwater (1987) and Atwater *et al.* (1987) have shown that the SW Washington coast has experienced episodic sudden subsidence. Rapid subsidence appears to have occurred about 300, 1600(?), 1700, 2700, and 3100 years ago. This is interpreted as representing coseismic subsidence associated with repeated subduction zone earthquakes. This investigation is ongoing. Observations along longer stretches of coast combined with better dating control and correlation of proposed subsidence events will provide additional information with which to test various tectonic models and their impact on seismic hazard evaluation in this region.

LONG-TERM EARTHQUAKE POTENTIAL

One goal of geologic studies is to quantify seismic hazard in such a way that it can be used for engineering decisions regarding seismic design. For the seismic design of dams, power plants, hospitals, and schools it has been common practice to use deterministic design criteria. That is, the design is based on the assumption that a particular earthquake magnitude, level of ground motion, or amount of displacement on a fault will occur during the life of the facility. In a particular situation this may, in fact, be the best approach. In recent years, probabilistic seismic hazard analysis (PSHA) has become increasingly used. This provides estimates of the probability of future earthquake occurrence or level of ground motion through the use of earthquake hazard models, which express assumptions regarding the timing and size of events based on a physical and statistical understanding of the earthquake process. PSHA can be used for relatively straightforward estimates such as the probability of an earthquake on a specific segment of a fault during a future period of time when the average recurrence interval, its uncertainty, and the elapsed time are known. For example, what is the probability of a M 7.5 earthquake on the Mojave segment of the San Andreas fault during the next 30 years? It can also be used to evaluate levels of ground motion at sites that are complicated by, and must take into account, multiple sources, different size earthquakes and recurrence intervals, and different possible attenuation relationships. The entire question of PSHA is treated in a National Academy report (1988) on probabilistic seismic hazard analysis. The report discusses, among other things, needs of the user, needs of earth science information, treatment of uncertainties, applications, and future directions. It is likely that as the understanding of fault behavior advances, the use of probabilities based on more refined geologic input will take on greater importance in guiding judgments regarding seismic design parameters. Some of the more common earthquake hazard models are discussed below. They range from very simple models that require very few data constraints to very complex models that, because of their large number of data constraints, have rarely been applied.

Poisson–Exponential Model

The most commonly used hazard model (see Cornell, 1968) is based on the assumption that earthquakes follow a Poisson process. That is, along a fault or within a seismic source zone, earthquakes are assumed to occur randomly in time and space. Coupled with this assumption is the exponential distribution of earthquake magnitudes. The Poisson-exponential model assumes that the times between earthquake occurrences are exponentially distributed and there is some time between occurrences of particular magnitudes. Therefore, the time of occurrence of the next earthquake is independent of the elapsed time since the prior one. Also, the Poisson process has no "memory" in that the magnitude of the next earthquake will not depend on the magnitude of any past events. Finally, the magnitude, locations, and times-of-occurrence of earthquakes along the fault are independent. This means, for example, that a long period of quiescence does not imply anything about the size of the next earthquake. Also, the next event is just as likely to occur on a segment of a fault that recently ruptured as on any other segment. Where data on faults and fault behavior is lacking, the Poisson-exponential model may be necessary and useful. However, in many cases the assumptions of the model may not be compatible with our understanding of the physical processes of earthquake generation.

For the Poisson-exponential model few data constraints are required. The probability of occurrence of x number of events during time t is only a function of the rate ν (the average number of events per unit time or the average recurrence interval):

$$P(x) = \frac{(\nu t)^x e^{-t}}{x!}$$

The rate ν may come directly from geologically derived estimates of earthquake recurrence.

Time-Predictable Model

The time-predictable model, as proposed by Shimazaki and Nakata (1980), is based on assumptions of constant rates of stress and strain accumulation, and that stress accumulates to some relatively constant threshold at which failure occurs. From these assumptions, given the size of the most recent strain release (usually expressed as coseismic fault slip) and the rate of strain accumulation (slip rate), one can predict the time to the next earthquake (Figure 21). In this regard, the time-predictable model is relatively deterministic, although some uncertainty may be introduced in the model parameters. For example, stochastic models of earthquake occurrence have been developed based on the time-predictable model (Anagnos and Kiremidjian, 1984).

The evaluation of seismic hazards would be greatly simplified if all faults followed a time-predictable behavior. However, this is not the case. Rather, time-predictable behavior may be strongly dependent on tectonic environment. Along plate boundaries, such as major transform fault zones like the San Andreas, or subduction zones, where the rate and source of stress are relatively constant and the rate of strain accumulation is high, major faults and fault segments may approach a generally uniform behavior. In these cases, a time-predictable model may provide a very reasonable approach to quantifying fault-specific hazard, even if there is some uncertainty in the precision regarding the regularity of recurrence. Geodetic observations in Japan suggest that the rate of strain accumulation between large earthquakes is not constant, but if the true variations in rate can be measured, the time-predictable model may still be applicable (Thatcher, 1984). However, the time-predictable model does not appear to be applicable to intraplate environments where, as described, repeat times for the same-size earthquake on a fault can be highly variable.

Renewal Models

Renewal models, which are also referred to as real-time models, also imply a time-dependent accumulation of energy between major earthquakes. As opposed to the Poisson model, renewal

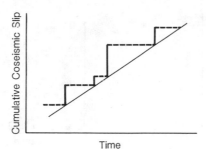

FIGURE 21. The time-predictable recurrence model. With information on the amount of the most recent coseismic fault slip (heavy line) and the assumption of linear strain accumulation (thin line), the time to the next earthquake (dashed line) can be estimated. (From Shimazaki and Nakata, 1980).

models have a one-step "memory" that considers the time since the most recent event (Figure 22). That is, the likelihood of earthquake occurrence during a particular future period of interest, which is referred to as a conditional probability, is related to the elapsed time since the most recent event and the average recurrence interval between major earthquakes. To use this model, the parameters required are the elapsed time, the average recurrence interval and its standard deviation, σ, and the appropriate probability density function for the expected recurrence interval. Figure 22 shows a probability density function for earthquake recurrence. The probability of an earthquake in the interval $(t, t + \Delta t)$ is given by the area of dark shading under the probability density curve. The probability, conditional on the earthquake not having occurred prior to t, is the ratio of the area of dark shading to the sum of the areas with dark and light shading. Different types of probability density functions have been used in a number of time-dependent probability studies. These include Gaussian, Weibull, and log normal probability density functions (Lindh, 1983; Sykes and Nishenko, 1984; Nishenko, 1985; Nishenko and Buland, 1987).

Renewal models have been widely used to describe earthquake occurrence (Veneziano and Cornell, 1974; Kameda and Ozaki, 1978; Savey et al., 1980; Grandori et al., 1984). More complex models that are based on an assumed renewal process have also been proposed. However, additional parameters are required to specify these models. For example, the semi-Markov two-step memory that relates the probability of future earthquakes of particular sizes to both the elapsed time since the most recent event and the magnitude of the prior event. This model has been used to assess probabilities of earthquake occurrence in Alaska (Patwardhan et al., 1980) and the Wasatch fault zone (Cluff et al., 1980).

When there are sufficient data time-dependent models have the potential to provide the most realistic estimates of seismic hazard on individual faults or segments. Figure 23 is an example of 30-year conditional probabilities calculated for major segments of the San Andreas fault (Lindh, 1987). To develop these probabilities, Lindh (1987) used elapsed times since historical events, recurrence intervals based on actual intervals from paleoseismicity data or calculated using estimates of slip rate and slip per event, and a Gaussian probability density function.

Treating Uncertainties and Alternatives in Geologic Data

It should be obvious by this point that there is natural variability in the earthquake process as well as uncertainty in our ability to quantify that process. This uncertainty must be accounted for in a seismic hazard analysis. One approach that is becoming increasingly more widespread is the use of logic trees (Power et al., 1981; Kulkarni et al., 1984; Youngs et al., 1987). This involves specifying discrete alternatives for input parameter values, fault behavior models, or states of nature and assigning a weight or likelihood that each alternative is the correct one. Figure 24 is a logic tree from a study of ground shaking hazard along the Wasatch front (Youngs et al., 1987). Note that this analysis tries to account for alternative attenuation relationships, exponential versus characteristic

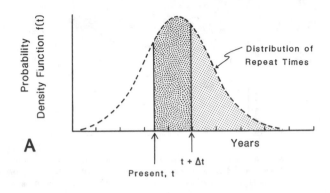

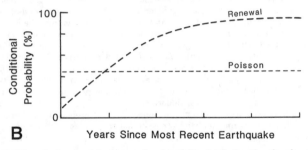

FIGURE 22. Schematic diagram of simple renewal model. A) The distribution of earthquake repeat times (recurrence intervals) is represented by the probability density function (dashed curve). The future time period of interest is shown from present t to $t + \Delta t$ (dark shading). The conditional probability of earthquake occurrence is defined as the ratio of the area of dark shading under the density function to the sum of the areas with dark and light shading. B) For the renewal model the conditional proability varies as a function of the elapsed time since the most recent earthquake, whereas the Poisson estimate is independent of the elapsed time.

recurrence models, the appropriateness of using segmentation and the correct segmentation model, variations in fault geometry (dip and length) that affect estimates of maximum magnitude, variability in the magnitude of the maximum earthquake, and the range of possible recurrence intervals. A positive feature of the logic tree is that the sensitivity of the outcome to the selection of different weights for each alternative can be analyzed.

MOVING INTO THE 1990s

What lies ahead for the geologic characterization of seismic sources in the 1990s? Certain trends are already established. Fault segmentation, with all of its implications for fault behavior, will continue to be a focus of research and application. Perhaps by the end of the decade the answer to the question "Can we identify ahead of time what part of a fault will rupture, where will the rupture start, and where will it end?" will, for many sources, be yes. One important aspect of segmentation that deserves careful consideration is the relation between fault zone structure, dynamic rupture, and the resulting ground motions. This is a promising area of investigation that requires the integration of strong motion data from historical events with detailed information on the surface structure of faults and their structure and physical properties at depth.

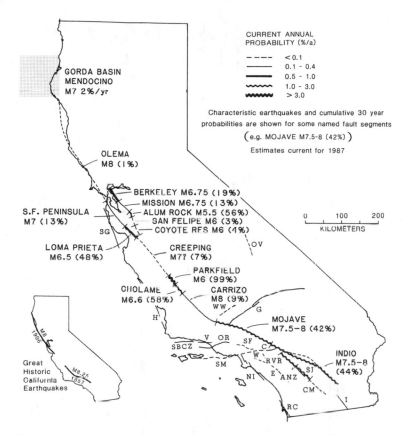

FIGURE 23. Annual and 30-year probabilities for selected segments of the San Andreas fault. (From Lindh, 1987.)

Obtaining more precise earthquake recurrence intervals will also be an area of expanding research. The ability to work backward in time and decipher past patterns of earthquake activity is critical for understanding the future behavior of faults. This is being aided by new technology and techniques in age dating that are allowing us to determine the timing of paleoearthquakes with greater precision and to date deposits that were previously considered undatable. An intriguing allied idea in the early stages of discussion is that of the "megatrench." This would be a series of engineered, deep trenches excavated to expose longer intervals of the geologic record. These could greatly expand our understanding of patterns of earthquake recurrence and the seismic cycle.

There is also little doubt that a great deal of effort will be put into modifying existing and developing new hazard models and methodologies, largely probabilistic, that incorporate fault parameters (slip rate, recurrence intervals, segmentation models) and their uncertainties in estimates of long-term earthquake potential.

We often try to force the earth to obey patterns of behavior or work in a way we perceive as correct. The basis for this viewpoint is generally limited. It is quite apparent that there is a spectrum

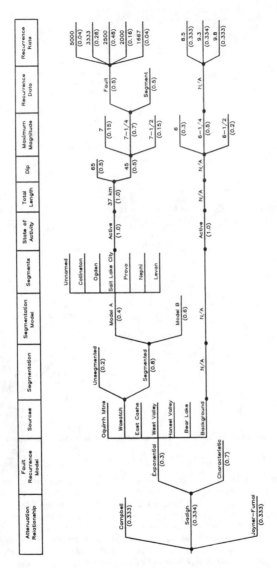

FIGURE 24. Hazard model logic tree. Parameter values are shown above the branches and assessed probabilities are shown in parentheses. (From Youngs *et al.*, 1987.)

of fault and earthquake behavior. Geological studies have and will continue to expand the window from which to view the occurrence of earthquakes in time and space. It is with this increasing perspective and understanding of the variability and uncertainty in the earthquake processes that earth scientists and engineers will be able to make the most reasonable decisions regarding seismic source characterization and seismic design. The 1990s will be an exciting time for the earthquake geologist.

ACKNOWLEDGMENTS

I would like to thank Bob Wallace and Bill Joyner for their insightful comments on the paper. Carol Sullivan prepared the manuscript. The responsibility for the views and opinions expressed in the paper is mine alone. Many of these have been developed from years of exchanging observations and points of view, often in strange and out-of-the-way places, with my colleagues in the fraternity of Holocene faulting geology.

REFERENCES

Aki, K., 1966, Generation and propagation of G-waves from the Niigata earthquake of June 19, 1964. 2. Estimation of earthquake movement, released energy, and stress strain drop from G-wave spectrum, *Bull. Earthquake Res. Inst., (Tokyo Univ.)* 44, 23–88.

Aki, K., 1979, Characterization of barriers on an earthquake fault, *J. Geophys. Res.* 84, 6140–6148.

Aki, K., 1984, Asperities, barriers, and characteristic earthquakes, *J. Geophys. Res.* 89, 5867–5872.

Allen, C. R., 1968, The tectonic environments of seismically active and inactive areas along the San Andreas fault system, in *Proceedings of the Conference on Geological Problems of the San Andreas System*, W. R. Dickinson and A. Granz, eds., Stanford University Publications Geological Science 11, 70–82.

Allen, C. R., 1986, Seismological and paleoseismological techniques of research in active tectonics, in *Active Tectonics*, National Academy Press, Washington, D.C., 148–154.

Anderson, J. G., 1979, Estimating the seismicity from geological structure for seismic-risk studies, *Bull. Seismol. Soc. Am.* 69, 163–168.

Anagnos, T., and A. S. Kiremidjian, 1984, Temporal dependence in earthquake occurrence, in *Proceedings of the Eighth World Conference on Earthquake Engineering* 1, 255–262.

Atwater, B. F., 1987, Evidence for Great Holocene earthquakes along the outer coast of Washington State, *Science* 236, 942–944.

Atwater, B. F., A. G. Hull, and K. A. Bevis, 1987, Aperiodic Holocene recurrence of widespread, probably coseismic subsidence in southwestern Washington (abstract), *EOS* 68, 1468.

Bakun, W. H., and A. G. Lindh, 1985, The Parkfield, California, earthquake prediction experiment, *Science* 229, 619–624.

Bakun, W. H., and T. V. McEvilly, 1984, Recurrence models and the Parkfield, California, earthquakes, *J. Geophys. Res.* 89, 3051–3058.

Båth, M., 1981, Earthquake recurrence of a particular type, *Pure Appl. Geophys.* 119, 1063–1076.

Båth, M., 1983, Earthquake frequency and energy in Greece, *Tectonophysics* 95, 1233–1252.

Birkeland, P. W., 1985, Quaternary soils of the western United States, in *Soils and Quaternary Landscape Evolution*, edited by J. Boardman, John Wiley, New York, 303–324.

Bonilla, M. G., and J. M. Buchanan, 1970, Interim report on worldwide historic surface faulting, *U.S. Geol. Surv. Open-File Rep.*, 32 pp

Bonilla, M. G., R. K. Mark, and J. J. Lienkaemper, 1984, Statistical relations among earthquake magnitude, surface rupture length, and surface fault displacement, *Bull. Seismol. Soc. Am.* 74, 2379–2411.

Brune, J. N., 1968, Seismic movement, seismicity and rate of slip along major fault zones, *J. Geophys. Res.* 73, 777–784.

Bruhn, R. L., P. R. Gibler, and W. T. Pamy, 1987, Rupture characteristics of normal faults: an example from the Wasatch fault zone, Utah, in *Continental Extensional Tectonics*, M. P. Coward, J. F. Dewey and P. L. Hancock, eds., *Geol. Soc. London Special Publication* 28, 337–353.

Campbell, K. W., 1977, The use of seismotectonics in Bayesian estimation of seismic risk, Report No. UCLA–ENG–7744, School of Engineering and Applied Sciences, University of California, Los Angeles.

Clark, M. M., K. K. Harms, J. J. Lienkaemper, J. A. Perkins, M. J. Rymer, and R. V. Sharp,

1983, The May 2, 1983 earthquake at Coalinga, California: the search for surface faulting, in *The Coalinga Earthquake Sequence Commencing May 2, 1983*, *U.S. Geol. Surv. Open-File Rep.* 83–511, 8–11.

Clark, M. M., J. J. Lienkaemper, D. S. Harwood, K. R. Lajoie, J. C. Matti, J. A. Perkins, M. J. Rymer, Sarna–Wojcicki, R. V. Sharp, J. D. Sims, J. C. Tinsley, and J. I. Ziony, 1984, Preliminary slip-rate table for late Quaternary faults of California, *U.S. Geol. Surv. Open-File Rep.* 84–106, 5 plates, 12 pp.

Cluff, L. S., A. S. Patwardhan, and K. J. Coppersmith, 1980, Estimating the probability of occurrences of surface faulting earthquakes on the Wasatch fault zone, Utah, *Bull. Seismol. Soc. Am.* 70, 463–478.

Colman, S. M., and K. Watson, 1983, Ages estimated from a diffusion equation model for scarp degradation, *Science* 221, 263–265.

Cornell, C. A., 1968, Engineering seismic risk analysis, *Bull. Seismol. Soc. Am.* 58, 1583–1606.

Crone, A. J., Introduction to directions in paleoseismology, in *Proceedings of Conference XXXIX, Directions in Paleoseismology*, A. J. Crone and E. M. Omdahl, eds., *U.S. Geol. Surv. Open-File Rep.* 87–673, 1–6.

Crone, A. J., and M. N. Machette, 1984, Surface faulting accompanying the Borah Peak earthquake, central Idaho, *Geology* 12, 664–667.

Crone, A. J., M. N. Machette, M. G. Bonilla, J. J. Lienkaemper, K. L. Pierce, W. E. Scott, and R. C. Bucknam, Surface faulting accompanying the Borah Peak earthquake and segmentation of the Lost River fault zone, central Idaho, *Bull. Seismol. Soc. Am.* 77, 739–770.

Donahue, D. J., T. H. Zabel, A. J. T. Jull, P. E. Damon, and K. H. Purser, 1983, Results of tests and measurements from the NSF regional accelerator facility for radioisotope dating, *Radiocarbon* 25, 719–728.

Doser, D. I., and R. B. Smith, 1982, Seismic moment rates in the Utah region, *Bull. Seismol. Soc. Am.* 72, 525–555.

Eaton, J. P., 1985, The May 2, 1983 Coalinga earthquake and its aftershocks: a detailed study of the hypocenter distribution and the focal mechanisms of the larger aftershocks, in *Proc. of Workshop XXVII, Mechanics of the May 2, 1983, Coalinga Earthquake*, M. J. Rymer and W. L. Ellsworth, eds., *U.S. Geol. Surv. Open-File Rep.* 85–44, 132–201.

Forman, S. L., 1987, The application of thermoluminescence (TL) dating to normal faulted terrain, in *Proceedings of Conference XXXIX, Directions in Paleoseismology*, A. J. Crone and E. M. Omdahl, eds., *U.S. Geol. Surv. Open-File Rep.* 87–673, 42–49.

Grandori, G., E. Guargenti, and V. Petrini, 1984, On the use of renewal processes in seismic hazard analysis, in *Proceedings of the Eighth World Conference on Earthquake Engineering* 1, 287–293.

Hait, M. J., Jr., and W. E. Scott, 1975, Holocene faulting, Lost River range, Idaho (abstract), *Geol. Soc. Am. Abstr. Prog.* 10, 217.

Hall, N. T., 1984, Holocene history of the San Andreas fault between Crystal Springs reservoir and San Andreas dam, San Mateo County, California, *Bull. Seismol. Soc. Am.* 74 281–300.

Hanks, T. C., 1985, The national earthquake hazards reduction program—scientific status, *U.S. Geol. Surv. Bull.* 1659, 40 pp.

Hanks, T. C., and H. Kanamori, 1979, A moment-magnitude scale, *J. Geophys. Res.* 84, 2981–2987.

Hanks, T. C., and D. P. Schwartz, 1987, Morphologic dating of the pre-1983 fault scarp on the Lost River fault at Doublespring Pass Road, Custer County, Idaho, *Bull. Seismol. Soc. Am.* 77, 837–846.

Hanks, T. C., and R. E. Wallace, 1985, Morphological analysis of the Lake Lahontan shoreline and beachfront fault scarps, Pershing County, Nevada, *Bull. Seismol. Soc. Am.* 75, 835–846.

Hanks, T. C., R. C. Bucknam, K. R. Lajoie, and R. E. Wallace, 1984, Modification of wavecut and faulting controlled landforms, *J. Geophys. Res.* 89, 5771–5790.

Harden, J. W., 1982, A quantitative index of soil development from field descriptions: examples from a chronosequence in central California, *Geoderma* 28, 1–28.

Harden, J. W., and E. M. Taylor, 1983, A quantitative comparison of soil development in four climatic regimes, *Quaternary Research* 20, 342–359.

Jacoby, G. C., P. R. Sheppard, and K. E. Sieh, 1987, Was the 8 December 1812 California earthquake produced by the San Andreas fault? Evidence from trees near Wrightwood (abstract), *Seismol. Soc. Am. Proc. 82nd Annual Mtg., Seismol. Res. Lett.* 58, 14.

Kameda, H., and H. Ozaki, 1979, A renewal process model for use in seismic risk analysis, *Mem. Faculty of Engineer., (Kyoto Univ.)* VXLI, 11–35.

King, G., and G. Yielding, 1984, The evolution of a thrust fault system: Processes of rupture initiation, propagation and termination in the 1980 El Asnam (Algeria) earthquake, *Geophys. J. R. astr. Soc.* 77, 913–933.

King, G. C. P., and J. Nabelek, 1985, Role of fault bands in the initiation and termination of earthquake rupture, *Science* 228, 984–987.

Kromer, B., M. Rhein, M. Bruns, H. Schoch–Fischer, K. O. Münnich, M. Stuiver, and B. Becker, 1986, Radiocarbon calibration data for the 6th to the 8th millenia, BC, *Radiocarbon* 28, 954–960.

Kulkarni, R. B., and A. S. Patwardhan, 1980, Characterization of temporally varying seismic exposure function, in *International Conference on Engineering for Protection from Natural Disaster*, Asian Inst. Tech., Bangkok, 415–425.

Kulkarni, R. B., R. R. Youngs, and K. J. Coppersmith, 1984, Assessment of confidence intervals for results of seismic hazard analysis, *Proceedings of the Eighth World Conference on Earthquake Engineering* 1, 263–270.

Kutschera, W., 1983, Accelerator mass spectrometry: from nuclear physics to dating, *Radiocarbon* 25, 677–691.

Lindh, A. G., 1983, Preliminary assessment of long-term probabilities for large earthquakes along selected segments of the San Andreas fault system in California, *U.S. Geol. Surv. Open-File Rep.* 83–63, 5 pp.

Lindh, A. G., 1987, Estimates of long-term probabilities for large earthquakes along selected fault segments of the San Andreas fault system in California, in *Proceedings of Symposium on Earthquake Prediction—Present Status*, University of Poona, in press.

Lund, W. R., and D. P. Schwartz, 1986, Fault behavior and earthquake recurrence at the Dry Creek site, Salt Lake segment, Wasatch fault zone, Utah (abstract), *EOS* 67, 1107.

Machette, M. N., 1985, Calcic soils of the southwestern United States, in *Soils and Quaternary Geology of the Southwestern United Stated*, edited by D. L. Weide, *Geol. Soc. Am. Bull.* 203, 1–21.

Machette, M. N., 1986, History of Quaternary offset and paleoseismicity along the La Jencia fault, central Rio Grande rift, New Mexico, *Bull. Seismol. Soc. Am.* 76, 259–272.

Machette, M. N., S. F. Personius, and A. R. Nelson, 1987, Quaternary geology along the Wasatch fault zone—Segmentation, recent investigations, and preliminary conclusions in *Assessment of Regional Earthquake Hazards and Risk along the Wasatch Front, Utah*, W. W. Hays and P. Gori, eds., *U.S. Geol. Surv. Open-File Rep.* 87–585, A1–A72.

Malde, H. E., 1987, Quaternary faulting near Arco and Howe, Idaho, *Bull. Seismol. Soc. Am.* 77, 847–867.

McFadden, L. D., and J. C. Tinsley, 1985, Rate and depth of pedogenic-carbonate accumulation in soils: formulation and testing of a compartment model, in *Soils and Quaternary Geology of the Southwestern United States*, edited by D. L. Weide, *Geol. Soc. Am. Bull.* 203, 23–41.

Molnar, P., 1979, Earthquake recurrence intervals and plate tectonics, *Bull. Seismol. Soc. Am.* 69, 115–133.

Nash, D. B., 1980, Morphologic dating of degraded normal fault scarps, *J. Geol.* 88, 353–360.

Nash, D. B., 1984, Morphologic dating of fluvial terrace scarps and fault scarps near West Yellowstone, Montana, *Geol. Soc. Am. Bull.* 95, 1413–1424.

National Academy Press, 1988, *Probabilistic Seismic Hazard Analysis*, in press.

Nishenko, S. P., 1985, Seismic potential for large and great interplate earthquakes along the Chilean and southern Peruvian margins of South America: a quantitative reappraisal, *J. Geophys. Res.* 90, 3589–3615.

Nishenko, S. P., and R. Buland, 1987, A generic recurrence interval distribution for earthquake forecasting, *Bull. Seismol. Soc. Am.* 77, 1382–1399.

Obermeier, S. R., G. S. Gohn, R. E. Weems, R. L. Gelinas, and M. Rubin, 1985, Geologic evidence for moderate to large earthquakes near Charleston, South Carolina, *Science* **227**, 408–412.

Ouchi, S., 1985, Response of alluvial rivers to slow active tectonic movement, *Geol. Soc. Am. Bull.* **95**, 504–515.

Papastamatiou, D., 1980, Incorporation of crustal deformation to seismic hazard analysis, *Bull. Seismol. Soc. Am.* **70**, 1321–1335.

Patwardhan, A. S., R. B. Kulkarni, and D. Tocher, 1980, A semi-Markov model for characterizing recurrence of great earthquakes, *Bull. Seismol. Soc. Am.* **70**, 323–347.

Pearson, G. W., and M. Stuiver, 1986, High-precision calibration of the radiocarbon time scale, AD1950–2500BC, *Radiocarbon* **28**, 839–862.

Pierce, K. L., 1986, Dating methods, in *Active Tectonics*, National Academy Press, Washington, D.C., 195–214.

Pierce, K. L., and S. M. Colman, 1986, Effect of height and orientation (microclimate) on geomorphic degradation rates and processes, late glacial terrace scarps in central Idaho, *Geol. Soc. Am. Bull.* **97**, 869–885.

Power, M. S., K. J. Copprsmith, R. R. Youngs, D. P. Schwartz, and E. H. Swan, 1981, Seismic exposure analysis for the WNP-2 and WNP-1/4 site: Appendix 2.5K to Amendment No. 18, Final Safety Analysis Report WNP-2, Washington Public Power Supply System.

Richter, C. F., 1958, *Elementary Seismology*, W. H. Freeman, San Francisco.

Russ, D. P., 1979, Late Holocene faulting and earthquake recurrence in the Redfoot Lake area, northwestern Tennessee, *Geol. Soc. Am. Bull.* **90**, 1013–1018.

Russ, D. P., 1982, Style and significance of surface deformation in the vicinity of New Madrid, Missouri, *U.S. Geol. Surv. Prof. Paper* **1236H**, 95–114.

Rymer, M. J., and W. L. Ellsworth, 1985, Mechanics of the May 2, 1983 Coalinga, California earthquake: an introduction, in *Proc. Workshop XXVII, Mechanics of the May 2, 1983, Coalinga Earthquake*, M. J. Rymer and W. L. Ellsworth, eds., *U.S. Geol. Surv. Open-File Rep.* **85–44**, 1–3.

Rymer, M. J., K. K. Harms, J. J. Lienkaemper, and M. M. Clark, 1985, Rupture of the Nuñez fault during the Coalinga earthquake sequence, in *Proc. Workshop XXVII, Mechanics of the May 2, 1983, Coalinga Earthquake*, M. J. Rymer and W. L. Ellsworth, eds., *U.S. Geol. Surv. Open-File Rep.* **85–44**, 294–312.

Savy, J. B., H. C. Shah, and D. Boore, 1980, Nonstationary risk model with geophysical input, *J. Struc. Div. ASCE* **106**, 145–164.

Schumm, S. A., 1986, Alluvial river response to active tectonics, in *Active Tectonics*, National Academy Press, Washington, D.C., 80–94.

Schwartz, D. P., 1987, Earthquakes of the Holocene, *Reviews of Geophysics* **25**, 1197–1202.

Schwartz, D. P., and K. J. Coppersmith, 1984, Fault behavior and characteristic earthquakes: Examples from the Wasatch and San Andreas faults, *J. Geophys. Res.* **89**, 5681–5698.

Schwartz, D. P., and K. J. Coppersmith, 1986, Seismic hazards: new trends in analysis using geologic data, in *Active Tectonics*, National Academy Press, Washington, D.C., 215–230.

Schwartz, D. P., and A. J. Crone, 1985, The 1983 Borah Peak earthquake: A calibration event for quantifying earthquake recurrence and fault behavior on Great Basin normal faults, in *Proceedings of Workshop XXVIII on the Borah Peak, Idaho, Earthquake*, R. S. Stein and R. C. Bucknam, eds., *U.S. Geol. Surv. Open-File Rep.* **85–290**, 153–160.

Schwartz, D. P., and A. J. Crone, 1988, Paleoseismicity of the Lost River fault zone, Idaho: earthquake recurrence and segmentation (abstract), *Geol. Soc. Am. Abstr. Prog.* **20**, 228.

Schwartz, D. P., K. J. Coppersmith, and F. H. Swan, III, 1984, Methods for estimating maximum earthquake magnitudes, in *Proceedings of the Eighth World Conference on Earthquake Engineering* **1**, 279–285.

Scott, W. E., K. L. Pierce, and M. H. Hait, Jr., 1985, Quaternary tectonic setting of the Borah Peak earthquake, central Idaho, *Bull. Seismol. Soc. Am.* **75**, 1053–1056.

Shimazaki, K., and T. Nakata, 1980, Time-predictable recurrence model for large earthquakes, *Geophys. Res. Lett.* **7**, 279–282.

Sibson, R. H., 1985, Stopping of earthquake ruptures at dilatational fault jogs, *Nature* **316**, 248–251.

Sibson, R. H., 1986, Rupture interaction with fault jogs, in *Earthquake Source Mechanics, Maurice*

Ewing Ser., S. Das, J. Boatwright, and C. H. Scholz, eds., AGU, Washington, D.C., 6, 156–167.

Sieh, K. E., Slip rate across the San Andreas fault and prehistoric earthquakes at Indio, California (abstract), *EOS* 67, 1200.

Sieh, K. E., 1978, Prehistoric large earthquakes produced by slip on the San Andreas fault at Pallett Creek, southern California, *J. Geophys. Res.* 83, 3907–3939.

Sieh, K. E., 1984, Lateral offset and revised dates of large earthquakes along the San Andreas fault at Pallett Creek, southern California, *J. Geophys. Res.* 89, 7641–7670.

Sieh, K. E., and R. H. Jahns, 1984, Holocene activity of the San Andreas fault at Wallace Creek, California, *Geol. Soc. Am. Bull.* 45, 883–896.

Sieh, K. E., M. Stuiver, and D. Brillinger, 1988, A very precise chronology of earthquakes produced by the San Andreas fault in southern California, paper submitted to *J. Geophys. Res.*

Singh, S. K., M. Rodriguez, and L. Esteva, 1983, Statistics of small earthquakes and frequency of large earthquakes along the Mexico subduction zone, *Bull. Seismol. Soc. Am.* 73, 1779–1796.

Slemmons, D. B., 1977, State-of-the-art for assessing earthquake hazards in the United States, Report 6: Faults and earthquake magnitude, U.S. Army Corps of Engineers, Waterways Experiment Station, Mis. Paper s–73–1, 129 pp.

Snay, R. A., R. B. Smith, and T. Soler, 1984, Horizontal strain across the Wasatch front near Salt Lake City, Utah, *J. Geophys. Res.* 89, 113–122.

Smith, R. B., and R. L. Bruhn, 1984, Intraplate extensional tectonics of the eastern Basin–Range: Inferences on structural style from seismic reflection data, regional tectonics, and thermal-mechanical models of brittle-ductile deformation, *J. Geophys. Res.* 89, 5733–5762.

Smith, S. W., 1976, Determination of maximum earthquake magnitude, *Geophys. Res. Lett.* 3, 351–354.

Stein, R. S., and G. C. P. King, 1984, Seismic potential revealed by surface folding: 1983 Coalinga, California earthquake, *Science* 224, 869–872.

Stuiver, M., and G. W. Pearson, 1986, High-precision calibration of the radiocarbon time scale, AD1950–500BC, *Radiocarbon* 28, 805–838.

Swan, F. H., III, D. P. Schwartz, and L. S. Cluff, 1980, Recurrence of moderate to large magnitude earthquakes produced by surface faulting on the Wasatch fault zone, Utah, *Bull. Seismol. Soc. Am.* 70, 1438–1462.

Swan, F. H., 1987, Temporal clustering of paleoseismic events on the Oued Fodda fault, Algeria, 1987, in *Proceedings of Conference XXXIX, Directions in Paleoseismology*, A. J. Crone and E. M Omdahl, eds., *U.S. Geol. Surv. Open File Rep.* 87–673, 239 248.

Sykes, L. R., and S. Nishenko, 1984, Probabilities of occurrence of large plate rupturing earthquakes for the San Andreas, San Jacinto, and Imperial faults, California, 1983–2003, *J. Geophys. Res.* 89, 5905 5927

Talwani, P., and J. Cox, 1985, Paleoseismic evidence for recurrence of earthquakes near Charleston, South Carolina, *Science* 229, 379–381.

Thatcher, W., 1984, The earthquake deformation cycle, recurrence, and the time-predictable model, *J. Geophys. Res.* 89, 5674–5680.

Thorson, R. M., W. S. Clayton, and L. Seeber, 1986, Geologic evidence for a large pre-historic earthquake in eastern Connecticut, *Geology* 14, 463–467.

Utsu, T., 1971, Aftershocks and earthquake statistics (III), *J. Fac. Sci. (Hokkaido Univ.) Ser. VII (Geophys.)* 3, 1337–1346.

Veneziano, D., and C. A. Cornell, 1976, Earthquake models with spatial and temporal memory for engineering seismic risk analysis, Department of Civil Engineering, MIT, R74–18, Cambridge, MA.

Wallace, R. E., 1970, Earthquake recurrence inervals on the San Andreas fault, California, *Geol. Soc. Am. Bull.* 81, 2875–2890.

Wallace, R. E., 1981, Active faults, paleoseismology and earthquake hazards in the Western United States, in *Earthquake Prediction: An International Review*, D. W. Simpson and P. G. Richards, eds., Maurice Ewing Series 4, American Geophysical Union, Washington, D.C., 209–216.

Wallace, R. E., 1984b, Patterns and timing of late Quaternary faulting in the Great Basin province and relation to some regional tectonic features, *J. Geophys. Res.* **89**, 5763–5769.

Wallace, R. E., 1987, Grouping and migration of surface faulting and variations in slip rates on faults in the Great Basin province, *Bull. Seismol. Soc. Am.* **77**, 868–876.

Weldon, R. J., II, and K. E. Sieh, 1985, Holocene rate of slip and tentative recurrence interval for large earthquakes on the San Andreas fault, Cajon Pass, southern California, *Geol. Soc. Am. Bull.* **95**, 793–812.

Wesnousky, S. G., 1986, Earthquakes, Quaternary faults, and seismic hazard in California, *J. Geophys. Res.* **91**, 12,587–12,631.

Wesnousky, S., C. H. Scholz, K. Shimazaki, and T. Matsuda, 1983, Earthquake frequency distribution and the mechanics of faulting, *J. Geophys. Res.* **89**, 9331–9340.

Wheeler, R. L., and K. B. Krystinik, 1987, Persistent and nonpersistent segmentation of the Wasatch fault zone, Utah—statistical analyses for evaluation of seismic hazard, in *Assessment of Regional Earthquake Hazards and Risk Along the Wasatch Front, Utah*, Volume 1, P. L. Gori and W. W. Hays, eds., *U.S. Geol. Surv. Open-File Rep.* **87–585**, B1–124.

Wintle, A. G., and D. J. Huntley, 1982, Thermoluminescence dating of sediments, *Quaternary Ser. Rev.* **1**, 31–53.

Wyss, M., 1979, Estimating maximum expectable magnitude of earthquakes from fault dimensions, *Geology* **7**, 336–340.

Yeats, R. S., 1986, Active faults related to folding, in *Active Tectonics*, National Academy Press, Washington, D. C., 63–79.

Yielding, G., J. A. Jackson, G. C. P. King, H. Sinuhal, C. Vita–Finzi, and R. M. Wood, 1981, Relations between surface deformation, fault geometry, seismicity and rupture characteristics during the El Asnam (Algeria) earthquake of 10 October 1980, *Earth Planet. Sci. Lett.* **56**, 287–304.

Youngs, R. R., and K. J. Coppersmith, 1985, Implications of fault slip rates and earthquake recurrence models to probabilistic seismic hazard estimates, *Bull. Seismol. Soc. Am.* **75**, 939–964.

Youngs, R. R., F. H. Swan, M. S. Power, D. P. Schwartz, and R. K. Green, 1987, Probabilistic analysis of earthquake ground shaking hazard along the Wasatch front, Utah, in *Assessment of Regional Earthquake Hazards and Risk Along the Wasatch Front, Utah*, Volume 1, P. L. Gori and W. W. Hays, eds., *U.S. Geol. Surv. Open-File Rep.* **87–585**, N1–48.

Zoback, M. L., 1983, Structure and Cenozoic tectonism along the Wasatch fault zone, *Geol. Soc. Am. Mem.* **157**, 3–27.

MEASUREMENT, CHARACTERIZATION, AND PREDICTION
OF STRONG GROUND MOTION

WILLIAM B. JOYNER AND DAVID M. BOORE*

The estimation of ground motion in future earthquakes for engineering purposes is one of the primary motivations for the measurement and processing of strong-motion data. The response spectrum is the best representation of ground motion because it takes account of the natural frequencies of structures. The conventional practice of using peak acceleration to scale standard response spectral shapes is likely to lead to serious error, except at high frequencies, because the shapes of response spectra depend strongly on magnitude and local geologic site conditions. Magnitude, distance, and site conditions are the principal variables used in predicting future ground motions. A number of predictive relationships derived from regression analysis of strong motion data are available for horizontal peak acceleration, velocity, and response spectral values. Theoretical prediction of ground motion calls for stochastic source models because source heterogeneities control the amplitude of ground motion at most, if not all, frequencies of engineering interest. Stochastic source models have been used for predicting ground motion in regions such as eastern North America where little recorded data are available. Ground motion predictions for large earthquakes have also been made by summation of recordings of smaller earthquakes. This technique does not take proper account of directivity except in very special circumstances. Aside from directivity, it is possible to do the summation in such a way that the low-frequency and high-frequency limits of the spectrum of the predicted motion obey appropriate scaling laws, but the spectrum may be deficient in amplitude at intermediate frequencies.

INTRODUCTION

There has been great progress in the last ten years in the study of strong earthquake ground motion and its engineering applications. New data and analysis have provided the basis for more reliable empirical estimates of ground motion in future earthquakes. Theoretical methods have been developed for estimation of ground-motion parameters and simulation of ground-motion time series. These methods are particularly helpful for regions such as eastern North America where strong-motion data are sparse. In what follows we survey the field, first reviewing developments in ground-motion measurement and data processing. We then consider the choice of parameters for characterizing strong ground motion and describe the wave-types involved in strong ground motion and the factors affecting ground-motion amplitudes. We conclude by describing methods for predicting ground motion.

Our intent was to make this paper a comprehensive and self-contained manual describing the newer methods of ground-motion prediction. We feel that this goal justifies the added detail and complexity in the paper and the introduction of much material from seismology that many engineers may find unfamiliar.

---

* U.S. Geological Survey, 345 Middlefield Road MS 977, Menlo Park, CA 94025

43

MEASUREMENT AND PROCESSING OF STRONG GROUND-MOTION DATA

Measurement

Since the time the first strong-motion record was obtained in 1933 most strong-motion data have been recorded on accelerographs with photographic recording. These instruments are triggered by the motion itself, and some part of the initial motion is therefore lost. Numerical calculations with the data require that the photographic film or paper record be digitized. At first this was done manually; later the process was automated. Now, digital recording instruments using force-balance accelerometers are coming into wider use (Anderson et al., 1983). Digital recording eliminates the delays and loss of accuracy associated with digitizing film or paper records and also permits recovery of the initial portion of the signal. Borcherdt et al. (1984) give a tabulation and discussion of the characteristics of digital recording systems available in the United States. As more experience is gained with digital recorders their use can be expected to increase further, but conventional recorders are still predominant, and it will be many years before they can be replaced. Of particular interest among new developments in digital instrumentation is the GEOS recording system (Borcherdt et al., 1985), a broad-band system with 16-bit dynamic range that facilitates the simultaneous recording of large and small motions at the same site. GEOS is especially suited for aftershock studies (Borcherdt et al., 1983; Boatwright, 1985).

Special attention has recently been devoted to the collection and analysis of data from spatially distributed arrays and networks of recording instruments. By analyzing data recorded on the El Centro differential array during the 1979 Imperial Valley, California, earthquake Spudich and Cranswick (1984) were able to show that the ground motion consisted of waves radiated from a compact region around the rupture front and that the rupture front progressed more or less coherently from the hypocenter to the limits of the rupture. Data from the SMART 1 digital array in Taiwan have been used to examine the spatial coherence of ground motion (Bolt et al.,1984; Abrahamson, 1985). Because of the earthquake that has been predicted at Parkfield, California, a network of strong-motion instruments has been established there by the California Strong-Motion Instrumentation Program (McJunkin and Shakal, 1983), and special strong-motion arrays arrays have been installed there by the Electric Power Research Institute and the California Strong-Motion Instrumentation Program and by the U.S. Geological Survey. At Anza, California, a broad-band, wide-dynamic-range, digital network of short-period seismometers supplemented by accelerographs has been set up to study earthquake source processes in a seismic gap along the San Jacinto fault (Berger et al., 1984).

One of the most important data sets in strong-motion seismology is that recorded in the 1979 Imperial Valley, California, earthquake because it gives good coverage of the near-source region for a shallow earthquake of moment magnitude as large as 6.5. The outstanding need now is for near-source data from shallow earthquakes of larger magnitude. Other significant data sets recently recorded are listed in Table 1. The 1985 earthquakes in Chile and Mexico are of particular importance because they contributed the first and only extensive data sets from the near-source regions of large subduction-zone earthquakes.

Processing

Early methods of strong-motion data processing consisted of subtracting a best-fitting parabola from the accelerogram before integrating to velocity and displacement (references on early methods are given by Trifunac, 1971). Subtraction of the parabola—called baseline correction—had no physical basis, but served to remove long-period errors that would be grossly magnified by double integration. Modern data-processing methods are derived from proposals by Trifunac (1971, 1972) that a high-pass digital filter be used for the baseline correction and a centered finite-difference approximation be used to correct for the effect of the instrument.

TABLE 1.   RECENT STRONG-MOTION DATA SETS

Earthquake	Day Month Year GMT	Magnitude	References
Coalinga, California	2 5 83	M_L 6.7	Shakal and McJunkin, 1983 Maley et al., 1983
Borah Peak, Idaho	28 10 83	M_S 7.3	Jackson and Boatwright, 1985
Morgan Hill, California	24 4 84	M_L 6.2	Brady et al., 1984a, b Huang et al., 1985 Shakal et al., 1986a
Central Chile	3 3 85	M_S 7.8	Saragoni, 1985
Michoacan, Mexico	19 9 85	M_S 8.1	Anderson et al., 1986
Nahanni, Canada	23 12 85	M_S 6.9	Weichert et al., 1986
Palm Springs, California	8 7 86	M_L 5.9	Huang et al., 1986 Porcella et al., 1987a
Chalfant Valley, California	21 7 86	M_L 6.0	Maley et al., 1986
San Salvador, El Salvador	10 10 86	M_S 5.4	Shakal et al., 1986b
Whittier Narrows, California	1 10 87	M_L 6.1	Etheredge and Porcella, 1987 Shakal et al., 1987
Superstition Hills, California	24 11 87	M_L 5.8 M_L6.0	Huang et al., 1987 Porcella et al., 1987b

The processing scheme initially developed from Trifunac's proposals (Trifunac and Lee, 1973) was used successfully on hundreds of seismograms, but there are two problems with it. The finite-difference approximation used for the instrument correction is satisfactory if the sampling rate is high enough, but at 50 samples per second, the rate used for most records processed before 1975, the approximation is poor for frequencies above about 10 Hz (Raugh, 1981; Shyam Sunder and Connor, 1982). Figure 1 compares the response of the finite-difference instrument-correction filter for sample rates of 50 and 200 samples per second with the exact response of the ideal filter for a natural frequency of 25 Hz and damping of 0.6 critical. Most present-day record processing is done at 200 samples per second, a rate at which the finite-difference approximation is adequate as shown in Figure 1. Other methods of instrument correction, however, are preferable. The correction can be applied exactly in the frequency domain, or it can be implemented to arbitrarily high accuracy with a time-domain convolution filter (Raugh, 1981; Shyam Sunder and Connor, 1982).

Shyam Sunder and Connor (1982) pointed out a second problem with the method described by Trifunac and Lee (1973). In the method an acausal high-pass filter is applied and then the result is integrated without including the filter transient that precedes the start time of the unfiltered data. Integrating an acausally filtered record without including the leading filter transient has the effect of shifting the result by a constant amount and can be thought of as reintroducing the low frequencies that were removed by the filter. In the method described by Trifunac and Lee the effects of excluding the leading filter transient were removed later by filtering the velocity and filtering the displacement. For that and other reasons, however, the method was unnecessarily complex. If the filter transients are included, it is possible to use a simple processing scheme consisting only of instrument correction and high- and low-pass filtering of the acceleration, followed by integration to velocity and displacement. The filter transients should also be included in the computation of response spectra; otherwise, spurious long periods will be incorporated into the result. The low-pass filter is necessary if instrument correction is to be performed because the instrument correction strongly amplifies any high-frequency noise present in the record (Figure 1). Because the accelerograph response is flat up to about half of the natural frequency and the natural frequency is typically about 25 Hz for modern instruments, instrument correction is necessary only for the sake

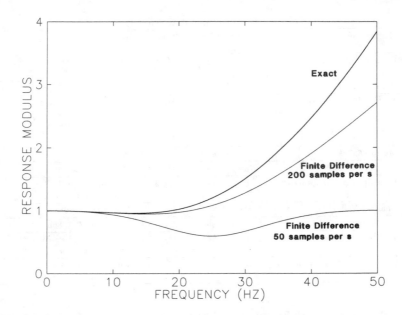

FIGURE 1. The exact response of the ideal instrument-correction filter for a natural frequency of 25 hz and damping of 0.6 critical compared with the response of the finite-difference approximation for 50 and 200 samples per s.

of the high-frequency components of the record. The high-pass filter is generally necessary because any low-frequency noise present in the record is strongly amplified by the double integration to displacement and by the computation of response spectral values at long periods. Direct use of time-domain convolution for the high-pass filter operation requires excessive computer time, much more than recursive filters or frequency-domain filters. Each of the processing stages may be done in a number of different ways. To insure satisfactory results, however, care must be taken with respect to certain key points. If filtering is done in the frequency domain, the record must be padded with enough zeros to avoid wrap-around error. If bidirectional recursive filtering is done, the record must be padded with enough zeros to accomodate the filter transient at the end of the record. If acausal filters are used, as stated previously, subsequent integration and the calculation of response spectra must include the filter transient that precedes the origin time of the original record. It is advisable to avoid using high-pass filters with too sharp a cutoff. Such filters may produce oscillations at frequencies near the cutoff frequencies, as shown by Fletcher et al. (1980).

In the last few years the U.S.Geological Survey has completely revised its standard processing scheme (Raugh, 1981; Converse et al., 1984). Previously the Survey used the methods described by Trifunac and Lee (1973). The revised scheme combines the low-pass filter and instrument correction and offers the option of implementing the combination either in the frequency domain or by time-domain convolution. An advantage of combining the low-pass filter and instrument correction is a substantial reduction in the number of filter weights required for specified accuracy in the time-domain convolution for the instrument correction. For routine operations the high-pass filter is a bidirectional Butterworth recursive filter, although other options are available. The current version contains provision for including the leading filter transient in subsequent computations, but that feature is not described in the latest published manual (Converse, 1984). The California

Strong-Motion Instrumentation Program uses a processing scheme that is a modified version (Shakal and Ragsdale, 1984) of the methods described by Trifunac and Lee (1973). Shyam Sunder and Connor (1982) proposed a scheme in which the instrument correction is performed by time-domain convolution and the low- and high-pass filters are combined and implemented by an elliptical recursive filter. Khemici and Chiang (1984) describe a system in which instrument correction, high- and low-pass filtering, and integration are all combined into one operation in the frequency domain.

Record-processing procedures are available which, if properly applied, are entirely adequate for the ranges of frequency and amplitude that have engineering significance. The most critical problem in applying the record-processing procedures is choosing the cutoff frequency for the high-pass filter. The best recent approach to that problem (Shakal and Ragsdale, 1984; Lee and Trifunac, 1984; Gerald Brady, written communication, 1987; Kenneth Campbell, oral communication, 1987) is to digitize one of the fixed traces which are present on the records from most modern analog strong-motion instruments and use the results to estimate the spectrum of the noise in the processing system. The proper cutoff frequency is chosen on the basis of a comparison between the spectrum of the noise and the spectrum of the accelerograph trace. Whether a filter with a given cutoff frequency removes a significant portion of the signal can be judged with the help of some results from seismology. For earthquakes large enough to be of engineering interest, the Brune (1970, 1971) spectrum of acceleration, which has been shown to be a good description of strong ground motion (Hanks and McGuire, 1981; Boore, 1983; Anderson and Hough, 1984), is approximately flat at intermediate frequencies down to a corner frequency f_0 below which the spectrum is proportional to frequency squared. The corner frequency is given by

$$f_0 = 4.9 \times 10^6 \beta (\Delta\sigma/M_0)^{1/3}$$

where the units of f_0, β, $\Delta\sigma$, and M_0 are Hz, km/s, bars, and dyne-cm, respectively. For $\beta = 3.2$ km/sec and $\Delta\sigma = 100$ bars (Boore, 1983) the formula simplifies to $f_0 = 10^{-(M-5)/2}$, where M is moment magnitude. Unless the cutoff frequency is significantly less than the corner frequency, application of a high-pass filter should be expected to modify the signal substantially. For example, an earthquake of magnitude 6 has a corner frequency of about 0.3 Hz according to the formula. Unless the cutoff frequency is much smaller than 0.3, we should expect the signal from a magnitude 6 earthquake to be significantly changed by filtering.

Most strong-motion accelerograph data include low-frequency noise sufficient to make high-pass filtering mandatory, and the permanent displacement is thereby lost. If data of sufficient accuracy are obtained, and if the instrument is not subjected to tilt, then it should be possible to forego high-pass filtering and doubly integrate the acceleration to displacement, retaining the permanent displacement. In the case of triggered accelerographs this feat would require determination of the initial velocity. The processing system used by the U.S. Geological Survey (Converse, 1984) contains a procedure for obtaining the initial velocity. The accelerogram is first integrated to velocity without high-pass filtering. The final portion of the velocity trace after the strong motion has subsided is then fit by a least-squares straight line, on the assumption that the velocity must be zero after the strong motion ceases. In fitting the straight line, tapered weights are applied at the beginning and end of the fitted segment to prevent partial cycles of low-level motion from biasing the determination of the line. The tapered weights are cosine half bells. The fitted line is extended to the start time of the record and subtracted, point by point, from the velocity, which is then integrated to give displacement. The slope of the line fitted to velocity is subtracted from the acceleration.

Digital recorders preserve the initial motion, simplifying the task of determining permanent displacement. Iwan et al. (1985) have proposed a scheme for processing digitally recorded data that replaces the high-pass filtering operation by corrections to make the average of both the acceleration and velocity zero over the final portion of the record. The permanent displacement is preserved in this scheme. A constant correction is applied to the portion of the record corresponding to the time of strong shaking, and a different constant correction is applied to the remainder of the record after the end of strong shaking. The values of the two different corrections are chosen so as to achieve the desired zero values for average velocity and average acceleration over the final portion of the record. Anderson et al. (1986) applied this scheme to determine permanent displacement at accelerograph sites in the epicentral region of the 1985 Michoacan, Mexico earthquake, including one site where the resulting vertical displacement of nearly 1 m was confirmed by observations of coastal uplift. The scheme will only work, of course, with data of sufficient relative accuracy.

CHARACTERIZATION OF STRONG GROUND MOTION

A number of different parameters may be used to characterize strong ground motion for purposes of seismic design. These include peak acceleration, peak velocity, response spectral values, and Fourier spectral values. Peak displacement has also been suggested, but peak displacement is too sensitive to the somewhat arbitrary choice of high-pass filter cutoff used in record processing. The most useful of these parameters are response spectral values. The response spectrum is the basis, either directly or indirectly, of most earthquake-resistant design. It may be used in the dynamic analysis of structures, and it is the basis for the relation, in building codes, between the lateral-design-force coefficient and the period of the building (Structural Engineers Association of California, 1980; Applied Technology Council, 1978). The response spectrum is useful because it represents the response, to a given ground motion, of a set of simple mathematical models of structures. It can be defined as the maximum response, to the given motion, of a set of single-degree-of freedom oscillators (for example, mass-spring systems) having different natural periods and damping. The response spectrum as customarily defined represents the response of a damped elastic system and does not incorporate the nonlinear response to be expected from real structures at high levels of motion. Nonlinear response spectra can be computed, but they would be different for different kinds of structures. As a practical matter, linear response spectra, coupled with response-reduction factors calculated by nonlinear analysis of particular structural types, are more likely to be used than nonlinear response spectra (Cornell and Sewell, 1987).

Although response spectral values may be the most useful of the parameters describing ground motion, the most emphasis in engineering practice, at least in the past, has been placed on peak horizontal acceleration. The conventional method for estimating response spectral values uses peak horizontal acceleration to scale some normalized spectral shape such as the Nuclear Regulatory Commission's Regulatory Guide 1.60 spectrum (U.S. Atomic Energy Commission, 1973). Such a procedure would be generally valid only if the shape of the response spectrum were independent of earthquake magnitude, source distance, and recording site conditions. In fact, a number of independent studies (McGuire, 1974; Mohraz, 1976; Trifunac and Anderson, 1978, Joyner and Boore, 1982) have shown that the shape of the response spectrum is strongly dependent on magnitude and site conditions, and, at frequencies less than about 3 Hz, large errors can result from the practice of scaling fixed spectral shapes by peak acceleration. These errors can be partially avoided by Newmark and Hall's (1969) method, in which the short-period portion of the spectrum is proportional to peak acceleration, the intermediate portion (about 0.3 to 2.0 sec) to peak velocity, and the long-period portion to peak displacement. The proportionality factor between velocity and intermediate period response, however, varies significantly with magnitude and site conditions as indicated by our predictive equations for response and for peak velocity (Joyner and Boore, 1982). Peak displacement, moreover, is, as previously mentioned, a quantity that is highly sensitive to the choice of high-pass filter cutoff. The obvious solution is to predict response values directly (e.g. Joyner and Boore, 1982). Another approach is the use of peak acceleration to scale a normalized spectral shape that varies with magnitude and site conditions (Idriss, 1985, 1987).

The search for a single parameter to characterize ground motion is clearly doomed to failure. Because the shape of the spectrum changes with magnitude and site conditions, a single parameter that represents ground motion well at one frequency must necessarily fail to do so at others.

In general, the study of ground motion has been focused on the horizontal component because of its greater engineering significance, and this review will not deal explicitly with the vertical component. If needed, estimates of vertical motion can be obtained from procedures similar to those discussed in this paper for estimating the horizontal component of motion.

PREDICTION OF STRONG GROUND MOTION

Factors That Affect Strong Ground Motion

Wave Types Involved in Strong Ground Motion. Horizontal ground motion is produced by S body waves (shear waves that travel through the earth) and by surface waves (waves that propagate along the surface). Where velocity increases with depth, which is the usual case, fundamental-mode surface waves tend to arrive later than body waves and may be distinguished by their later arrival. Higher-mode surface waves may arrive at more nearly the same time as body waves and may therefore be difficult to distinguish from body waves; in fact the distinction between body waves and higher-mode surface waves is not always meaningful.

The engineering importance of surface waves is sometimes overstated. The strong motion with frequency in the range of about 2–10 Hz seen on horizontal accelerograms recorded within a few tens of km of the source can adequately be described in terms of S body waves, perhaps modified to some degree by scattering and reflection. Surface waves are also recorded by strong-motion instruments (Hanks, 1975). Typically they are recorded at sites on deep sedimentary basins, they have periods in the general range of 3–10 sec, and they arrive later than the S body waves. In some cases, perhaps in most cases, these waves are generated at the margins of the sedimentary basins by conversion from body waves in the high-velocity material bounding the basin (Vidale and Helmberger, 1988). In such basins fundamental-mode surface waves at frequencies of engineering interest are confined to shallow depths, a few hundred meters or so, and sources at seismogenic depths are ineffective in exciting them. They can more readily be excited by body waves incident at basin margins.

Concerning the role of surface waves in strong ground motion, there is a widespread misconception that surface-wave amplitude decays with distance r as $r^{-1/2}$ whereas body waves decay as r^{-1} and surface waves will therefore dominate on distant records. This concept is wrong in at least two different ways. In the first place, while Fourier spectral amplitudes of surface-wave ground motion do in fact decay as $r^{-1/2}$, time-domain amplitudes do not, because of dispersion. Well-dispersed surface waves have a time-domain amplitude decay of r^{-1}, and Airy phases, which correspond to stationary points on the group-velocity dispersion curve, have a decay of $r^{-5/6}$ (Ewing et al., 1957, p. 358). Furthermore, fundamental-mode surface waves are confined to the shallower layers, and are therefore subject to greater anelastic attenuation than body waves.

At distances of about 100 km and greater the dominant phase on strong-motion records is the L_g phase, which is a superposition of multiply-reflected S waves trapped in the crust by supercritical reflection (Bouchon, 1982; Kennett, 1985). It is an Airy phase and therefore has a theoretical decay with distance of $r^{-5/6}$. It is dominant over direct body waves at distance probably because direct body waves decay with distance more rapidly than r^{-1} for realistic distributions of velocity with depth. In contrast to the direct body wave whose duration is fixed by the source duration, the L_g phase has a duration that increases with distance—an important point in ground-motion modeling (Herrmann, 1985).

Magnitude. Earthquake ground motions depend on the size of the earthquake, the most common measure of which is magnitude. Of the many different kinds of earthquake magnitude that have been defined, the two most commonly cited in earthquake engineering are the Richter local magnitude M_L and the surface-wave magnitude M_S. M_L is determined from the trace amplitude on a record made by a particular kind of seismograph, the Wood-Anderson seismograph, located within a few hundred km of the earthquake. M_S is determined from the ground motion associated with surface waves of 20 s period recorded anywhere in the world. For earthquakes in California with M_L less than about 6.5 the commonly cited magnitude is M_L. For earthquakes worldwide with M_S greater than about 6.5 the most commonly cited magnitude is M_S.

The recently defined moment magnitude M (Hanks and Kanamori, 1979) has the advantage that it corresponds to a well-defined physical property of the earthquake source. It is defined in terms of the seismic moment M_0, which is the product of three factors, the area of the rupture surface, the average slip, and the modulus of rigidity in the source zone. Moment magnitude is thus a measure of the size of an earthquake in a very specific sense. The equation for computing moment magnitude is

$$M = \frac{2}{3} \log M_0 - 10.7 \tag{1}$$

where the units of M_0 are dyne-cm. Use of moment magnitude has the advantage of making it easier to relate earthquake occurrence rates to geologically determined fault slip rates. It is sometimes stated that because M_L is determined from an instrument with a natural period in the period range of greatest engineering interest M_L should be preferred as the measure of earthquake size to use in making ground-motion estimates for engineering purposes. Catalog values of M_L for large earthquakes, however, are commonly poorly determined (Hutton and Boore, 1987; Boore, 1988), and moment magnitude is the better measure for such use. We compiled moment magnitudes for most of the earthquakes in the western North American strong-motion data set prior to 1981 (Joyner and Boore, 1981). Ekström and Dziewonski (1985) gave moments for 35 earthquakes in western North America occuring between 1977 and 1983. Moment determinations are available for most past earthquakes with important strong-motion records and can be expected for most, if not all, such earthquakes in the future.

In developing equations for the empirical estimation of earthquake ground motion we use moment magnitude as the measure of earthquake size (Joyner and Boore, 1981, 1982). Campbell (1981, 1988) and Idriss (1985) used M_S for earthquakes with M_L and M_S greater than or equal to 6.0 and M_L for earthquakes with M_L and M_S less than 6.0. Generally speaking, below a moment magnitude of about 8.0, moment magnitudes are approximately the same as magnitudes assigned by Campbell's rule, which in turn are the commonly cited magnitudes. The 1979 Imperial Valley, California, earthquake, however, was assigned a magnitude of 6.5 by us and 6.9 by Campbell. This discrepancy has a significant effect on the end results because the 1979 Imperial Valley earthquake contributes a large share of the near-source data points to both data sets.

Distance. Because the rupture surface for earthquakes may extend over tens or hundreds of kilometers, there is ambiguity in defining the source distance for a strong-motion record. Various measures of source distance have been used in the development of relationships for estimating ground motion. Some of these are illustrated in Figure 2. The early analyses tended to use epicentral distance because it was readily available. Obvious problems arise with the use of either epicentral or hypocentral distance in the case of earthquakes like the 1966 Parkfield, California, earthquake or the 1979 Imperial Valley, California, earthquake, which have very long rupture zones with the epicenter at one end and recording stations at the other. For some stations the epicenter and hypocenter are many times more distant than the closer portions of the rupture which are in fact the sources of the peak motions. Similar problems arise with the use of distance to the centroid of the rupture. Some stations may be far from the centroid but close to the rupture. In general it must be expected that different parts of the fault rupture will produce the peak motion at different recording stations. It might seem that the distance measure to use is the distance to the part of the rupture producing the peak motion. Where that part of the rupture is located, however, is unknown for many past earthquakes and all future earthquakes. Most recent work has used some variation on the closest distance to the rupture. Where that part of the rupture to the vertical projecton of the rupture on the surface of the earth (Joyner and Boore, 1981, 1982). Campbell (1981, 1988) used the closest distance to the rupture. In his 1981 paper he interpreted that as the distance to the surface rupture in the case of ruptures that broke the surface. In his 1988 paper he changed the interpretation to the closest distance to the "zone of seismogenic rupture." The top of that zone was identified from

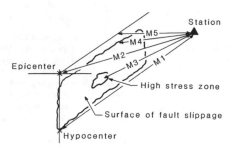

Distance Measures (from recording station)
M1 - Hypocentral
M2 - Epicentral
M3 - Dist. to energetic zone
M4 - Dist. to slipped fault
M5 - Dist. to surface projection of fault

FIGURE 2. Diagram illustrating different distance measures used in predictive relationships (from Shakal and Benneuter, 1981).

the aftershock distribution if possible, otherwise by the intersection of the fault surface with the surface of the basement rock.

For a point source in a uniform medium, geometric spreading of a direct body wave produces a distance dependence of $1/r$ for ground-motion amplitudes, where r is distance. Anelastic attenuation and/or scattering further reduce each frequency component by a factor of $\exp(-\pi fr/Q\beta)$, where f is frequency, β is the propagation velocity, and Q is the quality factor, which may depend on frequency ($1/Q$ is equal to twice the fraction of critical damping for the material). In a nonuniform medium geometric spreading may not be well-represented by the $1/r$ factor. In cases where spreading of $1/r$ is assumed, analysis commonly indicates Q increasing with frequency. As Frankel and Wennerberg (1987) point out, however, the apparent frequency dependence of Q may be simply the consequence of underestimating the effect of geometric spreading.

Burger et al. (1987) and Barker et al. (1988) have shown that, in a layered medium, horizontal ground motion at the shortest ranges is dominated by the upgoing direct S wave, which typically falls off more rapidly than $1/r$. At somewhat greater distances depending on the crustal structure, the motion is dominated by the interaction between the upgoing S wave and the S wave that initially heads downward and is later refracted upward toward the surface. In the vicinity of 100 km the motion is dominated by supercritical reflections from the base of the crust and in some cases from interfaces within the crust. Beyond 100 km the motion is dominated by the L_g phase, which, as noted above, is a superposition of multiple supercritically reflected S waves and has a falloff of $r^{-5/6}$. These complications can be modeled but only if the crustal structure is well-known and only for specific source depths. In most cases simpler methods will be relied upon for estimating ground motion.

There is no clear-cut basis for choosing a simple functional form to represent the dependence of ground-motion amplitudes with distance. We have used $exp(kr)/r$, where k is chosen to fit the data, except for certain long-period response ordinates for which the value of k turned out to be positive. In those cases we used r^d, where d is chosen to fit the data (Joyner and Boore, 1981, 1982). Most other workers have used r^d with d generally less than -1.0. It is probably not appropriate to attach much physical significance to the parameter values obtained by fitting these simple functions to the data.

It is of interest to examine the distance dependence of the Wood-Anderson amplitudes used in assigning Richter local magnitude M_L because many more data are available than in the strong-motion data set. Wood-Anderson amplitudes in northern California were investigated by Bakun and Joyner (1984) and in southern California by Hutton and Boore (1987). There were sufficient data in both data sets to determine both d and k in the function $r^d \exp(kr)$ used to model the distance dependence. The value of d determined for central California was statistically indistinguishable from -1.0, and for southern California it was -1.11. The distance dependence of Wood-Anderson amplitudes found in both studies is very similar to that found for peak horizontal acceleration and velocity from strong-motion data recorded principally in California (Joyner and Boore, 1981, 1982). This similarity indicates that data collected by high-sensitivity seismograph networks can be used in developing locally applicable equations for estimating strong motion in regions where few strong-motion data are available (see Rogers et al., 1988).

A significant issue in estimating future earthquake ground motion is the question of whether or not the shape of the curves relating peak ground-motion amplitudes to distance are dependent on magnitude. This issue is illustrated in Figure 3, which shows the curves for mean peak horizontal acceleration given by Campbell (1981), which have a magnitude-dependent shape, compared with ours, which do not (Joyner and Boore, 1981). Both his equations and ours contain a parameter that takes on the value zero if the shape is magnitude-independent and not otherwise. He finds that the parameter is significantly different from zero in his equations; we do not in ours. The reasons for the conflicting conclusions are not clear. Differences in the definitions of distance may be a factor. Lacking statistical evidence for significance we would include the parameter only if we believed that theoretical considerations called for it. We do give a theoretical argument in the appendix of our paper for a small degree of magnitude-dependence, but too small to justify modifying the equations. We remain unconvinced by other theoretical justifications for magnitude-dependence. As to the finite-source argument, we acknowledge that the whole rupture surface may be large compared to the distance to the recording site, but we would argue that the source of the peak motion is some restricted portion of the whole rupture and may well be small compared to the distance to any recording site on the surface of the earth. As we will show later, the question of magnitude-dependent shape does not make an important difference for predicting peak horizontal acceleration or response spectra at periods near 0.2 s, the difference being small compared to the standard error of an individual prediction. For peak horizontal velocity or response spectra at periods of about 1.0 s or longer, however, the difference is important.

Site Conditions. Significant site effects include the effects of topography and local site geology and the effects on the ground motion of structures within which or near which the recording instruments are installed. The effects of topography are the subject of a large literature (e.g. Boore, 1972, 1973; Bouchon, 1973; Tucker et al., 1984; Brune et al., 1984; Bard and Tucker, 1985; Geli et al., 1988; Andrews et al. 1988), but since the effects are negligible at most sites we will not consider them further here. The effect of local site geology has also been widely discussed (e.g. Kanai, 1952; Gutenberg, 1957; Idriss and Seed, 1968; Borcherdt, 1970; Rogers et al., 1984).

In considering local geological effects it is necessary to distinguish between two kinds of site amplification. The more widely recognized but actually less common of the two is the resonant amplification resulting from reinforcing mutiple reflections within low-velocity layers near the surface. This kind of amplification is highly sensitive to frequency. The second kind does not require sharp, reflecting interfaces, is not sensitive to frequency, and results from low velocity, or, more precisely, low impedance near the surface, impedance being the product of density and velocity. If we consider a tube of rays and neglect losses due to reflection, scattering and anelastic attenuation, the energy along a tube of rays is constant. If the effect of changes in the cross-sectional area of the tube can be neglected, which is commonly the case, then amplitude is inversely proportional to the square root of impedance (Aki and Richards, 1980, p. 127).

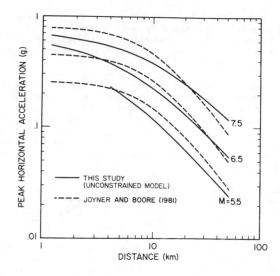

FIGURE 3. Curves of mean peak horizontal acceleration given by Campbell (1981), which have a magnitude-dependent shape, compared with those of Joyner and Boore (1981), which do not. The Joyner and Boore curves for the larger peak of two horizontal components have been reduced by 12 percent so that they may be compared with the mean peak. (Illustration modified from Campbell, 1981.)

Material with low velocity (and thereby low impedance) tends to have low Q, so, depending on the frequency of the motion and the thickness of the material, the amplifying effect of low-velocity material may be partly or totally offset by greater attenuation.

Largely because the data on site conditions at strong-motion recording sites are limited, the means used to incorporate site effects into the analysis of strong-motion data sets tend to be rather crude. We divided recording sites into two categories, rock and soil (Joyner and Boore, 1981, 1982). Sites with less than 5 m of soil overlying rock were put into the rock category. We found no statistically significant difference between the rock and soil sites in peak horizontal acceleration or response spectral values at periods less than 0.3 s, suggesting that attenuation in the soils offsets the amplification for high frequencies. Joyner and Fumal (1984) found no correlation at soil sites between peak horizontal acceleration and depth to rock for a range in depth to rock from nearly zero to nearly one km. This suggests that most of the attenuation in soils occurs at shallow depth, perhaps less than something like 100–200 m. For peak horizontal velocity and response at periods longer than about 0.3 s, we found significantly higher values on soil than on rock, higher by a factor of 1.5 for peak horizontal velocity and by a factor of 1.9 for one-second horizontal pseudovelocity response. If the effects of attenuation just offset the effects of amplification for high-frequency measures of ground motion, it would be expected that amplification would dominate for lower-frequency measures, since the limited available data suggest that Q in soils is not dependent on frequency.

Campbell (1981) separated sites with less than 10 m of soil overlying rock into a special category of shallow soil sites. For other sites he, as we, found no significant difference between peak horizontal acceleration on rock and soil. The shallow soil sites, however, had a higher peak horizontal acceleration than the other categories by a factor of 1.8 on the average. Though we might argue about the criterion for delimiting the category of shallow soil sites, we do not disagree with the designation of such a category; our data set simply did not include a sufficient number of shallow soil sites to support a separate category. Results from analysis of Italian strong-motion data (Sabetta and Pugliese, 1987) support higher horizontal accelerations at shallow soil sites. Mohraz (1976) showed that horizontal response spectra at shallow soil sites were preferentially richer in high frequencies than at sites where the soil was more than 60 m thick.

In an attempt to achieve a more physically based method of predicting site effects, Joyner and Fumal (1984) used downhole shear-velocity data at 33 strong-motion sites to develop equations for site effects in terms of the shear velocity averaged from the surface to a depth equal to one-quarter wavelength at the period of interest. Campbell (1988) took a different approach to the site effect on peak horizontal velocity and, like Trifunac and Lee (1979), expressed the site effect in terms of the depth of sediments beneath the recording site. No physical interpretation was given of this representation of the site effect. Unlike Trifunac and Lee, Campbell expressed the site effect in terms of a nonlinear function of sediment depth.

The methods described above do not incorporate the effect of resonance. To do so requires more detailed site information than is often available. For cases where the required information is available, methods of modeling assuming body waves in plane layers can be used. Pioneer work with such methods was done by Idriss and Seed (1968; Schnabel et al., 1972). They used the equivalent-linear method to incorporate nonlinear behavior in the soils, that is, they used a linear analysis with values of the dynamic soil properties chosen by an iterative procedure to correspond with the average strain level determined by the analysis. Joyner and Chen (1975) presented a truly nonlinear method and showed that for motion of short period and high amplitude on thick soil deposits the equivalent-linear method does not give a satisfactory approximation to the nonlinear results. Silva (1976) described a general system for the linear modeling of plane-layered systems based on the Thompson-Haskell method. All these kinds of analyses depend on the assumption of plane layers and, if nonlinearity is presumed, on laboratory data for dynamic soil properties. The adequacy of the plane-layer approximation has been demonstrated at a number of sites (e.g. Joyner et al., 1976; Johnson and Silva, 1981), though there are sites for which it is not appropriate (King and Tucker, 1984). There is some uncertainty, however, over the degree to which laboratory data are representative of the properties of soils in situ and in particular over the level of shaking at which nonlinearity becomes a significant factor for soils in situ.

An extreme example of the effect of site conditions is afforded by the damage in Mexico City from the 1985 Michoacan, Mexico, earthquakes, whose source zones were 300 km or more from the city. More than 300 buildings in Mexico City were destroyed or badly damaged. Nearly all of the buildings that collapsed in the city were located in an old lake bed, characterized by very soft clay soils. Response spectra at 5 percent damping showed amplification of approximately ten times at 2 s period on the lake bed relative to surrounding areas (Anderson et al., 1986; Singh et al., 1988). Materials similar to the soft clays of the old lake bed are found elsewhere in the world, and changes have been made in the upcoming editions of the Recommended Lateral Force Requirements of the Structural Engineers Association of California (1988) and the Uniform Building Code (International Conference of Building Officials, 1988) in an attempt to accomodate situations of this kind.

Many strong-motion records come from instruments located at the ground floor or in the basements of buildings or on the abutments of dams. Such records are affected by the response of the structure in which or near which the instrument is located. Even so-called "free-field" sites may be affected at high frequencies by the response of the instrument shelter (Bycroft, 1978; Crouse et al., 1984), though for the typical shelters used by the U.S. Geological Survey and the California Division of Mines and Geology the effects are at frequencies too high to be of concern (Crouse and Hushmand, 1987). We attempted to minimize the effects of structures by excluding all data recorded on the abutments of dams or in the basements of buildings three or more stories in height (Joyner and Boore, 1981, 1982). Campbell (1981, 1988) included the effect of the building as a parameter to be determined by analysis of the data.

Fault Type, Depth, and Repeat Time. Fault type, depth, and repeat time have been suggested as important in determining ground-motion amplitudes because of their presumed relation to the stress state at the source or to stress changes associated with the earthquake. McGarr (1984) argues that ground motion should increase with depth and that ground motion should be greater from reverse faults than from normal faults, with strike-slip faults having intermediate values. Kanamori and Allen (1986) present data showing that faults with longer repeat times have shorter lengths for the same magnitude, indicating a larger average stress drop and, presumably, higher ground motion.

Empirical attempts to correlate fault type, depth, and repeat time with measures of strong ground motion do not support clear-cut conclusions. Campbell (1988) finds that peak horizontal acceleration and velocity in shallow reverse-slip earthquakes are larger on the average by factors of about 1.4 and 1.6, respectively, than in strike-slip earthquakes. Reexamination of data we used in developing our published ground-motion predictive equations indicates values of peak horizontal acceleration higher on the average by a factor of about 1.25 in reverse-slip earthquakes than in strike-slip earthquakes. We cannot tell, however, whether the difference is due to fault type or repeat time. The reverse-slip faults are generally the ones with long repeat times and the strike-slip faults the ones with short repeat times. The conspicuous exception is the 1979 St. Elias, Alaska, earthquake, a reverse-slip event which lies on the boundary between the Pacific and North American plates, and therefore, presumably has a short repeat time. The peak horizontal acceleration values for the St. Elias earthquake on the average fall below the predictions of the equations developed from the whole data set, suggesting that repeat time and not fault type is the controlling variable, but no definite conclusions should be drawn based on a single earthquake.

Data on the amplitude of ground motion from normal-slip earthquakes compared to other types is subject to conflicting interpretations. McGarr (1984; see also McGarr, 1986) found a large difference between normal- and reverse-slip events. For a selected data set with few observations of large events he examined the effect of fault type and focal depth on peak acceleration and peak velocity. To remove the effect of distance he multiplied the peak motions by the hypocentral distance. He assumed that peak acceleration was independent of moment M_0, and to remove the effect of M_0 on peak velocity he divided by $M_0^{1/3}$. The resulting peak values were approximately proportional to depth. On the average the peak accelerations were about a factor of 3 greater and the peak velocities a factor of 2 greater for reverse- than for normal-slip events. Values for strike-slip events were intermediate. Other studies do not show such large differences. Westaway and Smith (1987) compared peak horizontal acceleration from normal-slip earthquakes with our equations (Joyner and Boore, 1981) and with Campbell's (1981) equations. Both sets of equations are based primarily on data from reverse- and strike-slip events. After considering data from more than 600 records of normal-slip events in the western United States, the Mediterranean region, and New Zealand, Westaway and Smith concluded that there is no significant difference between normal-slip events and others. The data set included some records with source distances greater than 100 km, and the argument could be made that regional differences in attenuation may have obscured differences due to fault type, but the data set also included data, such as the record from the base of the Matahina Dam in New Zealand, at distances short enough so that one would not expect a significant effect due to differences in attenuation (New Zealand Geological Survey, 1987).

The Italian strong-motion data set (Sabetta and Pulgliese, 1987) is particularly pertinent to the comparison between normal- and reverse-slip earthquakes. The data from the Friuli region are all from reverse-slip events and the rest of the data are all from normal-slip events. Sabetta and Pugliese developed predictive equations for peak horizontal acceleration from the whole data set of 95 records and also for the truncated data set of 52 records formed by omitting the records from the Friuli region. The equations developed from the truncated data set gave values only about 10 percent less than the equations developed for the whole data set for a stiff or deep soil site 10 km from a magnitude 6.5 event, indicating that the effect, if any, of fault type was small. Most of the Friuli records were from aftershocks, however, so that this conclusion holds in the general case only if aftershocks can be presumed characteristic of main shocks.

The data discussed up to this point has all come from shallow crustal earthquakes, for which the fault ruptures are confined to a zone within about 20 km of the earth's surface. Another important class of earthquakes is subduction-zone earthquakes such as occur off the Pacific coasts of Japan, Alaska, and Central and South America. Such earthquakes occur over a range of depths of a few hundred kilometers. In an analysis of subduction-zone data from the Northern Honshu zone Crouse *et al.* (1988) found no significant differences in the horizontal response spectral values from reverse, normal and strike-slip events. Their regression analysis showed a linear dependence of the logarithm of spectral response on focal depth. For periods less than about 1.0 s deeper events gave larger response than shallower events. At 0.1 s period the difference was the largest and corresponded to a factor of 2.5 for an increase of 100 km in depth. At periods greater than about 1.0 s the deeper events gave response that was smaller than shallower events. The difference at 4.0 s was the largest and corresponded to a factor of 1.7 for an increase of 100 km in depth. Crouse *et al.* suggested that the deeper events might have higher amplitude at short period because of higher stress drop or lower anelastic attenuation. They suggested that the deeper events might have lower amplitude at long period because they are less effective in generating surface waves. They did note, however, that the inclusion of the depth term in the regression did not substantially improve the fit to the data.

Directivity and Radiation Pattern. If the angle between the source-to-recording-site vector and the direction of rupture propagation is small, the recorded ground motion may be substantially increased in amplitude. This effect, called directivity (Ben-Menahem, 1961), can be expected to occur for incoherent as well as coherent ruptures (Boore and Joyner, 1978). The work of Boatwright and Boore (1982) shows that strong ground motion can be significantly affected by directivity, but it is not clear how to incorporate directivity into schemes for predicting ground motion in future earthquakes. The variable of importance for directivity is the angle θ between the rupture direction and the source-to-recording-site vector, and θ is not known in general for future earthquakes. Furthermore, for sites close to large earthquake ruptures, where reliable prediction is most important, θ changes during rupture propagation. Araya and Der Kiureghian (1986) have suggested an approach for including directivity in ground-motion predictive equations. Campbell (1988) included directivity as a parameter in his analysis, but he applied it to only 3 recordings out of a total of 134, acknowledging that other recordings in the data set might also be "affected to some degree" by directivity. Most authors of ground-motion predictive equations have not explicitly included a variable representing directivity. To the extent that directivity affects the records in the data set, however, the equations and the estimates of variability will reflect the effect. Near-source recording sites are more likely to be affected by directivity, and the predictive equations will give higher near-source values as a result.

Even without rupture propagation, the ground motion from a point dislocation will depend on the direction the ray leaves the source. This dependence, referred to as radiation pattern, is mentioned here only for the sake of completeness. The radiation pattern, which is relatively simple in a homogeneous medium, is complicated in the real world by scattering and refraction caused by variations in propagation velocity. The fault zone itself probably represents a low-velocity zone, and rays leaving the fault zone at a large angle to the normal may undergo significant refraction. Under these circumstances attempts to incorporate the radiation-pattern effect in ground-motion prediction equations are unlikely to yield much benefit, and no one, to our knowledge, has proposed doing so.

Empirical Prediction

The Method of Representative Accelerograms. Guzman and Jennings (1976) suggested a method, later elaborated by Heaton *et al.* (1986), for establishing design ground motions. The method begins with the selection of a suite of accelerograms recorded for magnitudes, distances, local site conditions and other factors similar to the postulated design earthquake. Each accelerogram and its response spectrum are multiplied by a constant scaling factor to account for differences in magnitude and source distance between the design earthquake and the accelerogram. The collection of response spectra is used to portray the range of ground motion to be expected at the site. The method has the advantage of directness and of maintaining a close tie to the basic data. The necessity for deciding which accelerograms represent conditions similar to the design earthquake and which do not, however, introduces an undesirable element of subjectivity. In our view, well-designed regression analyses of the whole data set make more efficient use of the available information than the method of representative accelerograms.

Development of Predictive Relationships. A comprehensive review of ground-motion predictive relationships developed before the 1979 Imperial Valley, California, earthquake is given by Idriss (1979). The 1979 Imperial Valley earthquake marked a major change in the strong-motion data base by providing many more near-source data points than had been available previously. More recent reviews have been written by us (Boore and Joyner, 1982) and Campbell (1985). Predictive relationships for ground motion may be expressed in graphical form or as mathematical equations. An example of graphical relationships is provided by the widely-used Schnabel and Seed (1973) curves, which are the basis of the seismic risk maps by Algermissen and Perkins (1976) and Algermissen *et al.* (1982) as well as the ATC-3 seismic zone maps (Applied Technology Council, 1978). In the case of predictive relationships expressed as mathematical equations, the equations contain parameters that are adjusted in some way so as to fit the available strong-motion data. Brillinger and Preisler (1984, 1985) demonstrated that there is a significant between-earthquake component of variance in ground-motion data in addition to the within-earthquake component. Such being the case, ordinary least squares is not strictly correct as a method for determining the parameters of the predictive equations. Brillinger and Preisler describe maximum-likelihood methods applicable to this case. The need to take account of the between-earthquake variance was recognized by us in our use of a two-stage regression procedure (described below) to develop predictive relationships (Joyner and Boore, 1981). The two-stage procedure gives virtually the same values as the methods of Brillinger and Preisler for each of the two components of the variance, indicating that it is equivalent. Campbell (1981, 1988) uses weighted least squares in a scheme "designed to give each earthquake an equal weight in the analysis in each of nine distance intervals."

In our view it is important to choose a form for the predictive equations based as much as possible on physical grounds. We consider this guideline important because data are sparse or nonexistent for important ranges of the predictive variables. If data were plentiful, it would matter less what form were chosen; either the form would fit the data, or the lack of fit would be obvious. Only those predictive variables should be used whose inclusion can be justified by physical reasoning. Furthermore, the number of adjustable parameters should be kept as small as possible.

A key feature of the data set for shallow earthquakes is the scarcity of data points for distances less than about 20 km and magnitudes greater than 7.0. Confident predictions can simply not be made in that range of magnitude and distance, which is, unfortunately, where predictions are most needed.

Much of the early strong-motion record processing was done on data sampled at 50 samples per second using a finite-difference instrument-correction algorithm. Response values derived from these data may be inaccurate at periods less than about 0.1 s. The peak accelerations obtained from such data may also be in error. Many workers have used uncorrected rather than corrected peak acceleration values to avoid the bias introduced by the early record processing (*e.g.* Joyner and Boore, 1981, 1982; Campbell, 1981, 1988).

Examples of Predictive Equations for Shallow Earthquakes. We now proceed to give examples of a number of predictive equations that have been proposed in the last 10 years. We have attempted to give all the information needed to make predictions of ground motions from the various equations. At the expense of easy visual comparisons of the equations themselves, we have retained the author's form of the equations, units, and preference for natural or common logarithms in order to forestall errors of conversion.

Joyner and Boore. Figure 4a shows the distribution in magnitude and distance of the strong-motion data we used in developing predictive equations for peak horizontal acceleration, and Figure 4b shows the distribution we used in developing equations for peak horizontal velocity and pseudovelocity response at 5 percent damping (Joyner and Boore, 1981, 1982). There are more points on Figure 4a than Figure 4b because record processing is necessary to obtain velocity and response spectra, and not all of the records represented on Figure 4a have been processed.   The data sets are restricted to earthquakes in western North America with moment magnitude greater than 5.0 and to shallow earthquakes, defined as those for which the fault rupture lies mainly above a depth of 20 km. To minimize the effect of structures we exclude from the data sets all records made at the base of buildings three or more stories in height and on the abutments of dams.   We exclude all earthquakes for which the data were in our opinion inadequate for estimating the source distance to an accuracy better than 5 km. Bias may be introduced into a strong-motion data set if low amplitude ground motions are preferentially excluded because they fall below the trigger level of operational instruments or, if recorded, the peak motions are excluded from the data set for any reason having to do with amplitude (*e.g.* a record is not digitized because of its low amplitude). To avoid this bias we exclude, for each earthquake, data recorded at a distance greater than the smallest distance at which an operational instrument did not trigger or at which data were excluded from the data set for a reason related to amplitude.

As previously mentioned, we use a two-stage regression procedure to develop the predictive equations. The precedure is illustrated in Figure 5. In the first stage the equation

$$\log y = A_i + d \log r + kr + s$$
$$r = (r_0^2 + h^2)^{1/2} \tag{2}$$

is fit to the data, where y is the ground-motion parameter to be predicted, r_0 is the shortest distance (km) from the recording site to the vertical projection of the earthquake fault rupture on the surface of the earth, and s is the site correction for soil sites (sites with 5 m or more of soil). Initially d is set equal to -1.0. Values of k and s and a value A_i for each earthquake i are determined by linear regression for a trial value of h. The final value of h is determined by a simple search procedure to minimize the sum of squares of the residuals, unless the corresponding value of k is positive. In that case k is set to zero and the process repeated this time with variable d.

In the second stage a first- or second-order polynomial in magnitude is fit by least squares to the values A_i.

$$A_i = a + b(\mathrm{M_i} - 6) + c(\mathrm{M_i} - 6)^2 \tag{3}$$

where $\mathrm{M_i}$ is moment magnitude of earthquake i. In our early papers we used ordinary least squares for the second-stage regression, and, for peak acceleration, excluded all of the earthquakes for which only one recording was available (Joyner and Boore, 1981, 1982). In our 1982 paper, for peak velocity and response spectra, we did not exclude any earthquakes because so few were available. Actually, the correct way to do the second-stage regression is by weighted least squares (Bevington, 1969). The weighting factor, $1/\sigma_i^2$, is the inverse of the variance of A_i:

$$\sigma_i^2 = \sigma_a^2/N_i + \sigma_b^2 \tag{4}$$

where σ_a^2 is the within-earthquake component of the total variance and is given by the variance of data points in the first-stage regression, N_i is the number of records in the data set for earthquake i, and σ_b^2 is the between-earthquake component of the total variance. σ_b^2 is determined iteratively in the second-stage regression by using zero as a starting value. Use of weighted least squares for the second-stage regression does not produce a large change in the resulting equations.

The final predictive equation is

$$\log y = a + b(\mathrm{M} - 6) + c(\mathrm{M} - 6)^2 + d \log r + kr + s$$
$$5.0 \le \mathrm{M} \le 7.7$$
$$r = (r_0^2 + h^2)^{1/2} \tag{5}$$

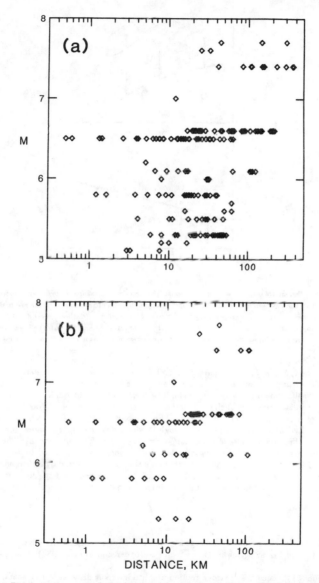

FIGURE 4. Distribution in magnitude and distance of the strong-motion data used by Joyner and Boore (1982) to develop predictive relationships for (a) peak horizontal acceleration and (b) peak horizontal velocity and response spectra.

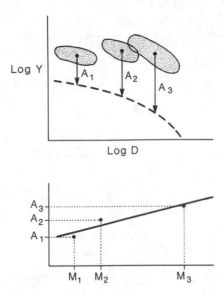

FIGURE 5. Schematic diagram showing the two-stage regression procedure of Joyner and Boore (1981, 1982). The upper part of the figure illustrates the first stage, in which the sum of the squares of the residuals is minimized by varying the shape of the curve representing the dependence on distance (dashed curve) and by shifting all the data points from the ith earthquake (contained within a closed curve) by an amount A_i. The lower part of the figure illustrates the second stage, in which the offset factors A_i are regressed against magnitude to determine the magnitude dependence.

where values of a, b, c, d, k, s, and h are given in Table 2 for predicting parameters corresponding to the randomly oriented horizontal component and in Table 3 for predicting parameters corresponding to the larger of the two horizontal components. The coefficients for the randomly oriented horizontal component are determined by including in the regression both horizontal components from each site as if they were independent data points. That procedure gives the correct coefficients for the randomly oriented horizontal component, though it would not give correct values for the standard errors of the coefficients, because the two components are not in fact independent. The values given in Table 2 for peak horizontal acceleration and velocity were not determined by regression. They were obtained from the values in Table 3 by changing the constant term a by an amount determined by averaging the difference between the logarithm of the larger peak and the average logarithm of the two peaks over a selected subset of the data. An alternative for the soil effect is

$$s = e \log \left(\frac{V_S}{V_{S0}} \right) \qquad (6)$$

where V_S is the site shear velocity averaged to a depth of one-quarter wavelength at the period of interest, and e and V_{S0} are given in Tables 2 and 3 (Joyner and Fumal, 1984).

Equation (5) gives curves of strong-motion amplitude versus distance with a shape independent of magnitude. A magnitude-dependent shape can be obtained by taking

$$h = h_1 \exp[h_2(\mathbf{M} - 6)] \qquad (7)$$

TABLE 2. PARAMETERS IN THE PREDICTIVE EQUATIONS OF JOYNER AND BOORE (1982) FOR THE RANDOMLY ORIENTED HORIZONTAL COMPONENT OF PSEUDOVELOCITY RESPONSE (CM/S) AT 5 PERCENT DAMPING AND OF PEAK ACCELERATION (g) AND VELOCITY (CM/S)

Period (s)	a	b	c	h	d	k	s	V_{S0}	e	$\sigma_{\log y}$
				Pseudovelocity response						
0.1	2.16	0.25	-0.06	11.3	-1.0	-0.0073	-0.02			0.28
0.15	2.40	.30	- .08	10.8	-1.0	- .0067	- .02			.28
0.2	2.46	.35	- .09	9.6	-1.0	- .0063	- .01			.28
0.3	2.47	.42	- .11	6.9	-1.0	- .0058	.04	590	-0.28	.28
0.4	2.44	.47	- .13	5.7	-1.0	- .0054	.10	830	- .33	.31
0.5	2.41	.52	- .14	5.1	-1.0	- .0051	.14	1020	- .38	.33
0.75	2.34	.60	- .16	4.8	-1.0	- .0045	.23	1410	- .46	.33
1.0	2.28	.67	- .17	4.7	-1.0	- .0039	.27	1580	- .51	.33
1.5	2.19	.74	- .19	4.7	-1.0	- .0026	.31	1620	- .59	.33
2.0	2.12	.79	- .20	4.7	-1.0	- .0015	.32	1620	- .64	.33
3.0	2.02	.85	- .22	4.7	-0.98	.0	.32	1550	- .72	.33
4.0	1.96	0.88	-0.24	4.7	-0.95	0.0	0.29	1450	-0.78	0.33
				Peak acceleration						
	0.43	0.23	0.0	8.0	-1.0	-0.0027	0.0			0.28
				Peak velocity						
	2.09	0.49	0.0	4.0	-1.0	-0.0026	0.17	1190	-0.45	0.33

where h_1 and h_2 are chosen to fit the data. As previously mentioned, however, we find that the difference between h_2 so chosen and zero is not statistically significant (in particular, see Figure 16 in Boore and Joyner, 1982).

Figure 6 gives the curves for peak acceleration and velocity for the randomly oriented horizontal component, and Figures 7, 8, and 9 give the corresponding pseudovelocity response spectra at 5 percent damping. Figure 7 shows the strong dependence, previously mentioned, of the shape of response spectra on magnitude and site conditions. Figures 8 and 9 show that, between 10 and 40 km at least, there is very little dependence of the shape on distance—in sharp contradiction with conventional wisdom.

Figures 8 and 9 do show that long-period response decreases more than short-period response in going from 0 to 10 km. This reflects the fact that the values of h for short periods are greater than for long periods; the values for peak acceleration are greater than for peak velocity (Tables 2 and 3). It is not completely clear why this should be the case. The larger value of h for peak acceleration and short-period response may represent a limitation in acceleration near the source by the limited strength of near-surface materials.

Tables 2 and 3 give estimates of $\sigma_{\log y}$, the standard deviation of an individual prediction of $\log y$ using the equations. The estimates for peak horizontal acceleration and velocity agree quite well with those of McGuire (1978), who used a data set specially constructed to avoid bias in the estimate of residuals caused by multiple records from a single event or by multiple records from the same site of different events. Some other workers, however, as will be seen, have obtained smaller values for $\sigma_{\log y}$.

Crouse et al. The following equation for peak horizontal acceleration and horizontal pseudovelocity

TABLE 3. PARAMETERS IN THE PREDICTIVE EQUATIONS OF JOYNER AND BOORE
(1982) FOR THE LARGER OF TWO HORIZONTAL COMPONENTS
OF PSEUDOVELOCITY RESPONSE (CM/S) AT 5 PERCENT DAMPING
AND OF PEAK ACCELERATION (g) AND VELOCITY (CM/SEC)

Period (s)	a	b	c	h	d	k	s	V_{S0}	e	$\sigma_{\log y}$
				Pseudovelocity response						
0.1	2.24	0.30	-0.09	10.6	-1.0	-0.0067	-0.06			0.27
0.15	2.46	.34	- .10	10.3	-1.0	- .0063	- .05			.27
0.2	2.54	.37	- .11	9.3	-1.0	- .0061	- .03			.27
0.3	2.56	.43	- .12	7.0	-1.0	- .0057	.04	650	-0.20	.27
0.4	2.54	.49	- .13	5.7	-1.0	- .0055	.09	870	- .26	.30
0.5	2.53	.53	- .14	5.2	-1.0	- .0053	.12	1050	- .30	.32
0.75	2.46	.61	- .15	4.7	-1.0	- .0049	.19	1410	- .39	.35
1.0	2.41	.66	- .16	4.6	-1.0	- .0044	.24	1580	- .45	.35
1.5	2.32	.71	- .17	4.6	-1.0	- .0034	.30	1780	- .53	.35
2.0	2.26	.75	- .18	4.6	-1.0	- .0025	.32	1820	- .59	.35
3.0	2.17	.78	- .19	4.6	-1.0	.0	.29	1620	- .67	.35
4.0	2.10	0.80	-0.20	4.6	0.98	0.0	0.24	1320	-0.73	0.35
				Peak acceleration						
	0.49	0.23	0.0	8.0	-1.0	-0.0027	0.0			0.28
				Peak velocity						
	2.17	0.49	0.0	4.0	-1.0	-0.0026	0.17	1190	-0.45	0.33

response at 5 percent damping was developed from data recorded at deep soil sites (generally greater than 60 m in thickness) during shallow crustal earthquakes in southern California by C. B. Crouse (written communication, 1987; Vyas *et al.*, 1988):

$$\ln y = a + bM_S + cM_S^2 + d\ln(r + 1) + kr \tag{8}$$

where y is peak horizontal acceleration (gal) or horizontal pseudovelocity response (cm/s), M_S is surface-wave magnitude, r is closest distance (km) from rupture surface to recording site, and the coefficients a, b, c, d, and k are given in Table 4 along with $\sigma_{\ln y}$, the standard deviation of a single prediction of $\ln y$. The values of standard deviation are similar to those in Tables 2 and 3 when the conversion is made from natural to common logarithm (multiply the numbers in Table 4 by 0.43). Both horizontal components were used so that the values of y predicted by equation (8) correspond to the randomly oriented horizontal component.

Sadigh et al. Equation (9) for peak horizontal acceleration and horizontal pseudoacceleration response at 5 percent damping was developed using data from the western United States supplemented by significant recordings of earthquakes at depth less than 20 km from other parts of the world (Sadigh, written communication, 1987; Sadigh *et al.*, 1986). Both horizontal components were used.

$$\ln y = a + bM + c_1(8.5 - M)^{c_2} + d\ln[r + h_1 \exp(h_2M)] \tag{9}$$

where y is peak horizontal acceleration (g) or horizontal pseudoacceleration (g), M is moment magnitude, r is the closest distance (km) to the rupture surface, and the values of a, b, c_i, d, and h_i are given in Table 5 along with expressions for $\sigma_{\ln y}$, the standard deviation of an individual prediction of $\ln y$. Note that $\sigma_{\ln y}$ depends on M and is significantly smaller for M $\geq$ 6.5 than the values in Table 4 and smaller than the values in Tables 2 and 3 if they were converted to natural logarithms. Equation (9) was derived for strike-slip earthquakes; to obtain values corresponding to reverse-slip events, the value of y from equation (9) should be increased by 20 percent.

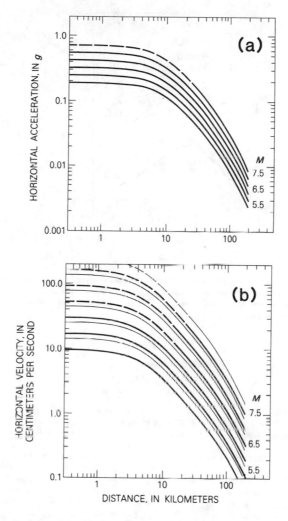

FIGURE 6. Predictive relationships from Joyner and Boore (1982) for shallow earthquakes giving (a) peak horizontal acceleration, reduced by 13 percent so as to approximate the value for the randomly oriented horizontal component, and (b) peak horizontal velocity at rock sites (heavy lines) and soil sites (light lines), reduced by 17 percent so as to approximate the value for the randomly oriented horizontal component. Curves are dashed where not constrained by data. Distance is the closest distance to the vertical projection of the rupture on the surface of the earth.

Donovan and Borstein. Equation (10) was developed for peak horizontal acceleration from data from the western United States (Donovan and Bornstein, 1978). Both horizontal components were

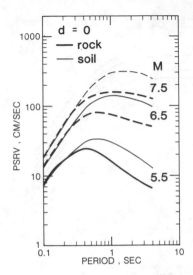

FIGURE 7. Predicted pseudovelocity response spectra of shallow earthquakes for 5 percent damping at rock sites (heavy lines) and soil sites (light lines) for zero distance and the indicated values of moment magnitude (Joyner and Boore, 1982). Spectra correspond to the randomly oriented horizontal component. Curves are dashed where not constrained by data. Distance is as defined for Figure 6.

used.

$$y = a \exp(bM)(r + 25)^d$$
$$a = 2,154,000(r)^{-2.10}$$
$$b = 0.046 + 0.445 \log(r)$$
$$d = -2.515 + 0.486 \log(r)$$

(10)

where y is peak horizontal acceleration (gal), M is magnitude, and r is distance (km) to the energy center, presumed to be at a depth of 5 km. Table 6 gives $\sigma_{\ln y}$, the standard deviation of the natural logarithm of an individual prediction of y, as it is presumed to vary with y.

Campbell. Campbell (1988) developed equations for predicting peak horizontal acceleration and velocity from a selected worldwide data set meeting the following criteria: "(1) the largest horizontal component of peak acceleration was at least 0.02 g; (2) the accelerograph triggered early enough to record the strongest phase of shaking; (3) the magnitude of the earthquake was 5.0 or larger; (4) the closest distance to seismogenic rupture was less than 30 or 50 km, depending on whether the magnitude of the earthquake was less than or greater than 6.25; (5) the shallowest extent of seismogenic rupture was no deeper than 25 km; and (6) the recording site was located on unconsolidated deposits." Records from instruments on the abutments or toes of dams were excluded. Two equations are given, equation (11) for the unconstrained relationship and equation (12) for the constrained relationship, which includes an anelastic attenuation term to permit extrapolation beyond the near-source region.

$$\ln y = a + bM + d \ln[r + h_1 \exp(h_2 M)] + s$$

(11)

$$\ln y = a + bM + d \ln[r + h_1 \exp(h_2 M)] + kr + s$$

(12)

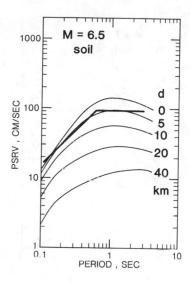

FIGURE 8. Predicted pseudovelocity response spectra (light lines) of shallow earthquakes for 5 percent damping at soil sites for a moment magnitude of 6.5 and the indicated distances (Joyner and Boore, 1982) compared to the ATC-3 spectrum (heavy line) for soil type S_2 in the highest seismic zone (Applied Technology Council, 1978). Predicted spectra correspond to the randomly oriented horizontal component. Distance is as defined for Figure 6.

where y is the mean of the peak acceleration (g) or velocity (cm/s) values for the two horizontal components, M is surface wave magnitude M_S if both local magnitude M_L and M_S are greater than or equal to 6.0 or M_L if both M_S and M_L are less than 6.0, and r is the shortest distance (km) to the zone of seismogenic rupture, identified where possible from the aftershock distribution, otherwise from other data, particularly the intersection of the fault surface with the surface of basement rock (Campbell does not state what magnitude to use if the relative sizes of M_S and M_L do not fall into one of the categories above). A value of 0.0059 was assumed for k for the regression analysis of the data base; in using equation (12) for predicting ground motion, a value appropriate for the region should be chosen. For peak horizontal acceleration

$$s = e_1 K_1 + e_2 K_2 + e_3 K_3 + e_4 K_4 + e_5 K_5 + e_6 (K_4 + K_5) \tanh(e_7 r) \qquad (13)$$

and for peak velocity

$$s = e_1 K_1 + e_2 K_2 + e_3 K_3 \tanh(e_4 D) + e_5 (1 - K_3) \tanh(e_6 D) \qquad (14)$$

where s incorporates the effects of fault type, directivity, soil type, building size, and building embedment. D is depth (km) to crystalline basement rock. Values of a, b, d, h_i, and e_i are given in Table 7 along with values of $\sigma_{\ln y}$, the standard deviation of an individual prediction of $\ln y$. The standard deviations are less than one-half of those in Tables 2 and 3 after conversion from natural to common logarithms. Values of K_i are given in Table 8.

Idriss. Equation (15) was developed for the randomly oriented horizontal component of peak horizontal acceleration by Idriss (1985, 1987).

$$\ln y = \ln a + d \ln(r + 20) \qquad (15)$$

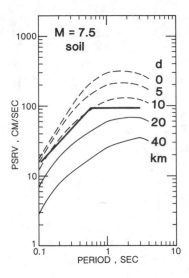

FIGURE 9. Predicted pseudovelocity response spectra (light lines) of shallow earthquakes for 5 percent damping at soil sites for a moment magnitude of 7.5 and the indicated distances (Joyner and Boore, 1982) compared to the ATC-3 spectrum (heavy line) for soil type S_2 in the highest seismic zone (Applied Technology Council, 1978). Curves are dashed where not constrained by data. Predicted spectra correspond to the randomly oriented horizontal component. Distance is as defined for Figure 6.

TABLE 4. PARAMETERS IN THE PREDICTIVE EQUATIONS
OF CROUSE (WRITTEN COMMUNICATION, 1987) FOR THE RANDOMLY ORIENTED
HORIZONTAL COMPONENT OF PSEUDOVELOCITY RESPONSE (CM/S)
AT 5 PERCENT DAMPING AND OF PEAK ACCELERATION (GAL)

Period (s)	a	b	c	d	k	$\sigma_{\ln y}$
Pseudovelocity response						
0.05	- 2.44178	0.84826	-0.02579	-0.52916	-0.00961	0.59914
0.10	- 0.61623	0.62660	- .00999	- .50106	- .01199	.68673
0.20	- 4.47801	2.00876	- .11673	- .32102	- .01423	.64716
0.40	- 1.35559	1.17453	- .04411	- .47398	- .00782	.62089
0.60	- 6.02161	2.66493	- .15619	- .52586	- .00548	.62275
1.00	- 5.89916	2.48235	- .13036	- .52261	- .00405	.62745
2.00	-11.48576	4.01914	- .23152	- .56791	- .00280	.63277
2.50	-12.33454	4.15828	- .23359	- .56280	- .00320	.66459
4.00	-14.90528	4.54962	- .24999	- .32351	- .00738	.73830
6.00	-14.77796	4.33959	-0.23491	-0.20849	-0.00791	0.79595
Peak acceleration						
	2.48456	0.73377	-0.01509	-0.50558	-0.00935	0.58082

TABLE 5. PARAMETERS IN THE PREDICTIVE EQUATIONS
OF SADIGH (WRITTEN COMMUNICATION, 1987) FOR THE RANDOMLY ORIENTED
HORIZONTAL COMPONENT OF PSEUDOACCELERATION RESPONSE (g)
AT 5 PERCENT DAMPING AND OF PEAK ACCELERATION (g)

Period (s)	a	b	c_1	c_2	d	M < 6.5			M ≥ 6.5		
						h_1	h_2	$\sigma_{\ln y}$	h_1	h_2	$\sigma_{\ln y}$
						Pseudoacceleration response at soil sites					
0.1	-2.024	1.1	0.007	2.5	-1.75	0.8217	0.4814	1.332 − 0.148M	0.3157	0.6286	0.37
0.2	-1.696	1.1	.0	2.5	-1.75	.8217	.4814	1.453 − 0.162M	.3157	.6286	.40
0.3	-1.638	1.1	- .008	2.5	-1.75	.8217	.4814	1.486 − 0.164M	.3157	.6286	.42
0.5	-1.659	1.1	- .025	2.5	-1.75	.8217	.4814	1.584 − 0.176M	.3157	.6286	.44
1.0	-1.975	1.1	- .060	2.5	-1.75	.8217	.4814	1.62 − 0.18M	.3157	.6286	.45
2.0	-2.414	1.1	- .105	2.5	-1.75	.8217	.4814	1.62 − 0.18M	.3157	.6286	.45
4.0	-3.068	1.1	-0.160	2.5	-1.75	0.8217	0.4814	1.62 − 0.18M	0.3157	0.6286	0.45
						Peak acceleration at soil sites					
	-2.611	1.1	0.0	2.5	-1.75	0.8217	0.4184	1.26 − 0.14M	0.3157	0.6286	0.35
						Pseudoacceleration response at rock sites					
0.1	-0.688	1.1	0.007	2.5	-2.05	1.353	0.406	1.332 − 0.148M	0.579	0.537	0.37
0.2	-0.479	1.1	- .008	2.5	-2.05	1.353	.406	1.453 − 0.162M	.579	.537	.40
0.3	-0.543	1.1	- .018	2.5	-2.05	1.353	.406	1.486 − 0.164M	.579	.537	.42
0.5	-0.793	1.1	- .036	2.5	-2.05	1.353	.406	1.584 − 0.176M	.579	.537	.44
1.0	-1.376	1.1	- .065	2.5	-2.05	1.353	.406	1.62 − 0.18M	.579	.537	.45
2.0	-2.142	1.1	- .100	2.5	-2.05	1.353	.406	1.62 − 0.18M	.579	.537	.45
4.0	-3.177	1.1	-0.150	2.5	-2.05	1.353	0.406	1.62 − 0.18M	0.579	0.537	0.45
						Peak acceleration at rock sites					
	-1.406	1.1	0.0	2.5	-2.05	1.353	0.406	1.26 − 0.14M	0.579	0.537	0.35

TABLE 6. STANDARD DEVIATION GIVEN BY DONOVAN AND BORNSTEIN (1978)
FOR THE NATURAL LOGARITHM OF AN INDIVIDUAL PREDICTION
OF PEAK HORIZONTAL ACCELERATION

Peak acceleration	0.01	0.05	0.10	0.15
Standard deviation of natural logarithm of peak acceleration	0.50	0.48	0.46	0.41

where y is peak horizontal acceleration (g), M is surface-wave magnitude for M greater than or equal to 6 and local magnitude otherwise, and r is the closest distance (km) to the source for M greater than 6 and hypocentral distance otherwise. Values of a and d are given in Table 9 along with $\sigma_{\ln y}$, the standard deviation of an individual prediction of ln y, which is treated as a function of M. Idriss proposed that peak acceleration from equation (15) be used to scale the response spectral shapes shown in Figure 10 for different site conditions with magnitude- and period-dependent correction factors shown in Figure 11. Figures 10 and 11 constitute another demonstration of the dependence of the shape of response spectra on magnitude and site conditions.

Comparisons of Predictive Equations. To compare the different relationships properly, adjustments must be made for the different definitions of distance. Figure 12a compares peak horizontal acceleration for the randomly oriented horizontal component at magnitude 6.5 as predicted by Donovan and Bornstein (1978), Joyner and Boore (1982), Idriss (1987), and Campbell (1988).

TABLE 7.   PARAMETERS IN THE PREDICTIVE EQUATIONS OF CAMPBELL (1988)
FOR MEAN PEAK HORIZONTAL ACCELERATION (g) AND VELOCITY (CM/S)

	Acceleration		Velocity	
	Unconstrained	Constrained	Unconstrained	Constrained
a	-2.817	-3.303	-0.798	-1.584
b	0.702	0.850	1.02	1.18
d	-1.20	-1.25	-1.26	-1.24
h_1	0.0921	0.0872	0.0150	0.00907
h_2	0.584	0.678	0.812	0.951
e_1	0.32	0.34	0.47	0.49
e_2	0.52	0.53	0.95	0.99
e_3	0.41	0.41	0.63	0.53
e_4	-0.85	-0.86	0.39	0.41
e_5	-1.14	-1.12	0.72	0.60
e_6	0.87	0.89	0.75	0.88
e_7	0.068	0.065	—	—
$\sigma_{\ln y}$	0.30	0.30	0.26	0.27

TABLE 8.   PARAMETER K_i IN THE PREDICTIVE EQUATIONS OF CAMPBELL (1988)
FOR MEAN PEAK HORIZONTAL ACCELERATION AND VELOCITY

Peak acceleration		
Fault type	$K_1 =$	1 reverse 0 strike-slip
Source directivity	$K_2 =$	1 rupture toward site 0 other
Shallow soil	$K_3 =$	1 soils$\leq$10 m deep 0 other
Embedment	$K_4 =$	1 basements of buildings 3–9 stories 0 other
	$K_5 =$	1 basements of buildings 10 or more stories 0 other
Peak velocity		
Fault type	$K_1 =$	1 reverse 0 strike-slip
Source directivity	$K_2 =$	1 rupture toward site 0 other
Building size	$K_3 =$	1 shelters and buildings less than 5 stories 0 other

The definition of distance used in Figure 12a is the closest distance to the vertical projection of
the rupture on the surface of the earth. The curves of Donovan and Bornstein and Campbell were
adjusted assuming a source depth of 5 km. The curve shown for Idriss is that for deep soil sites. The
curve shown for Campbell is that for strike-slip earthquakes recorded at free-field sites with more
than 10 m of soil, and no allowance is made for directivity. In this and subsequent comparisons
there is no indication of where the curves are not constrained by data as there is in Figures 6, 7,
and 9. At short distance, where it matters the most, the different relationships agree to within a
fraction of the uncertainty of an individual prediction as given by any of the authors. This suggests
that the short-distance predictions at magnitude 6.5 are controlled by the data. The differences at

TABLE 9. PARAMETERS IN THE PREDICTIVE EQUATIONS OF IDRISS (1987) FOR
THE RANDOMLY-ORIENTED HORIZONTAL COMPONENT OF PEAK ACCELERATION

M	Rock and stiff soil sites		Deep soil sites		
	a	d	a	d	$\sigma_{\ln y}$
4.5	606	-2.57	189	-2.22	0.70
5.0	617	-2.46	195	-2.13	.58
5.5	452	-2.28	147	-1.97	.48
6.0	282	-2.07	98	-1.79	.42
6.5	164	-1.85	61.6	-1.60	.38
7.0	91.7	-1.63	37.2	-1.41	.35
7.5	49.8	-1.41	22	-1.22	.35
8.0	28.5	-1.21	13.7	-1.05	.35
8.5	15.9	-1.01	8.4	-0.88	0.35

large distance are not of much practical importance; they are due, at least in part, to the inclusion of different records in the different data sets. Figure 12b gives the same comparison for magnitude 7.5. The agreement at short distance is not as good as at magnitude 6.5, reflecting the scarcity of data points, but it is within the uncertainty of an individual prediction.

Figure 13a compares 0.2 s pseudovelocity response at 5 percent damping for the randomly oriented horizontal component as predicted for magnitude 6.5 by Crouse (written communication, 1987; Vyas et al., 1988), Idriss (1987), Sadigh (written communication, 1987; Sadigh et al., 1986) and us (Joyner and Boore, 1982). The definition of distance is the same as in Figure 12. The curves shown for Crouse and Idriss are those for deep soil sites; the other two are for soil sites. Figure 13b gives the same comparison for magnitude 7.5. The differences shown for 0.2 s response are somewhat larger than for peak horizontal acceleration but smaller than for longer-period measures of ground motion, shown in Figures 14a and 14b, which give the same comparisons for 1.0 s pseudovelocity response. The maximum differences for motions large enough to be of engineering concern are a little larger than a factor of two; they occur for magnitude 7.5 at distances less than 10 km, a magnitude and distance range for which very few data are available.

For the 1.0 s response at magnitude 7.5 shown on Figure 14b, all four curves at short distances are higher, by factors of 1.5 to 3, than the highest values of the ATC-3 spectrum for firm ground. Ninety cm/s is the highest value of the ATC-3 pseudovelocity response spectrum at 1 hz for 5 percent damping on soil type S_2 with a response modification factor of 1 (see Figure 9). Concern that response values near the source may greatly exceed the values implicit in building codes is the reason for the proposal that a zone 5 should be added to the four zones of existing building codes. Zone 5 would lie within 8 km (5 miles) of a fault considered capable of producing a magnitude 7.5 earthquake.

Predictive Equations Developed from Italian Data. The predictive equations for shallow earthquakes described and compared in previous sections are based primarily on data from the western United States. There is a large set of strong-motion data recorded from shallow earthquakes in Italy. Predictive relationships for peak horizontal acceleration and velocity based on this data set have been developed by Sabetta and Pugliese (1987). They included only data believed to represent free-field conditions and only data where the triggering earthquake was reliably identified, had a magnitude greater than 4.5, was recorded by at least two stations, had an epicenter determined to an accuracy of 5 km or less, and had a magnitude accurate to 0.3. They used three site categories, stiff, shallow soil (5–20 m thick) and deep soil (more than 20 m thick). The equation fitted to the data is

$$\log y = a + bM - \log(r_0^2 + h^2)^{1/2} + s \tag{16}$$

where y is the larger value from the two horizontal components for either peak acceleration (g) or

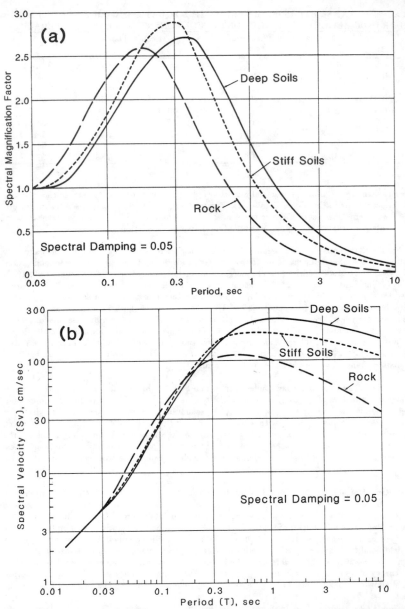

FIGURE 10. (a) Normalized spectral shapes for 5 percent damping for three different soil conditions and magnitudes in the range between 6.5 and 7.0 and (b) the same shape replotted in terms of pseudovelocity response for a zero-period acceleration of 1 *g* (adapted from Idriss, 1985).

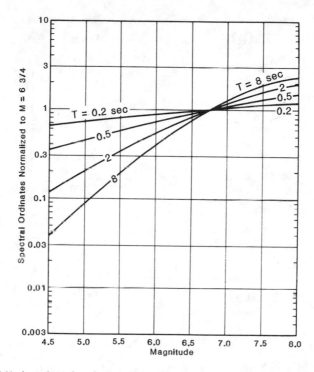

FIGURE 11. Approximate dependence of spectral ordinates on magnitude, normalized to a magnitude of 0.75 (from Idriss, 1987)

velocity (cm/s). M is given by surface-wave magnitude M_S when both M_S and local magnitude M_L are greater than or equal to 5.5, and M_L otherwise. r_0 is the closest distance (km) to the vertical projection of the rupture on the surface of the earth. Values of a, b, h, and s are given in Table 10 along with $\sigma_{\log y}$, the standard deviation of an individual prediction of $\log y$. Sabetta and Pugliese tried a magnitude-dependent h but did not find a statistically significant improvement in fit.

Equation (16) is similar in form to ours (Joyner and Boore, 1982), which facilitates comparison. Figures 15a and 15b give the comparison for peak horizontal acceleration and velocity, respectively. For distances less than 100 km the agreement is within a fraction of the standard deviation of an individual prediction for either study. In their paper Sabetta and Pugliese made the comparison using an earlier version of our curves (Joyner and Boore, 1981). Our earlier curve for peak horizontal velocity, which was based on fewer data than our 1982 equations, does not agree as well as the curve shown in Figure 15b. The differences at large distance between the two sets of curves in Figures 15a and 15b have little practical importance. They are probably due to our exclusion, for each earthquake, of data recorded at distances greater than the distance to the closest operational instrument that did not trigger.

Predictive Equations for Subduction-Zone Earthquakes. Data from subduction-zone earthquakes is generally treated separately because of presumed differences in source and/or propagation conditions. Jacob and Mori (1984) suggest that "the high variability of stress drops

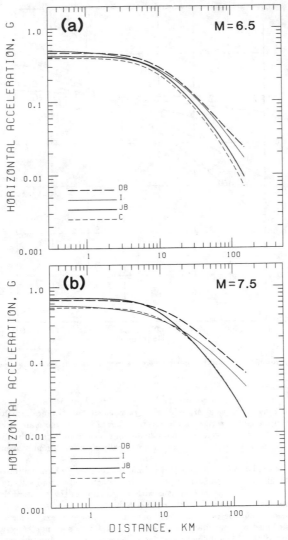

FIGURE 12. Comparison of different relationships for peak horizontal acceleration at magnitude 6.5 (a) and 7.5 (b). DB, from Donovan and Bornstein (1978); I, from Idriss (1987) for deep soil sites; JB, from Joyner and Boore (1982), reduced by 13 percent so as to approximate the value for the randomly oriented horizontal component; C, the constrained relationship of Campbell (1988) for a strike-slip earthquake recorded at a free-field site with soil more than 10 m deep and no allowance made for directivity. The distance plotted is the closest distance to the vertical projection of the rupture on the surface of the earth. The curves of Donovan and Bornstein and those of Campbell are adjusted assuming a source depth of 5 km.

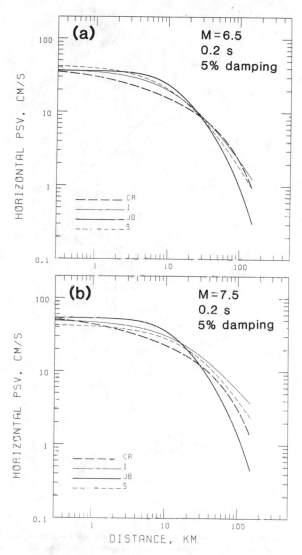

FIGURE 13. Comparison of different relationships for 0.2 s horizontal pseudovelocity response at 5 percent damping for magnitude 6.5 (a) and 7.5 (b). CR, from Crouse (written communication, 1987; Vyas *et al.*, 1988) for deep soil sites; I, from Idriss (1987) for deep soil sites; JB, from Joyner and Boore (1982) for soil sites; S, from Sadigh (written communication, 1987; Sadigh *et al.*, 1986) for soil sites. Distance is as defined for Figure 12.

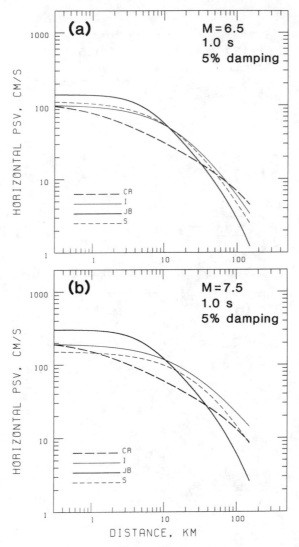

FIGURE 14. Comparison of different relationships for 1.0 s horizontal pseudovelocity response at 5 percent damping for magnitude 6.5 (a) and 7.5 (b). Curves are as defined for Figure 13, and distance is as defined for Figure 12.

typical for some subduction zones" is responsible for the apparently larger scatter in peak horizontal acceleration from Alaskan data compared with data from shallow earthquakes in the rest of the western United States. Boore (1986), however, was successful in simulating teleseismic P waves

TABLE 10. PARAMETERS IN THE PREDICTIVE EQUATIONS
OF SABETTA AND PUGLIESE (1987) FOR THE LARGER PEAK ACCELERATION (g)
AND VELOCITY (CM/S) OF TWO HORIZONTAL COMPONENTS

	a	b	h	s Stiff	s Shallow soil	s Deep soil	$\sigma_{\log y}$
Peak acceleration	-1.562	0.306	5.8	0.0	0.169	0.0	0.173
Peak velocity	-0.710	0.455	3.6	0.0	0.133	0.133	0.215

from subduction zone earthquakes with moment magnitudes up to 9.5, using the same stochastic source model with the same stress parameter as he used for simulating strong-motion data from shallow earthquakes in the western United States. This suggests a general similarity between the source processes of subduction-zone earthquakes and shallow crustal earthquakes. There are significant differences in geometry and propagation path between the two kinds of earthquake data. Of particular importance may be less anelastic attenuation along the deeper paths characteristic of data from subduction-zone earthquakes.

The M_S 7.8 Central Chile earthquake and the M_S 8.1 Michoacan, Mexico, earthquake, both in 1985, provided the first extensive sets of data from the epicentral regions of large subduction-zone earthquakes. Figure 16, taken from Anderson et al. (1986), shows peak accelerations from these data sets. The values for the Chilean earthquake are substantially higher at distances less than 100 km.

Crouse et al. Equation (17) was developed from pseudovelocity response data at soil sites from subduction earthquakes in the Northern Honshu zone (Crouse et al., 1988). Both horizontal components were used.

$$\ln y = a + b\mathrm{M} + d\ln r + qh \qquad (17)$$

where y is the pseudovelocity response value (cm/s) at 5 percent damping for the randomly oriented horizontal component, M is moment magnitude, r is distance (km) to the center of energy release, and h is focal depth (km). Values of a, b, d, and q are given in Table 11 along with $\sigma_{\ln y}$, the standard deviation of an individual prediction. The distance to the center of energy release was taken as the hypocentral distance for all earthquakes with M less than 7.5. For most of the larger events the distance was taken as the distance to the centroid of the fault plane defined by the aftershocks. If studies were available identifying the location of the greatest energy release, the distance to that point was used. Crouse et al. showed that equation (17) also fits data from stiff soil sites in the Kurile, Nankai, Alaskan, and Mexican subduction zones but appears not to fit data from stiff-soil sites in the Peru/North Chile and the New Britain/Bougainville zones.

Kawashima. Equation (18) for peak horizontal acceleration (gal), velocity (cm/s), and displacement (cm) and horizontal acceleration response (gal) for 5 percent damping was developed from data recorded in Japanese earthquakes with focal depth less than 60 km (Kawashima et al., 1984). Presumably most of the earthquakes are subduction-zone events. Data recorded on structures (including first floor and basement) were excluded.

$$y = a10^{bM}(\Delta + 30)^d \qquad (18)$$

where y represents the peak motion or the acceleration response which is the maximum over all possible azimuths of the horizontal component, M is JMA (Japanese Meteorological Agency) magnitude, and Δ is epicentral distance (km). Values of a, b, and d are given in Table 12 along with values of $\sigma_{\log y}$, the standard deviation of the common logarithm of an individual prediction of y. Different values of a, b, and $\sigma_{\log y}$ are given for three different site conditions defined in Table 13.

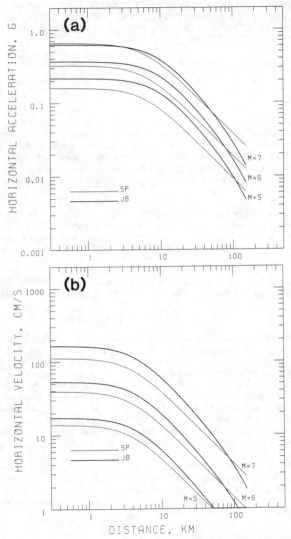

FIGURE 15. (a) Comparison of different relationships for the larger value of peak acceleration from the two horizontal components. SP, from Sabetta and Pugliese (1987) for Italian data at stiff and deep soil sites; JB, from Joyner and Boore (1982) for western North American data. (b) Comparison of different relationships for the larger value of peak velocity from the two horizontal components. SP, from Sabetta and Pugliese (1987) for Italian data at soil sites; JB, from Joyner and Boore (1982) for western North American data at soil sites. Distance is as defined for Figure 12.

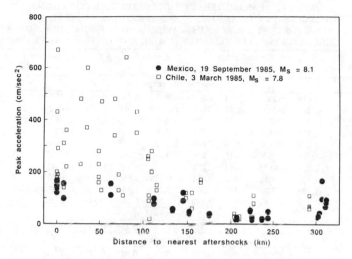

FIGURE 16. Peak horizontal acceleration in the 19 September 1985 Mexican earthquake and the 3 March 1985 Chilean earthquake plotted against distance outside the boundary of the aftershock zone (from Anderson et al., 1986). Copyright 1986 by the AAAS.

TABLE 11. PARAMETERS IN THE PREDICTIVE EQUATIONS OF CROUSE ET AL. (1988) FOR THE RANDOMLY ORIENTED HORIZONTAL COMPONENT OF PSEUDOVELOCITY RESPONSE (CM/S) AT 5 PERCENT DAMPING

Period (s)	a	b	d	q	$\sigma_{\ln y}$
0.1	1.86	0.48	-1.02	0.0093	0.668
0.2	3.19	0.44	-0.98	.0053	.672
0.4	1.29	0.68	-0.84	.0041	.597
0.6	0.67	0.85	-0.95	.0030	.674
0.8	-0.38	0.96	-0.87	.0017	.703
1.0	-1.13	1.00	0.03	.0	.713
1.5	-2.79	1.18	0.09	-.0007	.663
2.0	-3.04	1.26	-0.78	-.0008	.718
3.0	-3.46	1.34	-0.85	-.0046	.730
4.0	-4.09	1.39	-0.85	-0.0053	0.720

Theoretical Prediction

The usual method for calculating theoretical strong-motion seismograms is kinematic and requires specification of the distribution in space and time of slip on the fault and a means of calculating the Green's functions, which give the response to slip at a point. There are now available a number of methods for calculating complete Green's functions for earth models in which material properties vary only with depth (Spudich and Ascher, 1983; Ascher and Spudich, 1986; Olson et al., 1984). If lateral heterogeneity is a significant factor, ray methods (Spudich and Frazer, 1984) can be used to compute high-frequency ($> 1Hz$) ground motions at near-source distances. The methods for calculating complete Green's functions have been used to infer the distribution of slip

TABLE 12. PARAMETERS IN THE PREDICTIVE EQUATIONS
OF KAWASHIMA *ET AL.* (1984) FOR HORIZONTAL ACCELERATION
RESPONSE (GAL) AT 5 PERCENT DAMPING AND FOR PEAK HORIZONTAL
ACCELERATION (GAL), VELOCITY (CM/S), AND DISPLACEMENT (CM)

Period (s)	d	Ground group 1			Ground group 2			Ground group 3		
		a	b	$\sigma_{\log y}$	a	b	$\sigma_{\log y}$	a	b	$\sigma_{\log y}$
				Acceleration response						
0.1	-1.178	2420	0.211	0.262	848.0	0.262	0.256	1307	0.208	0.219
0.15	-1.178	2407	.216	.229	629.1	.288	.244	948.2	.238	.218
0.2	-1.178	1269	.247	.226	466.0	.315	.273	1128	.228	.211
0.3	-1.178	574.8	.273	.241	266.8	.345	.270	1263	.224	.217
0.5	-1.178	211.8	.299	.278	102.2	.388	.249	580.6	.281	.240
0.7	-1.178	102.5	.317	.239	34.34	.440	.245	65.67	.421	.243
1.0	-1.178	40.10	.344	.273	5.04	.548	.305	7.41	.541	.307
1.5	-1.178	7.12	.432	.254	0.719	.630	.288	0.803	.647	.305
2.0	-1.178	5.78	.417	.267	0.347	.644	.264	0.351	.666	.276
3.0	-1.178	1.67	0.462	0.249	0.361	0.586	0.248	0.262	0.635	0.263
				Peak acceleration						
	-1.218	987.4	0.216	0.216	232.5	0.313	0.224	403.8	0.265	0.197
				Peak velocity						
	-1.222	20.8	0.263	0.236	2.81	0.430	0.239	5.11	0.404	0.243
				Peak displacement						
	-1.254	0.626	0.372	0.262	0.062	0.567	0.258	0.070	0.584	0.262

TABLE 13. CLASSIFICATION OF SITE CONDITIONS FOR THE PREDICTIVE
EQUATIONS OF KAWASHIMA *ET AL.* (1984)

	Geological Definition	Definition by Natural Period
Group 1	Tertiary or older rock or diluvium less than 10m thick	Period less than 0.2 s
Group 2	Diluvium with thickness 10m or more or alluvium less than 25 m thick including soft layer less than 5 m thick	Period between 0.2 and 0.6 s
Group 3	Other than the above, usually soft alluvium or reclaimed land	Period more than 0.6 s

in past earthquakes by direct inversion (Olson and Apsel, 1982; Hartzell and Heaton, 1983) and by trial-and-error forward modeling (Archuleta, 1984). Such studies are important in advancing our understanding of earthquake source processes. The calculation of theoretical seismograms for future earthquakes, however, requires a means of specifying the distribution of slip in future earthquakes. Assuming uniform slip would not give realistic seismograms. We have done computations indicating that the amplitude of ground motions at all frequencies of engineering interest is controlled by the heterogeneities in the fault rupture (Boore and Joyner, 1986). Because there is no way to predict the heterogeneities of future ruptures, we turn to stochastic source models as the basis for theoretical ground-motion prediction.

Stochastic Source Models. Stochastic source models make possible what we consider to be the first realistic theoretical predictions of strong ground motion. We describe two stochastic source models, the barrier model of Papageorgiou and Aki (1983a, b) and the stochastic ω-square model of Hanks and McGuire (1981). In the barrier model a rectangular fault plane is covered by circular cracks of equal diameter separated by unbroken barriers. Individual cracks rupture independently and randomly, and their radiation is described by Sato and Hirasawa's (1973) equations. The Fourier spectrum of the resulting ground motion has random phase, which is the reason we classify the barrier model as a stochastic source model. The barrier model is specified by five basic parameters, fault length, fault width, maximum slip, rupture velocity, and barrier interval. A sixth parameter, the cohesive zone size, is introduced to explain the cutoff of high frequencies in the spectrum. The cutoff frequency f_m (Hanks, 1982) is considered by Papageorgiou and Aki (1983a) to be a source parameter that may vary with other source parameters. This contrasts with the view of advocates of the Hanks-McGuire model who, while admitting the possibility of source-controlled cutoff, generally consider the cutoff of high frequencies to be an effect of near-surface attenuation at the recording site. Papageorgiou and Aki (1983b) applied the barrier model to the analysis of six California earthquakes and showed that the barrier interval is strongly related to the maximum slip.

According to the Hanks McGuire (1981) model earthquake accelerations are band-limited white noise in the band between the corner frequency f_0 and f_m, and the spectral shape is given by the Brune (1970, 1971) spectrum. In addition to moment M_0, the model is specified by two parameters, the stress parameter $\Delta\sigma$ and f_m. Hanks and McGuire used this model with the aid of random vibration theory to predict horizontal peak acceleration and rms acceleration and obtained excellent agreement with empirical data over the magnitude range from 4.0 to 7.7. Boore (1983) made use of both stochastic simulations and random vibration theory to test the predictions of the Hanks-McGuire model for peak horizontal velocity and response spectra as well as peak acceleration. He showed agreement between model predictions and data covering a magnitude range from less than 1.0 to more than 7.0. Hanks and Boore (1984) showed that the model predictions reproduce the correlation between log moment and local magnitude M_L for California earthquakes in the M_L range from 0 to 7. Boore (1986) compared model predictions with peak teleseismic P-wave amplitudes given by Houston and Kanamori (1986) for earthquakes with moment magnitude up to 9.5 and showed good agreement.

A question with the Hanks-McGuire model arises in extending the model to magnitudes greater than the critical magnitude corresponding to rupture of the full width of the seismogenic zone. For larger earthquakes similarity must break down because rupture width can no longer increase as the moment increases. Joyner (1984) proposed a scaling law that accomplishes the extension of the Hanks-McGuire model to magnitudes for which similarity no longer applies.

Ground-Motion Prediction with Stochastic Source Models. Two methods are available for making ground-motion predictions with stochastic source models. One uses Monte Carlo simulations in the time domain and the other uses random vibration theory. The two methods complement each other. Calculations with random-vibration theory require less computer time, whereas Monte Carlo simulations are useful in applications that require time series and serve as a check on the assumptions underlying the random-vibration approach. Both methods are summarized below; more detailed descriptions are given by Boore (1983, 1986, 1987; Boore and Joyner, 1984a; Boore and Atkinson, 1987).

For both methods the spectrum of ground motion is given as a function of frequency f by

$$R(f) = CS(f)A(f)D(f)I(f) \tag{19}$$

Where the factors C, S, A, D, and I represent, respectively, a scaling factor, the source spectrum, an amplification factor, a diminution factor, and an instrument-response factor. Only $S(f)$ depends on seismic moment. C is given by

$$C = \frac{R_{\Theta\Phi}FV}{4\pi\rho_0\beta_0^3\mathcal{R}} \tag{20}$$

where $R_{\Theta\Phi}$ is the radiation pattern averaged over an appropriate range of azimuth and takeoff angle (e.g., Boore and Boatwright, 1984), F accounts for the free-surface effect, V represents the partition of energy from a vector into horizontal components (if needed), ρ_0 and β_0 are the density and shear velocity in the source region, and $\mathcal{R}$ is the geometric spreading factor. F and V are usually given values of 2 and $1/\sqrt{2}$, respectively. For body waves within about 100 km $\mathcal{R} = r$, where r is hypocentral distance. For distances beyond 100 km, where the L_g phase dominates, $\mathcal{R}$ will be proportional to $\sqrt{r}$. For teleseismic body waves a different treatment of geometric spreading is required (Boore, 1986).

For the barrier model the source factor $S(f)$ is given by Papageorgiou (1988). For the Hanks-McGuire model $S(f)$ is

$$S(f) = M_0/[1 + (f/f_0)^2] \tag{21}$$

where M_0 is the seismic moment and f_0 the corner frequency, given by

$$f_0 = 4.9 \times 10^6 \beta_0 (\Delta\sigma/M_0)^{1/3} \tag{22}$$

$\Delta\sigma$ is a parameter with the dimensions of stress, f_0 is in Hz, β_0 in km/s, $\Delta\sigma$ in bars, and M_0 in dyne-cm (Brune, 1970, 1971).

A modified version of $S(f)$ was given by Joyner (1984) to accommodate the breakdown of similarity which must occur when the moment exceeds the critical moment M_{0c} corresponding to rupture of the entire width of the seismogenic zone.

$$\begin{aligned} S(f) &= M_0/(1 + if/f_B)^{1/2} & f \le f_A \\ &= M_0 \left(\frac{f_A}{f}\right)^{3/2} /(1 + if/f_B)^{1/2} & f \ge f_A \end{aligned} \tag{23}$$

where

$$\begin{aligned} f_A &= 4.9 \times 10^6 \beta_0 \lambda^{-1/4} (\Delta\sigma/M_0)^{1/3} \\ f_B &= 4.9 \times 10^6 \beta_0 \lambda^{3/4} (\Delta\sigma/M_0)^{1/3} & M_0 \le M_{0c} \\[1em] f_A &= 4.9 \times 10^6 \beta_0 \lambda^{-1/4} \Delta\sigma^{1/3} M_{0c}^{1/6} M_0^{-1/2} \\ f_B &= 4.9 \times 10^6 \beta_0 \lambda^{3/4} (\Delta\sigma/M_{0c})^{1/3} & M_0 \ge M_{0c} \end{aligned} \tag{24}$$

λ is the ratio between the length and the width of the fault [assigned a typical value of 4 by Joyner (1984)], and the other symbols are as defined for equation (22).

The amplification factor can be given in different ways. Perhaps the most familiar is the frequency-dependent transfer function that results from wave propagation in a stack of layers (e.g. Boore and Joyner, 1984b). Amplification can also be represented in terms of site impedance, as discussed previously. Conservation of energy requires that amplitude increase as impedance decreases in going from the source region to the recording site. The amplification factor can be approximated by $\sqrt{\rho_0\beta_0/\rho_r\beta_r}$, where the subscripts r refers to material near the recorder and the subscript 0 refers to material near the source. Table 14 gives the logarithm of the correction factor as a function of frequency for a typical western North American strong-motion recording site on rock, as estimated by Boore (1986). The variations in density are expected to be minor for rock and were ignored in Table 14. The frequency dependence of the correction arises from the assumption that ρ_r and β_r are effective properties averaged over a depth equal to a quarter-wavelength (Joyner and Fumal, 1984). Note that the correction can be greater than a factor of two even at rock sites.

The diminution factor may be written

$$D(f) = \exp[-\pi f r/Q(f)\beta]P(f) \tag{25}$$

where Q is a frequency-dependent attenuation function, β is the propagation velocity averaged over the path, and P is a high-cut filter. In western North America we use for Q the following function that fits a number of observations collected by Aki (see Figure 2 in Boore, 1984):

$$Q = 29.4 \frac{1 + (f/0.3)^{2.9}}{(f/0.3)^2} \tag{26}$$

TABLE 14. AMPLIFICATION FACTORS FOR
ROCK SITES IN WESTERN NORTH AMERICA

$\log f$	$\log \sqrt{\beta_0/\beta_r}$
-1	0.01
-0.5	0.04
0.0	0.13
0.5	0.34
1.0	0.37

The high-frequency behavior of this function is consistent with our analysis of response spectral attenuation (Joyner and Boore, 1982), but, as referred to previously, we do not know to what extent the agreement indicates that the Q function represents the true frequency dependence of Q and to what extent the agreement is influenced by error in the assumed geometric spreading of $1/r$. The high-cut filter P is needed to account for the observation that acceleration spectra generally show an abrupt depletion of high-frequency energy above some frequency f_m (Hanks, 1982). This filter can be represented by a Butterworth filter [Boore, 1983, equation (4)] with a steep rolloff (approximately 24 db/octave) above the corner frequency f_m. Another form of the P filter is

$$P(f) = \exp(-\pi\kappa_0 f) \qquad (27)$$

The form and notion is from Anderson and Hough (1984). For small distances or large Q, the $P(f)$ factor contributes most of the attenuation, and the filter $D(f)$ is then roughly equivalent to a step high-cut filter with a cutoff frequency f_m equal to $1/\pi\kappa_0$ (a relation pointed out to us by J. Boatwright). A possible physical mechanism for the $P(f)$ factor would be near-site attenuation in the upper kilometer or so of the ray path.

Finally, the filter $I(f)$ is used to shape the spectrum so that the output time series corresponds to the particular ground-motion parameter of interest. For example, if pseudo-velocity response spectra are to be computed, I is the response to ground displacement of an oscillator of frequency f_r and damping ζ

$$I(f) = \frac{Vf^2}{(f^2 - f_r^2) - i(2\zeta f f_r)}, \qquad (28)$$

with the magnification V given by $2\pi f_r$. If Richter local magnitude M_L is to be computed, I is given by equation (28) with $f_r = 1.25$, $\zeta = 0.8$, and $V = 2800$, the values for the Wood-Anderson seismograph used in defining the M_L scale. If peak velocity or peak acceleration are the quantities of interest, then

$$I(f) = (2\pi f i)^n \qquad (29)$$

where $n = 1$ or 2, respectively, for velocity or acceleration   The uncorrected response of an accelerometer can be simulated by using equation (28) with appropriate f_r and ζ (typically, $f_r = 25$ Hz and $\zeta = 0.6$) and $V = (2\pi f_r)^2$. The response equations for other instruments, such as the World Wide Standardized Seismograph Network short-period instrument, can be used as desired.

Once the spectrum $R(f)$ is determined, application of the time-domain method of ground-motion prediction is very simple. Gaussian white noise is generated with a random-number generator. The noise is windowed by either a shaping function or a box car whose duration T_w is given by $1/f_0 + 0.05r$ for the Hanks-McGuire model and by $1/f_A + 0.05r$ for Joyner's (1984) modification, where r is the source distance (km). The distance-dependent term is included to account for the spreading out of the source energy due to scattering and wave-propagation effects (Herrmann, 1985). The amplitude of the window is chosen so that the mean level of the white spectrum is unity. The noise sample is then filtered by the filter $R(f)$ from equation (19). [This method differs from the conventional engineering method for stochastic simulation in which white noise is filtered before being windowed rather than after. Safak and Boore (1988) show that windowing filtered noise alters the spectrum, so windowing should be done first.] Fourier transformation back to the time domain gives the simulated time series from which the peak value is obtained. The process is then repeated with different seeds for the random-number generator. Between 20 and 100 simulations are generally sufficient to give a good estimate of the peak motion. A spectrum of a single realization of the process will not match the target spectrum, but the average of a number of realizations will match, as shown in Figure 17. The time series for one realization for each of two different magnitudes are shown in Figure 18; the only parameter that was changed between the simulations for the two earthquakes was the magnitude.

The method of random-vibration theory can be used to obtain peak values without doing simulations. The method is based on the work of Cartwright and Longuet-Higgins (1956). To estimate the peak value y_{max} of the ground-motion parameter y, we obtain the spectrum $R(f)$ of y from equation (19) as in the simulation method. The zeroth, second, and fourth moments, m_0, m_2, and m_4, of the energy density spectrum are calculated from the equation

$$m_k = \frac{1}{\pi} \int_0^\infty \omega^k |R(f)|^2 d\omega \qquad (30)$$

where $\omega = 2\pi f$. The rms value of the ground-motion parameter is given by

$$y_{rms} = (m_0/T_r)^{1/2} \qquad (31)$$

For predicting peak acceleration or velocity T_r is the same as T_w, the time duration of the motion determined as described for the simulation method. The determination of T_r for predicting response spectral values is described later. The expected value of the peak $[E(y_{max})]$ is calculated either by an exact or an asymptotic formula depending on the values of the bandwidth parameter ξ and the number of extrema N, where

$$\xi = m_2/(m_0 m_4)^{1/2} \qquad (32)$$

and N is the largest integer contained in the number $2\tilde{f}T_w$, where

$$\tilde{f} = \frac{1}{2\pi}(m_4/m_2)^{1/2} \qquad (33)$$

If N is less than $14.7/\xi - 8$ then the expected value of the peak is given by the exact formula

$$E(y_{max}) = y_{rms}\sqrt{\frac{\pi}{2}} \sum_{l=1}^N (-1)^{l+1} \frac{C_l^N}{\sqrt{l}} \xi^l \qquad (34)$$

where C_l^N are the binomial coefficients $(= N!/l!(N-l)!)$. For larger values of N there may be numerical problems with the exact formula and the asymptotic solution

$$E(y_{max}) = y_{rms}\{[2\ln(N)]^{1/2} + \gamma/[2\ln(N)]^{1/2}\} \qquad (35)$$

where $\gamma = 0.557216$ (Euler's constant), is recommended. For the asymptotic approximation, the quantities $\tilde{f}$ and N are recalculated as follows

$$\tilde{f} = \frac{1}{2\pi}(m_2/m_0)^{1/2}$$
$$N = 2\tilde{f}T_w \qquad (36)$$

The cutoff between the exact and asymptotic solutions was determined from numerical experiments, with the criteria being that at the transition the difference between $\log E(y_{max})$ from the two equations was less than 0.01 units.

When estimating response spectra a further refinement is required to the basic random-vibration theory described above. The random-vibration approach assumes a stationary time series. For small

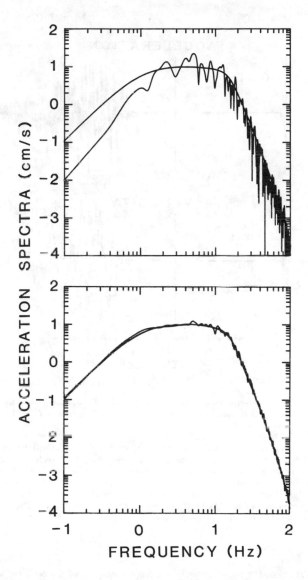

FIGURE 17. Fourier amplitude spectrum of acceleration at 10 km for magnitude 5.0. (Top) Smooth curve, given spectrum; jagged curve, spectrum for one realization of the simulation process. (Bottom) as above but averaged over 20 realizations (the average is the square root of the arithmetic mean of the squared moduli of the individual spectra).

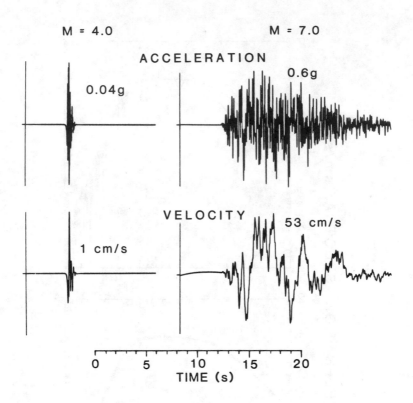

FIGURE 18. Simulated time series for magnitude 4 and 7 earthquakes at a distance of 10 km. Values given for peak motions are the average of peaks from 20 such time series. A low-cut filter with a cutoff frequency of 0.10 Hz has been applied to the velocity time series.

to moderate earthquakes, or low oscillator frequencies, or light damping the duration of motion may not be long enough to generate a pseudo-stationary response. We have developed an empirical correction to the duration of motion T_w to be applied in equation (31) [but not in the estimates of N] to account for the distribution of energy content beyond the ground-motion duration (Boore and Joyner, 1984a). We define the ground-motion duration T_w, as before, as $1/f_0 + 0.05r$ for the Hanks-McGuire model and as $1/f_A + 0.05r$ for Joyner's (1984) modification. The duration T_r to be used in equation (31) for computing y_{rms} is given by

$$T_r = T_w + \frac{T_0}{2\pi\zeta}\left(\frac{\gamma^3}{\gamma^3 + 1/3}\right) \tag{37}$$

where T_0 is the period of the oscillator in sec, ζ is the damping of the oscillator as a fraction of critical, and $\gamma = T_w/T_0$. Equation (37) was developed and verified by comparing the results of random-vibration theory with the results of Monte Carlo simulations.

As previously noted, the Hanks-McGuire model gives predictions in good agreement with the strong-motion data set collected in western North America. Encouraged by this agreement, Boore

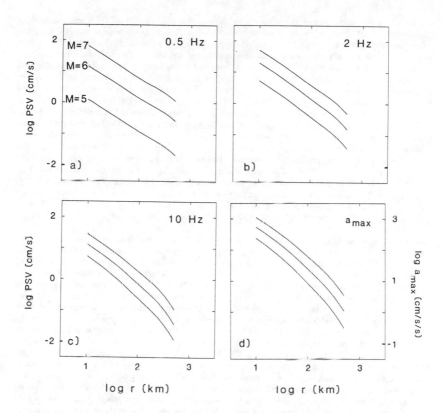

FIGURE 19. Predictive relationships for hard rock sites in eastern North America giving peak horizontal acceleration and pseudovelocity response at three periods for 5 percent damping (from Boore and Atkinson, 1987).

and Atkinson (1987) used the model to predict ground motion at rock sites in eastern North America, a region where there are few ground-motion recordings within 100 km of damaging earthquakes. They chose a Q function and a value for f_m appropriate for eastern North America. They ignored the amplification factor $A(f)$ on the grounds that near-surface shear velocity in rock in eastern North America is large enough to make the factor negligible. For a stress parameter $\Delta\sigma$ of 100 bars the model agrees with the limited available data. Figure 19 shows the predictions they made for peak horizontal acceleration and spectral response at 0.5, 2, and 10 Hz. Toro and McGuire (1987) also made a study applying the Hanks-McGuire model to the prediction of ground motion in eastern North America.

A limitation of the use of stochastic models for ground-motion predictions as described above is that the approach can be rigorously justified only for sites at distances large compared to the source dimensions. Joyner et al. (1988) have developed a method of stochastic simulation that is applicable at smaller distances. They start by generating functions over the rupture surface representing the total slip at each point. They do so by using Monte Carlo methods to obtain random-phase spectra in two-dimensional wave-number space. The spectra have constant amplitude (determined by the moment M_0) for wave-number vectors with modulus p less than a critical value and amplitudes proportional to $p^{-1.5}$ for p greater than the critical value. The critical value is controlled by the stress parameter $\Delta\sigma$ and is equal to f_A/v, where f_A is given by equation (24) and v is the rupture velocity. The function representing total slip is obtained by inverse Fourier transformation and multiplication by a window smoothly tapered to zero at the boundary of the rupture zone. Ground motion corresponding to a delta-function slip velocity at each point is calculated by assuming rupture propagation at uniform velocity from the hypocenter. The ground motion is Fourier transformed from the time to the frequency domain and multiplied by the filter $1/(1 + if/f_B)^{1/2}$ to obtain the spectrum $S(f)$ of ground motion corresponding to a slip-velocity function of the Kostrov (1964) type. The spectrum $S(f)$ is then multiplied by the factors C, A, D, and I in equation (19) and transformed back to the time domain so that the peak value can be obtained. In the actual computations the rupture surface is broken up into zones based on distance to the recording site so that the Q filter can be applied for the appropriate site distance. The process is repeated as many times as needed with different seeds for the Monte Carlo random-number generator. The ground motions resulting from this method have ω-square spectra at distant sites. Simulations of the 1979 Imperial Valley, California, earthquake agree reasonably well with observed data at sites near the source for a rupture velocity of 0.8 β_0. The results, however, are highly sensitive to the assumed rupture velocity, with amplitudes at the fault lower by about a factor of two for a rupture velocity of 0.7 β_0.

Comparison of the Hanks-McGuire Model and the Barrier Model. The differences between the Hanks-McGuire model and the barrier model represent essentially philosophical differences about how best to describe the seismic source and do not necessarily imply large differences in predicted ground motion. With the methods described in the preceding section it is possible to show just what the differences are. Fourier spectra of horizontal acceleration for the two models are shown in Figure 20. The solid lines in Figure 20 represent Joyner's (1984) modification of the Hanks-McGuire model for a critical magnitude M_{0c} larger than 8.0. The amplification factor is that in Table 14, corresponding to western North American rock sites, the Q function is given by equation (26), $P(f) = \exp(-\pi\kappa_0 f)$ with κ_0 chosen as 0.02 to correspond to rock sites, and $\Delta\sigma = 50$ bars. The dashed lines represent the spectra appropriate for the barrier model as specified by Papageorgiou (1988) with no amplification factor applied. The data used by Papageorgiou to obtain the spectral shapes were recorded at the surface of the earth and therefore implicitly contain the amplification factors. Using the random-vibration method described in the previous section, we computed peak acceleration and response spectral values for both models. The values for peak acceleration are compared in Figure 21 for both models with the results of regression analysis of strong-motion data from western North America (Joyner and Boore, 1982). For magnitudes above 6 the results from the two models agree well with each other and with the data. The values for the response spectrum at 5 Hz and 5 percent damping are shown in Figure 22. The agreement is not quite so good in this case but still reasonably good above magnitude 6 with the barrier model a little closer to the observed data. The values for the response spectrum at 1 Hz and 5 percent damping are shown in Figure 23, which shows two curves for data, the upper one for soil sites and the lower one for rock sites. The curve representing Joyner's modification of the Hanks-McGuire model agrees with data for rock sites, which it should, given that the values of the amplification factor and κ_0 correspond to rock sites. The curve representing the barrier model agrees reasonably well above magnitude 6 with data from soil sites. Papageorgiou (1988) makes no distinction between spectra for rock sites and soil sites, but most of the strong-motion data available for guiding the choices that control the shape of the spectrum were recorded at soil sites. It is therefore not surprising that the predictions of the barrier model agree better with data recorded at soil sites.

Hybrid Prediction

A method for predicting ground motion recently popular among seismologists is the summation of recordings of small earthquakes, considered as Green's functions, in an attempt to simulate the ground motion from larger events (Hartzell, 1978, 1982; Wu, 1978; Kanamori, 1979; Hadley and Helmberger, 1980; Mikumo *et al.*, 1981; Irikura and Muramatu, 1982; Hadley *et al.*, 1982;

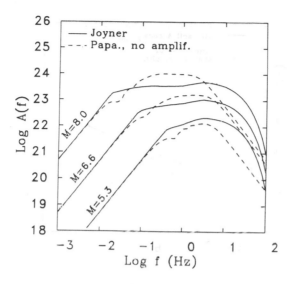

FIGURE 20. Spectra of horizontal acceleration at a distance of 20 km for the indicated magnitudes. Solid line, Joyner's (1984) modification of the Hanks-McGuire model with the amplification factor in Table 14, Q given by equation (26), $P(f)=\exp(-\pi\kappa_0 f)$, $\kappa_0=0.02$, and $\Delta\sigma=50$ bars; dashed line, the barrier model as specified by Papageorgiou (1988) with no amplification factor applied.

Irikura, 1983; Coats et al., 1984; Houston and Kanamori, 1984; Imagawa et al., 1984; Munguia and Brune, 1984; Hutchings, 1985; Heaton and Hartzell, 1986). The small earthquakes (henceforth called subevents) ideally are located near the hypothetical source and recorded at the site for which the large-event simulation is desired. If these ideal conditions are met, then the method incorporates wave-propagation effects over the whole path from source to recording site as well as local site effects. Generally, however, these conditions are not met. Users of the method generally postulate some distribution of subevents over a fault plane and sum them in accordance with an assumed geometry of rupture propagation. Most users include randomness of some sort in their methods for summing subevents. This randomness may be thought of as representing a degree of random heterogeneity characteristic of large earthquakes. It also performs an important function in preventing spurious periodicities in the simulated motion resulting from summing over uniform grids in space or over points equally spaced in time. Irikura (1983), who did not use any randomness in his summation, relied on a special smoothing technique to eliminate spurious periodicities.

In the spirit of the original concept of the subevent as a Green's function, the corner frequency of the subevent should be higher than any frequency of interest in the simulated motion. In that case the subevent record will be a true impulse response, and the spectrum of the simulated event will depend on how the subevents are distributed over the fault and in time. The quality of the simulation will depend, accordingly on how well the distribution of slip is represented over the fault and in time, in particular, how well the degree and kind of heterogeneity of faulting is represented. In the general case, however, because of limited dynamic range in the subevent records, it may not be possible to use subevents so small that their corner frequencies are higher than any frequency of engineering interest and still maintain the desired bandwidth in the simulated motion. For example, if we wanted to keep the subevent corner frequencies above 3 Hz, we could use subevents no larger than a moment magnitude of about 4; if we wanted to keep the subevent corner frequencies above

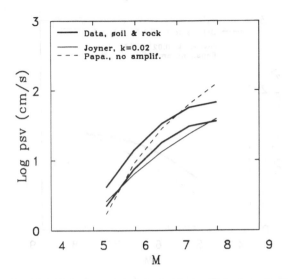

FIGURE 23. Pseudovelocity response at 1 Hz and 5 percent damping for the randomly oriented horizontal component as a function of magnitude for a distance of 20 km. Heavy lines, data from shallow earthquakes in western North America (Joyner and Boore, 1982) at soil sites (upper heavy line) and rock sites (lower heavy line); light line, predictions for Joyner's modification of the Hanks-McGuire model; dashed line, predictions for the barrier model (see caption of Figure 20 for details).

for the method of simulating large events by summing recordings of small ones. The problem is illustrated in Figure 24, where the light line represents the expected value of the spectrum obtained by summing identical subevents according to equation (39) and the heavy line represents the ω-square spectrum corresponding to the moment of the simulated event. The spectrum obtained by summing subevents falls significantly below the ω-square spectrum in the vicinity of the corner frequency of the simulated event. In our earlier paper, from which Figure 24 is taken, we did not treat this difference as a significant problem (Joyner and Boore, 1986). We now believe that it is significant and that it is inherent in any method of summing subevents distributed randomly with uniform probability.

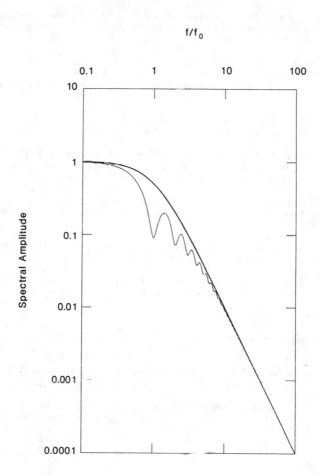

FIGURE 24. Spectrum (light line) of the simulated event obtained by random summation of identical subevents according to equation (39) for a difference of one unit in moment magnitude between simulated event and subevent compared to the ω-square spectrum (heavy line) corresponding to the moment of the simulated event. The axes have been normalized by the long-period level and corner frequency (f_0) of the simulated event.

REFERENCES

Abrahamson, N. A. (1985). Estimation of seismic wave coherency and rupture velocity using the SMART 1 strong-motion array recordings, *Rep. UCB/EERC-85/02*, 134 p., Earthquake Engin. Res. Cen., Univ. of Calif., Berkeley.

Aki, K. (1967). Scaling law of seismic spectrum, *J. Geophys. Res.* **72**, 1217–1231.

Aki, K. and P. G. Richards (1980). *Quantitative Seismology Theory and Methods* **1**, 557 p., W. H. Freeman and Company.

Algermissen, S. T., and D. M. Perkins (1976). A probabilistic estimate of maximum acceleration in rock in the contiguous United States, *U.S. Geol. Surv. Open-File Rept. 76-416*, 45 p.

Algermissen, S. T., D. M. Perkins, P. C. Thenhaus, S. L. Hanson, and B. L. Bender (1982). Probabilistic estimates of maximum acceleration and velocity in rock in the contiguous United States, *U.S. Geol. Surv. Open-File Rept. 82-1033*, 99 p.

Anderson, J. G., J. N. Brune, J. Prince, and F. L. Vernon, III (1983). Preliminary report on the use of digital strong motion recorders in the Mexicali Valley, Baha California, *Bull. Seism. Soc. Am.* **73**, 1451–1467.

Anderson, J. G. and S. E. Hough (1984). A model for the shape of the Fourier amplitude spectrum of acceleration at high frequencies, *Bull. Seism. Soc. Am.* **74**, 1969–1993.

Anderson, J. G., P. Bodin, J. N. Brune, J. Prince, S. K. Singh, R. Quass, and M. Onate (1986). Strong ground motion from the Michaocan, Mexico, earthquake, *Science* **233**, 1043–1049, 5 September.

Andrews, M. C., C. Dietel, T. Noce, E. Sembera, and J. Bicknell (1988). Preliminary analyses of digital recordings of the Superstition Hills, California aftershock sequence (abs.), submitted to *Earthquake Notes*.

Applied Technology Council (1978). *Tentative Provisions for the Development of Seismic Regulations for Buildings*, Applied Technology Council Publication ATC 3-06, 505 p.

Araya, R. and A. Der Kiureghian (1986). Seismic hazard analysis including source directivity effect, *Proc. of 3rd. U.S. Nat. Conf. on Earthquake Engin.* **1**, convened Charleston, South Carolina, August 24–28, 1986, 269–280.

Archuleta, R. J. (1984). A faulting model for the 1979 Imperial Valley earthquake, *J. Geophys. Res.* **89**, 4459–4585.

Ascher, U. and P. Spudich (1986). A hybrid collocation method for calculating complete theoretical seismograms in vertically varying media, *Geophys. J. R. Astron. Soc.* **86**, 19–40.

Bakun, W. H. and W. B. Joyner (1984). The M_L scale in central California, *Bull. Seism. Soc. Am.* **74**, 1827–1843

Bard, P.-Y. and B. E. Tucker (1985). Underground and ridge site effects: a comparison of observation and theory, *Bull. Seism. Soc. Am.* **75**, 905–922.

Barker, J. S., P. G. Somerville, and J. P. McLaren (1988). Modeling ground-motion attenuation in eastern North America, in *Proc. Symp. on Seismic Hazards, Ground Motions, Soil-Liquefaction and Engineering Practice in Eastern North America*, K. H. Jacob, ed., National Center for

Earthquake Engineering Research Tech. Report NCEER–87–0025, 339–352.

Ben-Menahem, A. (1961). Radiation of seismic surface-waves from finite moving sources, *Bull. Seism. Soc. Am.* **51**, 401–435.

Berger, J., L. M. Baker, J. N. Brune, J. B. Fletcher, T. C. Hanks, and F. L. Vernon, III, (1984). The Anza array: a high-dynamic-range, broadband, digitally radiotelemetered, seismic array, *Bull. Seism. Soc. Am.* **74**, 1469–1481.

Bevington, P. R. (1969). *Data Reduction and Error Analysis for the Physical Sciences*, 336 p., McGraw-Hill.

Boatwright, J. and D. M. Boore (1982). Analysis of the ground accelerations radiated by the 1980 Livermore Valley earthquakes for directivity and dynamic source characteristics, *Bull. Seism. Soc. Am.* **72**, 1843–1865.

Boatwright, J. (1985). Characteristics of the aftershock sequence of the Borah Peak, Idaho, earthquake determined from digital recordings of the events, *Bull. Seism. Soc. Am.* **75**, 1265–1284.

Bolt, B. A., N. Abrahamson, and Y. T. Yeh (1984). The variation of strong ground motion over short distances, in *Proc. of 8th World Conf. on Earthquake Engin. (San Francisco)* **2**, 183–189.

Boore, D. M. (1972). A note on the effect of simple topography on seismic SH waves, *Bull. Seism. Soc. Am.* **62**, 275–284.

Boore, D. M. (1973). The effect of simple topography on seismic waves: implications for the accelerations recorded at Pacoima Dam, San Fernando Valley, California, *Bull. Seism. Soc. Am.* **63**, 1603–1609.

Boore, D. M. and W. B. Joyner (1978). The influence of rupture incoherence on seismic directivity, *Bull. Seism. Soc. Am.* **68**, 283–300.

Boore, D. M. and W. B. Joyner (1982). The empirical prediction of ground motion, *Bull. Seism. Soc. Am.* **72**, S43–S60.

Boore, D. M. (1983). Stochastic simulation of high-frequency ground motions based on seismological models of the radiated spectra, *Bull. Seism. Soc. Am.* **73**, 1865–1894.

Boore, D. M. (1984). Use of seismoscope records to determine M_L and peak velocities, *Bull. Seism. Soc. Am.* **74**, 315–324.

Boore, D. M. and J. Boatwright (1984). Average body-wave radiation coefficients, *Bull. Seism. Soc. Am.* **74**, 1615–1621.

Boore, D. M. and W. B. Joyner (1984a). A note on the use of random vibration theory to predict peak amplitudes of transient signals, *Bull. Seism. Soc. Am.* **74**, 2035–2039.

Boore, D. M. and W. B. Joyner (1984b). Ground motions and response spectra at soil sites from seismological models of radiated spectra, in *Proc. of Eighth World Conf. on Earthquake Engin. (San Francisco)* **2**, 457–464.

Boore, D. M. (1986). Short-period P- and S-wave radiation from large earthquakes: implications for spectral scaling relations, *Bull. Seism. Soc. Am.* **76**, 43–64.

Boore, D. M. and W. B. Joyner (1986). Prediction of earthquake ground motion at periods of interest for base-isolated structures, in Proc. Seminar and Workshop on Base Isolation and Passive Energy Dissipation, March 12–14, 1986, San Francisco, Calif., 355–370, Applied Technology Council.

Boore, D. M. (1987). The prediction of strong ground motion, in *Strong Ground Motion Seismology*, M. Erdik and M. N. Toksöz, Editors, NATO Advanced Studies Institute series, D. Reidel Publishing Company, Dordrecht, The Netherlands, 109–141.

Boore, D. M. and G. M. Atkinson (1987). Stochastic prediction of ground motion and spectral response parameters at hard-rock sites in eastern North America, *Bull. Seism. Soc. Am.* **77**, 440–467.

Boore, D. M. (1988). The Richter scale—its development and use for determining earthquake source parameters, submitted to *Physics of the Earth and Planetary Interiors*, in press.

Borcherdt, R. D. (1970). Effects of local geology on ground motion near San Francisco Bay, *Bull. Seism. Soc. Am.* **60**, 29–61.

Borcherdt, R. D., G. L. Maxwell, C. S. Mueller, R. McClearn, G. Sembera, and L. Wennerberg (1983). Digital strong-motion data recorded by U.S. Geological Survey near Coalinga, California, in *The Coalinga Earthquake Sequence Commencing May 2, 1983*, 61–76, U.S. Geol. Surv. Open-File Rep. 83-511.

Borcherdt, R. D., J. G. Anderson, C. B. Crouse, N. C. Donovan, T. V. McEvilly, and A. F. Shakal (1984). *National Planning Considerations for the Acquisition of Strong-Ground-Motion Data*, Pub. 84-08 Earthquake Engin. Res. Inst., 57 p.

Borcherdt, R. D., J. B. Fletcher, E. G. Jensen, G. L. Maxwell, J. R. Van Schaak, R. E. Warrick, E. Cranswick, M. J. S. Johnston, and R. McClearn (1985). A general earthquake-observation system, *Bull. Seism. Soc. Am.* **75**, 1783–1825.

Bouchon, M. (1973). Effect of topography on surface motion, *Bull. Seism. Soc. Am.* **63**, 615–632.

Bouchon, M. (1982). The complete synthesis of seismic crustal phases at regional distances, *J. Geophys. Res.* **87**, 1735–1741.

Brady, A. G., R. L. Porcella, G. N. Bycroft, E. C. Etheredge, P. N. Mork, B. Silverstein, and A. F. Shakal (1984a). Strong-motion results from the main shock of April 24, 1984, in *The Morgan Hill, California Earthquake of April 24, 1984 (A Preliminary Report)* 1, 18–26, *U.S. Geol. Surv. Open-File Rep. 84-498A*.

Brady, A. G., R. L. Porcella, G. N. Bycroft, E. C. Etheredge, P. N. Mork, B. Silverstein, and A. F. Shakal (1984b). Strong-motion results from the main shock of April 24, 1984—computer plots, *The Morgan Hill, California earthquake of April 24, 1984 (A Preliminary Report)* 2, *U.S. Geol. Surv. Open-File Rep.84-498B*, 118 p.

Brillinger, D. R. and H. K. Preisler (1984). An exploratory analysis of the Joyner-Boore attenuation data, *Bull. Seism. Soc. Am.* **74**, 1441–1450.

Brillinger, D. R. and H. K. Preisler (1985). Further analysis of the Joyner-Boore attenuation data, *Bull. Seism. Soc. Am.* **75**, 611–614.

Brune, J. N. (1970). Tectonic stress and the spectra of seismic shear waves from earthquakes, *J. Geophys. Res.* **75**, 4997–5009.

Brune, J. N. (1971). Correction, *J. Geophys. Res.* 76, 5002.

Brune, J. N., R. Lovberg, R. Anooshehpoor, and L. Wang (1984). Measurements of topographic amplification on foam rubber models using newly designed quadrant position detectors (abs.), *Earthquake Notes* 55, 22.

Burger, R. W., P. G. Somerville, J. S. Barker, R. B. Herrmann, and D. V. Helmberger (1987). The effect of crustal structure on strong ground motion attenuation relations in eastern North America, *Bull. Seism. Soc. Am.* 77, 420–439.

Bycroft, G. N. (1978). The effect of soil-structure interaction on seismometer readings, *Bull. Seism. Soc. Am.* 68, 823–843.

Campbell, K. W. (1981). Near-source attenuation of peak horizontal acceleration, *Bull. Seism. Soc. Am.* 71, 2039–2070.

Campbell, K. W. (1985). Strong motion attenuation relations: a ten-year perspective, *Earthquake Spectra* 1, 759–804.

Campbell, K. W. (1988). Predicting strong ground motion in Utah, in *Evaluation of Regional and Urban Earthquake Hazards and Risk in Utah*, W. W. Hays and P. L. Gori, Editors, *U.S. Geol. Surv. Profess. Paper*, in preparation.

Cartwright, D. E. and M. S Longuet-Higgins (1956). The statistical distribution of the maxima of a random function, *Proc. R. Soc. London* 237, 212–232.

Coats, D. A., H. Kanamori, and H. Houston (1984). Simulation of strong motion from the 1964 Alaskan earthquake (abs.), *Earthquake Notes* 55, 18.

Converse, A. M. (1984). AGRAM: a series of computer programs for processing digitized strong-motion accelerograms, *U.S. Geol. Surv. Open File Rep. 84-525*.

Converse, A M., A. G. Brady, and W. B. Joyner (1984). Improvements in strong-motion data processing procedures, in *Proc. of 8th World Conf. on Earthquake Engin. (San Francisco)*, 2, 143–148.

Cornell, C. A. and R. T. Sewell (1987). Non-linear-behavior intensity measures in seismic hazard analysis, *Proc. Internat. Seminar on Seismic Zonation*, December, 1987, Guangzhou, China.

Crouse, C. B., G. C. Liang, and G. R. Martin (1984). Experimental study of soil-structure interaction at an accelerograph station, *Bull. Seism. Soc. Am.* 74, 1995–2013.

Crouse, C. B. and B. Hushmand (1987). Experimental investigations of soil-structure interaction at CDMG and USGS accelerograph statons (abs.), *Earthquake Notes* 58, 10.

Crouse, C. B., Y. K. Vyas, and B. A. Schell (1988). Ground motions from subduction-zone earthquakes, *Bull. Seism. Soc. Am.* 78, 1–25.

Donovan, N. C. and A. E. Bornstein (1978). Uncertainties in seismic risk procedures, *Proc. Am. Soc. Civil Eng., J. Geotech. Eng. Div.* 104, 869–887.

Ekström, G. and A. M. Dziewonski (1985). Centroid-moment tensor solutions for 35 earthquakes in Western North America (1977–1983), *Bull. Seism. Soc. Am.* 75, 23–29.

Etheredge, E. C. and R. L. Porcella (1987). Strong-motion data from the October 1, 1987 Whittier

Narrows earthquake, *U.S. Geol. Surv. Open-File Rep. 87-616,* 64 p.

Ewing, W. M., W. S. Jardetzky, and F. Press (1957). *Elastic Waves in Layered Media,* 380 p., McGraw-Hill.

Frankel, F. and L. Wennerberg (1987). On the frequency dependence of shear-wave Q in the crust from 1 to 15 Hz (abs.), *EOS, Trans. Am. Geophys. Union* 68, 1362.

Fletcher, J. B., A. G. Brady, and T. C. Hanks (1980). Strong-motion accelerograms of the Oroville, California, aftershocks: data processing and the aftershock of 0350 August 6, 1975, *Bull. Seism. Soc. Am.* 70, 243-367.

Geli, L., P.-Y. Bard, and B. Jullien (1988). The effect of topography on earthquake ground motion: a review and new results, *Bull. Seism. Soc. Am.* 78, 42-63.

Gutenberg, B. (1957). Effects of ground on earthquake motion, *Bull. Seism. Soc. Am.* 47, 221-250.

Guzman, R. A. and P. C. Jennings (1976). Design spectra for nuclear power plants, *Proc. Am. Soc. Civil Engin., J. Power Div.* 102, 165-178.

Hadley, D. M. and D. V. Helmberger (1980). Simulation of strong ground motions, *Bull. Seism. Soc. Am.* 70, 617-610.

Hadley, D. M., D. V. Helmberger, and J. A. Orcutt (1982). Peak acceleration scaling studies, *Bull. Seism. Soc. Am.* 72, 959-979.

Hanks, T. C. (1975). Strong ground motion of the San Fernando, California, earthquake: ground displacements, *Bull. Seism. Soc. Am.* 65, 193-225.

Hanks, T. C. and H. Kanamori (1979). A moment magnitude scale, *J. Geophys. Res.* 84, 2348-2350.

Hanks, T. C. and R. K. McGuire (1981). The character of high-frequency strong ground motion, *Bull. Seism. Soc. Am.* 71, 2071-2095.

Hanks, T. C. (1982). f_{max}, *Bull. Seism.Soc. Am.* 72, 1867-1879.

Hanks, T. C. and D. M. Boore (1984). Moment-magnitude relations in theory and practice, *J. Geophys. Res.* 89, 6229-6235.

Hartzell, S. H. (1978). Earthquake aftershocks as Green's functions, *Geophys. Res. Lett.* 5, 1-4.

Hartzell, S. H. (1982). Simulation of ground accelerations for the May 1980 Mammoth Lakes, California, earthquakes, *Bull. Seism. Soc. Am.* 72, 2381-2387.

Hartzell, S. H. and T. H. Heaton (1983). Inversion of strong ground motion and teleseismic waveform data for the fault rupture history of the 1979 Imperial Valley, California, earthquake, *Bull. Seism. Soc. Am.* 73, 1553-1583.

Heaton, T. H., and S. H. Hartzell (1986). Estimation of strong ground motions from hypothetical earthquakes on the Cascadia subduction zone, Pacific Northwest, *U.S. Geol. Surv. Open-File Rep. 86-328,* 40 p.

Heaton, T. H., F. Tajima, and A. W. Mori (1986). Estimating ground motions using recorded accelerograms, *Surveys in Geophys.* 8, 25-83.

Herrmann, R. B. (1985). An extension of random vibration theory estimates of strong ground motion to large distances, *Bull. Seism. Soc. Am.* **75**, 1447–1453.

Houston, H. and H. Kanamori (1984). The effect of asperities on short-period seismic radiation with application to the 1964 Alaskan earthquake (abs.), *Earthqukae Notes* **55**, 18.

Houston, H. and H. Kanamori (1986). Source spectra of great earthquakes: teleseismic constraints on rupture processes and strong motion, *Bull. Seism. Soc. Am.* **76**, 19–42.

Huang, M. J., A. F. Shakal, D. L. Parke, R. W. Sherburne, and R. V. Nutt (1985). Processed data from the strong-motion records of the Morgan Hill earthquake of 24 April 1984. Part II. Structural-response records, *Rep. OSMS 85-05*, 320 p., Office of Strong Motion Studies, Calif. Div. Mines and Geol.

Huang, M. J., R. W. Sherburne, D. L. Parke, and A. F. Shakal (1986). CSMIP strong-motion records from the Palm Springs, California earthquake of 8 July 1986, *Rep. OSMS 86-05*, 74 p., Office of Strong Motion Studies, Calif. Div. Mines and Geol.

Huang, M. J., T. Q. Cao, C. E. Ventura, D. L. Parke, and A. F. Shakal (1987). CSMIP strong-motion records from the Superstition Hills, Imperial County, California earthquakes of 23 and 24 November 1987, *Rep. OSMS 87-06*, 42 p., Office of Strong Motion Studies, Calif. Div. Mines and Geol.

Hutchings, L. (1985). Modeling earthquakes with empirical Green's functions (abs.), *Earthquake Notes* **56**, 14.

Hutton, L. K. and D. M. Boore (1987). The M_L scale in southern California, *Bull. Seism. Soc. Am.* **77**, 2074–2094.

Idriss, I. M. and H. B. Seed (1968). Seismic response of horizontal soil layers, *Proc. Am. Soc. Civil Engin., J. Soil Mech. and Found. Div* **94**, 1003–1031.

Idriss, I. M. (1979). Characteristics of earthquake ground motions, in *Earthquake Engineering and Soil Dynamics, Proc. Am. Soc. Civil Eng. Geotech. Eng. Div. Specialty Conf.*, June 19–21, 1978, Pasadena, California, **3**, 1151–1265.

Idriss, I. M. (1985). Evaluating seismic risk in engineering practice, *Proc. Eleventh Internat. Conf. on Soil Mech. and Foundation Eng.*, August 12–16, 1985, San Francisco, California, **1**, 255–320, A. A Balkema, Rotterdam.

Idriss, I. M. (1987). Earthquake ground motions, Lecture notes, Course on Strong Ground Motion, Earthquake Engin. Res. Inst., Pasadena, Calif., April 10–11, 1987.

Imagawa, K., N. Mikami, and T. Mikumo (1984). Analytical and semi-empirical synthesis of near-field seismic waveforms for investigating the rupture mechanism of major earthquakes, *J. Phys. Earth* **32**, 317–338.

International Conference of Building Officials (1988). *Uniform Building Code*, 1988 edition, in preparation.

Irikura, K. and I. Muramatu (1982). Synthesis of strong ground motions from large earthquakes using observed seismograms of small events, *Proc. of 3rd Internat. Microzonation Conf.* **1**, Seattle, 447–458.

Irikura, K. (1983). Semi-empirical estimation of strong ground motions during large earthquakes,

Bull. Disaster Prevention Res. Inst. (Kyoto Univ.) **33**, 63–104.

Iwan, W. D., M. A. Moser, and C.-Y. Peng (1985). Some observations on strong-motion earthquake measurement using a digital accelerograph, *Bull. Seism. Soc. Am.* **75**, 1225–1246.

Jackson, S. M. and J. Boatwright (1985). The Borah Peak, Idaho earthquake of October 28, 1983— strong ground motion, *Earthquake Spectra* **2**, 51–69.

Jacob, K. H. and J. Mori (1984). Strong motions in Alaska-type subduction zone environments, *Proc. Eighth World Conf. on Earthquake Engin. (San Francisco)* **2**, 311–317.

Johnson, L. R. and W. Silva (1981). The effects of unconsolidated sediments upon the ground motion during local earthquakes, *Bull. Seism. Soc. Am.* **71**, 127–142.

Joyner, W. B. and A. T. F. Chen (1975). Calculation of nonlinear ground response in earthquakes, *Bull. Seism. Soc. Am.* **65**, 1315–1336.

Joyner, W. B., R. E. Warrick, and A. A. Oliver, III, (1976). Analysis of seismograms from a downhole array in sediments near San Francisco Bay, *Bull. Seism. Soc. Am.* **66**, 937–958.

Joyner, W. B. and D. M. Boore (1981). Peak horizontal acceleration and velocity from strong-motion records including records from the 1979 Imperial Valley, California, earthquake, *Bull. Seism. Soc. Am.* **71**, 2011–2038.

Joyner, W. B. and D. M. Boore (1982). Prediction of earthquake response spectra, *Proc. 51st Ann. Convention Structural Eng. Assoc. of Cal.*, also *U. S. Geol. Surv. Open-File Rept. 82-977*, 16 p.

Joyner, W. B. (1984). A scaling law for the spectra of large earthquakes, *Bull. Seism. Soc. Am.* **74**, 1167–1188.

Joyner, W. B. and T. E. Fumal (1984). Use of measured shear-wave velocity for predicting geologic site effects on strong ground motion, *Proc. Eighth World Conf. on Earthquake Eng. (San Francisco)* **2**, 777–783.

Joyner, W. B. and D. M. Boore (1986). On simulating large earthquakes by Green's-function addition of smaller earthquakes, in *Earthquake Source Mechanics, Maurice Ewing Ser.* **6**, edited by S. Das et al., 269–274, Am. Geophys. Union.

Joyner, W. B., K. W. Campbell, and S. C. Harmsen (1988). Near-source simulation of earthquake ground motion based on the stochastic omega-square model (abs.), submitted to *Earthquake Notes*.

Kanai, K. (1952). Relation between the nature of surface layer and the amplitudes of earthquake motions, *Bull. Earthquake Res. Inst., Tokyo Univ.* **30**, 31–37.

Kanamori, H. (1979). A semi-empirical approach to prediction of long-period ground motions from great earthquakes, *Bull. Seism. Soc. Am.* **69**, 1645–1670.

Kanamori, H. and C. R. Allen (1986). Earthquake repeat time and average stress drop, in *Earthquake Source Mechanics, Maurice Ewing Ser.* **6**, edited by S. Das et al., 227–235, Am. Geophys. Union.

Kawashima, K., K. Aizawa, and K. Takahashi (1984). Attenuation of peak ground motion and absolute acceleration response spectra, *Proc. Eighth World Conf. on Earthquake Engin. (San Francisco)* **2**, 257–264.

Kennett, B. L. N. (1985). On regional S, Bull. Seism. Soc. Am. 75, 1077–1086.

Kostrov, B. V. (1964). Self-similar problems of propagation of shear cracks (translation), J. Appl. Math. Mech. 28, 1077–1078.

Khemici, O. and W.-L. Chiang (1984). Frequency domain corrections of earthquake accelerograms with experimental verifications, Proc. Eighth World Conf. on Earthquake Engin. (San Francisco) 2, 103–110.

King, J. L. and B. E. Tucker (1984). Observed variations of earthquake motion across a sediment-filled valley, Bull. Seism. Soc. Am. 74, 137–151.

Lee, V. W. and M. D. Trifunac (1984). Current developments in data processing of strong motion accelerograms, Rep. 84-01, 99 p., Dept. Civil Engin., Univ. So. Calif., Los Angeles.

Maley, R. P., A. G. Brady, E. C. Etheredge, D. A. Johnson, P. N. Mork, and J. C. Switzer (1983). Analog strong-motion data and processed main event records obtained by U.S. Geological Survey near Coalinga, California, in The Coalinga Earthquake Sequence Commencing May 2, 1983, 38–60, U.S. Geol. Surv. Open-File Rep. 83-511.

Maley, R. P., E. C. Etheredge, and A. Acosta (1986). U. S. Geological Survey strong-motion records from the Chalfant Valley, California, earthquake of July 21, 1986, U. S. Geol. Surv. Open-File Rep. 86-568, 19 p.

McGarr, A. (1984). Scaling of ground motion parameters, state of stress, and focal depth, J. Geophys. Res. 89, 6969–6979.

McGarr, A. (1986). Some observations indicating complications in the nature of earthquake scaling, in Earthquake Source Mechanics, Maurice Ewing Ser. 6, edited by S. Das et al., 217–225, Am. Geophys. Union.

McGuire, R. K. (1974). Seismic structural response risk analysis, incorporating peak response regressions on earthquake magnitude and distance, Research Report R74-51, Dept. Civil Eng., Mass. Inst. of Tech., Cambridge, Mass., 371 p.

McGuire, R. K. (1978). Seismic ground motion parameter relations, Proc. Am. Soc. Civil Eng., J. Geotech. Eng. Div. 104, 481–490.

McJunkin, R. D. and A. F. Shakal (1983). The Parkfield strong-motion array, Calif. Geol. 36, 27–34.

Mikumo, T., K. Irikura, and K. Imagawa (1981). Near-field strong-motion synthesis from foreshock and aftershock records and the rupture process of the main shock fault (abs.), IASPEI 21st General Assembly, London, Canada, July 20–30, 1981.

Mohraz, B. (1976). A study of earthquake response spectra for different geological conditions, Bull. Seism. Soc. Am. 66, 915–935.

Munguía, L. and J. N. Brune (1984). Simulations of strong ground motions for earthquakes in the Mexicali-Imperial valley, Proc. of Workshop on Strong Ground Motion Simulation and Earthquake Engineering Applications, Pub. 85-02 Earthquake Engin. Res. Inst., April 30–May 3, 1984, Los Altos, California, 21-1–21-19.

Newmark, N. M. and W. J. Hall (1969). Seismic design criteria for nuclear reactor facilities, Proc. Fourth World Conf. on Earthquake Eng. (Santiago) 2, B4-37–B4-50.

New Zealand Geological Survey (1987). The March 2, 1987, earthquake near Edgecumbe, North Island, New Zealand, *EOS, Trans. Am. Geophys. Union* 68, 1162–1171.

Olson, A. H. and R. J. Apsel (1982). Finite faults and inverse theory with applications to the 1979 Imperial Valley earthquake, *Bull. Seism. Soc. Am.* 72, 1969–2001.

Olson, A. H., J. A. Orcutt, and G. A. Frazier (1984). The discrete wavenumber/finite element method for synthetic seismograms, *Geophys. J. R. Astron. Soc.* 77, 421–460.

Papageorgiou, A. S. and K. Aki (1983a). A specific barrier model for the quantitative description of inhomogeneous faulting and the prediction of strong ground motion. I. Description of the model, *Bull. Seism. Soc. Am.* 73, 693–722.

Papageorgiou, A. S. and K. Aki (1983b). A specific barrier model for the quantitative description of inhomogeneous faulting and the prediction of strong ground motion. Part II. Applications of the model, *Bull. Seism. Soc. Am.* 73, 953–978.

Papageorgiou, A. S. (1988). On two characteristic frequencies of acceleration spectra: patch corner frequency and f_{max}, *Bull. Seism. Soc. Am.*, scheduled for the April issue.

Porcella, R. L., E. C. Etheredge, R. P. Maley, and J. C. Switzer (1987a). Strong-motion data from the July 8, 1986 North Palm Springs earthquake and aftershocks, *U.S. Geol. Surv. Open-File Rep. 87-155*, 37 p.

Porcella, R. L., E. C. Etheredge, R. P. Maley, and J. C. Switzer (1987b). Strong-motion data from the Superstition Hills earthquakes of 0154 and 1315 (GMT), November 24, 1987, *U.S. Geol. Surv. Open-File Rep. 87-672*, 56 p.

Raugh, M. R. (1981). Procedures for analysis of strong-motion records (abs.), *Earthquake Notes* 52, 17.

Rogers, A. M., R. D. Borcherdt, P. A. Covington, and D. M. Perkins (1984). A comparative ground response study near Los Angeles using recordings of Nevada nuclear tests and the 1971 San Fernando earthquake, *Bull. Seism. Soc. Am.* 74, 1925–1949.

Rogers, A. M., S. C. Harmsen, R. B. Herrmann, and M. E. Meremonte (1987). A study of ground motion attenuation in the southern Great Basin, Nevada-California, using several techniques for estimates of Q_S, log A_0, and coda Q, *J. Geophys. Res.* 92, 3527–3540.

Sabetta, F. and A. Pugliese (1987). Attenuation of peak horizontal acceleration and velocity from Italian strong-motion records, *Bull. Seism. Soc. Am.* 77, 1491–1513.

Sadigh, K., J. Egan, and R. Youngs (1986). Specification of ground motion for seismic design of long period structures (abs.), *Earthquake Notes* 57, 13.

Şafak, E. and D. M. Boore (1988). On low-frequency errors of uniformly modulated filtered white-noise models for ground motions, *Earthquake Engineering and Structural Dynamics* 16, in press.

Saragoni, R. H., M. B. Fresard, and S. Gonzales (1985). Analisis de los acelerogramas del terremoto del 3 de Marzo de 1985, I Parte, in Publicacion SES I 4/1985 (199), University of Chile.

Sato, T. and T. Hirasawa (1973). Body wave spectra from propagating shear cracks, *J. Phys. Earth* 21, 415–431.

Schnabel, P., H. B. Seed, and J. Lysmer (1972). Modification of seismograph records for effects of local soil conditions, *Bull. Seism. Soc. Am.* **62**, 1649–1664.

Schnabel, P. B. and H. B. Seed (1973). Accelerations in rock for earthquakes in the western United States, *Bull. Seism. Soc. Am.* **63**, 501–516.

Shakal, A. F. and D. L. Bernreuter (1981). *Empirical analyses of near-source ground motion, U.S. Nuclear Regulatory Commission Report NUREG/CR-2095.*

Shakal, A. F. and R. D. McJunkin (1983). Preliminary summary of CDMG strong-motion records from the 2 May 1983 Coalinga, California, earthquake, *Rep. OSMS 83-5.2*, 49 p., Office of Strong Motion Studies, Calif. Div. Mines and Geol.

Shakal, A. F. and J. T. Ragsdale (1984). Acceleration, velocity and displacement noise analysis for the CSMIP acceleration digitization system, *Proc. Eighth World Conf. on Earthquake Eng. (San Francisco)* **2**, 111–118.

Shakal, A. F., M. J. Huang, D. L. Parke, and R. W. Sherburne (1986a). Processed data from the strong-motion records of the Morgan Hill earthquake of 24 April 1984. Part I ground-response records, *Rep. OSMS 84-04*, 249 p., Office of Strong Motion Studies, Calif. Div. Mines and Geol.

Shakal, A. F., R. Linares, M J. Huang, and D. L. Parke (1986b). Processed strong motion data from the San Salvador earthquake of October 10, 1986, *Rep. OSMS 86-07*, 113 p., Office of Strong Motion Studies, Calif. Div. Mines and Geol.

Shakal, A. F., M. J. Huang, C. E. Ventura, D. L. Parke, T. Q. Cao, R. W. Sherburne, and R. Blasquez (1987). CSMIP strong-motion records from the Whittier, California earthquake of 1 October 1987, *Rep. OSMS 87-05*, 198 p., Office of Strong Motion Studies, Calif. Div. Mines and Geol.

Shyam Sunder, S. and J. J. Connor (1982). A new procedure for processing strong-motion earthquake signals, *Bull. Seism. Soc. Am.* **72**, 643–661.

Silva, W. (1976). Body waves in a layered anelastic solid, *Bull. Seism. Soc. Am.* **66**, 1539–1554.

Singh, S. K., E. Mena, and R. Castro (1988). Some aspects of source characteristics of the September 19, 1985, Michaocan earthquake and ground motion amplification in and near Mexico City from strong motion data, *Bull. Seis. Soc. Am.*, scheduled for the April issue.

Spudich, P. and U. Ascher (1983). Calculation of complete theoretical seismograms in vertically varying media using collocation methods, *Geophys. J. R. Astron. Soc.* **75**, 101–124.

Spudich, P. and L. N. Frazer (1984). Use of ray theory to calculate high-frequency radiation from earthquake sources having spatially variable rupture velocity and stress drop, *Bull. Seism. Soc. Am.* **74**, 2061–2082.

Spudich, P. and E. Cranswick (1984). Direct observation of rupture propagation during the 1979 Imperial Valley earthquake using a short baseline accelerometer array, *Bull. Seism. Soc. Am.* **74**, 2083–2114.

Structural Engineers Association of California (1980). *Recommended Lateral Force Requirements and Commentary*, Fourth Edition Revised.

Structural Engineers Association of California (1988). *Recommended Lateral Force Requirements and Commentary*, Fifth Edition, in preparation.

Toro, G. R. and R. K. McGuire (1987).   An investigation into earthquake ground motion characteristics in eastern North America, *Bull. Seism. Soc. Am.* **77**, 468–489.

Trifunac, M. D. (1971).   Zero baseline correction of strong-motion accelerograms, *Bull. Seism. Soc Am.* **61**, 1201–1211.

Trifunac, M. D. (1972).   A note on correction of strong-motion accelerograms for instrument response, *Bull. Seism. Soc Am.* **62**, 401–409.

Trifunac, M. D. and V. W. Lee (1973).   Routine processing of strong motion accelerograms, *Rep. EERL 73-03*, 360 p., Earthquake Engin. Res. Lab., Calif. Inst. Tech., Pasadena.

Trifunac, M. D. and J. G. Anderson (1978).   Preliminary empirical models for scaling pseudo relative velocity spectra, appendix A in *Methods for Prediction of Strong Earthquake Ground Motion, U.S. Nuclear Regulatory Commission Report NUREG/CR-0689*, A1–A90.

Trifunac, M. D. and V. W. Lee (1979).   Dependence of pseudo relative velocity spectra of strong motion acceleration on the depth of sedimentary deposits, *Report No. CE 79-02*, Dept. Civil Eng., Univ. of Southern Cal., Los Angeles, Cal., 67 p.

Tucker, B. E., J. L. King, D. Hatzfeld, and I. L. Nersesov (1984).   Observations of hard-rock site effects, *Bull. Seis. Soc. Am.* **74**, 121–136.

U.S. Atomic Energy Commission (1973).   Design response spectra for seismic design of nuclear power plants, *Regulatory Guide 1.60*, 4 p.

Vidale, J. E. and D. V. Helmberger (1988).   Elastic finite-difference modeling of the 1971 San Fernando, California earthquake, *Bull. Seism. Soc. Am.* **78**, 122–141.

Vyas, Y. K., C. B. Crouse, and B. A. Schell (1988).   Regional design ground motion criteria for the southern Bering Sea, *Conf. Offshore Mech. and Arctic Engin.*, Houston, Texas, February 7–12, 1988.

Weichert, D. H., R. J. Wetmiller, R. B. Horner, P. S. Munro, and P. N. Mork (1986).   Strong motion records from the 23 December 1985, M_S 6.9 Nahanni, NWT, and some associated earthquakes, *Geological Survey of Canada Open File Rept. 86-1-PGC*, 9 p.

Westaway, R. and R. B. Smith (1987).   Strong ground motion parameters for normal-faulting earthquakes (abs.), *EOS, Trans. Am. Geophys. Union* **68**, 1348.

Wu, F. T. (1978).   Prediction of strong ground motion using small earthquakes, *Proc. of 2nd Internat. Microzonation Conf.* **2**, San Francisco, 701–704.

LOCAL SITE EFFECTS ON STRONG GROUND MOTION

Keiiti Aki*

Abstract

This is a review of the state-of-the-art in evaluating site effects on strong ground motion. We start with examining the effectiveness of the broad classification of site conditions into soil and rock in strong motion prediction. After reviewing empirically determined site-specific amplification factors, we conclude that the conventional broad classification is not effective in characterizing the site effect especially for higher frequencies. We also review analytical approaches to the site effect, and find that we may have an adequate state-of-the-art in predicting the site effect for many realistic situations (except for the full three dimensional case), if we know input motion, velocity and density distribution, topography, sediment thickness and damping of sediment. We review the current practice in site characterization and conclude that the most realistic approach to the microzonation is to determine empirical site-amplification factors for as many sites as possible by the regression analysis of earthquake data, and correlate them with various geotechnical parameters of the site which are relatively easier to measure. Analytical studies on the causes of site effects will give helpful insight to the search for effective parameters.

Introduction

Aeons of weathering, erosion, deposition and other geological processes formed a great variety of topography and lithologic structures of irregular shapes and heterogeneous material near the surface of the Earth. Since all our engineering structures are constructed on these complex near-surface media the understanding of their effect on strong ground motion is essential for earthquake engineering.

The purpose of the present paper is to review the current knowledge and understanding of the effects of topographic and geologic conditions of a site on its strong ground motion and come up with recommendations on how to define and approach the problem.

*William M. Keck Foundation Professor, University of Southern California, Department of Geological Sciences, Los Angeles, CA 90089-0740.

In order to define the problem of site effect on strong ground motion, we must define what parameters of ground motion we are concerned, what the input wave field is and how the site conditions are characterized. These three questions are equally important for a clear definition of the site effect.

First, the same site would respond differently depending on the type of incident waves and the direction of their approach. The response will also depend on the coherency of incident wave field. For example, if the incident wave-field is always incoherent and considered to be composed of waves coming from various directions, the site effect would be stable and would not vary from one earthquake to another.

Secondly, the site conditions must be characterised properly to capture the essence of physical processes involved in the site effect. For example, the broad classification of a site into soil and rock generally leads to the conclusion that there is no need to consider the site effect for the peak ground acceleration or for the response spectra for frequencies higher than a few Hz. As shown later, the truth is not that the site effect does not exist for high frequencies, but that the current broad classification into soil and rock does not capture the site condition affecting high frequencies.

Finally, the site effect depends on what parameters of strong ground motion we are concerned with. For example, the peak ground velocity and displacement correlate with the broad classification of soil better than the peak ground acceleration. Some site characteristics such as absorption affect the amplitude of strong motion but not the duration, while others like the resonance of soft sediment will affect both amplitude and duration. The spatial variation of ground motion important for long structures such as bridges, on the other hand, would depend strongly on the nature of incident wave field.

Thus, a rich variety of site effects emerges by considering various choices of ground motion parameters, incident wave field and site characterics. We shall try, in our review of the current state of the art, to classify site effects systematically according to the orderly choice of the three factors.

One of the most useful way of synthesizing these various site effects is to construct a microzonation map. Such a map to be useful, however, the quantity plotted must vary from a place to another more than the range of its uncertainty at each site. Otherwise, the engineering significance of site effects is questionable as pointed out by Hudson (1972) with regard to the ground motion in Pasadena during the San Fernando earthquake of 1971.

Let us start our review with the conventional broad classification of site conditions such as soil and rock sites.

Broad Classification of Site Conditions

When Newmark et al. (1973) presented response spectra appropriate
for the design of critical structures such as nuclear power plant, the
spectral shape was considered to be site-independent, although they
noted the need for modification of the shape for the particular site
conditions for periods longer than 0.5 sec.

The site-dependent response spectra were published by Hayashi et al.
(1971) and Kuribayashi et al. (1972) in Japan, and by Seed et al.
(1976), Mohraz (1976) and Trifunac (1976a) among others in the U.S.

The classifications of geologic conditions used in the U.S. in the
past decade or so were summarized by Campbell (1985). For example,
Seed et al. (1976) use four classes of site conditions, namely, (1)
soft to medium clay and sand, (2) deep cohensionless soil, (3) stiff
soil, and (4) rock. On the other hand, Trifunac (1976a) follows
Trifunac and Brady (1975) and use three classes, namely, (1) soft
alluvium (2) intermediate and (3) hard basement or crystalline rock.

Different methods were used by various authors to estimate the
site effect. Seed et al. (1976) grouped observed spectral shapes into
the above four classes, and estimated their average as well as the
standard error. Trifunac (1976a) applied more formal regression
analysis to the observed Fourier spectra FS(T) by expressing the site
effect to the logarithm of FS(T) as

$$\log FS(T) = -d(T)S + \text{source and path effects} \tag{1}$$

where T is the period, S is 0 for alluvium, 2 for rock and 1 for the
intermediate site.

In spite of the difference in analysis method, the results obtained
by various authors both in the U.S. and Japan are remarkably
consistent. All of them show that soil sites have greater
amplification factors than rock sites for long periods, but the
relation tends to be reversed for short periods. For example, Trifunac
(1976a) found that basement rock sites show greater amplification than
alluvium sites for periods shorter than 0.2 sec as much as 1.5 times.
This is consistent with the trend of period dependence of the relative
amplification of granite and alluvium sites studied by Gutenberg
(1957). A similar cross-over at the period of about 0.2 sec can be
seen in the results of Mohraz (1976) and Seed et al. (1976) between the
response curves for soil and rock sites.

Recent study of the site effects at about 150 seismic stations in
the central California by the coda method (Phillips and Aki, 1986) also
revealed a similar frequency dependent site effect. They classified
the site into granite, Franciscan formation (Mesozoic), fault-zone
sediment and non-fault zone sediment, and found that granite sites have
the lowest amplification factor among all the sites at 1.5 Hz, but show
the highest amplification at 24 Hz. Thus, relative to the granite

site, other sites, show the cross-over as mentioned above. The
cross-over occurs at 4 Hz for Franciscan, 8 Hz for fault-zone sediment
and 20 Hz for non-fault-zone sediment.

The result from Japan is also similar. The response curves obtained
by Hayashi et al. (1971) show that the amplification factor for stiff
soils is lower than that for loose soils for periods longer than about
0.25 sec, but the relation is reversed for shorter periods. Likewise,
Katayama et al. (1978) found a similar cross-over for soft alluvium
site (type 4) and Tertiary or older sediment site (type 1) at about
0.25 sec.

With regard to the magnitude of amplification factor, soil sites
show up to a factor 2 to 3 greater amplification than rock sites for
periods longer than the cross-over period, while the amplification at
rock sites relative to soil sites for shorter periods is less than a
factor of 2.

The frequency dependence of site effect discussed above is
reflected in the difference in site effect among peak ground
acceleration, velocity and displacement. For example, Trifunac (1976b)
concluded that the influence of geological conditions at the recording
site appeared to be insignificant for peak acceleration but become
progressively more important for peaks of velocity and displacement.
This statement is consistent with the frequency dependent site effect,
because the predominant period in peak acceleration is in the general
range where the cross-over occurs, and there may be roughly equal
chance for amplification and deamplification, while the predominant
period in peak velocity and displacement is probably longer than the
cross-over period.

A similar conclusion about the site effect on peak acceleration,
velocity and displacement has been reported by Boore et al. (1980) and
Joyner and Boore (1981) who estimate the peak velocity at soil sites to
be a factor of 1.5 greater than that at rock sites. According to
Campbell and Duke (1974), the Arias intensity, which is the squared
acceleration spectra integrated over the whole frequency range, showed
higher value for alluvium sites than for rock sites, at least for the
data from the San Fernando earthquake of 1971.

Recent result from Japan obtained by Kawashima et al. (1986) who
analyzed 197 strong motion records by classifying them into three
groups of site conditions also supports the existence of cross-over.
They found that the peak acceleration is the lowest for the softest
site, while the peak velocity and displacement are the highest for the
softest site.

Let us now summarize what we found about average site effects
using the broad classification of site conditions.

(1) There exists a cross-over period; above it the soil site shows
higher amplification than the rock site and below it the relation is
reversed.

(2) The cross-over period is around 0.2 sec for both U.S. and Japan.

(3) The amplification of soil sites relative to rock sites for periods longer than the cross-over period amounts to a factor of 2 to 3, and the amplification of rock sites relative to soil sites for period shorter than the cross-over is less than a factor of 2.

(4) Peak ground accelerations are independent of site conditions.

(5) Peak velocity and displacement as well as the Arias intensity show higher values for soil sites than rock sites.

The above conclusions seemingly suggest that the variation in the site effect from a site to another may be at most a factor of 3, and the effect may decrease with decreasing period, becoming insignificant for short periods that prevail in the peak acceleration. We shall demonstrate, however, in the next section from a review of recent results on site-specific amplification factors determined by methods which do not require a site classification that the above suggestion is unfounded.

Site Specific Amplification Factor from Regression Analyses of Earthquake Data

A systematic study of site specific amplification factor of 26 strong motion seismograph sites was made by Kamiyama and Yanagisawa (1986) using 117 strong motion earthquake records in Japan which registered maximum accelerations greater than 0.02G. They fitted the observed velocity response spectra $V_{ij}(T)$ at the ith station for the jth earthquake by the following equation

$$\log V_{ij}(T) = a(T)M_j - b(T)\log(\Delta_{ij} + 30) - d(T) D_j + A_i(T) \qquad (2)$$

where T is the period, M_j and D_j are respectively the magnitude and focal depth of the jth earthquake, Δ_{ij} is the epicentral distance to the ith station from the jth earthquake, and $A_i(T)$ is the site amplification factor at the ith station. $a(T)$, $b(T)$, $d(T)$ and $A_i(T)$ are determined by the least squares method from horizontal component velocity response spectra for the period range 0.1 <T< 5 sec.

A unique determination of the site amplification factor requires an additional constraint. Kamiyama and Yanagisawa assumed that the amplification factor is 2 for a station on a hard rock, and chose Ofunato located on a rock site with the shear velocity of about 1 km/s as the reference station. This assumption is essentially equivalent to assume that Ofunato is located on a homogeneous half-space. However, the range of geographic variation of the amplification factor, which we are interested at this moment, is not affected by this assumption.

Fig. 1 and 2 show examples of the amplification factor $10A_i$. It is clear from these figures that the range of variation from a site to another is roughly independent of period in the range from 0.1 to 5 sec. In fact, from their results for all 25 stations, we obtain the following range of variation in A_i.

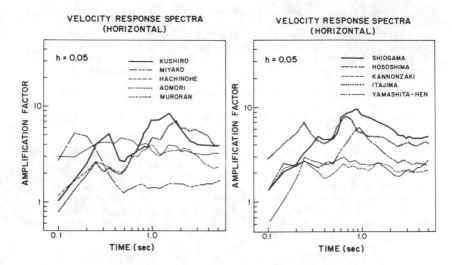

Fig. 1.   Site amplification factor as a function of frequency for
          various stations determined by regression analysis of
          observed velocity response spectra. (Reproduced from
          Kamiyama and Yanagisawa (1986).)

Fig. 2.   Site amplification factor as a function of frequency for
          various stations determined by regression analysis of
          observed velocity response spectra. (Reproduced from
          Kamiyama and Yanagisawa (1986).)

f in Hz	10	5	2	1	0.2
T in sec	0.1	0.2	0.5	1.0	5.0
range of A_1	0.92	0.64	0.70	1.13	0.58
10 (range of Ai)	8.3	4.4	5.0	13.5	3.8

    The above table shows that the variation is greatest around 1 Hz,
but there is no trend of decrease toward high frequency.

    As far as the author is aware, there have been no investigations of
site specific amplification factor by regression analysis of strong
motion data in the U.S. We have, however, extensive results for weak
motion data from the U.S. Geological Survey's seismic network in the
central California. Phillips and Aki (1986) determined the
amplification factor for coda waves at about 150 stations. Fig. 3
shows the amplification factor normalized to the mean of all the

stations for four groups of selected stations; namely those on granitic rocks, on the Franciscan formation, on the fault zone sediment, and on non-fault zone sediments. The amplification factor for coda waves has been shown to agree with the average of amplification factors for S waves over various directions of approach (Tsujiura, 1978; Tucker and King, 1984). This is consistent with the coda model of S to S backscattering proposed by Aki (1980).

The range of variation of amplification factor again show no strong decrease toward high frequency. In order to show the frequency dependence more clearly, we compare the histogram of amplification factor for 5 different frequencies; 1.5, 3, 6, 12 and 24 Hz in Fig. 4. The range of variation is somewhat greater for 1.5 Hz and smaller for 12 Hz, but the difference is slight. The effective range is about 2.5 in natural logarithm for 1.5 Hz and about 2 for 12 Hz, corresponding to the factor of 12 and 7.4 repectively. This is in good agreement with the result obtained from strong motion data in Japan; namely a factor of 13.5 for 1 Hz, and 8.3 for 10 Hz.

As mentioned in the introduction, for a meaningful microzonation map, the geographic variation of amplification factor must be significantly greater than the variation at a given site due to different incident wave field. Direct measurement of such variation is available from the comparative observations at a surface site and at a basement rock using a surface and borehole seismograph.

Kinoshita et al. (1986) calculated velocity response spectra for 27 earthquakes with magnitudes in the range from 5.5 to 7.4 in the east-central part of Japan, recorded at three borehole sites, namely, IWT (3.5 km deep), SHM (2.3 km) and FCH (2.8 km) in the Kanto region. The seismographs at the bottom are all located in the Tertiary bed rock

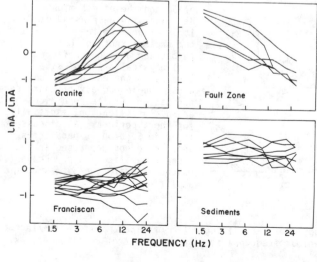

Fig. 3. Site amplification factor as a function of frequency for various sites determined by regression analysis of coda wave spectra. The natural logarithm of amplitude is shown relative to its mean over all stations used in the analysis. (Reproduced from Phillips and Aki (1986).)

at the depth of 600 to 700 meter below its upper boundary. The velocity response spectra are calculated for the seismographs located at the surface and in the bed rock at each site, and the ratio representing the site amplification was calculated for each earthquake. An example for the horizontal transverse component obtained at IWT is shown in Fig. 5, where the average amplification and the range of one standard error are shown for periods 0.1 to 5 sec. The standard error is about the same for the whole frequency range, and is a factor of a little less than 2. Nearly the same result is obtained for the horizontal radial component at IWT, as well as for both components at SHM and FCH.

Tucker and King's (1984) work on the amplification of three valley sites relative to rock site in the Garm region of USSR, and Tsujiura's (1978) work on the relative amplification of several sites at Dodaira also support the conclusion that the variation of amplification factor due to the variation of incident wave is frequency independent in the range from about 1Hz to 25 Hz and the standard error of variation is less than a factor of 2.

Since the geographical variation of site specific amplification factor obtained by regression analysis was 13 for 1 Hz and about 8 for 10 Hz as discussed earlier, we may conclude that a very meaningful microzonation map predicting the amplification factor with a standard error less than a factor of 2 can be constructed for the frequency range at least from 1 to 10 Hz.

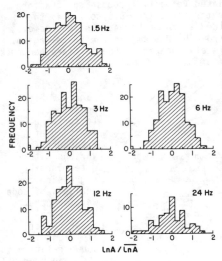

In the preceding section on the broad classification of site effects, it was suggested that the geographical variation in the site effect might be at most a factor of 3, and the effect might decrease with decreasing period. It is clear from the conclusion in the present section that the above suggestion does not reflect the real physical state of site effects on the earth, but only shows the inadequacy of characterization of site conditions by a broad classification, especially for short periods. As discussed later in Section 9, parameters which have not been used in the broad classification can be useful to characterize the site-specific amplification.

Fig. 4.   Histogram of natural logarithm of site amplification factor at various frequencies. (Reproduced from Phillips and Aki (1986).)

Use of Microtremor for Evaluating Site Effects

As demonstrated by Phillips and Aki (1986), the coda method is an
effective way of finding a frequency-dependent site-specific
amplification factor for S waves averaged over all directions of wave
approach. Two issues may be
raised regarding the coda
method. One is the problem
of non-linearity of soil at
a high strain level which
cannot be addressed by the
coda method using low strain
signals. We shall address
this problem in the next
section.

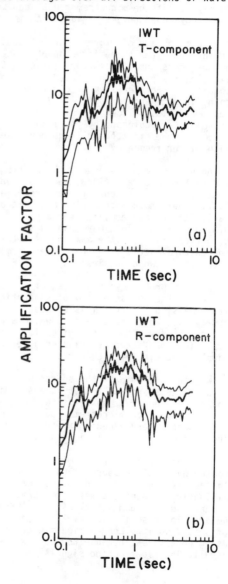

Fig. 5.   Amplification
factor for the surface site
relative to the borehole
bottom in the basement rock
with the range of one
standard error for station
1WT. (a) transverse
component; (b) radial
component. (Reproduced from
Kinoshita et al. (1986).)

The other issue is the relative complexity in applying the coda method as compared to the microtremor method which has been used for many years by Kanai and his colleagues in Japan as a means of seismic zoning. Unlike microtremor, the measurement of coda waves requires an instrument which can wait and record a small local earthquake. Thus, we may ask if the microtremor data can be used to evaluate the frequency-dependent, site-specific amplification factor.

The advantage of microtremor method is the simplicity of measurement, and there is no doubt that the spectral feature of microtremors shows gross correlation with the site conditions. For example, the predominant period of microtremor is an indicator of site condition. For Tertiary or older rock, the period tends to be shorter than 0.2 sec, and for soft alluvium or reclaimed land longer than 0.6 sec. For U.S., Alcock (1972) reports greater damage in the town of Grand Valley, Colorado by the 1969 Rulison underground explosion for area with the microtremor frequency lower than 12.5 Hz than for area with higher frequency.

An extensive measurement of microtremors in the U.S. was carried out by Tanaka et al. (1968) who reported amplitudes and frequency distribution of periods of microtremor at 309 sites in the Western U.S.. Their most interesting observation was that the range of variation of both amplitudes and periods was about the same between the U.S. and Japan. This may mean that the extent of variation of site conditions is about the same between the two countries, assuming that the cultural activities that generate microtremors are about the same between the two countries.

A major problem with the use of microtremor for the study of site effect is the impossibility of separating their source-path effects from the site effect. This situation is radically different from the coda waves, which have been shown to share the common source and path effects at any sites for a given local earthquake. Thus, the site amplification factor relative to a reference site can be obtained simply by taking amplitude ratio.

We cannot eliminate the source-path effect for microtremor, because the source of tremor at a site is usually different from the source of tremor at another site. For example, the relative site amplification for microtremor observed in Mexico city are different from those observed during the earthquake of 19 September 1985 by an order of magnitude (Celebi et al., 1987).

Long-period microtremor within an area, on the other hand, can be caused by a common source such as a distant oceanographic disturbance. In fact, the usefulness of microtremor with period 1 to 5 sec for evaluating site effect was demonstrated by Ohta et al. (1978) for Japan, and by Kagami et al. (1982, 1986) for California. They found a general increase in the level of long-period amplitude with thickness of soil deposit. Long-period microtremors would supplement the method using coda waves, because the coda waves from small local earthquakes usually lack long-period signals.

Another promising approach using microtremor is the determination
of shear velocity distribution with depth by the analysis of dispersion
curves extracted from microtremor data. Okada et al. (1987)
successfully determined shear velocity distribution to the depth of 2
km using microtremor in the frequency range from 0.2 to 1 Hz recorded
by a small array of a few km aperture.

They suggested that the spatial autocorrelation method due to Aki
(1957), which gives the same result as the frequency-wave number method
due to Capon (1969), may be simple enough for a real-time, on-site
determination of shear velocity structure using a microprocessor-
controlled microtremor measurement array.

Site Effects on Weak and Strong Motions

A vast amount of literature exists on the non-linearity of soil
including the liquefaction phenomena, and there is no question about
its importance to the understanding of site effect on strong ground
motion. The present author, however, is not qualified for reviewing
the soil mechanical aspect of the subject and must restrict himself to
the seismological literature on comparative studies of weak and strong
motions at a given site.

So far as the author is aware of, the only strong motion record
demonstrating the striking effect of non-linarity of soil is the record
of the Niigata earthquake of June 16, 1964 obtained at site 701 (No. 2
apartment house, Kawagishi-cho, Niigata). As shown in Fig. 6, about
8-10 sec after the beginning of recording, short-period motions
suddenly become small and are dominated by long-period (5.5 sec)
motions. The apartment No. 2 which housed the seismograph suffered a
large tilt, but the nearby No. 4 apartment completely fell on the
ground by the liquefaction of soil consisting of water saturated sand.

Except for the above example, it is usually not easy to clearly
recognize the non-linear effect on observed strong motion records. As
Esteva (1977) states, the influence of non-linearities is often
overshadowed by the overall patterns of shock generation and
propagation.

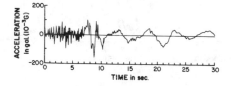

Fig. 6.   N-S component
acceleration record of
the Niigata earthquake
of June 16, 1964,
obtained at Site 701,
No. 2 apartment house,
Kawagishi-cho, Niigata.

As a matter of fact, seismologists tend to find a good correlation between weak and strong motions at a given site, namely, similar amplification factors for both, implying that non-linearities are not important as the first order effect in most cases.

Rogers et al. (1984) recorded seismic motions from the underground nuclear tests at the Nevada Test Sites at 28 sites in the Los Angeles area at which strong ground motions were also recorded during the 1971 San Fernando earthquake. They chose four reference sites on relatively harder rock in Pasadena, Hollywood, Van Nuys and Palos Verdes. They then computed the Fourier amplitude spectra for the NTS signal (about 329 sec duration) and the San Fernando record (41 sec duration), and obtained the ratio of the spectrum to the corresponding spectrum observed at the closest reference site. Fig. 7 shows the spectral ratio for the NTS signal plotted against that for the San Fernando earthquake for 4 different period bands over which the spectral ratio was averaged. Total period, short period, intermediate period and long period band correspond to the period range 0.2 to 10.0, 0.2 to 0.5, 0.5 to 1.0 and 1 to 3.3 sec, respectively. In spite of the large range in signal levels up to 10^{-3} for the strong motion data, the site amplification factor shows a good correlation between the weak and strong motion data. This correlation is remarkable in view of the

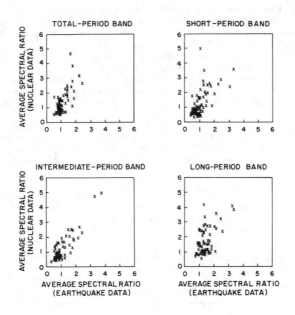

Fig. 7.   Relation between the amplification factor for nuclear test and that for the San Fernando strong motion. The amplification factor is relative to the reference site of each group. (Reproduced from Rogers et al. (1984).)

difference in the incident wave field between the two data sets. The
signal duration, incident direction, and wave composition are different
between the two, and one may expect the standard error of a little less
than a factor of 2 for each ratio as discussed earlier. Most of the
scattered points in Fig. 8 are indeed within a factor of 2 from the
line corresponding to the equality of two spectral ratios. We also
note for both NTS and earthquake data that the range of geographic
variation of amplification factor is about a factor of 10 in agreement
with our earlier conclusion.

A similar agreement of the amplification factor between
weak and strong motion was observed by Tucker and King (1984) for a
sediment-filled valley in Garm, USSR. Fig. 8 shows the spectral ratio
of the middle to the edge of the valley for weak (10^{-9} - 10^{-3}g) and
strong acceleration (.04-0.2g). The ratios are plotted for different
events to show scatter due to different incident directions. No
significant differences can be recognized on the average between the
weak motion group and the strong motion group.

Similarity, Murphy et al. (1971) concluded that a linear model can
explain the major features of the amplification effect at various sites
in the NTS for a wide range of ground motion (10^{-5} to 1g) caused by the
underground nuclear testing. Joyner et al. (1981) also found that the
effect of alluvium on strong ground motion observed during the Coyote
Lake, California, earthquake of 1979 can be explained without invoking
nonlinear soil response.

In a more qualitative study, Benites et al. (1987) found a good
correlation between the damage pattern for past large earthquakes and
the weak motion amplification for small earthquakes in Lima, Peru.

Thus, the comparison of amplification factor at a given site
between weak and strong motion generally supports a good correlation
between them. We may conclude, then, except for an obvious case of
liquefaction, that the amplification factor obtained for a given site
using the weak motion data can be used to predict the first order
effect on strong ground motion at the site.

Nature of Strong Motion Wave Field

The most direct way of finding the nature of strong motion wave
field is probably to make spatial-temporal correlation analysis or the
frequency-wave number analysis of data collected by a dense network of
seismographs such as the SMART-1 array in Taiwan and the differential
array in El Centro.

The first attempt along this line was made by Aki and Tsujiura
(1959) using the records of small local earthquakes obtained by an
array of six stations deployed within an area of 500 m diameter over
granitic rocks near Tsukuba, Japan. They analyzed records of 18
earthquakes in the frequency range from 0.3 to 17 Hz. Using an analog
computer for calculating correlation coefficient among all station
pairs, they estimated the fraction of power carried by the regular
plane waves coming from the earthquake source for consecutive time

windows (each 2.5 s long) from P waves to the coda. The probability of
finding the plane wave in the P wave part, S wave part, P to S interval
and post S-arrival was found to be 100%, 78%, 62% and 30% respectively.
Thus, the S wave part which constitute the maximum motion contained
greater fraction of plane waves coming from the source than the P-coda
or the S-coda part. The fraction of power carried by plane waves in
the time window containing P waves was found to be the highest; it was
up to 80% and about 40% on the average. In the time window containing
the S waves the fraction of power carried by these plane waves was 60%
in the largest case, and 20 to 30% in most cases.

The above low fraction of power carried by plane waves in the P
and S wave parts may be partly due to the long time window (2.5 s) with
respect to the source duration of earthquakes analyzed, which permits
contamination by scattered waves
into these parts. In fact, the
study by Spudich and Cranswick
(1984) on the data obtained by the
El Centro differential array (213 m
long linear array) during the 1979
Imperial Valley earthquake (M_S=6.9)
revealed much higher degree of
coherency in vertical acceleration
during the first 9 sec period, and
in horizontal acceleration during
the 6 to 11 sec period. They
attributed these coherent waves to
direct P and S waves from a small
region surrounding the propagating
rupture front.

The SMART-1 array in Taiwan has
produced valuable data for studying
the wave field of strong ground
motion for great ranges of
earthquake size, epicentral distance
and focal depth as described in a
recent review by Abrahamson et al.
(1987).

Using multi-station measures of
coherency, Abrahamson (1985) found,
for example, that across a 2 km
aperture subarray for a M-6.3 event
at an epicentral distance 20 km and
focal depth 25 km, S-wave coherency

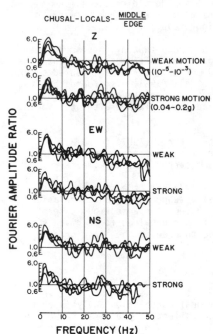

Fig. 8.   S-wave amplification factor at mid-valley sites relative to
            valley-edge in the Chusal Valley for local earthquakes.
            Results for weak motions and strong motions are shown
            separately. (Reproduced from Tucker and King (1984).)

decays rapidly with frequency above approximately 2 Hz while the P-wave coherency decays at frequencies above approximately 3 Hz. Since the aperture is ten times greater than the El Centro array, the loss of coherency above 2-3 Hz may not be a strong contradiction to what was observed during the Imperial Valley earthquake. The results from Tsukuba are certainly comparable to those from the SMART-1 array considering the aperture size about a quarter of the latter.

An extremely intriguing result on the magnitude dependence of the variability of peak acceleration within the SMART-1 array was reported by Abrahamson (1987). He found that the standard deviation of the natural logarithm of the peak ground acceleration within the array significantly decreases with the increasing magnitude. The standard deviation is about 0.35 at M=4 and decreases to about 0.17 at M=7. The above magnitude dependence may be attributed partly to the more coherent waves from larger earthquakes because of the tendency that larger earthquakes have longer predominant periods, and located further from the array on the average. If the magnitude dependence still remains after removing these wave-propagational effects, it must be attributed to the non-linear effect of soil. The above observation, thus, presents a future problem of the first order importance for geotechnical engineers and seismologists.

Causes of Local Variations in Ground Motion

Earlier we have concluded that the geographical variation of site specific amplification factor amounts to about a factor of 10 for the frequency range from 1 to 10 Hz, and the conventional broad classification of site conditions is inadequate for capturing the real site effect.

In order to find a better characterization of site conditions, it is essential to understand what causes local variations in ground motion. let us start with the simplest, namely, the effect of a flat free surface.

Flat Free Surface

As well known, the flat free surface doubles the vertically incident S waves. For SH waves polarized in the horizontal direction, the amplification of factor 2 applies to all incidence directions, and there will be no local variations caused by the free surface.

For SV waves, however, the flat free surface has an extremely complex effect. This is not a subject of academic interest but of practical concern of major importance, because the extremely localized damage pattern due to the recent Whittier Narrows earthquake of 1987 has been attributed by Sammis et al. (1987) to the free surface effect on SV waves incident near the critical angle θ_c. The critical angle is given by sin $\theta_{c}=\beta/\alpha$, where α is the P wave velocity and β is the S wave velocity. At the critical angle, the horizontal component of slowness of S waves matches the P wave slowness, and a strong coupling occurs, including the generation of SP waves (P waves converted from S waves and propagating along the surface).

Several surprising effects are expected for SV waves incident near the critical angle. First, the horizontal component displacement at the surface shows a sharp peak for plane SV waves incident at the critical angle as shown in Fig. 9 for the case of Poisson's ratio 0.25. The amplification factor amounts to about 5 for a narrow (~1°) range of incidence angle. The peak amplification depends on Poisson's ratio, and increases with decreasing Poisson's ratio as shown in Fig. 10 together with the critical angle. The range of incidence angle with high amplification narrows rapidly with decreasing Poisson's ratio. This suggests that the effect may be smoothed out for spherical waves composed of plane waves with distributed directions. In fact, the calculation by Pekeris and Lifson (1957) for a point source of vertical force does not appear to show the effect as strong as expected for plane waves.

We found, however, even more drastic effect in the result of Pekeris and Lifson. For a source varying as the step function in time, SV waves generate surface displacement of step function at a distance shorter than the critical distance and of a logarithmic singularity beyond the critical distance. In other words, the displacement is finite inside the critical distance, and becomes infinite outside. The horizontal radial component displacements at various distances (critical distance in this case of Poisson's ratio 0.25 is H/√2, where H is the focal depth) are shown in Fig. 11. Of course, the infinite displacement does not occur in reality because the source time function is band limited. In this case the displacement due to SV waves will have the same amplitude spectral shape but a π/2 phase shift across the critical distance.

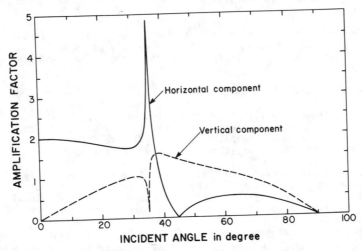

Fig. 9.   The amplitude of horizontal (solid line) and vertical (broken line) component displacement at the free surface due to incident SV waves plotted as a function of incidence angle, for the case of Poisson's ratio 0.25.

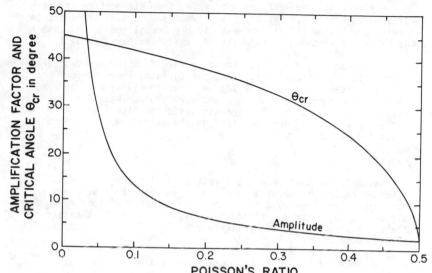

Fig. 10. The critical angle and the peak amplitude at the critical angle as a function of Poisson's ratio.

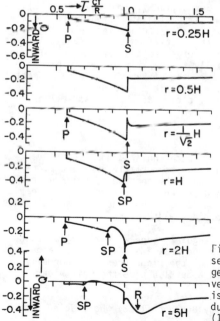

Fig. 11. Horizontal displacement seismogram at the surface of a homogeneous halfspace due to a buried vertical line force at depth H. r is the epicentral distance. (Reproduced from Pekeri s and Lifson (1957).)

The effect of free surface on SV waves from dislocation sources was studied by Kawasaki et al. (1973) who found a similar result to Pekeris and Lifton (1957), and was clearly identified in the case of the Kita-mino earthquake of August 19, 1961 (Kawasaki, 1975).

In addition to the drastic change in wave form for S waves, SP waves (S converted to P propagating along the surface) appear beyond the critical distance. Chapman (1972) and Bouchon (1978) made an interesting observation that SP waves have sharper wave form for lower Poisson's ratio. Bouchon further discussed the effect of finite fault size and low velocity surface layer on the seismic motion near the critical distance.

Topography

A natural item to follow the effect of a flat free surface is the effect of topography on seismic motion. In order to describe the topography effect, we need to specify the geometry of topography as well as the incident wave field. Let us start with the simplest case of a wedge-shaped ridge and valley where plane SH waves polarized in the direction of the axis of ridge or valley are incident.

SH Waves Incident on Wedge-Shaped Ridge or Valley: A Rule of Thumb.

A surprisingly simple exact solution exists for the motion at the vertex of a wedge due to incident SH waves polarized in the direction of vertex. As pointed by Sanchez-Sesma (1985), Macdonald's (1902) solution gives the displacement amplification at the vertex to be $2/\nu$ when the angle of wedge is $\nu\pi$ (for $0<\nu<2$), for any incidence angle. For example, the amplification by the flat free surface ($\nu=1$), is 2 as well known, and it is 4 for the case of a rectangular wedge. Although this amplification is not necessarily the maximum value and higher amplification is observed at the far side of the vertex with respect to the incidence direction, it gives a convenient rule of thumb for the rough estimate of topographic amplification at a ridge as well as deamplification at a valley.

SH Waves Incident on Ridges.

Boore (1972) calculated the seismic motion at a non-planar free surface of a homogeneous half space due to normally incident plane SH waves using a finite difference method. He considered triangular ridges with slopes 23° and 35°, and showed that the motion at the ridge crest can be amplified up to about 70% more than the flat surface case for wave length comparable to the ridge width.

Smith (1975) also studied a triangular ridge with 20° slope using a finite element method. he found the peak amplification of about 50% greater than the flat surface case at the ridge crest for normally incident SH waves with wave length about 1.6 time the half width of ridge.

A similar result was obtained by Bouchon (1973) who studied the effect of shape of a ridge on the surface motion for the normal and oblique (35°) incidence using the time-domain extension of the Aki-Larner (1970) method. Fig. 12 shows the shape of ridge and amplitude of surface motion relative to the flat surface case for various ratios of height (h) to half width (l). The wave length (λ) was chosen to be 5h for all cases. The amplification at the ridge crest amounts to a little greater than 50%.

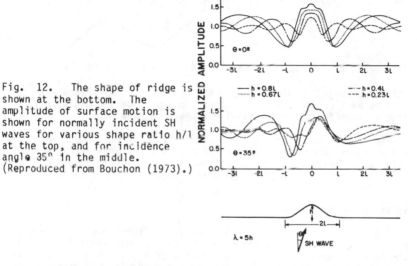

Fig. 12. The shape of ridge is shown at the bottom. The amplitude of surface motion is shown for normally incident SH waves for various shape ratio h/l at the top, and for incidence angle 35° in the middle. (Reproduced from Bouchon (1973).)

Bard (1982), refining the method used by Bouchon (1973), studied the details of wave scattering phenomena involved in the ridge crest amplification. He considered the mountain model of Sills (1978) given by the following equation for elevation,

$$\zeta(x) = h \ (1-a) \ \exp \ (-3a) \tag{3}$$

where $a = (X/\ell)^2$. This topography is completely defined by its half-width ℓ and its height h as shown in Fig. 13. SH waves of the form of a Ricker wavelet, $f(t) = (b-0.5) \exp (-b)$ with $b=(\pi t/t_p)^2$, were considered. Bard investigated the physics of ridge effect on SH waves by examining time, frequency as well as wave-number domain solutions for various shape ratios h/ℓ, incidence angles and characteristic periods t_p. He identified the following two effects operating in the phenomena; one is the local amplification associated with the convex curvature of ridge crest, and the other is the diffracted waves generated at and propagated away from the ridge crest. The local amplification shows a broad spectral peak for wave lengths comparable with or a slightly shorter than the mountain width, and generally decreases with increasing incidence angle. On the other hand, diffracted waves become stronger for the forward scattering and weaker for the back scattering as the incidence angle increases. Their

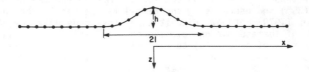

Fig. 13.   The shape of ridge used by Sills (1978) and Bard (1982).

Fig. 14.   One of the repeated ridge models used by Bard (1982).

lateral propagation along ridge slope and their interference with the
primary wave creates rapid variations in amplitude and phase, giving
rise to significant differential motions along the slope.

Bard (1982) and Bard and Tucker (1985) further considered models of
three ridges in parallel as shown in Fig. 14, and showed that a
relatively narrow band, additional amplification of about a factor of
1.5 relative to the single ridge case occurs due to a lateral
resonance.

SH Waves Incident on Canyons

The simple rule of thumb by Sanchez-Sesma (1985) discussed earlier
suggests amplification at the edge of a canyon and deamplification at
the bottom. Bouchon (1973) showed that such a pattern develops in the
case of normally incident SH waves for a canyon with the depth greater
than about 1/3 of the half-width. Both the amplification and
deamplification increases with the canyon depth. The Aki-Larner method
used by Bouchon, however, is limited to cases of relatively gentle
slope. A similar result was obtained by Sanchez-Sesma and Rosenblueth
(1979) who used the boundary integral equation method which is
applicable to topography of an arbitrary shape. Trifunac (1973), on
the other hand, gave an exact solution for the case of SH waves
incident on a semi-cylindrical canyon. His solution delineated the
detailed amplification-deamplification pattern for various frequencies
and incidence angles.

Fig. 15 and 16, for incidence angle 0° and 30° respectively, show
the amplitude of surface displacement plotted as a function of two
variables; one is the horizontal distance X normalized to the canyon
depth or radius a (the canyon occupies -1<X/a<1) and the other is the
normalized frequency η (=2a/λ) for the range from 0.25 (wavelength λ
equal to 8 times the canyon depth) to 2.0 (wave length equal to the
canyon depth). In the case of incidence angle 30°, waves are incident
from the nagative X axis, and Fig. 16 shows clearly the shadowing
effect of the canyon on its far side (positive X), and the strong
interference on its near side (negative X) between the incident and
reflected waves.

The spectral amplitude diagrams such as Fig. 17 and 18 do not convey
the whole picture of phenomena because the phase information is
completely missing. Thus, we need a time-domain solution to capture
the physics of wave scattering phenomena by a canyon. Recently, Kawase
(1987a) developed an efficient method for calculating the time-domain
solution in which the boundary element method (Brebbia, 1978) is
combined with the Green's function calculated by the discrete wave
number method (Bouchon and Aki, 1977). Kawase's results corresponding
to Trifunac's cases for the incident wave form of the Ricker wavelet
with characteristic frequency of η=2 are presented in Fig. 17 and 18
respectively for the incidence angles of 0° and 30°. It is clear from
these figures that the peaks and troughs in amplitude distribution in
the frequency domain do not necessarily mean actual large and small
amplitudes in the time domain. For example, the time domain solution
in Fig. 17 shows that both direct waves and reflectecd waves (marked by
arrows) from the canyon surface have amplitudes nearly constant over
the entire horizontal surface outside the canyon. Thus, the spectral

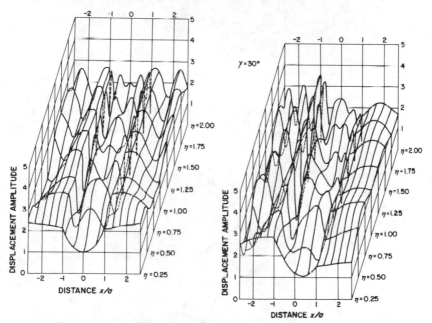

Fig. 15. The amplitude of surface motion across a cylindrical canyon
due to normally incident SH waves as a function of horizontal distance X normalized to the canyon depth and the normalized frequency η(=2a/λ) where λ is the wave length of
incident SH waves. (Reproduced from Trifunac (1973).)

Fig. 16. The same as Fig. 17 except that the waves are incident from
-X at an incidence angle of 30°. (Reproduced from Trifunac
(1973).)

amplitude variation over the same surface shown in Fig. 15 does not
mean the actual variation in amplitude along the surface, but only
means an apparent fluctuation in Fourier transform amplitude due to
contributions from both the direct and reflected waves. The diffracted

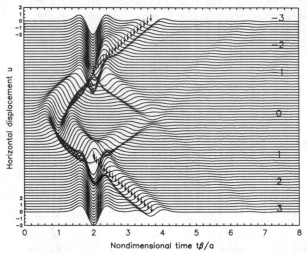

Fig. 17.   The time domain solution for the case corresponding to Fig.
           17. The incident waveform is the Ricker pulse with
           predominant frequency of η=2. (Reproduced from Kawase
           (1987a).)

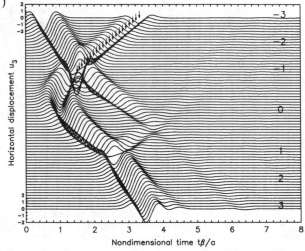

Fig. 18.   The time domain solution for the case corresponding to Fig.
           18. The incident wave form is the Ricker pulse with
           predominant frequency of η=2. (Reproduced from Kawase
           (1987a).)

waves from the edge of the canyon are difficult to distinguish at the surface outside the canyon from the wave reflected at the canyon.

On the other hand, the effect of shadow at the far-side edge of canyon (X=1) is clear both in the frequency domain and time domain, because the wavelet is isolated at this point. In general, we find that the scattered wave field observed at the surface consists of (1) direct (incident) waves, (2) reflected waves at the canyon surface and (3) diffracted waves generated at both edges of the canyon. The diffracted waves propating along the canyon surface ($|X/a|<1$) are the main motions observed inside the canyon after the arrival of direct wave. It is clear from Figures 17 and 18 that a large differential motion is expected near both edges of the canyon.

Trifunac's exact solutions have been extremely useful in serving as a classic test case for many approximate methods later developed for dealing with more general topographic geometries. Another set of exact solutions which has been used for testing approximate methods was obtained by Wong and Trifunac (1974a) for the canyon of elliptic cross-section.

In order to study the effect of canyon with an arbitrary shape, Wong and Jennings (1975) used the method of boundary integral equation to obtain solutions both in the frequency and time domain as well as response spectra. Considering a topography which simulates that near the Pacoima dam (the site of strong motion seismograph registering acceleration greater than 1g during the San Fernando earthquake of 1971), they found that the effect of canyon was strongest in the frequency-domain solution for wavelengths comparable to or shorter than the canyon width. The time domain solutions showed significant differences at different points, but not as large as seen in the frequency domain solution. The response spectra showed the smallest differences with significant effects only at high frequencies.

P and SV waves incident on ridges

The effects of a ridge on incident P and SV waves were studied by Bard (1982) for the same ridge geometry and using the same technique as for incident SH waves. He found that the amplification of displacement at the ridge relative to the flat case is weak for incident P waves, only 10% as compared to 38% for SH and 30% for SV waves for an identical ridge shape.

In the case of incident SH waves, the diffracted SH waves propagate from the ridge crest, as mentioned earlier. In the case of incident P waves, the diffracted waves are primarily Rayleigh waves because unlike SH waves P waves propagating along the free surface cannot satisfy the stress free condition and are quickly attenuated. In the case of incident SV waves, the diffracted waves are both Rayleigh and SP waves mentioned earlier in the section on the effect of flat free surface on SV waves. Because of the involvement of different kinds of waves, the pattern of surface motion is more complicated than the case of incident SH waves. The interference between the direct waves and diffracted waves again generates rapidly varying amplitude and phase along ridge slopes, giving rise to significant differential motions as in the case of SH waves.

When SV waves are incident at the critical angle θ_c, where $\sin\theta_c$ = (s velocity)/(p velocity), the anomalous amplification occurs at the

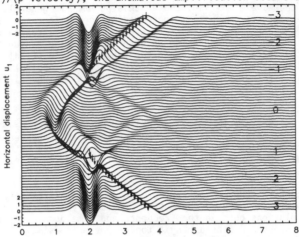

Fig. 19.   The time domain solution for horizontal component displacement at the surface of a cylindrical canyon with depth a due to normally incident SV waves with the same wave form as in Fig. 17. (Reproduced from Kawase (1987a).)

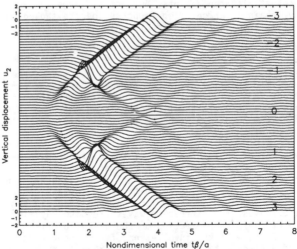

Fig. 20.   The same caption as Fig. 19, except that vertical displacement is shown. (Reproduced from Kawase (1987a).)

flat surface as mentioned earlier. Bard (1982) shows an extraordinary effect observed in this case that the surface motion at the ridge crest is reduced to about a half the case of the critical angle incidence. Further study is needed, however, to ascertain the existence of a similar effect for incident spherical waves from a localized source.

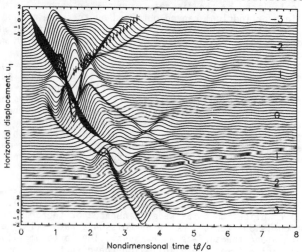

Fig. 21. The same caption as Fig. 19, except that the incidence angle is 30°; the critical angle for Poisson's ratio 1/3. (Reproduced from Kawase (1987a).)

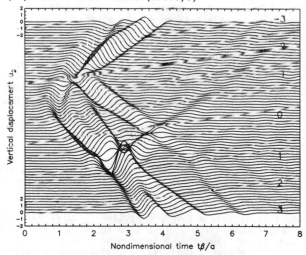

Fig. 22. The same caption as Fig. 21, except that vertical displacement is shown. (Reproduced from Kawase (1987a).)

P and SV Waves Incident on Canyons.

P and SV waves incident on canyons of various shapes have been
studied by Bouchon (1973) and Wong (1982) among others, but so far the
best demonstration of physics of the phenomena is given by Kawase
(1987a) who used the boundary element method combined with Green's
function calculated by the discrete wave number method. His result for
SH waves in a cylindrical canyon obtained by the same method described
earlier showed that the wave field consists of incident SH waves,
reflected SH waves at the canyon surface and diffracted SH waves
generated at both edges of the canyon.

In the case of vertically incident SV waves shown in Fig. 19
(horizontal component) and Fig. 20 (vertical component), we find a
similar result to the SH case except that diffracted waves now contain
P, SV and Rayleigh waves. The arrival times of SV waves reflected at
the canyon surface and observed on the surface outside the canyon are
again marked by arrows. They are difficult to distinguish from the
diffracted SV and Rayleigh waves generated at the canyon edge, although
the particle motion and apparent velocity supports that they are
probably of Rayleigh wave type.

For the case of SV waves with the incidence angle 30° shown in Fig.
21 (horizontal component) and Fig. 22 (vertical component), we find an
additional complication by SP waves generated at the critical incidence
angle, which is 30° for this case of Poisson's ratio (1/3) and
propagating along the surface as P waves. The large amplitude
horizontal motion immediately following the direct wave observed inside
the canyon is due to SP waves. This motion is not prominent in the
vertical component as expected.

The case of incident P waves is much simpler than the case of
incident SV waves, partly because of the absence of SP waves and partly
because of the relatively longer wave length of P waves for a given
period.

The conclusion of Bard (1982) after a comparative study of P, SV
and SH waves incident on a ridge, namely, "incident SV waves possess
the greatest scattering power and seem to be associated with the most
complicated diffraction scheme" appears to apply also to the case of a
canyon.

Rayleigh Waves Incident on Irregular Topographies

The effect of irregular topographies on Rayleigh waves in a
homogeneous half space has been studied by various reseachers. Here we
shall only describe some of the results which may have signficant
engineering application.

We found that canyons are very effective to block Rayleigh waves
with wave length comparable or less than the canyon depth. For
example, Wong (1982) showed by a least-squares approach for matching
boundary conditions that a semi-cylindrical canyon of depth (radius) a

will reduce the amplitude of transmitted Rayleigh wave down to 20% of that of incident Rayleigh waves for wave-length less than 1.5a, and to 50% for wave length between 1.5a and 2.5a. The corresponding time-domain solution obtained by Kawase (1987a) by the discrete wave-number boundary element method showed that the amplitude of transmitted Rayleigh waves is reduced to less than 10% of that of incident Rayleigh waves of the Ricker waveform with the effective wave-length of about 0.93a.

Another case of practical importance is the effect of a cliff on Rayleigh waves. Fuyuki and Nakano (1984) computed the effect of a step-like cliff on Rayleigh waves incident from the lower surface by a finite difference method, and measured the amplitude of transmitted Rayleigh waves observed at a horizontal distance from the cliff five times the wave length λ of Rayleigh waves. They found that the amplitude ratio of transmitted to incident Rayleigh waves decreases with the increasing height h of the step to about 30% at h=0.4λ, increases slightly to about 50% at h-0.7λ, and then again decreases to less than 10% for h greater than 1.2 λ. Their numerical results are in a good agreement with the theoretical prediction by Mal and Knopoff (1965) who omitted contributions of diffracted waves from the corners of cliff.

Fuyuki and Nakano (1984) also found significant S waves diffracted from the lower corner of cliff when Rayleigh waves are incident from the lower surface. The reciprocal problem of Rayleigh wave generation by a cliff due to incident SV waves are studied by Boore et al. (1981) using also a finite difference method. Their calculations revealed Rayleigh waves with amplitude as large as 0.4 times the amplitude of the surface motion of the incident waves in the absence of cliff, even for incident wave lengths several times the cliff height. Since Rayleigh waves have short horizontal wave lengths as compared with incident body waves, they play an important role in the differential motion.

Three Dimensional Topographies.

The theoretical study of the effect of three dimensional topographies is still a difficult task for the currently available computer, except for the case of cylindrical symmetry as investigated by Sanchez-Sesma (1983). An alternative approach is the physical modelling such as developed by King and Brune (1981) and Brune (1984) who used photographic recording of particle motion of foam-rubber models of realistic topographies. Brune (1984) modelled the topography around the Pacoima dam accelerograph site and found that for many angles of incidence the motion at the site is reduced rather than amplified relative to the flat area. This is explained as a consequence of two counteracting effect, namely the amplification of ridge and deamplification of canyon, because the ridge on which the accelerograph site is located is itself at the bottom of a canyon.

Flat Soft Surface Layer

The effect of soft surface layer on strong ground motion has been well recognized in Japan since early 1930's through pioneering observational studies by Ishimoto and theoretical studies by Sezawa. A simultaneous observation of ground motion due to the same earthquake at different sites with different geologic condition was already carried out by Takahashi and Hirano (1941) almost 50 years ago who was able to obtain the transfer function between two sites from observed seismograms and interpreted it in terms of a soft surface layer at one of the sites as reproduced in Aki and Richards (1980, p. 588). They are also probably the first to explicitly give the following well known formula for the amplification factor of surface displacement due to SH waves normally incident on a soft surface layer from underneath,

$$|U(w)| = 2\{\cos^2(\omega H/\beta_1) + (\rho_1\beta_1/\rho_2\beta_2)^2 \sin^2(\omega H/\beta_1)\}^{-1/2} \qquad (4)$$

where the incident wave is harmonic with unit amplitude and frequency ω. H, β_1, ρ_1, are the thickness, shear velocity and density of the surface layer, respectively and β_2 and ρ_2 are the shear velocity and density of the basement rock, respectively.

The above formula predicts the familiar factor of 2 amplification of the free surface effect for incident waves with wave length much longer than the layer thickness ($\omega H/\beta_1 \approx 0$). The amplification is peaked at incident wave lengths 4H, 4/3 H, 4/5 H,...at which the amplification factor is equal to twice the impedance ratio between the basement and the layer, namely, $2\rho_2\beta_2/\rho_1\beta_1$.

The peak amplification decreases with the increasing incidence angle as shown by Burridge et al. (1980). There still exists, however, the simple rule of peak amplification related to the impedance ratio if we extend the concept of impedance to non-vertical incidence case as $\rho\beta$ $\cos \theta$, where θ is the angle between the direction of wave propagation and the vertical. Since $\cos\theta$ approaches zero rapidly as θ approaches 90°, the peak amplification decreases rapidly as the wave incidence approaches grazing for plane SH wave incidence. Burridge et al. (1980) also calculated the amplification for incident P and SV waves. For the case of vertical incidence, there is no distinction between SV and SH, and the amplification factor for P waves is very similar to that for S waves. For non-vertical incidence cases, the situation becomes complicated because of coupling between P and SV waves. In particular, for incident SV waves with incidence angle (in the basement rock) greater than the critical angle θ_c, where $\sin\theta_c = \beta_2/\alpha_2$, a very sharply peaked amplification much beyond the impedance ratio occurrs at a frequency near the lowest resonant frequency ($\beta_1/4H$). Thus, the effect of a soft surface layer becomes drastically different between SH and SV waves as the incidence angle increases. As an example, Fig. 23 shows the amplification for vertically incident S waves for the case in which $\rho_2/\rho_1 = 1.2$, $\beta_2/\beta_1 = \alpha_2/\alpha_1 = 2.5$, and Poisson's ratio is 0.25. Fig. 24 shows the amplification for vertical (broken line) and horizontal (solid line) component for the same case but with the incidence angle 45°, which is beyond the critical angle for this Poisson's ratio, and a striking peak amplification as much as a factor of about 25 shows up.

An interesting and important effect of a soft surface layer is expected when the top of water table is contained in the layer, because the water table will be a strong discontinuity for P wave velocity but not for S wave velocity. As suggested by Cranswick and Mueller (1985),

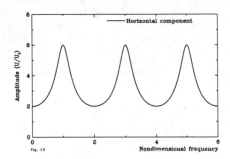

Fig. 23.   The amplification factor due to a surface layer for normally
            indicent S waves. The peak amplification is equal to twice
            the impedance contrast $(\rho_2/\beta_2)/\rho_1\beta_1)$, which is 6 in this
            case.

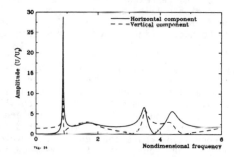

Fig. 24.   The amplification factor due to a surface layer for SV waves
            incident at 45°. The solid line for horizontal and the
            broken line for vertical component. (Recalculated using
            parameters in Burridge et al. (1980).)

P waves converted by the incident SV waves at the free surface may be
trapped in the layer above the water table and may become the cause of
high-frequency resonance often observed only in the vertical component
strong motion accelerograms obtained at soil sites.

Sediment-Filled Valley

    Actual soft surface layers are not flat but laterally confined in
the form of sediment-filled valley or basin. The finite lateral extent
of the surface layer introduces additional effects such as the
generation of surface waves at the edge and the resonance in lateral
direction, and tends to increase the amplitude as well as the duration
of ground motion. Numerous studies have been made on these effects by
various researchers using a variety of approaches. Let us first sample
some examples from studies by representative methods to gain some
perspective about available solutions.

Exact analytical solutions are available for the case of SH waves incident on a semi-cylindrical (Trifunac, 1971) and a semi-elliptical (Wong and Trifunac, 1974b) sediment-filled valley. They have been extremely useful for testing numerical methods applicable to more general geometries. Lee (1984) obtained solutions for the three-dimensional cases of P, SH and SV plane waves incident on a semi-spherical valley, matching boundary conditions by expanding the spherical wave functions into a power series.

Finite-difference (Harmsen and Harding, 1981) and finite-element (Ohtsuki and Harumi, 1983) have been used to show a strong generation of Rayleigh waves at the edge of a sediment-filled valley when P and SV waves are incident.

When the medium is composed of a basin imbedded in a homogeneous half-space, the problem can be reduced to a set of linear equations involving Green's function and unknown parameters describing the source distribution of scattered waves. The set of linear equations can be obtained starting with the representation theorem in the form of integral equation and discretizing the boundary surface (Brebbia, 1978), or it can be formed from the continuity of displacement and traction across the boundary using the complete systems of solutions (Herrera, 1981).

A great variety of the boundary method exists depending on how Green's function is calculated, how the boundary surface is discretized, how the complete systems of solutions are approximated and how the matching of displacement and traction is accomplished across the boundary. For example, Bouchon (1985), Campillo and Bouchon (1985) and Kawase (1987a) use the discrete wave number method for calculating Green's function. Wong (1982) uses the generalize inverse approach toward matching the boundary condition. Sanchez-Sesma and Esquival (1979) and Dravinski (1982, 1983) consider the sources of scattered waves distributed near but off the boundary to avoid the singularity of Green's function, while Kawase et al. (1982) eliminates the singularity by approximate integration over the segmented surface.

The Aki-Larner (1970) method based on the Rayleigh ansatz and the discrete wave number representation can be also considered as a particular case of approximation to the complete systems of solutions (Sanchez-Sesma et al., 1982). In the Aki-Larner method, the wave field in each layer is expressed as a superposition of plane harmonic waves including inhomogeneous plane waves, and the boundary condition is met in the horizontal wave-number domain taking advantage of the fast Fourier Transform. The method has been extended to the time domain by Bouchon (1973) and Bard and Bouchon (1980a, b), to three dimension by Niwa and Hirose (1985), to the case of multiple layers by Kohketsu (1987), and to the case of vertically inhomogeneous layers by Bard and Gariel (1986).

Ray methods (Hong and Helmberger, 1977) and their extension, Gaussian beam methods (Nowack and Aki, 1984) have also been used to study the ground motion in sediment-filled basins. They appear to give a surprisingly good result for the case of incident SH waves.

Recently, Sanchez-Sesma et al. (1987) presented a strikingly simple
representation of wave field as a sum of rays in a triangular basin
with a dip angle $\pi/2N$ (N=3,5,7---) under incident SH waves. Since ray
methods are the least time-consuming, the practical application to
three-dimensional earth model is possible for the deterministic
prediction of site effect using the present-day computer and has been
attempted (Ihnen and Hadley, 1987).

Ray methods, however, cannot deal with the cases in which
inhomogeneous plane waves, such as Rayleigh waves and beyond-critically
reflected waves, play major roles, as in the case of P and SV waves
incident on the sediment-filled basin.

In the following, we shall summarize major results on the effect of
sediment-filled basin on incident P, SV and SH waves obtained in the
literature.

SH Waves Incident on Sediment-Filled Valleys

The seismic motion of a sediment-filled valley due to incident SH
waves has been thoroughly studied by Bard and Bouchon (1980a, b, 1985),
Bard and Gariel (1985) and Bard (1983). Their approach of
investigating solutions in time, frequency and wave number domains was
particularly useful in clarifying the physical processes involved in
the complex phenomena. They considered two types of valley geometry as
shown in Fig. 25. Type 1 is a cosine-shaped valley with half-width D
and depth h. Type 2 has a flat bottom bounded by steep edges with half
width d_1 of the bottom part, d_2 of the edge part and depth h. The
density, shear velocity and rigidity of the sediment are ρ_1, β_1, and μ_1
and those of the basement rock and ρ_2, β_2, and μ_2 respectively. In the
examples reproduced here, ρ_1=2.0 g cm^{-3}, ρ_2=3.3 g cm^{-3}, β_1=0.7 km
sec^{-1}, β_2=3.5 km sec^{-1} and the damping is assumed to be 0 (or Q=∞).
Fig. 26 shows seismic motions at the surface of type 1 (cosine-shaped)
valley with h=200 m and D=5 km when a SH plane waves of Ricker waveform
with the characteristic period 0.732 s incident vertically from below.
Fig. 27, on the other hand, shows seismic motions at the surface of
type 2 (flat bottom) valley with h=500 m, d_1=4 km and d_2=1 km when a SH
plane waves of Ricker waveform with the effective period of 1.22S
incident vertically from below. Because of the symmetry, only one half

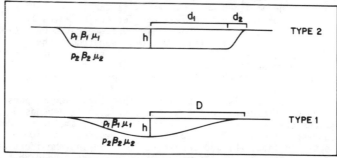

Fig. 25.   Two types of sediment-filled valley studied by Bard and
Bouchon (1980a,b).

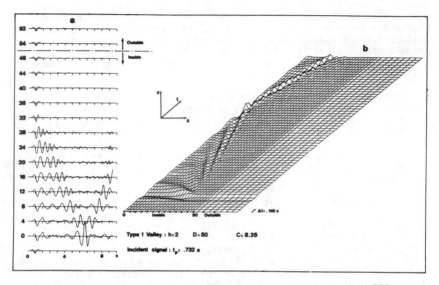

Fig. 26.   Response of a type 1 valley with maximum depth h=200 m,
            half-width D=5 km, to a vertically incident SH Ricker
            wavelet of characteristic period 0.732 sec. (a) The traces
            represent the displacement at surface receivers, spaced from
            0 to 6.2 km from the Valley Center. The bottom trace would
            be the surface displacement without the valley. (b) Diagram
            showing the spatial (x) and temporal evolution of the
            surface displacement in the valley and in its vicinity.
            The dots indicate the location of sites where the
            seismograms in (a) are computed. (Reproduced from Bard and
            Bouchon (1980a).)

of the valley is shown in these figures. In both cases, it is clear
that the main departure from the flat layer response is the Love waves
generated at the edge of valley and propagated back and forth between
both edges. The amplitude of Love waves is the largest at the valley
center because of the constructive interference of waves from both
edges. We find that stronger Love waves are generated by type 1
(cosine) valley than type 2 (flat bottom) even though the depth of
sediment is thicker for the latter. It is also apparent that seismic
motion in type 2 valley shows stronger flat-layer response because of
the broader width over which the sediment depth is constant.

    The departure of seismic motion from the flat-layer response due to
generation of Love waves in the sediment critically depends on the
damping chracteristic of the sediment. If, for example, the damping is
10% (corresponding to Q=5), most of Love waves seen in Fig. 27 will be
wiped out. In fact, Aki and Larner (1970) studying the identical
problem as in Fig. 27 but assuming 10% damping, concluded that the
flat-layer response is applicable to this case. Thus, the question of

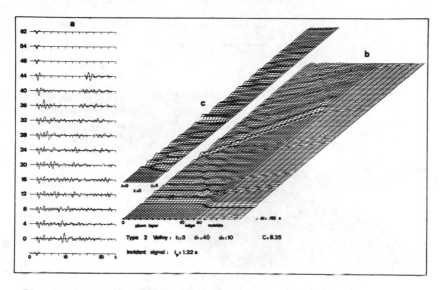

Fig. 27.   The same as Fig. 26 except that the SH wave with the
characteristic period 1.22 sec is incident on the type 2
valley with deeper depth (h=500 m). (Reproduced from Bard
and Bouchon (1980a).)

the applicability of the flat-layer response critically depends on the
damping of sediment. If the damping is strong, the flat-layer response
will give a satisfactory result in this case.

The effect of oblique incidence, however, may be quite different
between the flat-layer and the sediment-filled valley even if the
damping is strong. As mentioned earlier the peak amplification in the
flat layer case decreases with the incidence angle. Aki and Larner
(1970) showed that the surface motion of the type 1 valley is nearly
independent of the incidence angle. Bard and Bouchon (1980a) further
observes that the surface motion may be increased by the direct
transformation of obliquely incident SH waves into Love waves. An
example of strong Love waves generated at the near-source edge of
valley is shown in Fig. 28 for the same valley as shown in Fig. 27 but
for incidence angle of 45° and waveform of period 1.83s.

Bard and Bouchon (1985) recognized that the whole sediment-filled
valley begins to vibrate in phase with a single frequency when the
shape ratio h/P exceeds a certain critical value, where P is the total
width over which the sediment thickness is more than half its maximum.
This phenomenon was also seen in the semi-cylindrical valley studied by
Trifunac (1971). They called it "2-D resonance", and found that the
critical shape ratio depends on the velocity contrast between the
sediment and the basement. The critical shape ratio is smaller for the
greater velocity contrast as shown in Fig. 29.

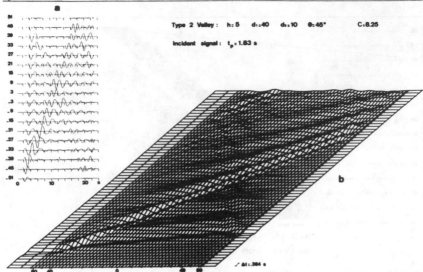

Fig. 28. The same as Fig. 27 except that the SH wave with the characteristic period 1.83 s is incident at an incidence angle of 45°. (Reproduced from Bard and Bouchon (1980a).)

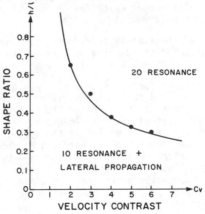

Fig. 29. The critical shape ratio for the 2-D resonance as a function of the velocity contrast between the sediment and basement for incident SH waves. (Reproduced from Bard and Bouchon (1985).)

The 2-D resonance, in general, shows higher frequency and higher peak amplification than the 1-D flat-layer resonance. Bard and Bouchon (1985) made a systematic study of nine cosine-shaped valleys with shape ratios ranging from 0.05 to 1.0, velocity contrast of 5, density contrast of 1.5 and damping of 2.5% (Q of 20). For each case, they

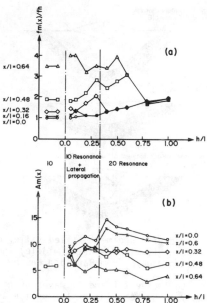

Fig. 30. Peak frequency and the corresponding amplification factor at
several horizontal locations x within the sediment as a
function of the shape ratio (h/ℓ). (Reproduced from Bard
and Bouchon (1985).)

measured the peak frequency and the corresponding amplification factor
at five surface sites equally spaced from X/P=0.0 (center) to X/P=0.64
(edge). They are shown in Fig. 30 together with the frequency and
amplification factor for the flat-layer case with the thickness equal
to the thickness below each site. It is extremely interesting to see
that as the shape ratio increases, the resonance frequency converges to
the single value for the 2-D resonance, while the amplification factor
diverges from the single value for the 1-D resonance (the impedance
ratio diminished slightly by the damping effect). Fig. 30 shows that
the 2-D amplification is up to 3 times the 1-D values near the valley
center.

Strong amplification of differential motions, such as strain, tilt
and rotation, by a sediment-filled valley is expected as demonstrated,
e.g., by Bouchon et al. (1982). As a rough estimate, the amplification
factor for differential motions would be inversely proportional to the
square of shear velocity in the sediment, because the peak displacement
amplification is proportional to the impedance contrast, and the wave
length is proportional to the shear velocity.

P and SV Waves Incident on Sediment-Filled Valley

Bard and Bouchon (1980b) extended their study of SH waves in

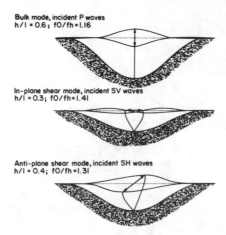

Fig. 31.   Vibration modes of a sediment-filled valley due to normally incident P waves (top), SV waves (middle) and SH waves (bottom). (Reproduced from Bard and Bouchon (1985).)

sediment-filled valley to incident P and SV waves using the same two types of valley geometry. The behavior of the motion is qualitatively similar to that for SH waves. The edge of valley generates surface waves (Rayleigh waves in this case) which are trapped between the two edges and increases the amplitude of the motion as well as its duration. The seismograms, however, are much more complicated than in the SH case because of interference among P, SV and Rayleigh waves.

The transition to the 2-D resonance occurs for both P and SV cases as in the SH case, but it appears to occur for SV at a smaller shape ratio than for SH and P. Bard and Bouchon (1980b) observed, for the type 1 valley with h=1 km, D=5 km, and velocity contrast of 5, the in-phase vibration of the whole valley that lasted far too long to be explainable by the flat-layer theory.

The fundamental modes of 2-D resonance excited by P, SV and SH waves are illustrated in Fig. 33 reproduced from Bard and Bouchon (1985).

The seismic motion in sediment-filled valleys due to incident P and SV waves has not been studied as extensively as for the SH case. We expect especially unusual phenomena for oblique incidence of SV waves as we have seen in the cases of flat free surface, ridge, canyon, and flat surface layer.

Comparison of Observation and Theory

Recently, Geli et al. (1986) made a comprehensive review of the effect of topography on seismic motion, thereby comparing observational results obtained by Davis and West (1973), Griffith and Bollinger (1979) and Tucker et al. (1984) with theoretical results obtained by

Boore (1972), Smith (1975), Sills (1978), Zhenpeng et al.(1980), Bard
(1982) and Zahradnik and Urban (1984).

They found that theoretical results are in agreement with results
from laboratory model experiments such as done by Rogers et al. (1974),
but they cannot explain the large amplification effect observed at some
ridge crests in the field. Some of the observed amplification are far
too large to be attributed to the simple topography effect considered
in the theoretical studies. Bard and Tucker (1985) and Geli et al.
(1986) suggested that combined effects of elevated topography, low
velocity layering at higher elevation, and laterally repeated ridges
may explain the observed high amplification.

A comparison of theory and observation becomes more difficult for
the effect of soft surface layer or sediment-filled valley because we
need to have not only the information on sub-surface velocity and
density distribution, but also a simultaneous observation of seismic
motions at the surface and at a reference point either in the basement
rock beneath the surface observation point or at a nearby exposed site.

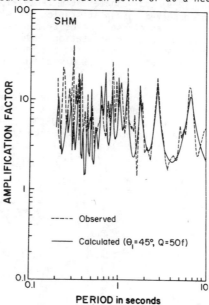

Fig. 32.   Comparison of the observed and calculated amplification
           factor at a borehole station SHM. The observed
           amplification is obtained from a simultaneous recording at
           the surface and borehole bottom. The calculated curve is
           based on the shear velocity, density, Q values, and
           thickness of layers above the basement rock known from
           borehole measurements. (Reproduced from Kinoshita et al.
           (1986).)

Examples from simultaneous observation at the surface and depth were given by Kinoshita et al. (1986) using several holes 2 to 3 km deep penetrated into the Tertiary basement rock near Tokyo, Japan. The shear wave velocity, density and thickness of layers above the basement rock as well as their Q values are known from various borehole seismic observations.

Fig. 32 shows a comparison of observed spectral ratio for surface and basement motion (dotted lines) at station SHM (depth of 2.3 km) and the theoretical amplification for SH waves with the incidence angle 45° in the basement rock. The earthquake used for calculating the amplitude ratio was M=7.0 earthquake of July 23, 1982 at a distance about 200 km from the station. The agreement between observation and flat-layer theory is very good both in the absolute level of amplification and peak frequencies.

Another example of good agreement comes from the Mexico earthquake of September 19, 1985. As shown in Fig. 33, Romo and Seed (1985), using the one-dimensional, vertical wave propagation analysis procedure of Seed and Idriss (1969) and Schnabel et al. (1972), compared the average spectrum of observed motion at station CAO with the spectrum computed for proper choices of shear velocity and damping of soil assuming that the motion observed at station UNAM corresponds to the incident wave beneath CAO. It was pointed out, however, by Kawase (1987b) that the duration at station CDAO was too long to be explained by the 1-D resonance.

A most impressive comparison of observed seismic motion in a sediment-filled valley and theoretical prediction was made by Bard and Tucker (1987) for the Chusal Valley, Garm, USSR. The thickness and seismic velocity of the sediments are known from the work of Sedova (1962) as shown in Fig. 34. At the top of Fig. 35, the NS component velocity seismogram band-passed between 1.5 and 4.5 Hz from the S wave of a local event is shown. The valley surface appears to move in phase and the amplitude is greatest at the valley center and decreases smoothly toward the valley edges.

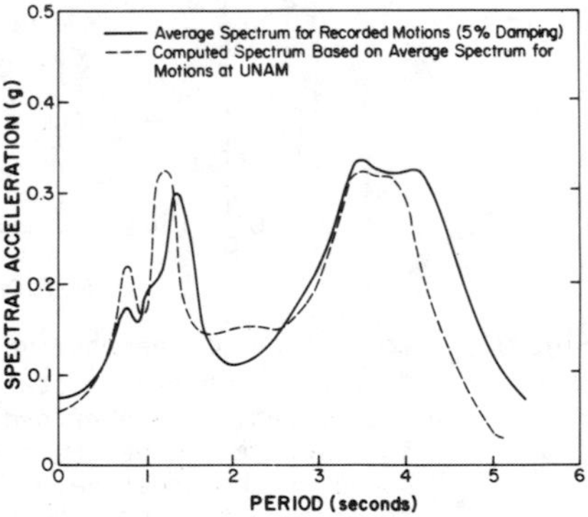

Fig. 33.   Comparison of spectra
for recorded and computed motions
at CAO site in Mexico City.
(Reproduced from Romo and Seed
(1986).)

Fig. 34.   Chusal Valley sediments
and seismometer sites. Plan view
sketch of Chusal Valley. The 5 m
- contours of sediment thickness,
together with the P and S wave
velocities as a function of depth
(inset, upper left) were
determined in a seismic reflection
study (Sedova, 1962). Also shown
are the seismometers sites 1
through 12 used in this study, and
the tunnel site (T) acting as a
common trigger and reference for
the valley stations. The dotted
line to the south indicates the
estimated boundary with the
saturated sediments of the Surkhob
Valley. (Reproduced from Bard and
Tucker (1987).)

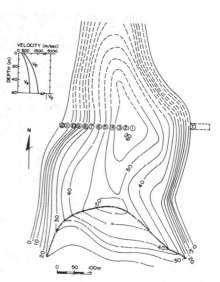

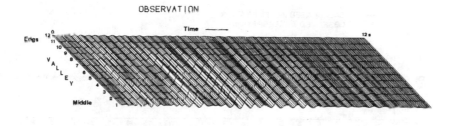

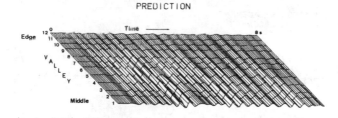

Fig. 35.   Two-dimensional anti-plane shear resonance across Chusal
            Valley. (Top) Observed space-time response of Chusal
            Valley. The 12 NS component seismograms were recorded
            during a very small local event on October 12, 1977, located
            3 km south of Chusal at a depth of 6 km. These records have
            been band-passed between 1.5 and 4.5 Hz in order to

142 EARTHQUAKE ENGINEERING

emphasize the fundamental resonance mode: the motion is
in-phase across the whole valley, and its amplitude
decreases from the center to the edge. (Bottom) Predicted
space-time SH response of Chusal Valley to the same event.
The incoming signal was chosen in such a way the predicted
signal at the valley edge would be the same as the recorded
one. The incoming wave is a vertical plane SH wave. These
seismograms are also band-passed between 1.5 and 4.5 Hz.
(Reproduced from Bard and Tucker (1987).)

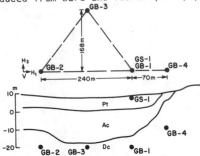

Fig. 36. Plan view (top) and cross-section (bottom) of seismograph
sites in the sediment and basement rock. (Reproduced from
Ohtsuki et al. (1984).)

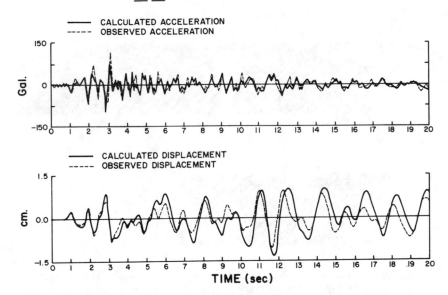

Fig. 37. Comparison of observed (dashed line) and calculated (solid
line) accelerograms and displacements at the GS-1 site (see
Fig. 38). (Reproduced from Ohtsuki et al. (1984).)

Using the Aki-Larner method extended to the layer with a vertical velocity gradient, Bard and Tucker (1987) calculated the seismic motion on the basis of Sedova's model as shown at the bottom of Fig. 35. We find that main features of observed motion are reproduced in the predicted.

Another example of good agreement between observed and calculated motions was obtained by Ohtsuki et al. (1984). They recorded simultaneously earthquake motions at the surface of a low velocity sediment-filled valley near its edge and at points in the basement rock, as shown in Fig. 36. The observed motion at the surface site (GS-1) was compared with the theoretical motion in Fig. 37 calculated by the hybrid method which combined a particle model (with the input motion given by the observed basement motion) and finite element method. The agreement is quite satisfactory both for acceleration and displacement.

The above examples suggest that we may have an adequate state-of-the-art in predicting the site effect on ground motion for many realistic situations, if we know (1) input motion, (2) velocity and density distribution, (3) topography, (4) sediment thickness, and (5) damping of sediment.

Our analysis method still need development for application to more general 3-D, heterogeneous and anisotropic cases, but the real difficulty lies in gaining the information about input motion and structural parameters mentioned above.

The analytical approach described in the preceeding several sections is expensive in terms of computer time, necessary input and structural information. Earlier, we described an alternative approach of empirically determining site-specific amplification using the regression analysis of earthquake data. There is an intermediate approach between these two extremes, namely, trying to correlate empirically determined amplification factor with the characteristic of site condition that can be measured relatively easily.

Characterization of Site Conditions

The single most important parameter affecting the site amplification is probably the near-surface shear wave velocity as can be found in our detailed discussions on the causes of local variations in ground motion. For example, the resonant peak amplification of a flat surface layer is proportion to the impedance contrast, which is inversely proportional to the near-surface shear wave velocity assuming that basement velocity is constant. If the shear wave velocity varies smoothly in both lateral and vertical directions, the resonance disappears, but the amplitude will be inversely proportional to the square root of impedance as shown in Aki and Richards (1980, p. 116), and therefore to the square root of near-surface shear wave velocity. In fact, a comprehensive study of empirical site amplification factor and various geologic and geotechnical parameters made by Rogers et al. (1985) for Los Angeles and San Francisco revealed that the most significant factor controlling site amplification is mean void ratio which strongly correlates (inversely) with the mean shear wave velocity.

Void ratios (e) are computed from dry density (GD) data obtained from the foundation engineering data by using the relation $e=(GS/GD)-1$, where GS is the density of the solid without voids. The void ratio data are more readily available than the shear wave velocity data. They are generally obtained from engineering boreholes as the depth-weighted mean for the upper 8 m.

Fig. 38 from Rogers et al. (1985) shows the short period spectral ratio smoothed over 0.2 to 0.5 sec at sites in the Los Angeles basin as a function of the void ratio of the site. It is remarkable that the range of variation reduced from a factor of 7 to a factor of 2 by specifying the void ratio.

Rogers et al. studied 9 other geotechnical parameters; (1) mean percentage of silt and clay, (2) thickness of Quaternary, (3) age, (4) thickness of Holocene, (5) depth to water table, (6) textural type, (7) depth to crystalline basement, (8) depth to cementation, and (9) mean shear wave velocity. They found that in addition to the void ratio and shear wave velocity, the thickness of unconsolidated sediment (principally Holocene) and the depth to basement rock are also significant parameters controlling the amplification for periods 0.2-0.5 s. At periods longer than 0.5s, depth to basement rock and the thickness of Quarternary sediments were found to be controlling factors. They found, at least for Los Angeles, that the depth to water table is not a reliable predictor of site amplification.

A geotechnical site parameter calculated from blow-count (N-value) profiles from standard penetration test has also been shown to have a significant relation with the amplification factor by Goto et al. (1982).

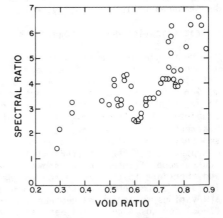

Fig. 38.   Spectral ratio relative to a rock site smoothed over the period range 0.2 to 0.5 sec at sites in the Los Angeles basin plotted against the void ratio (Reproduced from Rogers et al. (1985).)

All the above characterizations are anticipating that softer the material higher the amplification. There is, however, a definite trend that the relation may be reversed for frequencies higher than a few Hz as discussed in detail in an earlier section. In fact, Anderson and Hough (1984) presented a strikingly simple frequency dependence of the observed S-wave acceleration which clearly manifest the importance of absorption in site effects on high frequency seismic motion. They found that the acceleration spectrum can be characterized by a trend of exponential decay e-πkf, and the coefficient k is systematically smaller for sites on rock than for sites on alluvium.

Thus, for higher frequencies, the geotechnical parameters related to the softness of soil will have a relation with the site amplification factor opposite to the one for lower frequencies. Our future problem in this area would be to find an effective geotechnical parameter which can characterize this frequency dependent behavior of site amplification-deamplification effect.

The testing of site amplification predictions based on various geotechnical measurements being coordinated by the California Division of Mines and Geology as a part of the Parkfield Earthquake Prediction experiment is a useful starting point for finding such an effective parameter.

Conclusions

In this review of site effects on strong ground motion, we started with the broad classification of site conditions into soil and rock and found there exists a cross-over period above which the soil site shows higher amplification than the rock site and below which the relation is reversed. The cross-over period is around 0.2 sec for both U.S. and Japan. The amplification of soil sites relative to rock sites for periods longer than the cross-over period amount to a factor of 2 to 3, and the amplification of rock sites relative to soil sites for periods shorter than the cross-over is less than a factor of 2. No significant difference was found in the peak ground acceleration between soil and rock sites.

The above results do not mean that the site effect decreases with decreasing period, rather means that the broad classification fails to capture the essential factor controlling the site effect for shorter periods. This conclusion was obtained from the review of empirical site specific amplification obtained from regression analysis of strong motion and weak motion data.

Observations in both Japan and U.S. indicate that the geographic variation of site specific amplification factor obtained by regression analysis ranges over a factor of about 10 for frequencies between 1 and 10 Hz. Since the standard error of the observed variation of amplification factor for different directions of incident waves is less than a factor of 2, we may conclude that a very meaningful microzonation map predicting the amplification factor can be constructed for the frequency range at least from 1 to 10 Hz.

There are two alternative approaches toward the meaningful microzonation. One is to measure the site-specific amplification factor empirically using the data from large and small earthquakes. The other is to improve the characterization of site conditions to capture the frequency dependent amplification effect.

The microtremors are easier to observe than earthquakes and useful for a broad classification of site conditions, but cannot give accurate estimation of amplification factor because of the unknown source effect.

Numerous observations are presented to support that the amplification factors for weak and strong motions are similar to each other to the first order, except for the obvious case of liquefaction, although a latest result from SMART-1 array suggests an intriguing possibility of non-linear effect on peak ground acceleration.

In order to improve the site characterization, we reviewed theoretical studies on the causes of local variations in ground motion including the effects of flat free surface, topography, flat soft surface layer, sediment-filled valley. We present several successful comparisons of observation and theory, and suggest that we may have an adequate state-of-the-art in predicting the site effect on ground motion for many realistic situations, if we known (1) input motion, (2) velocity and density distribution, (3) topography (4) sediment thickness, and (5) dampling of sediment. The analysis method still need development for application to more general 3-D, heterogeneous and anisotropic cases, but the real difficulty lies in gaining information about input motion and structural parameters mentioned above.

The most realistic approach to the microzonation is then to determine empirical site-amplification factors for as many sites as possible by the regression analysis of earthquake data, and correlate them with various geotechnical parameters of the site which are relatively easier to measure. Analytical studies on the causes of site effects will give helpful insight to the search for effective parameters.

Acknowledgement

It is my pleasure to acknowledge numerous stimulating discussions with Pierre Yves Bard, Michel Campillo, Hiroshi Kawase and Francisco Sanchez-Sesma. My thanks are also due Jacqueline Le Falle, Cindy Waite and Hiroshi Kawase for preparing the camera-ready copy. This work was supported by the National Science Foundation under grant ECE-8616457.

REFERENCES

Abrahamson, N. A., B. A. Bolt, R. B. Darragh, J. Penzien, and Y. B. Tsai, The SMART 1 accelerograph array (1980-1987), Earthquake Spectra, $\underline{3}$, 263-288, 1987.

Abrahamson, N. A., Estimation of seismic wave coherency and rupture velocity using SMART 1 strong motion array recordings, EERC Report No. UCB/EERC-85/02, 1985.

Abrahamson, N. A., Some statistical properties of peak ground accelerations, submitted to Bull. Seis. Soc. Am., 1987.

Aki, K., and P. G. Richards, Quantitative Seismology: Theory and Methods, Freeman and Co., 1980.

Aki, K., Scattering and attenuation of shear waves in the lithosphere, J. Geophys. Res., $\underline{85}$, 6496-6504, 1980.

Aki, K., Space and time spectra of stationary stochastic waves with special reference to microtremors, Bull. Earthq. Res. Inst., $\underline{35}$, 415-456, 1957.

Aki, K., and K. L. Larner, Surface motion of a layered medium having an irregular interface due to incident plane SH waves, J. Geophys. Res., $\underline{75}$, 933-954, 1970.

Aki, K., and M. Tsujiura, Correlational study of near earthquake waves, Bull. Earthq. Res. Inst., Tokyo University, $\underline{37}$, 207-232, 1959.

Alcock, E. D., Grand Valley Colorado: A microzonation case history, Proc. of 1st Int. Conf. on Microzonation, 299-306, 1972.

Anderson, J. G., and S. E. Hough, A model for the shape of the Fourier amplitude spectrum of acceleration at high frequencies, Bull. Seis. Soc. Am., $\underline{74}$, 1969-1994, 1984.

Bard, P. Y., Les effets de site d'origine structurale en sismologie, modelisation et interpretation, application au risque sismique, these d'Etat, Universite Scientifique et Medicale de Grenoble, France, 1983.

Bard, P. Y., Diffracted waves and displacement field over two dimensional elevated topographies, Geophys. J. R. Astr. Soc., $\underline{71}$, 731-760, 1982.

Bard, P. Y., and M. Bouchon, The seismic response of sediment-filled valleys, Part I. The case of incident SH waves, Bull. Seis. Soc. Am., $\underline{70}$, 1263-1286, 1980a.

Bard, P. Y., and M. Bouchon, The seismic response of sediment-filled valleys, Part II. The case of incident SH waves, Bull. Seis. Soc. Am., $\underline{70}$, 1263-1286, 1980b.

Bard, P. Y., and M. Bouchon, The two-dimensional resonance of sediment-filled valleys, Bull. Seis. Soc. Am., 75, 519-541, 1985.

Bard, P. Y., and J. C. Gariel, The seismic response of two-dimensional sedimentary deposits with large vertical velocity gradients, Bull. Seis. Soc. Am., 76, 343-346, 1986.

Bard, P. Y., and B. E. Tucker, Ridge and tunnel effects: comparing observations with theory, Bull. Seis. Soc. Am., 75, 905-922, 1985.

Bard, P. Y., and B. E. Tucker, Predictability of sediment site amplification: a case study, preprint, 1987.

Benites, R., B. Tucker and J. Kuroiwa, A comparison of historical damage and response to weak motion in La Molina valley, Lima, Peru.

Boore, D. M., A note on the effect of simple topography on seismic SH waves, Bull. Seis. Soc. Am., 62, 275-284, 1972.

Boore, D. M., W. B. Joyner, A. A. Oliver, III, and R. A. Page, Peak acceleration, velocity, and displacement from strong motion records, Bull. Seis. Soc. Am., 70, 305-321, 1980.

Boore, D. M., S. C. Harmsen, and S. T. Harding, Wave scattering from a step change in surface topography, Bull. Seis. Soc. Am., 71, 117-125, 1981.

Bouchon, M., A simple, complete numerical solution to the problem of diffraction of SH waves by an irregular surface, J. Acoust. Soc. Am., 77, 1-5, 1985.

Bouchon, M., Effect of topography on surface motion, Bull. Seis. Soc. Am., 63, 615-632, 1973.

Bouchon, M., The importance of the surface or interface P wave in near-earthquake studies, Bull. Seis. Soc. Am., 68, 1293-1311, 1978.

Bouchon, M., and K. Aki, Discrete wavenumber representation of seismic source wavefield, Bull. Seis. Soc. Am., 67, 259-277, 1977.

Bouchon, M., K. Aki, and P. Y. Bard, Theoretical evaluation of differential ground motions produced by earthquakes, Proc. 3rd Microzonation Conf., G2-G12, Seattle, 1982.

Brebbia, C. A., The boundary element method for engineers, Pentech Press, London, 1978.

Brune, J. N., Preliminary results on topographic seismic amplification effect on a foam rubber model of the topography near Pacoima dam, Proc. 8th WCEE, 663-670, 1984.

Burridge, R., F. Mainardi, and G. Servizi, Phys. Earth Planet. Int., 22, 122-136, 1980.

Campbell, K. W., Strong ground motion attenuation relations: a ten-year perspective, Earthquake Spectra, 1, 759-804, 1985.

Campbell, K. W., and C. M. Duke, Bedrock intensity attenuation and site factors from San Fernando earthquake records, Bull. Seis. Soc. Am., 64, 173-185, 1974.

Campillo, M., and M. Bouchon, Synthetic SH seismograms in a laterally varying medium by the discrete wave-number method, Geophys. J. R. Astr. Soc., 83, 307-317, 1985.

Capon, J., High-resolution frequency-wave number spectrum analysis, Proc. IEEE, 57, 1408-1418, 1969.

Celebi, M., C. Dietel, J. Prince, M. Onate and G. Chavez, Site amplification in Mexico City (determined from 19 September 1985 strong-motion records and from recordings of weak motions), in Ground Motion and Engineering Seismology, ed. A. S. Cakmak, 141-152, Elsevier, 1987.

Chapman, C. H., Lamb's problem and comments on the paper "On leaking modes" by Usha Gupta, Pure Appl. Geophys., 94, 233-247, 1972.

Cranswick, E., and C. S. Mueller, High-frequency vertical particle motions in the Imperial Valley and SV to P coupling: The site response of differential array on the "Greenhouse effect," EOS, 66, no. 46, 968, 1985.

Davis, L. L., West, L. R., Observed effects of topography on ground motion, Bull. Seis. Soc. Am., 63, 283-298, 1973.

Dravinski, M., Amplification of P, SV and Rayleigh waves by two alluvial valleys, Soil Dynamics and Earthquake Eng., 2, 66-77, 1983.

Dravinski, M., Influence of interface depth upon strong ground motion, Bull. Seis. Soc. Am., 72, 597-614, 1982.

Esteva, L., Microzoning: models and reality, Proc. of 6th WCEE, New Dehli, 1977.

Fuyuki, M., and M. Nakano, Finite difference analysis of Rayleigh wave transmission past an upward step change, Bull. Seis. Soc. Am., 74, 893-911, 1984.

Geli, L., P. Y. Bard, and B. Jullien, The effect of topography on earthquake ground motion: A review and new results, submitted to Bull. Seis. Soc. Am., 1986.

Goto, H., H. Kameda, and M. Sugito, Use of N-value profiles for estimation of site dependent earthquake motions (in Japanese), Collected Papers 317, Japanese Society of Civil Engineering, 69-78, 1982.

Griffith, D. W., and G. A. Bollinger, The effect of Appalachian
Mountain topography on seismic waves, Bull. Seis. Soc. Am., 69,
1081-1105, 1979.

Gutenberg, B., Effects of ground on earthquake motion, Bull. Seis. Soc.
Am., 47, 221-250, 1957.

Harmsen, S. C., and S. T. Harding, Surface motion over a sedimentary
valley for incident plane P and SV waves, Bull. Seis. Soc. Am., 72,
655-670, 1981.

Hayashi, S., H. Tsuchida, and E. Kurata, Average response spectra for
various subsoil conditions, Third Joint Meeting, US-Japan Panel on
Wind and Seismic effects, UJNR, Tokyo, May 10-12, 1971.

Herrera, I., Boundary methods for fluids, in Finite Elements in Fluids
IV, R. H. Gallagher, Editor, John Wiley and Sons, New York, 1981.

Hong, T. L., and D. V. Helmberger, Glorified optics and wave
propagation in non planar structures, Bull. Seis. Soc. Am., 68,
1313-1330, 1977.

Hudson, D. E., Local distribution of strong earthquake ground motions,
Bull. Seis. Soc. Am., 62, 1765-1786, 1972.

Ihnen, S. M. and D. M. Hadley, Seismic hazard maps for Puget Sound,
Washington, Bull. Seis. Soc. Am., 77, 1091-1109, 1987.

Joyner, W. B., and D. M. Boore, Peak horizontal acceleration and
velocity from strong motion recording including records from the
1979 Imperial Valley, California, earthquake, Bull. Seis. Soc.
Am., 71, 2011-2038, 1981.

Joyner, W. B., R. E. Warrick and T. E. Fumal, The effect of Quaternary
alluvium on strong ground motion in the Coyote Lake, California,
earthquake of 1979, Bull. Seis. Soc. Am., 71, 1333-1350, 1981.

Kagami, H., C. M. Duke, G. C. Liang, and Y. Ohta, Observation of 1- to
5-second microtremors and their application to earthquake
engineering, Part 2: Evaluation of site effect upon seismic wave
amplification due to extremely deep soil deposits, Bull. Seis.
Soc. Am., 72, 987-998, 1982.

Kagami, H., S. Okada, K. Shiono, M. Oner, M. Dravinski, and A. K. Mal,
Observation of 1- to 5-second microtremors and their application
to earthquake engineering. Part 3: A two-dimensional study of
site effects in the San Fernando valley, Bull. Seis. Soc. Am., 76,
1801-1812, 1986.

Kamiyama, M. and E. Yanagisawa, A statistical model for estimating
response spectra of strong earthquake ground motions with emphasis
on local soil conditions, Soils and Foundations, 26, 16-32, 1986.

Katayama, T., T. Iwasaki, and M. Saeki, Statistical analysis of earthquake acceleration response spectra, Collected Papers, 275, Japanese Society of Civil Engineering, 29-40, 1978.

Kawasaki, I., Y. Suzuki, and R. Sato, Seismic waves due to a shear fault in a semi-infinite medium, Part I, Point source, J. Phys. Earth, 21, 251-284, 1973.

Kawasaki, I., The focal process of the Kita-mino earthquake of August 19, 1961, and its relationship to quaternary fault, the Hatogayu-Koike fault, J. Phys. Earth, 24, 227-250, 1975.

Kawase, H., K. Yoshida, S. Nakai, and Y. Koyanagi, Dynamic response of structure on a layered medium - a dipping layer and a flat layer -, Proc. 6th Japan Earthq. Eng. Symp., 1641-1648, 1982.

Kawase, H., Time-domain response of a semicircular canyon for incident SV, P and Rayleigh waves calculated by the discrete wave number boundary element method, submitted to Bull. Seis. Soc. Am., 1987a.

Kawase, H., Irregular ground analysis to interpret time-characteristics of strong motion recorded in Mexico City during 1985 Mexico earthquake, in Ground Motion and Engineering Seismology, A. S. Cakmak, editor, Elsevier, 467-476, 1987b.

Kawashima, K., K. Aizawa, and K. Takahashi, Attenuation of peak ground acceleration, velocity and displacement based on multiple regression analysis of Japanese strong motion records, Earthquake Engineering and Structural Dynamics, 14, 199-215, 1986.

King, J. L., and J. N. Brune, Modeling the seismic response of sedimentary basins, Bull. Seis. Soc. Am., 71, 1469-1487, 1981.

Kinoshita, S., T. Mikoshiba, and T. Hoshino, Estimation of the average amplification characteristics of a sedimentary layer for short period S-waves, Zisin (Journ. Seis. Soc. Japan), Ser. 2, 39, 67-80, 1986.

Kohketsu, K., 2-D reflectivity method and synthetic seismograms for irregularly layered structure, I. SH-wave generation, Geophys. J. R. Astr. Soc., 89, 821-838, 1987.

Kuribayashi, E., T. Iwasaki, Y. Iida and K. Tuji, Effects of Seismic and subsoil conditions on earthquake response spectra, Proc. Int. Conf. on Microzonation, 499-512, 1972.

Lee, V. W., Three-dimensional diffraction of plane P, SV and SH waves by a hemispherical alluvial valley, Soil Dynamics and Earthquake Engineering, 3, 133-144, 1984.

Macdonald, H. M., Electric waves, Cambridge University Press, Cambridge, England, 1902.

Mal, A. K., and L. Knopoff, Transmission of Rayleigh waves past a step change in elevation, Bull. Seis. Soc. Am., 55, 319-334, 1965.

Mohraz, B., A study of earthquake response spectra for different geological conditions, Bull. Seis. Soc. Am., 66, 915-935, 1976.

Murphy, J. R., A. H. Davis and N. L. Weaver, Amplification of seismic body waves by low-velocity surface layers, Bull. Seis. Soc. Am., 61, 109-146, 1971.

Newmark, N. M., J. A. Blume, and K. Kapur, Design response spectra for nuclear power plants, paper presented at the Structural Engineers ASCE Conference, San Francisco, Calif., April, 1973.

Niwa, Y., and S. Hirose, Three-dimensional analysis of ground motion by integral equation method in wave-number domain, Proc. 5th Int. Conf. Numerical Methods in Geomechanics, 1985.

Nowack, R., and K. Aki, The two-dimensional Gaussian beam synthetic method; testing and applications, J. Geophys. Res., 89, 7797-7819, 1984.

Ohta, Y., H. Kagami, N. Goto, and K. Kudo, Observation of 1- to 5-second microtremors and their application to earthquake engineering. Part 1: Comparison with long-period accelerations at the Tokachi-Oki earthquake of 1968, Bull. Seis. Soc. Am., 68, 767-779, 1978.

Ohtsuki, A., and K. Harumi, Effect of topographby and subsurface inhomogeneities on seismic SV waves, Int. J. Earthquake Engrg. Struct. Dyn., 11, 441-462, 1983.

Ohtsuki, A., H. Yamahara and T. Tazoh, Effect of lateral inhomogeneity on seismic waves, II observations and analyses, Earthquake Engineering and Structural Dynamics, 12, 795-816, 1984.

Okada, H., T. Matsushima, and E. Hidaka, Comparison of spatial autocorrelation method and frequency-wave number spectral method of estimating the phase velocity of Rayleigh waves in long-period microtremors, Geophysical Bulletin of Hokkaido Univ., Sapporo, Japan, No. 49, 53-62, 1987.

Pekeris, C. I. and H. Lifson, Motion of the surface of a uniform elastic half-space produced by a buried pulse, J. Acoust. Soc. Am., 29, 1233-1238, 1957.

Phillips, W. S., and K. Aki, Site amplification of coda waves from local earthquakes in central California, Bull. Seis. Soc. Am., 76, 627-648, 1986.

Rogers, A. M., R. D. Borcherdt, P. A. Covington, and D. M. Perkins, A comparative ground response study near Los Angeles using recordings of Nevada nuclear tests and the 1971 San Fernando earthquake, Bull. Seis. Soc. Am., 74, 1925-1949, 1984.

Rogers, A. M., L. J. Katz, and T. J. Benett, Topographic effect on ground motion for incident P waves: a model study, Bull. Seis. Soc. Am., 64, 437-456, 1974.

Rogers, A. M., J. C. Tinsley, and R. D. Borcherdt, Predicting relative ground response, in Evaluating Earthquake Hazards in the Los Angeles region, ed. J. I. Ziony, U.S.G.S. Prof. Paper 1360, 221-248, 1985.

Romo, M. P., and H. B. Seed, Analytical modeling of dynamic soil response in the Mexico earthquake of Sept. 19, 1985, in the Mexico Earthquakes - 1985, ed. M. A. Cassaro and E. M. Romero, 148-162, Am. Soc. Civil Eng., 1968.

Sammis, C. G., K. Aki and H. Kawase, Damage pattern due to the Whittier-Narrows earthquake of October 1, 1987: A ring of destruction by SV waves at critical incidence, submitted to Bull. Seis. Soc. Am., 1987.

Sanchez-Sesma, F. J., Diffraction of elastic SH waves by wedges, Bull. Seis. Soc. Am., 75, 1435-1446, 1985.

Sanchez-Sesma, F. J., Diffraction of elastic waves by three-dimensional surface irregularities, Bull. Seis. Soc. Am., 73, 1621-1636, 1983.

Sanchez-Sesma, F. J., F. J. Chavez-Garcia and M. A. Bravo, Seismic response of a class of alluvial valleys for incident SH waves, in press, Bull. Seis. Soc. Am., 1987.

Sanchez-Sesma, F. J., I. Herrera, and J. Aviles, A boundary method for elastic wave diffraction: application to scattering of SH waves by surface irregulariteies, Bull. Scis. Soc. Am., 72, 473-490, 1982.

Sanchez-Sesma, F. J., and E. Rosenblueth, Ground motion at canyons of arbitrary shape under incident SH waves, Int. J. Earthquake Eng. Struct. Dyn., 7, 441-450, 1979.

Sanchez-Sesma, F. J., and J. A. Esquivel, Ground motion on alluvial valleys under incident plane SH waves, Bull. Seis. Soc. Am., 69, 1107-1120, 1979.

Schnabel, P. B., J. Lysmer, and H. B. Seed, SHAKE: A computer program for earthquake response analysis of horizontally layered sites, Report No. EERC.72-12, University of California, Berkeley, December, 1972.

Sedova, E. N., Correlation of dynamic features of weak earthquakes with ground conditions (in Russian), Trudi Inst. Phys. Earth, U.S.S.R., 25, 211-225, 1962.

Seed, H. B., and I. M. Idriss, The influence of soil conditons on ground motions during earthquakes, J. Soil Mech. Found. Eng. Div., ASCE, 94, SM1, 93-137, 1969.

Seed, H. B., C. Ugas, and J. Lysmer, Site-dependent spectra for earthquake-resistant design, Bull. Seis. Soc. Am., 66, 221-243, 1976.

Sills, L. B., Scattering of horizontally polarized shear waves by surface irregularities, Geophys. J. R. Astr. Soc., 54, 319-348, 1978.

Smith, W. D., The application of finite element analysis to body wave propagation problems, Geophys. J. R. Astr. Soc., 42, 747-768, 1975.

Spudich, P., and E. Cranswick, Direct observation of rupture propagation during the 1979 Imperial Valley earthquake using a short baseline accelerometer array, Bull. Seis. Soc. Am., 74, 2083-2114, 1984.

Takahasi, R., and K. Hirano, Seismic vibrations of soft ground, Bull. Earthq. Res. Inst., Tokyo University, 19, 534-543, 1941.

Tanaka, T., K. Kanai, K. Osada, and D. J. Leeds, Observation of microtremors, XII, Bull. Earthq. Res. Inst., Tokyo Univ., 46, 1127-1147, 1968.

Trifunac, M. D., Preliminary empirical model for scaling Fourier amplitude spectra of strong ground acceleration in terms of earthquake magnitude, source-to-station distance and recording site condition, Bull. Seis. Soc. Am., 66, 1343-1373, 1976a.

Trifunac, M. D., Preliminary analysis of the peaks of strong earthquake ground motion-dependence of peaks on earthquake magnitude, epicentral distance and recording site conditions, Bull. Seis. Soc. Am., 66, 189-219, 1976b.

Trifunac, M. D., Scattering of plane SH waves by a semi-cylindrical canyon, Int. J. Earthquake Eng. Struct. Dyn., 1, 267-281, 1973.

Trifunac, M. D., Surface motion of a semi-cylindrical alluvial valley for incident plane SH waves, Bull. Seis. Soc. Am., 61, 1755-1770, 1971.

Tsujiura, M., Spectral analysis of the coda waves from local earthquakes, Bull. Earthq. Res. Inst., Tokyo Univ., 53, 1-48, 1978.

Tucker, B. E., and J. L. King, Dependence of sediment-filled valley response on the input amplitude and the valley properties, Bull. Seis. Soc. Am., 74, 153-165, 1984.

Tucker, B. E., J. L. King, D. Hatzfeld, and I. L. Nersesov, Observations of hard rock site effects, Bull. Seis. Soc. Am., 74, 121-136, 1984.

Wong, H. L., and M. D. Trifunac, Scattering of plane SH waves by a semi-elliptical canyon, Int. J. Earthquake Eng. Structr. Dyn., 3, 157-169, 1974a.

Wong, H. L., and M. D. Trifunac, Surface motion of a semi-elliptical alluvial valley for incident plane SH waves, Bull. Seis. Soc. Am., 64, 1389-1408, 1974b.

Wong, H. L., and P. C. Jennings, Effect of Canyon topography on strong ground motion, Bull. Seis. Soc. Am., 65, 1239-1257, 1975.

Wong, H. L., Effect of surface topography on the diffraction of P, SV and Rayleigh waves, Bull. Seis. Soc. Am., 72, 1167-1183, 1982.

Zahradnik, J., and L. Urban, Effect of a simple mountain range on underground seismic motion, Geophys. J. R. Astron. Soc., 79, 167-183, 1984.

Zhenpeng, L. Y., L. Baipo and Y. Yifan, Effect of three dimensional topography on earthquake ground motion, Proc. 7th World Conference on Earthquake Engineering, Instanbul, 2, 161-168, 1980.

USE OF DETAILED GEOLOGIC DATA IN REGIONAL PROBABILISTIC SEISMIC HAZARD ANALYSES: AN EXAMPLE FROM THE WASATCH FRONT, UTAH

R.R. Youngs* M. ASCE, F.H. Swan*, and M.S. Power* M. ASCE

ABSTRACT
Probabilistic assessments of earthquake hazards due to strong ground shaking are usually based solely on historical seismicity. A seismic hazard analysis conducted for the urban corridor along the Wasatch Front, Utah demonstrates how detailed geologic data can be integrated with seismicity data to better define the overall hazard and to evaluate local variations in the hazard based on proximity to active faults and to differences in the rate of activity on the faults. Using a logic tree methodology, geologic data and the uncertainty in these data were formally incorporated in the probabilistic hazard assessment. This approach results in more local variation in the hazard than assessments based solely on seismicity data. The geologic factors that had the greatest impact on the computed hazard along the Wasatch fault zone were the seismic source model (faults vs. source zones), the form of the fault-specific magnitude distribution (characteristic vs. exponential), and the frequency of large-magnitude earthquakes.

INTRODUCTION
Regional assessment of seismic ground motion hazard commonly involves development of hazard maps displaying the results of probabilistic seismic hazard analyses. In most cases these studies rely heavily on seismicity data and use geologic information only in general terms to delineate seismic source zones or to identify specific active faults. However, detailed geologic information, if available, can be used extensively in seismic hazard analyses to constrain the parameters of the seismic hazard model. The most common uses of more detailed geologic data have been to constrain maximum earthquake magnitudes using empirical relationships between earthquake rupture dimensions and magnitude, and to constrain earthquake recurrence estimates using fault slip rates (e.g. Anderson, 1979; Molnar, 1979; Greensfelder and others, 1980; Doser and Smith, 1982; Campbell, 1983; Youngs and Coppersmith, 1985a; Wesnousky, 1986). In well-studied tectonically active regions even greater use can be made of geologic information. Fault segmentation can be identified to further refine estimates of maximum magnitude and relative seismic activity. Paleoseismic data can be used to better constrain earthquake recurrence estimates based on Holocene slip rates and/or on the timing of pre-historic events. These data can be used to develop fault-specific models for earthquake recurrence.
Detailed geological information was incorporated into a regional seismic ground motion hazard assessment for the Wasatch Front area of north-central Utah (Figure 1). Extensive geological investigations have been conducted during the past two decades for the specific purpose of identifying and evaluating the potential earthquake hazards along the

*Geomatrix Consultants, One Market Plaza, Spear St. Tower, Suite 717, San Francisco, California 94105

Wasatch Front (Utah Geological Association, 1972; U.S. Geological Survey, 1980, 1984). The studies have focused on the Wasatch fault zone which displays evidence of repeated large magnitude earthquakes throughout the Holocene (Swan and others, 1980; Hanson and others, 1981; Schwartz and others, 1984; Schwartz and Coppersmith, 1984).

In contrast to the geologic data, the record of historical seismicity is short compared to the long repeat time between major earthquakes. Instrumental monitoring of the Wasatch Front dates back to only

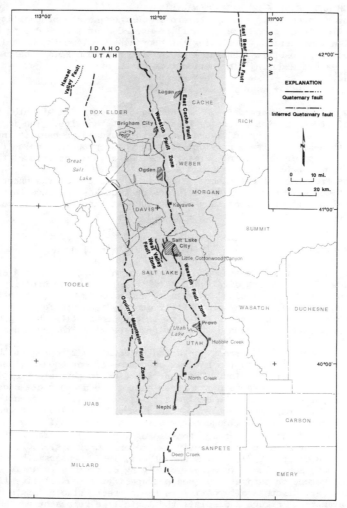

Figure 1 - Map showing location of the study area (shaded region) and the Quaternary faults modeled in the seismic hazard analysis.

1962 (Arabasz and others, 1980) and extensive telemetered seismic arrays have been in existence only since 1974. Prior to that, the earthquake catalog is limited primarily to shaking intensity data reported in the towns along the Wasatch Front. An important part of the approach is the inclusion of multiple interpretations of the available data in the model to represent the uncertainty in the hazard estimates. The complete hazard analysis is described in detail in Youngs and others (in press).

APPROACH AND RESULTS OF HAZARD ANALYSIS

For this analysis seismic hazard is defined as the probability that various levels of ground motion will be exceeded at a site during a specified time period. Following the approach developed by Cornell (1968), the probability that at a given site a ground motion parameter, Z, will exceed z during time period t is given by:

$$P(Z > z \mid t) = 1 - \exp[-\nu(z) \cdot t] \leq \nu(z) \cdot t \qquad (1)$$

where $\nu(z)$ is the average frequency during time period t at which the level of ground motion parameter Z exceeds z at the site. The inequality in Equation 1 is valid regardless of the assumed probability model for earthquake occurrence provided $\nu(z)$ is the appropriate value for the time period of interest, and $\nu(z) \cdot t$ provides a good estimate of the hazard for probabilities less than about 0.1.

The frequency of exceedance, $\nu(z)$, is a function of the uncertainty in the time, size and location of future earthquakes and uncertainty in the level of ground motions they may produce at the site. It is computed by the expression

$$\nu(z) = \sum_n \alpha_n(m^\circ) \int_{m^\circ}^{m^u} \int_{r=0}^{r=\infty} f(m) \cdot f(r) \cdot P(Z > z \mid m, r) \ dr \ dm \qquad (2)$$

where $\alpha_n(m^\circ)$ is the frequency of earthquakes on source n above a minimum magnitude of engineering significance, m°; $f(m)$ is the probability density function for event sizes between m° and a maximum event size for the source, m^u; $f(r)$ is the probability density function for distance to the earthquake rupture; and $P(Z > z \mid m, r)$ is the probability that, given a magnitude m earthquake at a distance r from the site, the ground motion exceeds level z.

The probability functions contained in Equation 2 represent the uncertainties inherent in the natural phenomena of earthquake generation and seismic wave propagation. The approach used in this study explicitly incorporates the uncertainties in selecting the appropriate models and model parameters into the analysis to assess their impact on the estimate of the expected level of seismic hazard as well as the uncertainty in that estimate. The uncertainty in modeling is incorporated into the hazard analysis through the use of logic trees. The logic tree formulation for seismic hazard analysis (Power and others, 1981; Kulkarni and others, 1984; Youngs and others, 1985; Coppersmith and Youngs, 1986; EPRI, 1987) involves specifying discrete alternatives for states of nature or parameter values and specifying the relative likelihood that each discrete alternative is the correct value or state of the input parameter. Figure 2 displays the logic tree representing the seismic hazard model developed for this study. The logic tree is laid out to provide a logical progression from general aspects/hypotheses

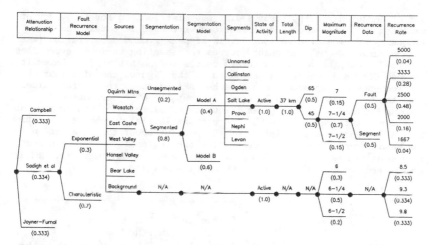

Figure 2 - Hazard model logic tree used for evaluating seismic hazards in the study. Parameter values are shown above the branches and assessed probabilities are shown in parentheses below the branches. (Note not all branches for all sources are shown.)

regarding the characteristics of seismicity and seismic wave propagation in the region to specific input parameters for individual faults and fault segments. The parameter values and their relative likelihoods were based on the best judgment of the authors (Youngs and others, in press) as to the uncertainty in defining input parameters and thus represent their confidence in the estimated hazard.

The first node of the logic tree represents the uncertainty in selecting the appropriate strong ground motion attenuation relationship. Attenuation was placed first in the tree because it is felt that a single relationship (whichever relationship may be "correct") is applicable to all earthquake sources in the region. As indicated in Figure 2, three attenuation relationships developed for soil site conditions were used: Campbell (1987), Sadigh and others (1986, presented in Youngs and others, in press), and Joyner and Boore (1982) as modified by Joyner and Fumal (1985) to be applicable to the random horizontal component. These three relationships were developed using magnitude definitions that are equivalent to local magnitude, M_L in the range of 3 to 7 and surface wave magnitude, M_s, in the range of 5 to 7.5. In applying these relationships to Utah we have assumed that ground motions are more directly correlated with magnitude than with seismic moment. The seismic moment-earthquake magnitude relationships used in the analysis were developed specifically for the Utah region (Doser and Smith, 1982).

The remaining nodes of the logic tree define the characteristics of the seismic sources. Specific aspects relating to the use of geologic data in defining the seismic sources and source parameters are described in the following section of this paper.

The annual frequencies of exceeding various levels of peak ground acceleration were developed by performing hazard computations using Equation 2 with the input parameters defined by each end branch of the logic tree shown in Figure 2. At each ground motion level, the complete

set of computed results forms a discrete distribution for frequency of
exceedance. The computed distributions were used to obtain the mean
frequency of exceeding various levels of peak ground acceleration (mean
hazard curve) as well as hazard curves representing various percentiles.
Figure 3 shows the resulting hazard curves for points at Salt Lake City,
Brigham City and Logan. As indicated in the figure, the distributions
in frequency of exceedance are somewhat skewed about the median value,
with the mean hazard curve located at about the 60th percentile.
Reading horizontally at a given frequency of exceedance, the 90-percent
confidence interval (5th to 95th percentile levels) in peak acceleration
typically represent a variation of ±50 percent about the level given by
the mean hazard curve. Maps presenting the regional variation in mean
hazard estimates along the Wasatch front for three different probability
levels are presented in Youngs and others (in press).

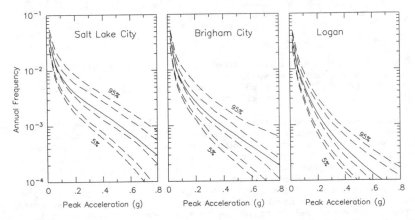

Figure 3 - Peak ground acceleration hazard curves for three locations in
Utah. Shown are the mean hazard curves (solid) and hazard curves
corresponding to the 5th, 15th, 50th, 85th, and 95th percentiles
(dashed) of the distributions for frequency of exceedence.

    Figure 4 shows the contributions of various seismic sources to the
hazard at the three sites. At each of the sites the hazard at low
acceleration levels is dominated by the background seismicity. Because
the rate of occurrence of moderate magnitude events is assumed to be
uniform throughout the background zone of seismicity, similar hazard
levels are obtained at all three sites. At higher acceleration levels,
the hazard is dominated by contributions from the identified faults, and
their rela-tive rates of seismic activity result in differences in the
computed hazard depending on proximity to the faults.

USE OF GEOLOGIC DATA
    Seismic Source Model: The first level of integration of geologic
data is in the definition of the potential sources of future earthquakes. The historical seismicity data, shown in Figure 5, define a
zone of con-centrated seismic activity, the Intermountain Seismic Belt.

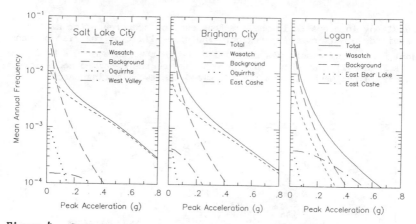

Figure 4 - Contributions of various sources to the computed mean hazard at three locations in Utah. Shown for each location are the mean hazard curves for various sources and the total mean hazard curve.

One approach to source modeling would be to use this zone (outlined in Figure 5) as the seismic source for the region with the occurrence of future events assumed to be uniformly distributed within its boundary. The use of such a model would imply relatively uniform seismic hazard within the region encompassed by the source zone.

A number of faults have been identified within the Intermountain Seismic Belt that have had repeated late Pleistocene and Holocene displacements (Figures 1 and 5). These faults are the most likely locations of future large earthquakes. An alternative source model would be to use the identified faults as the sources of future damaging earthquakes. Use of this model would imply variable seismic hazard within the study region. Figure 6 illustrates the differences in hazard at two points that result from the use of these two source models. Sites close to the Wasatch fault zone, such as Salt Lake City, have a much higher hazard for the fault-specific source model than for the source zone model, whereas a site located at some distance form the nearest fault but still within the source zone, has a significantly lower hazard for the fault-specific model compared to the source zone model.

The source model developed for this study used a combination of the two models described above. The mapped faults are considered to be the likely source of future large earthquakes ($M_L > 6$), but many of the small- to moderate-magnitude events cannot be directly associated with mapped Quaternary faults (Arabasz and others, 1980; Arabasz, 1984). Fault-specific sources were used to represent the mapped active faults that are the likely sources of large magnitude earthquakes ($M_L \geq 6$); and an area source was used to model the background seismicity of smaller-magnitude earthquakes that may be occurring on unknown faults or faults that are not known to be active during the Quaternary. On the fault-specific sources only the occurrence of events of magnitude 6 or larger was modeled. While smaller magnitude events may also occur on the mapped active faults, they do not appear to occur with a frequency discernible from the background rate. Therefore, the background source was used to model the occurrence of all events of magnitude $\leq M_L$ 6.

resulting in a spatially uniform frequency of occurrence for smaller magnitude events over the region. The upper limit on the size of randomly occurring events was assessed to be in the range of $6 \leq M_L \leq 6.5$ (see Figure 2). As shown in Figure 6, use of the combined model results in hazard estimates intermediate between the two limiting cases.

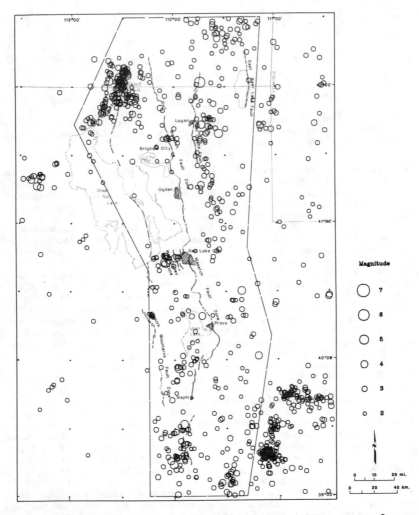

Figure 5 - Instrumental earthquake catalog of independent events of magnitude $M_L \geq 2$ occurring in time period July 1, 1962 to April 1, 1986. Boundary defines limits of background seismic source.

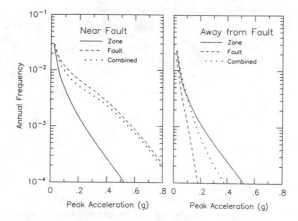

Figure 6 - Comparison of computed hazard using fault-specific sources only with hazard based on background zone source only.

Earthquake Magnitude Distribution: While the exponential magnitude distribution generally fits regional seismicity data, recent studies (e.g. Schwartz and Coppersmith, 1984) have suggested that the magnitude distribution for individual faults and fault segments may be better represented by a "characteristic" size that implies relatively more frequent large events than intermediate size events. The characteristic earthquake model of Schwartz and Coppersmith (1984) was developed on the basis of data from the Wasatch fault zone and the San Andreas fault in California. Studies of these faults indicate a discrepancy between the recurrence rate of large events on the faults based on paleoseismicity data and the rate obtained by extrapolation of an exponential distribution fit to the recorded seismicity that could be reasonably attributed to the faults. For this study, both the exponential magnitude distribution and the characteristic magnitude distribution, the latter as developed by Youngs and Coppersmith (1985a,b), were used to model earthquake recurrence for the fault-specific sources. The characteristic model was favored over the exponential model (probability 0.7 versus 0.3) as comparisons of the geologic and seismicity data from the Wasatch fault zone originally formed the basis for the model.

Figure 7 illustrates the effect of the choice of recurrence model on the computed hazard. At Salt Lake City, where the hazard is largely dominated by the Wasatch fault (see Figure 4), the choice of recurrence model has a significant effect on the hazard. At Logan, where the background source zone is a major contributor to the hazard, the computed hazard is less sensitive to the choice of recurrence model. Only an exponential recurrence model was used for the background source as it represents smaller magnitude regional seismicity.

Fault Segmentation: When individual fault zones are modeled as seismic sources in hazard analyses, they are commonly considered to produce earthquakes at a uniform rate along their length. However, recent studies (Schwartz and Coppersmith, 1984, 1986; Aki ,1979, 1984; Slemmons and Depolo, 1986) have suggested that fault zones often consist of separate segments that may rupture completely and independently

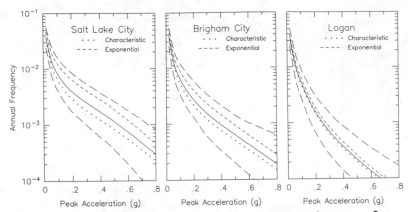

Figure 7 - Contribution to uncertainty in exceedance frequency from uncertainty in recurrence model. The long dashed curves are the 5th and 95th percentiles considering uncertainty in all parameters (see Figure 3) and the conditional mean hazard curves obtained using the two recurrence models are labeled.

during successive earthquakes of characteristic sizes, perhaps at different frequencies. This implies that the individual segments of a fault should be modeled as separate sources in hazard analyses. Fault segmentation not only affects the distribution of ruptures along the fault zone, it also provides constraints on the maximum rupture length and, therefore, on maximum earthquake magnitude, and may also affect recurrence estimates.

Studies by several investigators (Schwartz and Coppersmith, 1984; Mayer and MacLean, 1986; Machette and others, 1986; Personius, 1986) report evidence for distinct segments along the Wasatch fault zone. However, detailed analysis of some historical surface-faulting earthquakes suggests that faulting can rupture across and/or skip over barriers between adjacent segments. Examples of possible multiple segment ruptures include the 10 October 1980 El Asnam, Algeria earthquake (Deschamps and others, 1982; King and Yielding, 1984) and the 28 October 1983 Borah Peak, Idaho earthquake (Crone and Machette, 1984).

Both an unsegmented model and a segmented model were included for each fault. The unsegmented model allows for rupture across the segment boundaries during events and for the simultaneous rupture of two or more adjacent segments. Rupture during successive earthquakes is assumed to be uniformly distributed along the entire length of the fault zone and the maximum length of rupture is not necessarily constrained by boundaries between segments. The segmented model assumes that rupture will not occur across the segment boundaries and that the segments behave independently with the maximum length of rupture on a given segment equal to the segment length. Based on the results of paleoseismic investigations, the segmented fault model is considered to be more representative of fault-rupture behavior of normal slip Basin and Range faults. Accordingly, the unsegmented and segmented conditions for all the faults were assigned weights of 0.2 and 0.8, respectively. For the Wasatch fault zone two segmented fault models were included to account for uncertainty in the actual number of segments.

Figure 8 shows the effect of considering fault segmentation on the computed hazard. As could be expected, those sites where the hazard is dominated by the fault-specific sources exhibit the greatest sensitivity to the treatment of fault segmentation. The variations in computed hazard resulting from the use of a segmented or unsegmented fault zone result primarily from variations in the estimated recurrence rates for the various segments, as will be discussed subsequently. The two alternative segmentation models included in the analysis for the Wasatch fault zone yielded similar hazard estimates.

Maximum Earthquake Magnitude Assessment: Maximum earthquake magnitudes for fault-specific sources are typically assessed by estimating the physical parameters of rupture length and/or rupture area for the maximum size event and then relating these parameters to earthquake magnitude using published correlations for magnitude as a function of rupture size. In addition to these techniques, the paleoseismicity studies along the Wasatch fault zone have provided data on the amount of displacement that occurred during prior events. These data enable estimates to be made of the average displacement that may occur during future large earthquakes. The estimates of average displacement together with rupture area can be used to estimate the seismic moment for the largest events that may occur in the future, and from seismic moment obtain estimates of magnitude. The use of seismic moment to estimate magnitude is more appealing than the use of a single dimension such as rupture length or maximum displacement, because seismic moment incorporates essentially all the physical factors controlling earthquake size - fault rupture area, amount of slip and rigidity of the rock.

Three techniques were used in the analysis to estimate maximum magnitudes for the fault-specific sources. rupture length vs. magnitude (Slemmons, 1982; Bonilla and others, 1984); rupture area vs. magnitude (Wyss, 1979); and seismic moment vs. magnitude (modified from Doser and Smith, 1982 by Youngs and others, in press). As was the case for the selected attenuation relationships, these relationships were developed using magnitude definitions that are equivalent to local magnitude, M_L, in the range of 3 to 7 and surface wave magnitude, M_s, in the range of 5 to 7.5. In general, all three techniques gave estimates within 0.5 magnitude units of each other. For a segmented Wasatch fault zone, the maximum magnitudes for individual segments range from M_s 6-1/2 to 7-1/2. If the fault is unsegmented, somewhat different maximum rupture lengths are expected and the estimated maximum earthquake is in the range of about M_s 7 to 7-3/4. Figure 9 indicates the level of uncertainty in the computed hazard resulting from the range of estimates for maximum magnitude. As can be seen, the uncertainty in maximum magnitude contributes moderately to the uncertainty in the computed hazard.

Earthquake Recurrence: Recurrence estimates for the fault-specific sources were based on geologic data using the following two approaches. Where displacement data for individual past events were available, they were used to estimate the recurrence rate for the largest events. The form of the characteristic or exponential magnitude distribution was used to define the frequency of events down to magnitude 6.0. If only slip-rate data were available for a fault or fault segment, then a moment-rate approach was used with the seismicity rate defined by the relationship given by Anderson (1979) for the exponential model and by the relationship given by Youngs and Coppersmith (1985a,b) for the characteristic model. In these approaches, the recurrence estimates for individual faults are anchored to the mean repeat times of the largest

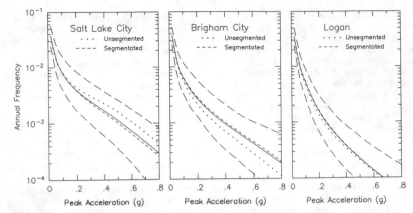

Figure 8 - Contribution to uncertainty in exceedance frequency from uncertainty in source segmentation. The long dashed curves are the 5th and 95th percentiles considering uncertainty in all parameters (see Figure 3) and the conditional mean hazard curves obtained for unsegmented and segmented faults are labeled.

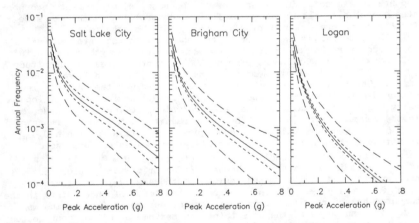

Figure 9 - Contribution to uncertainty in exceedance frequency from uncertainty in maximum magnitude. The long dashed curves are the 5th and 95th percentiles considering uncertainty in all parameters (see Figure 3) and the short dashed curves are the 5th and 95th percentiles considering only uncertainty in maximum magnitude.

events estimated from geological evidence rather than an extrapolation of the observed frequency of small events. The geologically-based estimates of recurrence of large-magnitude events are considered more reliable than the extrapolation of small-magnitude seismicity rates because of the difficulty in attributing instrumentally recorded seismicity to specific faults and the observed mismatch between the geologically-based rates for large events and extrapolation of the historical seismicity rate described above.

Figure 10 compares the cumulative earthquake recurrence for the seismic sources with the observed regional seismicity rate. For events of magnitude $\geq M_L$ 6, the aggregate recurrence estimates for all the fault-specific sources developed on the basis of geologic data are consistent with the observed seismicity rate based on the historical record. For smaller events, the historical seismicity rate was used to define the recurrence relationship for the background source.

Earthquake recurrence estimates for the Wasatch fault were made on the basis of the geologic data gathered in detailed fault trenching studies. The data provide estimates of the Holocene to late Pleistocene slip rate for most of the fault zone on the basis of the cumulative displacement of features ranging in age from 5,000 to 21,000 years B.P. These data together with data on the slip per event at the same locations provide direct estimates of the frequency of large, surface-rupturing earthquakes. The mean repeat times for large events on individual fault segments range from 1,350 to 2,550 years along the more active central portion of the Wasatch fault zone and from 7,000 to 10,000 years along the less active northern and southern ends of the zone. Uncertainty estimates on the age of features, cumulative displacement, and slip per event typically yielded coefficients of variation of 0.25 for the mean repeat time of large events. These uncertainty estimates were included in the analysis (see Figure 2) and, as indicated in Figure 11, represented a significant source of uncertainty in the estimated hazard. For the case when the fault zone was considered to be unsegmented, recurrence estimates were based on a moment-rate approach and a combined estimate for fault slip rate from the data for individual segments. In general, slip rates were used for earthquake recurrence estimates for the other fault-specific sources as displacement-per-event data were usually not available.

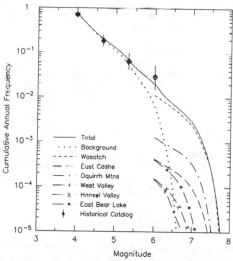

Figure 10 - Comparison of predicted recurrence rates for individual sources and total of all sources with observed frequency of earthquakes.

The variability of the recurrence frequencies for large events
along the central part of the Wasatch fault had only a small effect on
the computed hazards as the rate for individual segments were within
about 30 percent of the average rate for all segments. Those segments
having a shorter than average repeat time for the largest events have a
correspondingly higher hazard than estimated using the average repeat
time and those having a longer than average repeat time have a lower
hazard.

Real-Time Recurrence Estimates: The hazard analyses were conducted
using a memoryless, or Poisson, model for earthquake recurrence, that
is the frequency of events in the near future (e.g. the expected number
of events in the next 50 years) is a function of the long term earthquake recurrence rate without consideration of the elapsed time since
the most recent event. While the Poisson model is almost exclusively
used in seismic hazard analyses in practice, it is at odds with the
generally accepted physical model that earthquake occurrence on a fault
is a process of the gradual accumulation of strain, followed by sudden
release. The physical model implies that for individual faults undergoing relatively constant loading, the repeat time for those events that
release the builtup stress should be more or less cyclic in nature.
However, on a regional basis, the Poisson model for earthquake occurrence has been shown to be appropriate both on the basis of observation
(e.g. Gardner and Knopoff, 1974) and on the basis of theory as the
limiting result of a number of independent or nearly independent processes of whatever type on individual features (e.g. Brillinger, 1982). In
addition, Cornell and Winterstein (1986) have shown that the Poisson
model is conservative as long as the elapsed time since the most recent
renewing event is less than the mean repeat time.

The renewal model provides the simplest approach for making real
time estimates of earthquake occurrence. The renewal model specifies
that, neglecting the occurrence of multiple events, the probability of a
renewing event in the next t years given to years since the last renewing event is given by the expression:

$$P(\text{event in } t|t_o) = [F(t+t_o) - F(t_o)]/[1 - F(t_o)] \qquad (3)$$

where F() is the cumulative density function for interarrival time of
events. Investigators have employed different models for the distribution of interarrival times, such as the normal, lognormal, gamma, and
Weibull distributions. Cornell and Winterstein (1986) indicate that the
Weibull distribution provides the desired characteristics of a simple
analytical form and a hazard function that increases monotonically with
increasing t_o. (The hazard function is a measure of the likelihood of
an event in the next increment of time, dt, given no events in time t_o.)
They give a simple approximation for the Weibull distribution as:

$$F(t) \sim 1 - \exp(-\lambda t^{1/V[T]}) \qquad (4)$$

in which

$$\lambda \sim \{1 - 0.5V[T](1-V[T])\}/E[T] \qquad (5)$$

with E[T] and V[T] being the mean and coefficient of variation of interarrival times, respectively. Equations 4 and 5 are good approximations
for $0 \le V[T] \le 1$ and have the advantage of degenerating to the Poisson model
when V[T]=1 (Cornell and Winterstein, 1986).

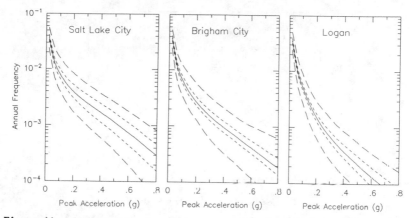

Figure 11 - Contribution to uncertainty in exceedance frequency from uncertainty in recurrence rate. The long dashed curves are the 5th and 95th percentiles considering uncertainty in all parameters and the short dashed curves are the 5th and 95th percentiles considering only uncertainty in recurrence rate.

Equations 3 through 5 were used to obtain real time estimates of the frequency of future large events at the six locations along the Wasatch fault zone for which data are available for time since the most recent event (Schwartz and others, 1984). The coefficient of variation of interarrival time, V[T], was assumed to be 0.7, reflecting the large variations observed in the limited data for interarrival times for large earthquakes on the Wasatch fault zone. At all locations the estimated elapsed times since the most recent event were less than the estimated mean return periods resulting in a range of 0.67 to 0.9 for the ratio of the estimated number of renewing events during the next 50 years to that given by the Poisson model. These ratios were assumed to apply only to the largest or characteristic events on the appropriate fault segments. Smaller events are not likely to result in substantial stress release and were considered to occur randomly (Poissonian) in time. The effect of the real time estimates of the near future rate of large events on the computed hazard along the Wasatch fault zone is shown in Figure 12. The effect is a minor decrease in hazard, with the largest decrease at long return periods where the hazard is dominated by contributions from the largest events.

SUMMARY

The example presented herein indicates how the result of detailed geological studies can impact the regional assessment of seismic hazards. As indicated, the sensitivity of the hazard at any one site to various interpretations of the data may vary depending on the controlling contributions to the hazard and the effects of the various interpretations on the assessment of the basic parameters of earthquake location and frequency. The geological factors that had the greatest impact on the computed hazard along the Wasatch fault zone were the source model (faults vs. zones), the fault-specific magnitude distribution (characteristic vs. exponential), and the frequency of large magni-

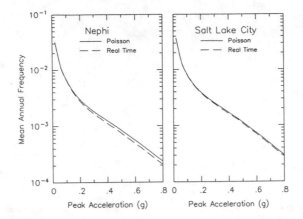

Peak Acceleration (g)   Peak Acceleration (g)

Figure 12 - Effect of use of real time estimates of earthquake occurrence probabilities on computed hazard at two locations along the Wasatch fault zone.

tude earthquakes. It is important that the uncertainties in interpreting the data be considered fully in the analysis in order that the uncertainty in the estimated hazard be expressed.

ACKNOWLEDGMENTS
The work presented in this paper was sponsored by the U.S. Geological Survey under contracts 14-08-0001-21914 and 14-08-0001-G1189. The authors wish to acknowledge David P. Schwartz of the U.S. Geological Survey and Robert K. Green of Woodward-Clyde Consultants for their contributions to the seismic hazard analysis.

APPENDIX-REFERENCES
Aki, K., 1979, Characterization of barriers on an earthquake fault: Journal of Geophysical Research, v. 84, pp. 6140-5148.
Aki, K., 1984, Asperities, barriers, and characteristic earthquakes: Journal of Geophysical Research, v. 89, n. B7, pp. 5867-5872.
Anderson, J.G., 1979, Estimating the seismicity from geological structure for seismic risk studies: Bulletin of the Seismological Society of America, v. 69, n. 1, pp. 135-158.
Arabasz, W.J., 1984, Earthquake behavior in the Wasatch front area-association with geologic structure, space-time occurrence, and stress state: in Proceedings of Conference XXVI - A Workshop on "Evaluation of Regional and Urban Earthquake Hazards and Risk in Utah," US Geological Survey Open-File Report 84-763, pp. 310-339.
Arabasz, W.J., R.B. Smith, and W.D. Richins, 1980, Earthquake studies along the Wasatch Front, Utah - network monitoring, seismicity, and seismic hazards: Bulletin of the Seismological Society of America, v. 70, n. 5, pp. 1479-1499.
Bonilla, M.G., R.K. Mark, and J.J. Lienkaemper, 1984, Statistical relations among earthquake magnitude, surface rupture length, and surface fault displacement: Bulletin of the Seismological Society of America, v. 74, n. 6, pp. 2379-2411.
Brillinger, D.R., 1982, Some bounds for seismic risk: Bulletin of the Seismological Society of America, v. 72, n. 4, pp. 1403-1410.
Campbell, K.W., 1983, Bayesian analysis of extreme earthquake occurrences, Part II, application to the San Jacinto fault zone of Southern California: Bulletin of the Seismological Society of America, v. 73, n. 4, pp. 1099-1115.
Campbell, K.W., 1987, Predicting strong ground motion in Utah: Assessment of Regional Earthquake hazards and Risk Along the Wasatch Front, Utah, U.S. Geological Survey Open File Report 87-585, v. II, pp. L-1-90.

Coppersmith, K.C. and R.R. Youngs, 1986, Capturing uncertainty in probabilistic seismic hazard assessments within intraplate environments: in Proceedings of the 3rd National Conference on Earthquake Engineering, Charleston, South Carolina, August 24-28, v. I, pp. 301-312.

Cornell, C.A., 1968, Engineering seismic risk analysis: Bulletin of the Seismological Society of America, v. 58, n. 5, pp. 1583-1606.

Cornell, C.A. and S.R. Winterstein, 1986, Applicability of the Poisson earthquake-occurrence model: EPRI NP-4770 Research Project P101-38 Final Report, August, see also The seismic hazard implications of non-Poissonian recurrence models (abs.): Earthquake Notes, v. 57, n. 1, p. 25.

Crone, A.J., and M.N. Machette, 1984, Surface faulting accompanying the Borah Peak earthquake, central Idaho: Geology, v. 12, pp. 664-667.

Deschamps, A., Y. Guademer, and A. Cisternas, 1982, The El Asnam, Algeria earthquake of 10 October 1980 - Multiple-source mechanism determined from long-period records: Bulletin Seismological Society of America, v. 72, n. 4, pp. 1111-1128.

Doser, D.I. and R.B. Smith, 1982, Seismic moment rates in the Utah region: Bulletin of the Seismological Society of America, v. 72, n. 2, pp. 525-551.

EPRI, 1987, Seismic hazard methodology for the central and eastern United States - Volume 1: Methodology: Report NP-4726, Volume 1, prepared for Seismicity Owners Group and Electric Power Research Institute under research projects P101-38, -45, -46, 2256-14, Revised, February, 1987.

Gardner, J.K., and L. Knopoff, 1964, Is the sequence of earthquakes in Southern California, with aftershocks removed, Poissonian: Bulletin of the Seismological Society of America, v. 64, n. 5, pp. 1363-1367.

Greensfelder, R.W., R.C. Kintzer, and M.R. Somerville, 1980, Seismotectonic regionalization of the Great Basin, and comparison of moment rates computed from Holocene strain and historic seismicity: summary: Bulletin of the Geological Society of America, v. 97, pp. 518-523.

Hanson, K.L., F.H. Swan, and D.P. Schwartz, 1981, Study of earthquake recurrence intervals on the Wasatch fault, Utah: Fifth semiannual technical report prepared for the U.S. Geological Survey under Contract No. 14-08-0001-19115, 15 pp.

Joyner, W.B., and D.M. Boore, 1982, Prediction of earthquake response spectra: US. Geological Survey Open File Report 82-977.

Joyner, W.B., and T.E. Fumal, 1985, Predictive mapping of earthquake ground motion, in Evaluating earthquake hazards in the Los Angeles region: U.S. Geological Survey Professional Paper 1360, pp 203-220.

King, G, and G. Yielding, 1984, The evolution of a thrust fault system - Process of rupture initiation, propagation and termination in the El Asnam (Algeria) earthquake: Geophysical Journal Royal Astronomical Society, v. 77, pp. 913-933.

Kulkarni, R.B., R.R. Youngs, and K.J. Coppersmith, 1984, Assessment of confidence intervals for results of seismic hazard analysis: in Proceedings of the Eight World Conference on Earthquake Engineering, San Francisco, California, v. 1, pp. 263-270.

Machette, M.N., S.F. Personius, W.E. Scott, and A.R. Nelson, 1986, Quaternary geology for ten fault segments and large-scale changes in slip rate along the Wasatch fault zone: Paper presented at Workshop on "Earthquake Hazards Along the Wasatch Front, Utah," Salt Lake City, Utah, July 14-18, 1986.

Mayer, L., and A. MacLean, 1986, Tectonic geomorphology of the Wasatch Front, Utah, using morphologic discriminant analysis - preliminary implications for Quaternary segmentation of the Wasatch fault zone: Geological Society of America, Abstracts with Programs, v. 18, n. 2, pp. 155.

Molnar, P., 1979, Earthquake recurrence intervals and plate tectonics: Bulletin of the Seismological Society of America, v. 69, n. 1, pp. 115-133.

Personius, S.F., 1986, The Brigham City segment - A new segment of the Wasatch fault zone, northern Utah: Geological Society of America, Abstract with Programs, v. 18, n. 5, p. 402.

Power, M.S., K.J. Coppersmith, R.R. Youngs, D.P. Schwartz, and F.H. Swan III, 1981, Seismic exposure analysis for the WNP-2 and WNP-1/4 site: Appendix 2.5K to Amendment No. 18 Final Safety Analysis Report WNP-2, for Washington Public Power Supply System, Richland Washington, September.

Sadigh, K., J.A. Egan and R.R. Youngs, 1986, Specification of ground motion for seismic design of long period structures: Earthquake Notes, v. 57, n. 1, p. 13.

Schwartz, D.P., and K.J. Coppersmith, 1984, Fault behavior and characteristic earthquakes from the Wasatch and San Andreas faults: Journal of Geophysical Research, v. 89, n. B7, pp. 5681-5698.

Schwartz, D.P. and K.J. Coppersmith, 1986, Seismic hazards: new trends in analysis using geologic data: in Active Tectonics, Academic Press, Washington D.C., pp. 215-230.

Schwartz, D.P., F.H. Swan, and L.S. Cluff, 1984, Fault behavior and earthquake recurrence along the Wasatch fault zone: in Proceedings of Conference XXVI - A Workshop on "Evaluation of Regional and Urban Earthquake Hazards and Risk in Utah," US Geological Survey Open-File Report 84-763, pp. 113-125.

Slemmons, D.B., 1982, Determination of design earthquake magnitudes for microzonation: in Proceedings of the Third International Earthquake Microzonation Conference, v. 1., pp 119-130.

Slemmons, D.B., and G.M. Depolo, 1986, Evaluation of active faulting and associated hazards: in National Research Council, Studies in Geophysics - Active Tectonics, Academic Press, Washington, D.C., pp. 45-62.

Swan, F.H., III, D.P. Schwartz, and L.S. Cluff, 1980, Recurrence of moderate to large magnitude earthquakes produced by surface faulting on the Wasatch fault, Utah: Bulletin of the Seismological Society of America, v. 70, n. 5, pp. 1431-1462.

U.S. Geological Survey, 1980, Earthquake hazards along the Wasatch and Sierra-Nevada frontal fault zones: Proceedings of Conference X, National Earthquake Hazards Reduction Program, 29 July 1979 - 1 August 1979, U.S. Geological Survey Open-File Report 80-801, 679 pp.

U.S. Geological Survey, 1984, A workshop on "Evaluation of regional and urban earthquake hazards and risk in Utah": Proceedings of Conference XXVI, National Earthquake Hazards Reduction Program, U.S. Geological Survey Open-File Report 84-763, 674 pp.

Utah Geological Association, 1972, Environmental Geology of the Wasatch Front, 1971: Utah Geological Association Publication 1, Salt Lake City, Utah.

Wesnousky, S.G., 1986, Earthquakes, Quaternary faults, and seismic hazards in California: Journal of Geophysical Research, v. 91, n. B12, pp. 12,587-12,632.

Wyss, M., 1979, Estimating maximum expectable magnitude of earthquakes from fault dimensions: Geology, v. 7, n. 7, pp. 336-340.

Youngs, R.R. and K.J. Coppersmith, 1985a, Development of a fault-specific recurrence model: Earthquake Notes (abs.), v. 56, n. 1, p. 16.

Youngs, R.R. and K.J. Coppersmith, 1985b, Implications of fault slip rates and earthquake recurrence models to probabilistic seismic hazard estimates: Bulletin of the Seismological Society of America, v. 75, n. 4, pp. 939-964.

Youngs, R.R., K.J. Coppersmith, M.S. Power and F.H. Swan III, 1985, Seismic hazard assessment of the Hanford region, eastern Washington State: in Proceedings of the DOE Natural Phenomena Hazards Mitigation Conference, Las Vegas, Nevada, October 7-11, pp. 169-176.

Youngs, R.R., F.H. Swan, M.S. Power, D.P. Schwartz, and R.K. Green, in press, Probabilistic Analysis of Earthquake Ground Shaking Hazard Along the Wasatch Front, Utah: U.S. Geological Survey Professional Paper, preprinted in Assessment of Regional Earthquake hazards and Risk Along the Wasatch Front, Utah, U.S. Geological Survey Open File Report 87-585, v. II, pp. M-1-110.

Site Effects at McGee Creek, California

Sandra H. Seale[1] and Ralph J. Archuleta[2]

Abstract

At the McGee Creek, California, site, 3-component strong-motion accelerometers are located at depths of 166 m, 35.0 m and 0.0 m. The surface material is glacial moraine, to a depth of 30.5m, overlying hornfels. Accelerations were recorded from the Round Valley, California, event of November, 1984 M_L 5.8. By separating out the SH components of acceleration, we were able to determine the orientations of the downhole instruments. By separating out the SV component of acceleration, we were able to determine the approximate angle of incidence of the signal at 166 m. A constant phase velocity Haskell-Thomson model was applied to generate synthetic SH seismograms at the surface using the accelerations recorded at 166m. In the frequency band 0.0 - 10.0 Hz, we compared the filtered synthetic records to the filtered surface data. The onset of the SH pulse is clearly seen, as are the reflections from the interface at 30.5 m. The synthetic record closely matches the data in amplitude and phase. The fit between the synthetic accelerogram and the data shows that the seismic amplification at the surface is a result of the contrast of the impedances (shear stiffnesses) of the near surface materials.

Introduction

This work addresses the problem of whether the attenuation in near-surface material is more than compensated by the amplification induced by the lower impedance of near-surface material and the expected resonance effects of an equivalent layer or layers on a halfspace. The importance of this issue was dramatically emphasized in the events of March 3, 1985 Valparaiso, Chile, M_s 8.1, [*Celebi, 1987*] and the September 19, 1985, Michoacan, Mexico, M_s 8.1, [*Anderson et al., 1986; Celebi et al., 1987*]. In both earthquakes the ground motion was amplified by the local near-surface site conditions. This effect is not limited to very large earthquakes. Similar effects are seen for the May 2, 1983 Coalinga, M_L 6.5 [*Mueller, 1986*] and April 24, 1984, Morgan Hill, M_L 6.1 [*Celebi, 1987*], California earthquakes. In fact, a major conclusion from the 1985 Valparaiso earthquake is that the weak ground motion could be used to identify the frequency ranges for which amplification occurs during the strong ground motion [*Celebi, 1987*]. The problem of amplification/attenuation due to local site conditions has been reviewed by both the seismological and engineering communities [*Seed and Idriss, 1970; Joyner et. al., 1976; Mueller, 1986; Celebi et. al., 1987*]. Given that so many metropolitan areas are built on alluvial fans or basins, this problem is of major importance to any organization concerned with earthquake hazards.

This work is almost unique in addressing the problem of local site amplification/attenuation in the United States (Japan has a number of arrays of downhole and surface accelerometers that have recorded some of their larger earthquakes [*Chen, 1985*]). This experiment is the only operational one in a seismically active area in which strong ground motion is being recorded. With the exception of the Richmond Field Station operated by the University of California, Berkeley, which is distant from most earthquakes, this experiment is the first one that attempts to directly compare the difference between strong ground motion in the basement to ground motion at the surface. *Seed and Idriss , 1970,* compared weak motion at various depths in a soil strata; however, they had no instruments placed in bedrock. *Joyner et. al., 1976,* analyzed weak motion from a downhole array near San Francisco. A more detailed discussion of these works is presented below.

This work has provided direct evidence for the comparison of weak motion to strong motion; that is, we can test in a limited way the range over which linear attenuation of seismic waves is valid. These *in situ* data directly demonstrate the effects of near-surface material on ground motion. The results of our analysis have important implications for nonlinear attenuation, microzonation using weak

[1]Assistant Research Geophysicist, Institute for Crustal Studies, University of California, Santa Barbara, CA 93106

[2]Associate Professor of Seismology, Department of Geological Sciences, University of California, Santa Barbara, CA 93106

ground motion, resonances, and our general understanding of attenuation and amplification due to local site conditions.

Data

The McGee Creek site (Figure 1) is located in the Mammoth Lakes area of California. We installed 3-component accelerometers and velocity transducers at McGee Creek [*Archuleta, 1986*] in November, 1984. These instruments are at depths of 166 m, 35 m, and 0 m (Figure 2). (At this time the vertical velocity transducer at 35 m is the only velocity transducer operating correctly at depth. All of the accelerometers are currently operating.) The largest earthquakes recorded at McGee Creek are the July 21, 1986, M_L 6.5 Chalfant Valley (37° 32.52' N, 118° 26.84' W, depth 11.5 km) [*Maley et. al., 1986; Cockerham and Corbett, 1987*], and the November 23, 1984, M_L 5.8 Round Valley (37° 27.30' N, 118° 36.18' W, depth, 13.4 km) [*Priestly and Smith, 1987*]. We have recently received the acceleration records of the Chalfant Valley event and are at work on the analysis of these data. Peak accelerations from the Chalfant Valley earthquake are around 100 cm/s^2 at the surface with considerably smaller amplitudes (about a factor of 5) at both 35 m and 166 m [*Maley et. al., 1986*]. The Round Valley mainshock and aftershock produced similar accelerations with a peak acceleration at the surface of around 120 cm/s^2 and an amplification factor of 4 from downhole to surface. Results from the analysis of the data from the Round Valley earthquake (Figure 3) are presented here.

Analysis

Our primary goal in this analysis was to take the accelerations recorded at 166 m, apply our knowledge of the material properties, and use linear wave propagation methods to produce a synthetic accelerogram at the surface for comparison with the recorded data. The first step was to determine the orientations of the downhole horizontal components. Based on the locations of the Round Valley earthquakes [*Robert Cockerham, USGS, personal communication*], we determined the angles of arrival for McGee Creek. By systematically rotating the horizontal components to maximize SH motion, we determined the orientations of the downhole components. In Figure 4 we show the direction of arrival between the mainshock and the aftershock. We also show the direction of the "North" component at 166 m, 225 ° measured clockwise from North, and 201° for 35 m. (Because we did not have any known orientations for the downhole horizontals, the components were nominally assigned directions of "North" and "West.") The "West" directions are now 315° at 166 m and 291° at 35 m. Using these directions, we rotated the horizontal components into SH (tangential) and SV (radial) components. Figure 5a shows the SH accelerations at 35 m and at 166 m based on our determined orientations of the horizontal components. It is quite clear that we have good phase agreement between the two records. In fact, after correcting for the velocity of the hornfels, these two records have a correlation of 89% for 1.5 sec after the onset of the SH motion. Figure 5b compares the surface SH acceleration with the SH acceleration at 166 m at the same scale. The amplification is obvious.

Based on the material properties and the propagation of shear waves in a layered medium, we took the 166 m record and produced a synthetic accelerogram for comparison with the surface record. Knowing that we have the equivalent of two layers over a halfspace [*Fumal et. al., 1985*], we expected that resonances would exist. Figure 6 shows the spectral ratio of the surface SH accelerogram to the 166 m SH accelerogram. There are many resonant spikes at frequencies above 10 Hz; the two prominent spikes around 3 and 6 Hz are explicable from simple theory given the material parameters of the two layers and the halfspace. The lowest peak represents the resonance of the two surface layers together, the next peak is the resonance of the upper layer.

In our first attempt to model this spectral ratio we used the original material properties and the Haskell-Thomson propagation matrix for a vertically propagating SH wave (horizontal wavenumber k = 0). This matrix relates the anti-plane shear stress and displacement at the bottom of the layer to the stress and displacement at the top of the layer:

$$\left\{ \begin{array}{c} v_2 \\ {}^{\tau_2}\!/_{\omega} \end{array} \right\} = \left[\begin{array}{cc} \cos \eta & {}^{-1}\!/_{\rho C_s} \sin \eta \\ \rho C_s \sin \eta & \cos \eta \end{array} \right] \left\{ \begin{array}{c} v_1 \\ {}^{\tau_1}\!/_{\omega} \end{array} \right\} \qquad \eta = \omega \, {}^h\!/_{C_s} \qquad (1)$$

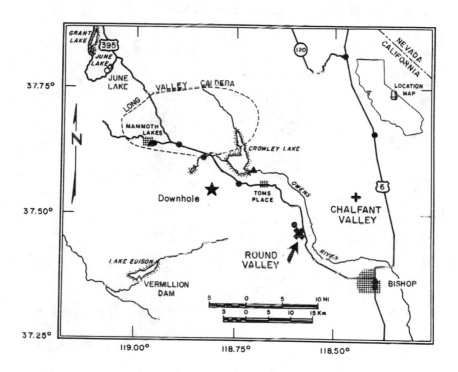

Figure 1. The location of the downhole in the Mammoth Lakes region of California.

McGEE CREEK RECORDING SITE

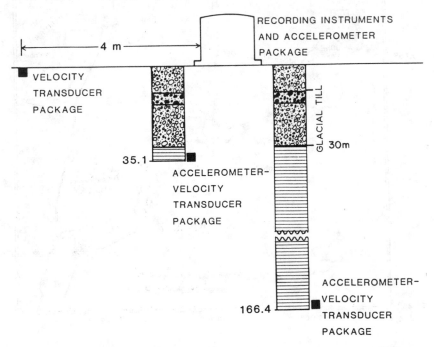

Figure 2. The instrumentation and geology at McGee Creek. The first 30.5 m are glacial till. At depths from 30.5 m to 166.4 m the medium is hornfels. The instruments are located at 0 m, 35 m, and 166 m.

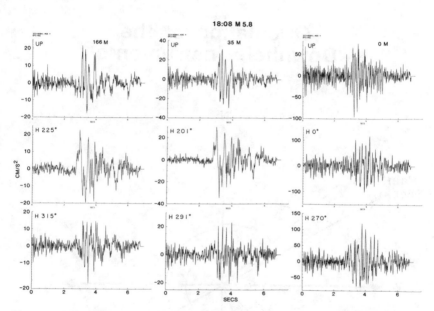

Figure 3. The time histories for three components of acceleration recorded at 166 m, 35 m and 0 m depths at McGee Creek from the M_L 5.8 November 23, 1984 Round Valley, California, earthquake. For each depth the components are arranged with the vertical component at the top and the two horizontals beneath.

Orientation of the
Downhole Instruments

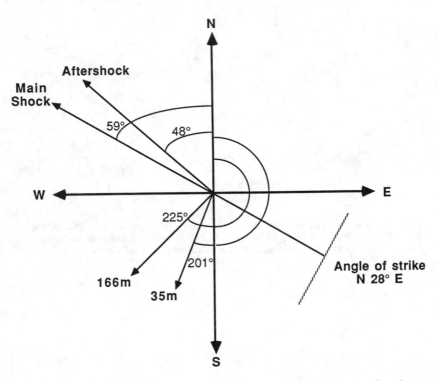

Figure 4. A schematic diagram showing in plan view the radial direction of seismic waves from the Round Valley mainshock (59° counterclockwise from North) and aftershock (48°), the strike of the mainshock, and the orientation of the "North" horizontal component at 166 m and 35 m.

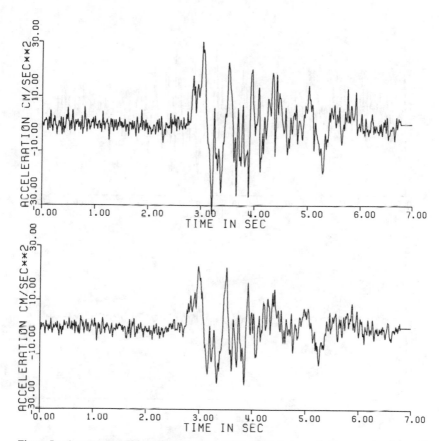

Figure 5a. A comparison between the SH acceleration time histories at 166 m (bottom trace) and at 35 m (top trace), plotted at the same scale. Based on the calculated orientations of the downhole horizontal components, the accelerograms were rotated into SH, horizontally polarized shear motion.

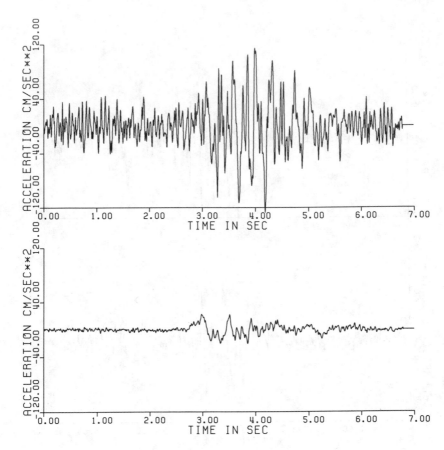

Figure 5b. A comparison between the SH acceleration time histories at 166 m (bottom trace) and at the surface (top trace), plotted at the same scale.

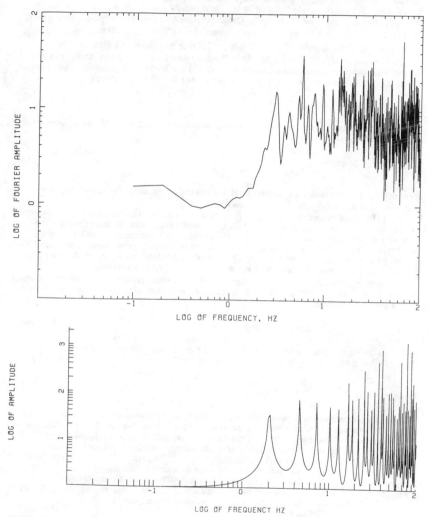

Figure 6. The spectral ratio of the SH acceleration at the surface to the SH acceleration at 166 m. Note the resonant peaks, particularly at about 3 and 6 Hz. Also note that the spectrum is fairly level for all frequencies greater than 20 Hz.

Figure 7. The spectral ratio of the surface to 166 m, computed using the Haskell-Thomson propagator matrices for vertically incident SH waves with the original material properties and no attenuation. Compared to the data, Figure 6, this spectral ratio is far too spiky, with large amplitudes and resonant peaks located at the wrong frequencies.

where v is the displacement, τ is the shear stress, ω is the frequency of the wave, h is the layer thickness and C_S is the shear velocity of the layer. The subscripts 1 and 2 indicate the state variables at the top and bottom of the layer, respectively. Our model consists of three layers: glacial till from 0 m to 14 m (C_S= 330 m/s, ρ = 2.0 g/cm^3), glacial till from 14 m to 30 m (C_S = 620 m/s, ρ = 2.1 g/cm^3), and hornfels from 30 m to 166 m (C_S = 1320 m/s, ρ = 2.5 g/cm^3). The velocity data were obtained from logs [Fumal et. al., 1985, Archuleta, 1986]. When the three layer matrices are multiplied together, we obtain the anti-plane shear stress and displacement at 166 m in terms of the stress and displacement at 0 m. Since the stress at the surface is zero, we can directly relate the displacements.

$$v_{166} = K\, v_0 \tag{2}$$

The amplitude spectrum is obtained by computing the absolute value of the ratio of the displacements as a function of frequency.

$$\left| v_0 \right| \Big/ \left| v_{166} \right| = 1 \Big/ \left| K \right| \tag{3}$$

(Note that the spectral ratio for displacements and accelerations are the same, since differentiation by time twice introduces $-\omega^2$ on both sides of equation (2).)

In the resulting spectral ratio (Figure 7) the resonant peaks occur at frequencies lower than in the data. We determined that the shear wave velocity in the halfspace (hornfels) was incorrect. The value of 1320 m/s, determined from logs, was much too small. A value consistent with the spectral ratio between 166 m and 35 m is 2800 m/s. Cross-correlation of the SH pulses between 166 m and 35 m.confirms a shear wave speed of 2800 m/s. The original shear velocity gave a Poisson's ratio $v = 0.45$; with $C_S = 2800$ m/s, $v = 0.28$, which is a more reasonable value for hornfels.

In the spectral ratio (Figure 7), the amplitudes of the resonant peaks are much too large, thus requiring the addition of damping to the model in the upper two layers. Hysteretic damping in the form of a complex shear modulus was added:

$$\mu = \mu\,(\,1 + 2\iota\beta\,) \tag{4}$$

where μ is the shear modulus of the layer and β is the damping ratio. A damping ratio of $\beta = .025$ (2.5% of critical) gave the best fit to the data. Quality factor Q is inversely proportional to damping.

$$Q = 1 \Big/ 2\beta \tag{5}$$

Our next step was to improve our model by considering SH waves with incidence angles other than 0°. By systematically rotating the components in the vertical plane to maximize SV motion, we determined an angle of incidence of 56°, measured from the vertical. Because the Round Valley earthquake and the McGee Creek site are both within the Sierra granitic batholith, we assumed a straight line ray path. The straight line ray path incidence angle is 58°. The angle of 56°, compared to 58°, is further confirmation of a nearly uniform medium between the source and the receiver at 166 m. The angle of 56° gives a horizontal phase velocity of 3377 m/s. (At this phase, the incidence angle in the top layer is 5°. The impedance contrast of the materials is such that the SH incidence is nearly vertical at the surface.) The Haskell-Thomson matrix for a wave with constant phase is

$$\begin{Bmatrix} v_2 \\ \tau_2/k \end{Bmatrix} = \begin{bmatrix} \cosh ksh & -1/\mu s \sinh ksh \\ -\mu s \sinh ksh & \cosh ksh \end{bmatrix} \begin{Bmatrix} v_1 \\ \tau_1/k \end{Bmatrix} \qquad s = \sqrt{1 - \left(\frac{\omega}{k\,C_s}\right)^2} \tag{6}$$

where here k is the horizontal wavenumber.

Again we multiplied the three layer matrices to obtain displacements at 166 m in terms of displacements at 0 m. In order to compute synthetics at the surface, we first performed a Fast Fourier Transform (FFT) on the acceleration record at 166 m. We had approximately seven seconds of data with a sampling rate of 200 samples/second. The data were padded with zeros out to 10.24 sec, so that the resulting transform was computed out to 100 Hz (see Figure 7). The transformed record at 166 m is multiplied by the Haskell-Thomson coefficient to produce a record at 0 m in the frequency domain. We then apply an inverse FFT to this record, which produces a synthetic accelerogram at the surface.

The original acceleration record and the synthetic were lowpass filtered to 10 Hz. The material parameters of the model were adjusted to obtain the best fit to the data. We found that the upper 14 m should have a shear velocity of about 290 m/s, as compared to 330 m/s. We also found that a β of 5% in the upper two layers was necessary to obtain the correct amplitudes. This means that the quality factor Q is about 10 for the surface material (glacial till). Our best estimates of the material parameters at McGee Creek are shown in Figure 8.

In Figure 9 we plot the synthetic SH acceleration time history (dashed line) with the actual data (solid line), both lowpass filtered to 10 Hz. Although frequencies above 10 Hz have been filtered from the data, the acceleration amplitudes are still about 95% of the unfiltered data. The agreement in amplitude, phase and duration is quite remarkable, especially considering that we are modeling the data up to 10 Hz with a single phase. We think that the extra beat that appears in the data is from a wave with a different phase.

Figure 10 shows the spectral ratio of the computed surface accelerogram to the data at 166 m. Comparison of this spectral ratio with the data (Figure 7) shows an immediate discrepancy. The spectral ratio in Figure 10 shows a pronounced attenuation of the spectral amplitudes for frequencies above 15 Hz, much more severely than was seen in the data. Yet for frequencies less than 15 Hz, the fit between the two is very good in the spectral amplitudes of the main resonant peaks (3 and 6 Hz) and in their location, i.e., the peaks occur at the same frequency in the data as in the synthetic. This would be expected based on the very good fit in the time domain of the surface synthetic to data. The discrepancy for frequencies greater than 15 to 20 Hz suggests that a constant Q may not be appropriate for all frequencies.

Discussion

This work can be compared to that of Seed and Idriss, 1970 and Joyner et. al., 1976. Seed and Idriss used a lumped-mass, variable damping model for the soil profile at Union Bay, Seattle. They applied recorded motions in glacial till to compute motions in the upper layers of clay and peat. Their solution was obtained iteratively to account for the nonlinear behavior of clay and peat. The synthetic seismograms in the peat were a good match, particularly for the long period motion. Joyner et. al. applied modeling techniques similar to ours to downhole data from the San Francisco Bay. They found very good agreement between their model predictions and the weak motion data at low frequencies (< 5 Hz). Joyner et. al. anticipated that the linear plane-layer model would be insufficient to predict strong motion behavior.

Our results show that the linear model does capture the amplification of strong motion in the surface layers (up to 10 Hz). The value of damping that we used - 5% - is reasonable for small strains in the glacial till. Clearly the amplification of accelerations in the surface is due to the impedance contrast of the surface material to the hornfels.

The issue of a frequency-dependent Q for higher frequencies needs to be investigated. The constant Q of our model introduces a trend in the spectral ratio above 15 Hz that does not exist in the data. This implies that the attenuation mechanism in the material may depend on frequency; however, this dependence will decrease the attenuation for higher frequencies.

This research is supported by the Office of Nuclear Regulatory Research, U.S. Nuclear Regulatory Commission Grant No. NRC-G-04-85-004.

Material Properties

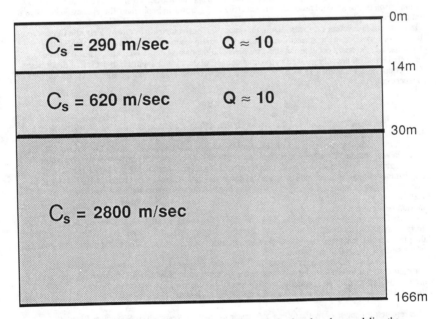

Figure 8. The material parameters that are most consistent with the data, based on modeling the observed time history and the spectral ratio. The major changes are the shear wave velocity of 2800 m/s in the hornfels compared with 1320 m/s; 290 m/s in the upper 14 m compared to 330 m/s and $Q \approx 10$, which was undetermined.

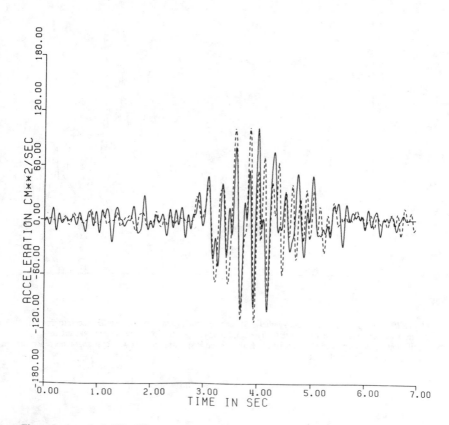

Figure 9. A synthetic SH accelerogram (dashed line), computed for frequencies up to 10 Hz using the Haskell-Thomson matrices to propagate the SH acceleration at 166 m to the surface. The material parameters shown in Figure 8 and an incidence angle of 56° at 166 m depth are used in the constant phase model. The actual SH accelerogram (solid line), recorded at the surface, is lowpass filtered to 10 Hz for comparison. The similarity in peak amplitudes, phase and duration is obvious. The data still show a reverberation that is slightly different from the synthetic.

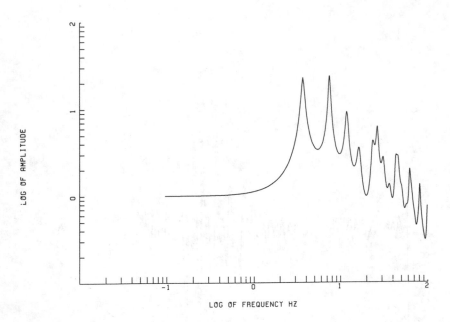

LOG OF FREQUENCY HZ

Figure 10. The spectral ratio of the synthetic SH accelerogram (0 m depth) to the SH accelerogram at 166 m. Compare with the data, Figure 6. The amplitude and position of the resonant peaks are in excellent agreement up to 15 Hz. Above 15 Hz the synthetic spectral ratio shows a systematic decrease due to Q not observed in the data.

References

Anderson, J. G., P. Bodin, J. N. Brune, J. Prince, S. K. Singh, R. Quaas and M. Onate (1986). Strong ground motion from the Michoacan, Mexico, earthquake, *Science*, 233, 1043-1049.

Archuleta, R. J. (1986). Downhole recordings of seismic radiation, *Earthquake Source Mechanics, Geophys. Monograph 37 (Maurice Ewing 6)*, S. Das, J. Boatwright and C. Scholz, Eds., American Geophysical Union, Washington, D. C., 319-329.

Celebi, M.(1987). Distant resonance effects of the earthquake - a preliminary study, *The Morgan Hill, California, Earthquake of April 24, 1984*, S. W. Hoose, Ed., U. S. Geological Survey Bulletin 1639, Washington, D. C., 105-110.

Celebi, M. (1987). Topographical and geological amplifications determined from the strong-motion and aftershock records of the 3 March 1985 Chile earthquake, *Bull. Seismol.Soc. Amer.*, 77, 1147-1167.

Celebi, M., J. Prince, C. Dietel, M. Onate and G. Chavez (1987). The culprit in Mexico City -- amplification of motions, *Earthquake Spectra*, 3, 315-328.

Chen, P. C. (1985). Vertical distribution of peak subsurface horizontal earthquake accelerations, *Report to the National Science Foundation*.

Cockerham, R. S. and E. J. Corbett (1987). The July 1986 Chalfant Valley, California, earthquake sequence: preliminary results, *Bull. Seismol. Soc. Amer.*, 77, 280-289.

Fumal, T. E., R. E. Warrick, E. C, Etheridge and R. J. Archuleta (1985). Downhole geology, seismic velocity structure and instrumentation at the McGee Creek, California, recording site, *Earthquake Notes*, 55, 5.

Joyner, W. B., R. E. Warrick and A. A. Oliver, III (1976). Analysis of seismograms from a downhole array in sediments near San Francisco Bay, *Bull. Seismol. Soc. Amer.*, 66, 937 - 958.

Maley, R. P., E. C. Etheridge and A. Acosta (1986). U. S. Geological Survey strong-motion records from the Chalfant Valley, California, earthquake of July 21, 1986, *Open File Report 86-568*, U. S. Geological Survey, Menlo Park, California.

Mueller, C. S. (1986). The influence of site conditions on near-source high-frequency ground motion: case studies from earthquakes in Imperial Valley, Ca., Coalinga, Ca., and Miramichi, Canada, *Ph.D. Thesis*, Stanford University, Stanford, California.

Priestly, K. F. and Smith, K. D, (1987). The 1984 Round Valley, California earthquake sequence, submitted to *Jour. Geophys. Res.*

Seed, H. B. and I. M. Idriss (1970). Analysis of ground motions at Union Bay, Seattle, during earthquakes and distant nuclear blasts, *Bull. Seismol. Soc. Amer.*, 60, 125 - 136.

SITE-SPECIFIC ESTIMATION OF SPATIAL INCOHERENCE
OF STRONG GROUND MOTION

Paul G. Somerville, James P. McLaren, Chandan K. Saikia[1],
and Donald V. Helmberger[2]

ABSTRACT

The large spatial variability that is observed in strong ground
motion has significant effects on the seismic response of structures
such as dams and bridges that are supported on large foundations. We
have formulated and tested a site-specific procedure for the
estimation of spatial incoherence of strong ground motions. The
procedure entails the estimation and combination of two contributions
to spatial incoherence. One contribution represents the effect of a
nearby extended source, which causes interference between simultaneous
arrivals from different regions of the fault. It is estimated from
closely-spaced simulated acceleration time histories of design basis
earthquakes. The other contribution represents complex aspects of
wave propagation and local site response that are not modeled in the
simulations. It is measured from closely-spaced recordings at the
site of a small earthquake or explosion. The procedure has been
tested by showing that the incoherence measured from Differential
Array recordings of an aftershock of the 1979 Imperial Valley
earthquake, when combined with the incoherence measured from
closely-spaced strong motion simulations of the mainshock, gives an
incoherence function that is consistent with that derived from the
mainshock accelerograms recorded at the Differential Array. The
simulated accelerograms of the 1979 Imperial Valley mainshock show
good agreement with the recorded accelerograms, thereby validating the
procedure used to simulate the mainshock recordings.

INTRODUCTION

The analysis of soil-structure interaction for the seismic design
of critical structures requires a realistic description of the wave
field that is incident at the foundation of the structure. An
important aspect of this wave field is its spatial variability from
one location on the foundation to another. If the structure is
situated on a rigid foundation, then the effect of spatial incoherence
is generally to reduce translational components of motion within the
structure (in comparison with those induced by coherent input motion),
and to induce torsion and rocking. If the structure is supported on
multiple foundations (as in a bridge) or has a very large foundation
(as in an arch dam), then the effect of spatial incoherence may be to
increase translational as well as rotational motions of the structure.

[1]Woodward-Clyde Consultants, 566 El Dorado St., Pasadena, CA 91101
[2]Prof., Seismological Lab., Calif. Inst. of Tech., Pasadena, CA

When estimating the spatial incoherence of ground motion at a specific site for use in design, we would ideally use closely-spaced strong motion recordings at the site from design basis earthquakes to directly measure the spatial incoherence of ground motions at the site. However, such strong motion recordings seldom exist, and it is therefore necessary to use some other procedure. One procedure is to assume that the spatial incoherence at the site in question is similar to that at some other site such as the Differential Array in the Imperial Valley (Smith et al, 1982; and this paper) or the SMART 1 Array at Lotung, Taiwan (Loh, 1985; Harichandran and Vanmarcke, 1986). However, since we expect the spatial incoherence of ground motions (as well as other aspects of site response) to be strongly site-dependent, this procedure may be difficult to justify in many cases.

In this paper, we describe an alternative procedure for the estimation of spatial incoherence of ground motions that can be used in a site-specific manner. Our approach is to separate ground motion incoherence into two distinct contributions, and thereby estimate a spatial incoherence function by individually estimating these contributions and then combining them. We have tested this procedure using data from the 1979 Imperial Valley earthquakes recorded at the El Centro differential array, and shown that the estimated spatial incoherence function for the mainshock is consistent with that measured directly from the mainshock recordings.

Although we use the term *spatial incoherence* to describe the spatial variations of ground motions, the formulation and measurement of the effect is most readily done in terms of spatial coherence. In the time domain, spatial coherence is defined by the cross correlation function of a pair of time histories recorded at adjacent stations. Estimates for a series of frequencies can be obtained by bandpass filtering of the cross correlation function. In the frequency domain, coherence is defined by the normalized cross-power spectrum of the two time histories, which can be obtained by Fourier transforming the cross correlation function. This spectrum is smoothed using a Parzen window (Jenkins and Watts, 1969) having a 2 Hz bandwidth. To illustrate the computational method, the spatial coherence of a pair of Differential Array recordings of the 1979 Imperial Valley mainshock is illustrated in Figure 1. The estimation of spatial coherence of ground motions is discussed in further detail by Abrahamson and Bolt (1987); Spudich and Cranswick (1984); and Smith et al (1982). The latter two papers describe measurements of spatial coherence of Differential Array recordings of the 1979 Imperial Valley earthquake.

From a set of pairs of strong motion recordings having different spatial separations, we can derive a spatial coherence function for use as input into soil-structure interaction analyses using methods such as that of Luco and Wong (1986). This function describes the manner in which spatial coherence of ground motions decreases with frequency and with spatial separation of observation points. For mathematical convenience, we have assumed that the dependence of coherence on station separation R and on frequency ω has the form:

$$C(\omega, R) = e^{\left((-a - b\omega^2)R\right)}$$

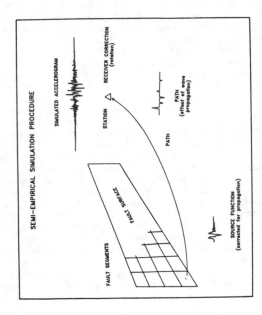

Figure 2. Schematic diagram of the semi-empirical ground motion simulation procedure. For each fault element, empirical source functions having appropriate radiation pattern are convolved with the Green's function for the appropriate range and depth and corrected for the receiver function. The ground motion contributions of the fault elements are lagged and summed to simulate rupture propagation over the fault surface.

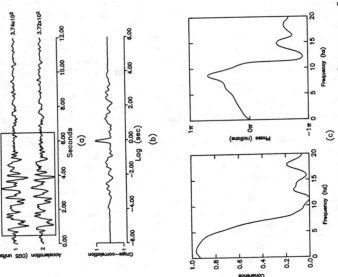

Figure 1. (a) South component accelerograms of the 1979 Imperial Valley mainshock recorded at stations 1 and 3 of the Differential Array. (b) the cross correlation function of (a) over the time window indicated by box. (c) the smoothed amplitude and phase spectra of (b).

where a and b are constants. Coherence is unity at all frequencies at zero separation, and decays as R increases. The dependence of coherence on separation R contains the product of frequency-dependent and frequency-independent terms. The ω^{-2} form of the frequency dependence is consistent with observed spectral measurements of coherence such as those shown in Figure 1.

OUTLINE OF THE GROUND MOTION SIMULATION PROCEDURE

The procedure for site-specific estimation of spatial incoherence requires the use of simulated accelerograms of design basis earthquakes. In the following, we briefly describe the simulation method, which is based on Hadley et al. (1982). A more complete description is given by Wald et al. (1988a).

The simulation procedure is summarized in Figure 2. The fault rupture surface is divided into a grid of elements, and the ground motion at a given site is obtained by lagging and summing the contributions from these elements in such a way as to simulate the propagation of rupture over the fault surface. While gross aspects of the source rupture process are treated kinematically as a spreading rupture front, stochastic aspects are included to simulate the irregularity in rupture velocity and slip velocity. Gross aspects of wave propagation are modeled by theoretical Green's functions calculated for a horizontally layered medium using generalized rays (Helmberger and Harkrider, 1978). Detailed aspects of the source radiation at high frequencies, as well as unmodeled propagational aspects such as scattering, are included empirically by using multiple recordings of a small earthquake In an extension of the Hadley et al. (1982) approach, the radiation from the fault is represented in a manner that empirically includes the observed breakdown in coherence of the radiation pattern at high frequencies. The method has been used successfully in the simulation of the extensive strong motion recordings of the 1979 Imperial Valley earthquake (Wald et al., 1988a) and of the 1987 Whittier Narrows earthquake (Wald et al., 1988b), as well as numerous less well recorded earthquakes.

Although considerable progress has been made in simulating the complexity of high frequency ground motions at a single station using the above methods, the spatial incoherence of these motions is not completely simulated when closely spaced time histories are generated. This is because the simulations generally use the same source function for a given fault element but with slightly different timing at adjacent locations. Thus the simulations do not contain spatial variations in the waveform of motions across the site from a given fault element that are expected to arise from diffraction and scattering. To obtain a complete estimate of spatial incoherence, these effects must therefore be included using another method that is described below.

ANALYSIS OF SPATIAL INCOHERENCE DUE TO WAVE PASSAGE

The passage of a wavefront across a site causes spatial incoherence of ground surface motions. We first analyse the case of waves originating from a point source, and then the case of waves originating from an extended source.

Point Source. If the separation of two locations at the site is much less than the source-to-site distance, and we consider a point source in a laterally homogeneous medium, then the wave field at the site can be approximated by a plane wave having particular azimuth and incidence angles. For vertical incidence, all locations at the site will experience the wave motion simultaneously and there will be no spatial incoherence. However, the incidence angle will generally not be vertical. In this case, while the waveform at adjacent locations may be identical, there will be a time (or phase) shift between the waves at adjacent locations resulting from the finite phase velocity of the waves across the site. This effect, which we call the wave passage effect, gives rise to spatial incoherence of ground surface motions. This wave passage contribution is included with other contributions to spatial incoherence, such as those due to scattering, when the unlagged cross correlation is measured. The unlagged cross correlation is the value of the cross correlation function obtained when the two time histories being correlated are aligned in absolute time, that is, without any time lag.

The wave passage contribution to spatial incoherence can be removed completely for the point source case by measuring the peak cross correlation, which is the cross correlation value obtained after one of the time histories has been lagged in time to produce the maximum cross correlation value. This time lagging is required to offset the time shift caused by the wave passage effect, as illustrated in the lower part of Figure 3.

Extended Source. We may represent an extended source as a series of point sources distributed over a fault plane. The waves from the different point sources will in general have different incidence angles and different azimuths, producing different time shifts between arrivals at adjacent locations, as shown in the upper part of Figure 3. This causes a reduction in the spatial coherence of ground motions relative to that of a single point source. Each additional point source contribution will in general further reduce the coherence of ground motions between adjacent locations. The effect of the superposition of plane waves from multiple sources is to superpose the wave passage contribution to spatial incoherence from each source.

As in the point source case, the wave passage contribution to spatial incoherence in the extended source case is included with other contributions when the unlagged cross correlation is measured. However, since the spatial incoherence effect from an extended source is three-dimensional, it cannot be completely described (or removed) using the simplifying assumption of a plane wave having a specified azimuth and incidence angle. Thus peak cross correlation measurements will remove some but not all of the wave passage contribution to spatial incoherence of ground motions from extended sources.

SITE-SPECIFIC PROCEDURE FOR ESTIMATING SPATIAL INCOHERENCE

The spatial incoherence estimate for a specific site is synthesized from two principal components which are illustrated schematically in Figure 3 and in the right column of Table 1. The first of these, estimated from ground motion simulations, incorporates the effect of an extended source (one whose dimensions are large compared with the source-to-site distance), which gives rise to

Source

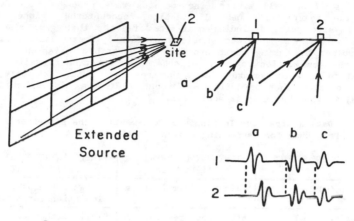

Extended
Source

Scattering

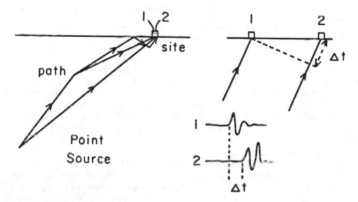

Point
Source

Figure 3. Schematic diagram of two contributions to spatial incoherence. Ray paths extending from multiple locations on an extended fault through a homogeneous medium to two adjacent locations at a site are shown in the top left corner. The ray paths at the two locations are shown in more detail in the top right corner, and below these are shown schematically the seismograms at the two sites. The three pairs of arrivals have similar waveforms at the two locations but different phase delays (wave passage effects) due to different angles of incidence (shown) and azimuthal angles (not shown). Ray paths extending from a point source through a scattering medium to the site are shown in the bottom left corner. The effect of scattering is represented by differences in waveform at the two locations, and the wave passage effect is shown as a difference in arrival time (bottom right corner).

interference between simultaneous arrivals from many different regions
of the fault. The second component, measured from ground motion
recordings of a small source, incorporates wave propagation effects
such as scattering and diffraction encountered along the
source-to-site path, including near-site effects, that give rise to
spatial variations in ground motion waveforms. These contributions
are combined by taking their product, as described in Appendix II.
Since the estimation of the incoherence function entails the
combination of two separate contributions, we must take care to
include each incoherence effect only once. In the discussion that
follows, we show how this is achieved.

TABLE 1. Data Sources for Testing and Implementing the Site-Specific
Procedure for Estimating Spatial Incoherence

INCOHERENCE EFFECT	TEST USING IMPERIAL VALLEY DATA	IMPLEMENTATION AT A SPECIFIC SITE
Source $C_s(\omega)$	Mainshock simulations	Design earthquake simulations at site
Scattering $C_p(\omega)$	Aftershock recordings	Small earthquake recordings at site
Combined Effect $C(\omega)$ (should match direct measurement)	$C_s(\omega) \times C_s(\omega)$	$C_s(\omega) \times C_p(\omega)$
Direct Measurement	Mainshock recordings	Not available

    In the simulated time histories, simultaneous arrivals from
different regions of the fault interfere in a complex manner, giving
rise to spatial incoherence. This spatial incoherence is due to the
non-vertical incidence of the waves (the wave passage effect), and
will be present even though the wave groups from individual fault
elements have the same waveform at different locations on the
foundation.
    Ground motions simulated using the above method do not include
wave propagation effects such as multipathing, scattering and complex
site response that might give rise to spatial variations in waveforms.
We therefore combine a separate estimate of these effects, which we
shall refer to as scattering, with the effects of the source as
estimated from the simulations. We estimate this scattering
contribution from peak cross correlation measurements of site
recordings from small earthquakes or explosions. By using events with
small source dimensions, we avoid including for a second time the
effect of an extended source. By measuring the peak coherence, we
also avoid including for a second time the effect of wave passage due
to a point source. Both the extended source and point source wave
passage effects are already included in our estimate of the source
contribution to spatial incoherence derived from the simulations.
    In our simulation procedure, the contribution at the site from
each fault element has a different path and is therefore expected to
have different scattering effects. Ideally, we would like to have
site recordings from a small event centered on each fault element with
which to make estimates of the scattering contribution to spatial

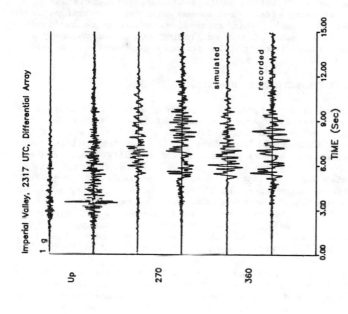

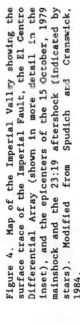

Figure 5. Recorded and simulated accelerograms of the 1979 Imperial Valley mainshock at station EDA, whose location is shown in Figure 4. For each component, the upper trace is simulated and the lower trace is recorded.

Figure 4. Map of the Imperial Valley showing the surface trace of the Imperial Fault, the El Centro Differential Array (shown in more detail in the inset), and the epicenters of the 15 October, 1979 mainshock and the 23:19 aftershock (indicated by stars). Modified from Spudich and Cranswick, 1984.

incoherence. However, we do not typically have such recordings, and must therefore use available recordings at appropriate distances to represent the scattering effect, and assume that the scattering effect is the same for each path. This assumption may be quite realistic, since we expect that most of the scattering effect occurs near the site, and that scattering may not be strongly dependent on the total path length.

TEST OF PROCEDURE FOR ESTIMATING SPATIAL INCOHERENCE

We have tested this procedure using strong motion recordings from the El Centro Differential Array of the October 15, 1979 Imperial Valley mainshock (M=6.5) and its aftershock (M=5.1) at 23:19 on the same day. A map showing the locations of the Imperial Valley fault rupture, the epicenters of the mainshock and the aftershock, and the Differential Array is shown in Figure 4. Since relative timing between the stations was lost, it is not possible to completely include the wave passage effect in this test by using unlagged cross correlation measurements. However, the wave passage effect can be partially included in the test by using measurements of peak cross correlation. By using unlagged cross correlation measurements of our simulations, we are also able to make a complete estimate of spatial incoherence at the Differential Array, but lack the data that would verify this estimate.

The test procedure, outlined schematically in the central column in Table 1, is to obtain separate estimates of source coherence and scattering coherence and combine them to give coherence estimates of the recorded mainshock ground motions. We then compare these estimates with those measured from the mainshock recordings.

The source contribution to incoherence of the mainshock is estimated from ground motion simulations of the mainshock at the Differential Array stations. The simulation procedure has been validated by showing that it produces accelerograms at the Differential Array (Figure 5) and at other stations of the El Centro array that are in good agreement with the recorded accelerograms (Wald et al, 1988a).

The peak coherence of the source contribution measured from the simulations of the mainshock is shown in Figure 6. Peak cross correlation measurements over 6-sec windows containing the strongest S wave arrivals of both horizontal components were made in three frequency bands: 2.5 to 5 Hz; 5 to 10 Hz, and 10 to 20 Hz. The best fit spatial coherence function assuming the form described above provides a fairly good fit to the measurements. In this and the following figures, the coherence function is used to illustrate gross trends in the measurements, and is not central to the test that we wish to perform.

The scattering contribution to coherence of the mainshock is represented by peak coherence measurements of the aftershock of 23:19, shown in Figure 7. The combined coherence (Figure 8) is obtained by multiplying the spectral coherence values of the two contributions, as described in Appendix II. This combined coherence, which was synthesized from a source contribution (derived from simulations) and a scattering contribution (derived from aftershock recordings), is very similar to the coherence measured from the mainshock recordings (Figure 9). The combined coherence estimate gives slightly higher

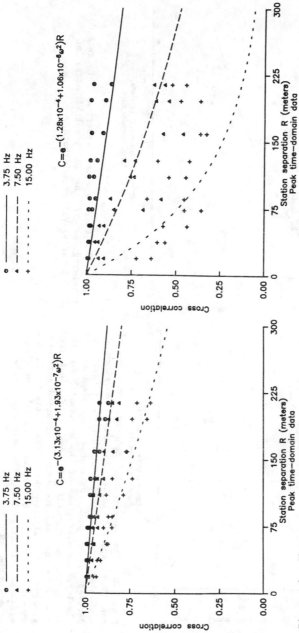

Figure 6. The source contribution to the peak spatial incoherence estimate of the 1979 Imperial Valley mainshock at the Differential Array, represented by peak coherence measurements of simulated accelerograms of the mainshock. Measurements are shown for three frequency bands (2.5-5, 5-10, and 10-20b Hz), labelled by their center frequencies. The best fit spatial coherence function given by the equation is shown by the curves.

Figure 7. The scattering contribution to the spatial incoherence estimate of the 1979 Imperial Valley mainshock at the Differential Array, represented by peak coherence measurements of recorded accelerograms of the 23:19 aftershock.

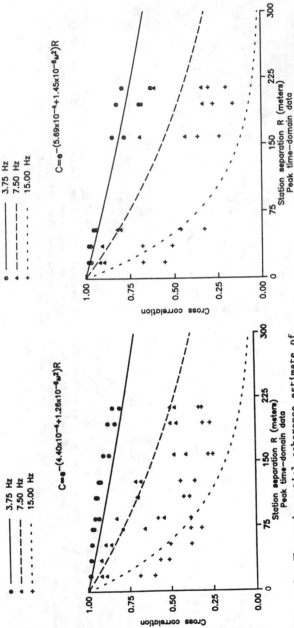

Figure 9. Peak spatial coherence of recorded accelerograms at the Differential Array of the 1979 Imperial Valley mainshock.

Figure 8. The peak spatial coherence estimate of the 1979 Imperial Valley mainshock, obtained by combining the source contribution (Figure 6, derived from peak coherence measurements of simulated accelerograms of the mainshock) and the scattering contribution (Figure 7, derived from peak coherence measurements of recorded accelerograms of the aftershock). This estimate is for comparison with the peak spatial coherence measured directly from the mainshock accelerograms, shown in Figure 9.

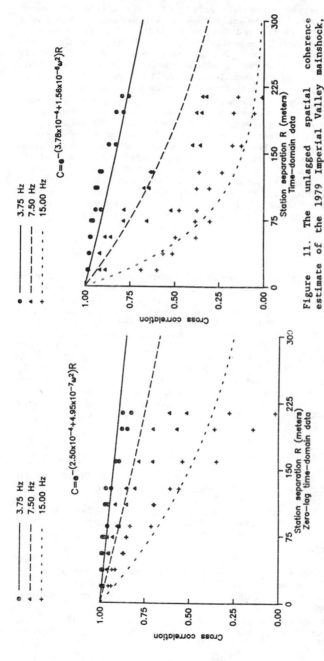

Figure 11. The unlagged spatial coherence estimate of the 1979 Imperial Valley mainshock, obtained by combining the source contribution (Figure 6, derived from simulated accelerograms of the mainshock) and the scattering contribution (Figure 7, derived from recorded accelerograms of the aftershock). The unlagged spatial incoherence estimate includes the wave passage effect completely, in contrast with the peak coherence estimate of Figure 8 and the direct peak coherence measurements of Figure 9.

Figure 10. The source contribution to the unlagged spatial coherence estimate of the 1979 Imperial Valley mainshock at the Differential Array, represented by unlagged coherence measurements of simulated accelerograms of the mainshock. The unlagged coherence estimate includes the wave passage effect completely, in contrast with the peak coherence estimate of Figure 6 which only partially includes the wave passage effect.

values than are observed. The data in Figure 9 are sparse because of
data dropouts in the recorded mainshock accelerograms at station 4
(Smith et al., 1982; Spudich and Cranswick, 1984). The large scatter
in the mainshock data, especially at 7.5 Hz, reflect the
systematically higher coherence in the east component than in the
south component. The measured coherence values show considerable
departures from the assumed functional form in Figures 7-9, but no
convenient alternative functional form is apparent in the data.

As indicated above, we have of necessity used peak coherence
measurements in this test because of the lack of absolute time in the
Differential Array data. However, our simulations of the mainshock do
not suffer from this limitation. By using unlagged cross correlation
measurements, shown in Figure 10, in place of the peak cross
correlation measurements of Figure 6, we can obtain an estimate of the
source contribution to spatial incoherence that completely includes
the wave passage effect. Combining this source estimate with the
scattering contribution of Figure 7, we are then able to make a
complete estimate of spatial incoherence of the mainshock at the
Differential Array (Figure 11), but lack the data that would verify
this estimate.

CONCLUSIONS

We have shown that our procedure of combining separate estimates
of coherence due to the source effect and to the scattering effect
provides an accurate estimate of the spatial incoherence of ground
motions of a large earthquake in the case of the Imperial Valley
Differential Array data. This procedure may therefore provide useful
estimates of spatial incoherence at other sites for which direct
estimates of spatial incoherence from large earthquakes are not
available. The application of the procedure at a specific site is
illustrated schematically in the right column of Table 1. The
procedure entails the combination of estimates of the source
contribution to incoherence from simulated accelerograms of design
basis earthquakes, and of the scattering contribution to incoherence
from closely-spaced site recordings of a nearby small earthquake or
explosion.

ACKNOWLEDGMENTS

This work was sponsored by Pacific Gas and Electric Company under
the direction of Dr Y.B. Tsai. The authors are grateful to the Branch
of Engineering Seismology and Geology of the U. S. Geological Survey
for providing the Differential Array recordings. The authors are also
grateful to an anonymous reviewer for very helpful comments.

APPENDIX I.- REFERENCES

Abrahamson, N.A. and B.A Bolt (1987). Array analysis and synthesis
    mapping of strong seismic motion, In: B.A. Bolt (editor), Strong
    Motion Synthetics: Computational techniques series, Academic
    Press, New York.
Hadley, D.M., D.V. Helmberger, and J. A. Orcutt (1982). Peak
    acceleration scaling studies, Bull. Seism. Soc. Am. 72, 959-979.

Harichandran, R.S. and E.H. Vanmarcke (1986). Stochastic variation of earthquake ground motion in space and time, *J. Eng. Mech.* 112, 154-174.

Helmberger, D.V. and D.G. Harkrider (1978). Modeling earthquakes with generalized ray theory, In: *Proceedings of IUTAM Symposium: Modern Problems in Elastic Wave Propagation*, John Wiley & Sons, Inc., New York, 499-518.

Jenkins, G.M. and D.G. Watts (1968). Spectral analysis and its applications, Holden-Day, Oakland, California.

Loh, C.-H. (1985) Analysis of the spatial variation of seismic waves and ground movements from SMART-1 array data, *Earthq. Eng. Struct. Dyn.* 13, 561-581.

Luco, J.E. and H.L. Wong (1986). Response of a rigid foundation to a spatially random ground motion, *Earthq. Eng. Struct. Dyn.* 14, 891-908.

Smith, S.W., J.E. Ehrenberg, and E.N. Hernandez (1982). Analysis of the El Centro differential array for the 1979 Inperial Valley earthquake, *Bull. Seism. Soc. Am.* 72, 237-258.

Spudich, P. and E. Cranswick (1984). Direct observation of rupture propagation during the 1979 Imperial Valley earthquake using a short baseline accelerometer array, *Bull. Seism. Soc. Am.* 74, 2083-2114.

Wald, D.J., L.J. Burdick, and P.G. Somerville (1988a). Simulation of acceleration time histories close to large earthquakes, In: *Proceedings of the Earthquake Engineering and Soil Dynamics II Conference*, American Society of Civil Engineers, this volume.

Wald, D.J., P.G. Somerville, and L.J. Burdick (1988b). Simulation of recorded accelerograms from the 1987 Whittier Narrows earthquake, *Earthquake Spectra*, in press.

APPENDIX II.- FORMULATION OF SPATIAL INCOHERENCE ESTIMATION PROCEDURE

The physical basis of a procedure for estimating the spatial incoherence of ground motions close to an extended source by combining the contributions from two effects was outlined above. We now describe the formulation of the procedure in more rigorous terms, and show how the contributions to spatial incoherence from the source and from scattering are combined. The simulated acceleration time history S(t) at a given location is constructed from the sum of suitably lagged contributions from each of n fault elements E(t):

$$S(t) = \sum_{i=1}^{n} E_i(t)$$

The coherence function that we wish to construct is derived conceptually from time histories A(t) that are formed by convolving each fault element time history E(t) with a scattering operator P(t). As discussed above, we assume that P(t) at a given location is the same for all fault elements:

$$A(t) = \sum_{i=1}^{n} \left[E_i(t) * P(t) \right]$$

where * denotes convolution. Time histories $A_j(t)$ at adjacent locations will be different because both the fault element contributions $E(t)$ and the scattering operator $P(t)$ are different at different locations. This can be written as

$$A(t) = \left[\sum_{i=1}^{n} E_i(t) \right] * P(t)$$

or

$$A(t) = S(t) * P(t)$$

The coherence function of two adjacent time histories $A_1(t)$ and $A_2(t)$ is given by their cross correlation function $C(\tau)$:

$$C(\tau) = \frac{\frac{1}{N} \sum_{i=1}^{N-\tau} A_1(t) A_2(t+\tau)}{\sqrt{\frac{1}{N} \sum_{i=1}^{N} A_1{}^2(t) \frac{1}{N} \sum_{i=1}^{N} A_2{}^2(t)}}$$

For brevity, we represent cross correlation by the symbol $\otimes$:

$$C(\tau) = A_1(t) \otimes A_2(t)$$

Expressing each time history as a convolution of the simulated time history $S(t)$ and the scattering operator $P(t)$, we obtain

$$C(\tau) = \left[S_1(t) * P_1(t) \right] \otimes \left[S_2(t) * P_2(t) \right]$$

or

$$C(\tau) = \left[S_1(t) * P_1(t) \right] * \left[S_2(-t) * P_2(-t) \right]$$

since correlation is identical to convolution with the time axis of one component reversed. This is equivalent to

$$C(\tau) = \left[S_1(t) \otimes S_2(t) \right] * \left[P_1(t) \otimes P_2(t) \right]$$

$$= C_s(\tau) * C_p(\tau)$$

This relation states that the cross correlation $C(\tau)$ of the two time histories $A_1(t)$ and $A_2(t)$ can be obtained from the convolution of separate estimates of the cross correlation functions of the simulated source $C_s(\tau)$ and of the scattering operator $C_p(\tau)$. In the frequency domain, the cross-power spectrum $C(\omega)$ of the two spectra $A_1(\omega)$ and $A_2(\omega)$ can therefore be obtained from the product of separate estimates $C_s(\omega)$ and $C_p(\omega)$ of the cross-power spectra of the source $S(\omega)$ and the scattering operator $P(\omega)$:

$$C(\omega) = C_s(\omega) \times C_p(\omega).$$

LOCAL SPATIAL VARIATION OF EARTHQUAKE GROUND MOTION

Ronald S. Harichandran[1], A.M. ASCE

INTRODUCTION

Consideration of the temporal and spatial variation of earthquake ground motion is important in the design of large structures (such as dams and containment structures for nuclear power plants), long structures with widely-spaced multiple supports (such as bridges and surface pipelines), and long buried structures (such as gas and oil pipelines, etc.). Recent data available from closely-spaced arrays of seismographs have been indispensable in the study of the space-time variation of ground motion. One such digital seismograph array is the SMART-1 array located in Lotung, Taiwan, which has recorded a large number of events. Some studies have been initiated based on the data obtained from this array (Bolt et. al. 1982; Loh et. al. 1982; Loh 1985; Harichandran & Vanmarcke 1986).

An initial model based on the preliminary study of one far-field event (Event 20) recorded by the SMART-1 array, which characterized the ground motion as a random field in space and time, was proposed by Harichandran and Vanmarcke (1986). In the present work, the results of additional extensive analyses on two far-field events (Events 20 and 24) are presented. Only 15-second long strong motion parts of the accelerograms dominated by S-waves were used in the analyses. The main characteristics of these event are given in Table 1. The model proposed earlier is verified, and a simpler model which may suffice for some applications is suggested. Kanai-Tajimi and the "double-filter" models for the point spectral density function of ground acceleration are compared.

The regularity between the gross time delays for different pairs of stations (also observed in the earlier study) are used to derive relative arrival times at each station through a multiple linear regression model. These arrival times are then used to estimate the direction of gross wave propagation. It is found that for both events the direction of propagation is away from the epicenter.

COHERENCY ANALYSIS

The SMART-1 array, shown in Figure 1, is a two-dimensional surface array that consists of a center station, C-00, and 12 stations on each

Table 1: Main Characteristics of the Two Events

Event	Magnitude	Azimuth (Deg)	Epicentral Dist. (km)	Depth (km)	Max. Acceleration (gal)	
					Radial	Transverse
20	6.9	105.7	116.6	30.6	60.6	64.0
24	7.2	130.0	92.3	25.0	68.7	47.0

---

[1] Assistant Professor, Department of Civil & Environmental Engineering, Michigan State University, East Lansing, MI 48824.

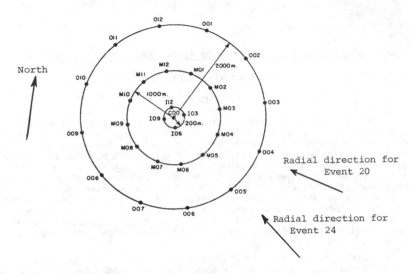

Figure 1: The SMART-1 Seismograph Array

of three concentric circles (inner I, middle M and outer O) with radii of 200, 1000 and 2000 m, respectively. Only the horizontal components, radial (direction away from the epicenter) and transverse (perpendicular to radial), recorded at the center station and stations on the inner and middle rings were used in the analyses.

For each component, the auto-spectra $S_j(f)$ and $S_k(f)$, cross correlation function, and the cross-spectrum $S_{jk}(f)$, of the accelerograms $x_j(t)$ and $x_k(t)$ recorded at two stations j and k, were estimated using conventional spectral estimation. A Hamming window with a bandwidth of 1 Hz was used (Jenkins & Watts 1968; Harichandran 1987). From these the complex-valued coherency spectrum, $\gamma_{jk}(f)$, and real-valued phase spectrum, $\phi_{jk}(f)$, were obtained through:

$$\gamma_{jk}(f) = \frac{S_{jk}(f)}{[S_j(f)S_k(f)]^{\frac{1}{2}}} \tag{1}$$

and
$$\phi_{jk}(f) = \tan^{-1}\left[\frac{\text{Im}[S_{jk}(f)]}{\text{Re}[S_{jk}(f)]}\right] \tag{2}$$

The gross propagation delay, d_{jk}, for each pair of accelerograms was estimated by removing any linear trend from the aligned phase

$$\phi'_{jk}(f) = \phi_{jk}(f) + 2\pi d_{jk}f \tag{3}$$

In most cases d_{jk} corresponded to the delay at which the cross correlation function had its peak.

Although approximately 150 station pairs with separations less than 1000 m were available, for computational expediency only a subset of these were considered in the analyses. In order to achieve a uniform representation over the separation range of interest, station pairs were selected arbitrarily to give approximately 6 pairs at every 100 m separation spanning from 0 to 1000 m (i.e., 6 pairs with separations around 100 m, 6 pairs with separations around 200 m, etc.). In total 56 pairs were used for event 20 and 53 for event 24. Initial processing and prior experience indicated that the accelerograms had an approximately isotropic (i.e., direction independent) coherency variation which was a function of frequency and station separation. Estimates of this average coherency variation were obtained by non-parametric smoothing of the m absolute value of coherency spectra, $\{|\gamma(\nu_p,f_q)|, p=1, 2,\ldots,m, q=1,2,\ldots,n\}$ (where ν_p is the station separation for the pth pair, and f_q is the qth discrete frequency), according to

$$|\hat{\gamma}(\nu,f)| = \frac{\displaystyle\sum_{p=1}^{m}\sum_{q=1}^{n} |\gamma(\nu_p,f_q)|\, w\!\left[\frac{\nu - \nu_p}{\Delta\nu}\right] w\!\left[\frac{f - f_q}{\Delta f}\right]}{\displaystyle\sum_{p=1}^{m}\sum_{q=1}^{n} w\!\left[\frac{\nu - \nu_p}{\Delta\nu}\right] w\!\left[\frac{f - f_q}{\Delta f}\right]} \tag{4}$$

in which $w(x) = e^{-x^2/2}$ is a smoothing window, and $\Delta\nu$ and Δf are smoothing parameters. Parameter values of $\Delta\nu = 33.3$ m and $\Delta f = 0.1$ Hz were used.

Cross sections of $|\hat{\gamma}(\nu,f)|$ at various separations and frequencies are shown in Figures 2 and 3. It can be seen that in general the coherencies decay with increasing separation and frequency. The coherencies tend not to decay below a value of about 0.2. This is expected because as the true coherency approaches zero the variance in the coherency estimate becomes large (Jenkins & Watts 1968).

A different smoothing procedure was used previously by Harichandran and Vanmarcke (1987). In that preliminary study the complex-valued coherencies were first aligned to remove the gross propagation effects. The aligned complex-valued coherencies, rather than the absolute values of coherency, were then smoothed according to (4). It was assumed there that the phase spectra were approximately linear for all accelerogram pairs. This is not true in general. Although the estimated phase is approximately linear for small station separations and low frequencies, it becomes more random at large separations and high frequencies. This is mainly due to the variance of the phase estimator in (2) tending to infinity as the true coherency approaches zero (Jenkins & Watts 1968). Therefore, in the method employed previously, a significant bias can be introduced when the estimated phase becomes more random, with the smoothed coherency becoming smaller and smaller as more and more station pairs are used. This problem is averted in the current method. Also, since the absolute values are now used for smoothing, there is no need to align the coherencies. (Alignement only changes the phase and not the absolute values of coherency.)

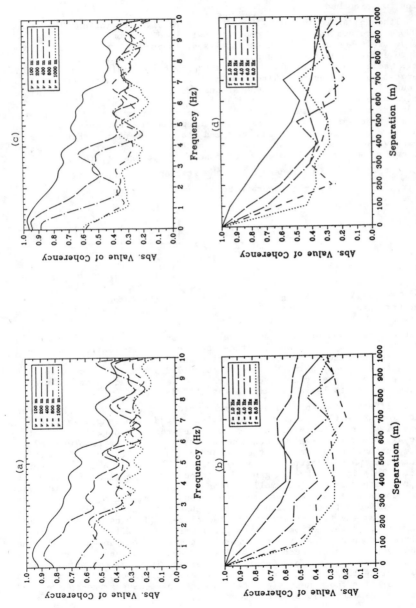

Figure 2: Smoothed Coherencies for Components of Event 20: (a-b) Radial; (c-d) Transverse

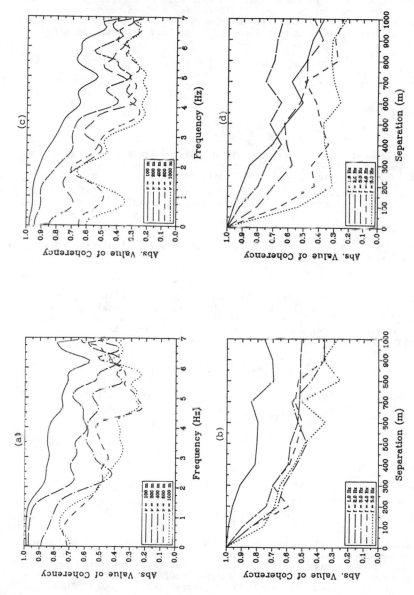

Figure 3: Smoothed Coherencies for Components of Event 24: (a-b) Radial; (c-d) Transverse

DIRECTION OF WAVE PROPAGATION

The time delays estimated between each pair of accelerograms showed a regular property of "closure". That is, for a given component of acceleration and a triplet of stations j,k and ℓ, the delay between station j and ℓ was roughly related to the delays between stations j and k, and k and ℓ, through

$$d_{j\ell} \approx d_{jk} + d_{kl} \qquad (5)$$

This property implied that it was possible to obtain a relative arrival time t_j at each station and that the delay for a pair of stations j and k could then be expressed as

$$d_{jk} = t_k - t_j \qquad (6)$$

In general, the number of station pairs analyzed were in excess of the number of stations, and therefore the relative arrival times were estimated through multiple linear regression. Collecting the delays d_{jk} estimated by pairwise analysis into a vector $\mathbf{d}$, and the relative arrival times t_j into a vector $\mathbf{t}$, the regression model may be expressed as

$$\mathbf{d} = \mathbf{Ht} + \epsilon \qquad (7)$$

in which ϵ is the observation and measurement error. The ith equation in the above set is of the form

$$d_{jk} = t_k - t_j + \epsilon_i \qquad (8)$$

Thus, each row of the matrix $\mathbf{H}$ has only two non-zero elements, of which one is 1 and the other is -1. Matrix $\mathbf{H}$ is not of full rank, reflecting the fact that the relative arrival time at one station can be arbitrarily chosen. Choosing station 1 to be the reference and the arrival time t_1 to be zero, equation (7) may be written in partitioned form as

$$\mathbf{d} = \left[\mathbf{H_1} \mid \mathbf{H_{n-1}} \right] \begin{bmatrix} t_1 \\ t_{n-1} \end{bmatrix} + \epsilon \qquad (9)$$

in which $\mathbf{H_1}$ is the first column and $\mathbf{H_{n-1}}$ the remaining part of $\mathbf{H}$, and the vector $\mathbf{t_{n-1}}$ contains t_j, j=2,3,...,n. Since $t_1 = 0$ this is equivalent to

$$\mathbf{d} = \mathbf{H_{n-1}} \mathbf{t_{n-1}} + \epsilon_{n-1} \qquad (10)$$

The matrix $\mathbf{H_{n-1}}$ is now of full rank, and the solution of the above multiple linear regression problem that minimizes $\epsilon_{n-1}^T \epsilon_{n-1}$, the sum of the square of the errors, is

$$\mathbf{t_{n-1}} = (\mathbf{H_{n-1}^T} \mathbf{H_{n-1}})^{-1} \mathbf{H_{n-1}^T} \mathbf{d} \qquad (11)$$

Relative arrival times were estimated using the radial and transverse components separately. Both components yielded roughly the same arrival time estimates. The average of the times estimated from both components are tabulated in Table 2. The coefficient of correlation, R^2, for the regressions was 0.99 or greater for all cases. This vali-

Table 2: Estimates of Relative Arrival Times at Stations in
Hundredths-of-a-second

Station	Event 20 (Ref: I01)	Event 24 (Ref: C00)	Station	Event 20 (Ref: I01)	Event 24 (Ref: C00)
C00	*	0.0			
I01	0.0	1.1	M01	0.0	3.6
I02	0.2	*	M02	-4.3	-6.6
I03	-3.3	-3.0	M03	-12.9	-15.0
I04	-3.6	-6.1	M04	-23.3	-32.0
I05	-4.9	-7.3	M05	-30.7	-40.8
I06	-3.8	-5.2	M06	-26.0	-34.3
I07	-2.3	*	M07	-12.6	-24.0
I08	-2.1	1.8	M08	-1.8	-10.4
I09	5.2	*	M09	2.7	*
I10	-4.4	2.7	M10	6.2	16.5
I11	0.4	2.2	M11	8.6	34.7
I12	0.5	2.3	M12	*	0.8

Arrival times for station that were not triggered are marked with *.

dates the significance of the regression model.

The arrival times give an indication of the direction of wave propagation. Approximate contours of the average arrival times obtained from the radial and transverse components for each event are shown in Figure 4. Stations I01 and M01 are indicated on the figures and the other stations are numbered clockwise from these. These contours may be thought of as propagating "wavefronts" as far as phase effects are concerned. Note, however, that coherency decays along a given wavefront, and hence they should not be confused with wavefronts in a plane wave model for which all points on a wavefront are fully coherent. For both events waves seem to propagate diagonally from the south-east to the north-west and are roughly directed away from the appropriate epicenters. The apparent roughness in some of the contours may be attributed to errors in the arrival times arising from estimation and/or from the clock settings at some stations. For event 20, if the arrival time at I10 is increased by about 0.05 second and that at I09 is decreased by about 0.04 second then the roughness in the contours for the relative arrival times of 0 and 3 (hundredths-of-a-second) would disappear. Similarly, for event 24, if the arrival time at station M12 was increased by about 0.16 seconds then the sharp curvatures of the positive contours would disappear. For event 20 it appears that the waves propagate faster when approaching the center station C00, but slow down after passing it.

If the arrival times, t_j, could be characterized by assuming a constant velocity and straight wavefronts, then a plot of t_j against the component along the propagation direction of the station separations (measured from the reference station), $s_j(\theta)$, should show a linear trend. Here θ denotes the azimuth of approach of the waves. Another technique for estimating the gross direction of propagation could, therefore, be based on varying θ until the sum of the squares of the errors of the linear regression line (forced through the origin) fit to the data $(t_j, s_j(\theta))$ was minimized. This was done using the average arrival times (from the radial and transverse components), and azimuths of approach of E43.8°S and E38.4°S were obtained for events 20 and 24, respectively. These directions agree well with the propagation direc-

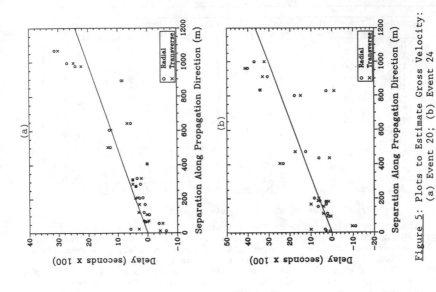

Figure 5: Plots to Estimate Gross Velocity:
(a) Event 20; (b) Event 24

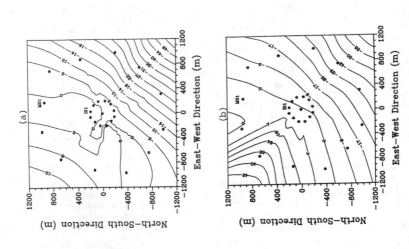

Figure 4: Contours of Arrival Times (sec/100):
(a) Event 20; (b) Event 24

tions in Figure 4. Plots of t_j vs. $s_j(\theta)$ (for the values of θ mentioned above) are shown, together with the regression lines, in Figure 5. The velocities based on the regression lines are approximately 4900 m/s and 3300 m/s for events 20 and 24, respectively. The substantial scatter in these plots, however, tend to suggest that the assumption of a constant velocity and straight wavefronts is too simplistic.

SPACE-TIME RANDOM FIELD MODELS

Preliminary analyses (Harichandran & Vanmarcke 1986) have shown that the space-time variation of any component of ground acceleration, over the dimensions of engineered structures, may be characterized as a homogeneous random field. The cross spectral density function of the accelerations at two stations j and k separated by a distance ν may be expressed as

$$S_{jk}(\nu,f) = S_x(f) \; \rho(\nu,f) \; e^{-i2\pi f d_{jk}} \qquad (12)$$

in which the autospectrum of acceleration, $S_x(f)$, is assumed to be the same at all points, $\rho(\nu,f)$ is a real-valued isotropic (i.e., direction independent) coherency function, d_{jk} is the time delay for waves to propagate from j to k, and $i = \sqrt{-1}$. $\rho(\nu,f)$ describes the coherency variation of the acceleration field, while the complex exponential accounts for the phases arising from wave propagation.

Harichandran and Vanmarcke (1986) proposed the following double exponential model for the coherency variation $\rho(\nu,f)$:

$$\rho(\nu,f) = A \exp\left[-\frac{2\nu}{\alpha\theta(f)} (1 - A + \alpha A)\right]$$

$$+ (1 - A) \exp\left[-\frac{2\nu}{\theta(f)} (1 - A + \alpha A)\right] \qquad (13)$$

in which the frequency variation is expressed through

$$\theta(f) = k\left[1 + \left[\frac{f}{f_0}\right]^b\right]^{-\frac{1}{2}} \qquad (14)$$

This model has five empirical parameters A, α, k, f_0 and b. A simpler single exponential model with the same frequency variation specified in equation (14), but with only three parameters k, f_0 and b, is

$$\rho(\nu,f) = \exp\left[-\frac{2\nu}{\theta(f)}\right] \qquad (15)$$

An alternative single exponential model used by Novak and Hindy (1980) is

$$\rho(\nu,f) = \exp\left[-\lambda(f\nu)^\mu\right] \qquad (16)$$

in which the variation with increasing frequency or separation is assumed to be similar.

The three models presented above were fitted to the unsmoothed coherencies of the aligned radial and transverse components of events 20 and 24. A nonlinear least squares algorithm was used to estimate the relevant parameters such that the sum of the squared differences between the estimated and theoretical coherencies was minimized. Typical

Table   3: Model Parameters for $\rho(\nu,f)$

Model	Parameter	Event 20		Event 24	
		Radial	Transverse	Radial	Transverse
Double	A	0.626	0.665	0.355	0.577
	α	0.022	0.019	0.086	0.037
	k	19700	39500	23100	16400
Exponential	f_0 (Hz)	2.02	0.52	0.55	2.33
	b	3.47	1.96	2.35	5.28
Single	k	2520	5160	11150	2980
Exponential	f_0 (Hz)	2.02	0.16	0.30	2.48
	b	1.86	1.03	1.54	3.27

cross sections of the first two models (equations (13) and (15)) are
shown in Figures 6 and 7, and the values of the model parameters obtained through the fitting procedure are presented in Table 3. A comparison of these figures with Figures 2 and 3 shows that the double
exponential model is able to capture the main features of the coherency
variation. The single exponential model agrees to some extent with the
double exponential model only for the radial component of event 24. In
general Novak's model does not fit the observed variations too well,
especially when the "corner frequency" f_0 is not close to zero. A comparison of the double exponential model and Novak's model for the radial component of event 20 is presented in Figure 8.

Two models that are commonly used for the point spectrum are the
Kanai-Tajimi spectrum

$$S_x(\omega) = \left[\frac{\{1 + 4\zeta_g^2[\omega/\omega_g]^2\}}{\{1 - [\omega/\omega_g]^2\}^2 + 4\zeta_g^2[\omega/\omega_g]^2} \right] S_0 \tag{17}$$

and the so-called "double-filter" spectrum (Hindy & Novak 1980)

$$S_x(\omega) = \left[\frac{\{1 + 4\zeta_g^2[\omega/\omega_g]^2\}}{\{1 - [\omega/\omega_g]^2\}^2 + 4\zeta_g^2[\omega/\omega_g]^2} \right] \left[\frac{[\omega/\omega_f]^4\}}{\{1 - [\omega/\omega_f]^2\}^2 + 4\zeta_f^2[\omega/\omega_f]^2} \right] S_0 \tag{18}$$

in which ω_g, ζ_g, ω_f, ζ_f, and S_0 are model parameters. For normalized
one-sided spectra (which have unit area) the intensity parameter S_0 is

$$S_0 = \left[\int_0^\infty S_x(\omega) \, d\omega \right]^{-1} \tag{19}$$

For the Kanai-Tajimi spectrum the above integral can be evaluated in
closed form, and $S_0 = 4\zeta_g/[\pi\omega_g(1 + 4\zeta_g^2)]$. The displacement spectrum
corresponding to the Kanai-Tajimi acceleration spectrum does not exist
(i.e., it has a singularity at f=0), while the double-filter spectrum
rectifies this problem. Thus, in studies where the displacement spectrum is required, the double-filter spectrum is more suitable.

The two spectra presented above were fitted to the average (over all
stations) normalized autospectrum for each component of each event. In
order to give good resolution, the autospectra were computed using a
bandwidth of 0.25 Hz. A nonlinear least squares fit was made on the
arithmetic values of the average spectra in the frequency range from
0.1 to 10 Hz. Typical fits are shown in Figure 9. The four-parameter
double-filter model gave a better fit in general, and the main differences in the two models occurred at very low frequencies. The estimated

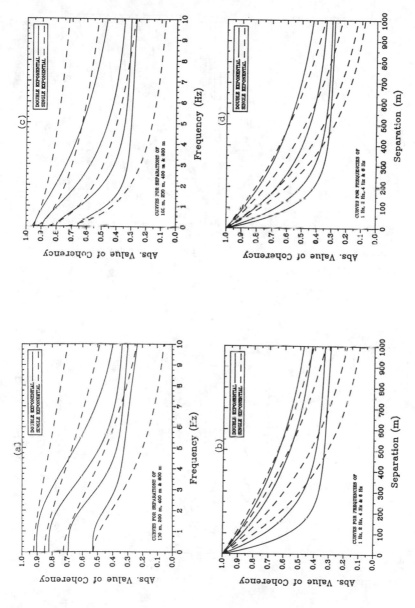

Figure 6: Fitted Coherency Models for Components of Event 20: (a-b) Radial; (c-d) Transverse

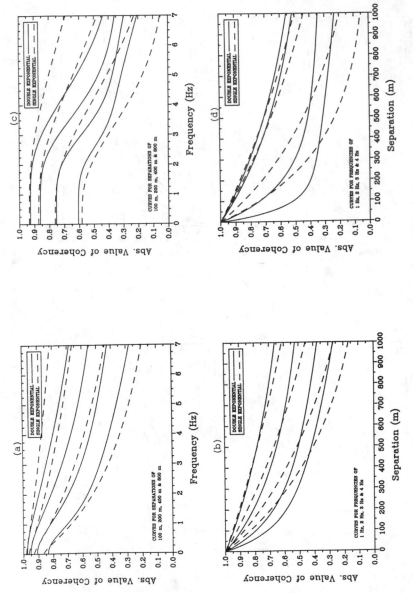

<u>Figure 7</u>: Fitted Coherency Models for Components of Event 24: (a-b) Radial; (c-d) Transverse

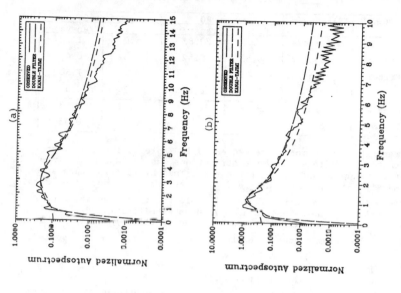

Figure 9: Fitted Autospectral Models for Transverse Components: (a) Event 20; (b) Event 24

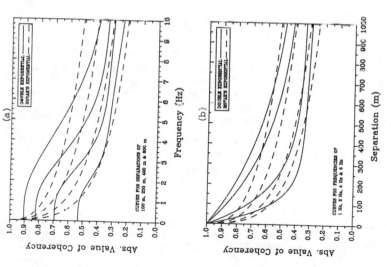

Figure 8: Comparison of Double Exponential and Novak's Model: Event 20 - Radial

Table 4: Model Parameters for Autospectra

Model	Parameter	Event 20		Event 24	
		Radial	Transverse	Radial	Transverse
Double-Filter	ω_g (rad/s)	20.22	20.82	5.05	1.92
	β_g	0.53	0.52	0.62	0.42
	ω_f (rad/s)	5.45	4.81	6.41	7.36
	β_f	0.46	0.61	0.27	0.37
Kanai-Tajimi	ω_g (rad/s)	19.29	21.31	7.21	8.53
	β_g	0.50	0.47	0.22	0.36

parameters for each model are presented in Table 4. The two events have substantially different power distributions, as is apparent from the spectral shapes and model parameters. The physical interpretation of the Kanai-Tajimi parameters f_g and ζ_g as ground frequency and damping characterizing the site is questionable for these events, since they are recorded at the same site and yet have significantly different values of f_g and ζ_g. The acceleration spectra would generally be influenced by the seismic source and ray path in addition to the soil properties.

The final part required to characterize the random field model is the specification of the propagation time delay d_{jk} between any two stations j and k. It was shown earlier that such delays can be characterized effectively through relative arrival times associated with each station. However, these arrival times are difficult to model accurately. A simple model, which assumes straight "wavefronts" and a constant propagation velocity vector **V**, yields the delays

$$d_{jk} = \frac{\mathbf{V} \cdot \boldsymbol{\nu}_{jk}}{V^2} \qquad (20)$$

in which $\boldsymbol{\nu}_{jk}$ is the separation vector between stations j and k, V^2 is the square of the magnitude of **V**, and the dot signifies the scalar dot product between the two vectors. In words, equation (20) says that the time delay d_{jk} is equal to the separation of stations j and k along the propagation direction (i.e., direction of **V**) divided by the velocity of propagation. This simple model is not able to represent the scatter of the arrival times in Figure 5, but characterizes their main trend. It appears that a more realistic approach would be to treat the arrival times as random variables, and to obtain the delays d_{jk} from these through equation (6). Such a characterization, however, would make the use of the model in studies of structural response more cumbersome.

CONCLUSIONS

Space-time random field models of ground acceleration, based on analyses of the horizontal components of two far-field earthquake events recorded by the SMART-1 seismograph array, are presented in this paper.

The gross propagation of seismic waves is characterized through the arrival times at each station. Time delays for waves to propagate from one station to another are estimated through correlation analysis, and the arrival times at stations are estimated from these using multiple linear regression.

Kanai-Tajimi and the so-called "double-filter" spectrum are used to model the autospectra of ground accelerations. The double-filter model, which has a higher number of parameters (four instead of two), yields a better fit to observed spectra, and is also more suitable for analyses in which the autospectrum of ground displacement is required.

Double and single exponential models are used to characterize the coherency of the acceleration fields. The double exponential model proposed by Harichandran and Vanmarcke (1986) is able to successfully characterize the coherency variation of these far-field events. The single exponential models, however, are not too satisfactory.

The phase components of the random field model are presently characterized by a single apparent wave velocity. Although this simplistic assumption is unable to fully characterize the arrival times at each station, it captures the essential trend in the arrival times.

ACKNOWLEDGMENTS

This research was supported by the National Science Foundation under Grant Number ECE-8605184. Any opinions, findings and conclusions are those of the author and do not reflect the views of the National Science Foundation.

The SMART-1 array data were made available by the Seismographic Station of the University of California at Berkeley and the Institute of Earth Sciences of the Academia Sinica in Taipei.

REFERENCES

1. B.A. Bolt, et. al. (1982). "Earthquake Strong Motions Recorded by a Large Near-Source Array of Digital Seismographs," *Earthquake Engineering and Structural Dynamics*, 10, 561-573.

2. Harichandran, R.S., and Vanmarcke, E. (1986). "Stochastic Variation of Earthquake Ground Motion in Space and Time," *Journal of Engineering Mechanics*, ASCE, 112(2), 154-174.

3. Harichandran, R.S. (1987). "Correlation Analysis in Space-Time Modeling of Strong Ground Motion," *Journal of Engineering Mechanics*, ASCE, 113(4), 629-634.

4. Hindy, A. and Novak, M. (1980). "Pipeline Response to Random Ground Motion," *Journal of Engineering Mechanics*, ASCE, 106(2), 339-360.

5. Jenkins, G.W. and Watts, D.G. (1968). *Spectral Analysis and its Applications*, Holden-Day, San Francisco.

6. Loh, C-H., Penzien, J., and Ysai, Y.B. (1982). "Engineering Analysis of SMART 1 Array Accelerograms," *Earthquake Engineering and Structural Dynamics*, 10, 575-591.

7. Loh, C-H. (1985). "Analysis of the Spatial Variation of Seismic Waves and Ground Movements from SMART-1 Array Data," *Earthquake Engineering and Structural Dynamics*, 13, 561-581 (1985).

FOAM RUBBER MODELING OF SOIL-STRUCTURE INTERACTION

A. Anooshehpoor*   J. N. Brunet   R. H. Lovberg‡

Results from a layered foam rubber model representing soil and structure at the El Centro Terminal Substation indicate that the response amplitude of the building at frequencies above 4 Hz is up to a factor of 3 lower than the response amplitude of the free field for vertically incident SH-waves. Two-dimensional computer model results by Shannon and Wilson, Inc. and Agbabian Associates (1980), predicted that the response amplitude in the building is up to a factor of 2 lower than the response amplitude of the free field for frequencies above 1.5 Hz. Our study also shows that at higher frequencies most of the reduced response is due to the energy being scattered by the shape of the foundation and not by its inertial mass. Results for incident angles of 50° and 90° also indicate high frequency filtering by the building. For horizontally incident waves the filtering of high frequencies by the building is less significant but it takes place for almost all frequencies. We use the vertically incident SH-wave results to estimate the free field ground acceleration during the May 18, 1940 El Centro earthquake for frequencies up to 10 Hz. The transfer function used to predict the free field motion is very similar in shape to the weighting function used by Munguia and Brune (1984), to match synthetic accelerograms with the 1940 El Centro earthquake. The results indicate that the free field acceleration was about a factor of two higher than recorded by the El Centro 1940 accelerogram.

1 Introduction

In recent years it has become obvious that many of the important problems relating to earthquake hazard are seismic engineering problems involving interaction between the structure and the local geological materials being shaken by the incoming waves. Although it is often possible to obtain numerical or analytical results for simplified models, it is often not clear what the effects of the necessary approximations are. Using physical models, very complex structures can be constructed and studied, and results used in conjunction with computer modeling to cross-check results and thus lend more confidence to both techniques. Furthermore, study of physical models can lead to physical insights which would not be evident from computer modeling or theoretical calculations.

Foam rubber (*polyurethane*) is lightweight and soft with a considerable range in density and elastic wave velocities. We use large blocks with dimensions of about $1m \times 2m \times 3m$. The light weight and low rigidity of foam rubber leads to modeling frequencies in the audio range. The attenuation properties of foam rubber corresponds to a Q of about 10, comparable to the Q values expected for Earth materials near the surface of the earth undergoing strong deformation during earthquakes.

The El Centro Terminal Substation Building (*TSB*) has been the site of accelerographs which recorded the May 1940 El Centro, and October 1979 Imperial Valley earthquakes. The Building has a massive foundation and rests on soft soil deposits. Under these conditions strong soil-structure interaction effects are likely to cause the recorded acceleration to differ from the true free-field ground motion. Shannon and Wilson, Inc. and Agbabian Associates (1980), based on *2-D* numerical modeling, concluded that at frequencies above 1.5 Hz, the motions in the building fall significantly

*Research Associate, Seismology Dept. Univ. of Nevada Reno, Reno NV 89557
†Professor of Geophysics, Seismology Dept., Univ. of Nevada Reno, Reno NV 89557
‡Professor of Physics, Physics Dept. B-019, Univ. of Calif. San Diego, La Jolla CA 92093

below the free field ground motions. The 3-D foam rubber modeling of this problem indicates similar strong soil-foundation interaction effects at frequencies above 4 Hz.

2 Experimental Procedure

2.1 Model Construction

The 1.8 lb/ft^3 (0.029 gr/cm^3) class foam rubber comes in a wide range of elastic wave velocities. To construct the model, foam types with the relative shear moduli—best representing the change in the earth's shear moduli at the TSB site—are selected using data obtained by Shannon and Wilson, Inc. and Agbabian Associates through their 1976 geotechnical investigation (*Figure 1*). The thickness of the layers is determined once the scale is decided:

$$scale = \frac{\lambda_m}{\lambda_e}$$

where λ_e is a typical wavelength in the *earth*, and λ_m is its corresponding wavelength in the *model*. With:

$$\lambda = \frac{\beta}{\nu}$$

we have:

$$scale = \frac{\beta_m}{\beta_e} \frac{\nu_e}{\nu_m}$$

where β and ν are the shear wave velocity and frequency respectively. The *scale* is $\frac{1}{47}$ which fixes the frequency in the *model* at 10 times the frequency in the *earth*.

Only the top six layers were modeled, in part to ensure that enough energy arrives at the surface when the bottom layer is excited (foam rubber has a low Q of about 10, and propagation through a thicker model attenuates the energy too much). To provide continuity, the layers are cemented together with a special glue that leaves a strong but flexible glue line.

The TSB model is made of *styrofoam* with a density of about 0.032 gr/cc, which was chosen because its density was very close to the appropriate density to represent the foundation of the TSB. The *styrofoam* is much more rigid than the *polyurethane* representing the soil, corresponding to the fact that the cement foundation of the TSB is much more rigid than the surrounding sediments.

2.2 Model Excitation

Mechanical impulse is applied to the foam models by a fast electromagnetic driver which is similar in its design to the driver of a large audio loudspeaker (*Figure 2*). The force is applied tangentially to a thin square fiberglass plate (56cm × 56cm) cemented to the rear of the foam block. The actual displacement function is a unidirectional pulse of about .5 mm amplitude, lasting about 8 milliseconds.

2.3 Recording of Motion

Displacement of the foam surface is measured by telescopic 2-*axis* optical position sensing detectors (PIN-$SC/4D$, by UNITED DETECTOR TECHNOLOGY, INC.), which are focused on small light emitting diodes (LED) embedded in the foam. Each of these units employs a solid-state quadrant-detector capable of resolving the centroid of the light spot image to about 2.5μm, depending on the brightness of the image. Six of these detectors are available for the simultaneous recording on the model surface. Normally one detector is reserved for monitoring the input motion.

2.4 Data Logging Facility

The data logger system consists of *COMPAQ* computer and *TECMAR* Labmaster A/D convertor. These records are converted to 12-*bit* digital words, and may be sampled at rates from 17 kHz for single channel, or $\frac{17}{n}$ kHz for n channels, to any sub-multiple of 1 kHz for all channels.

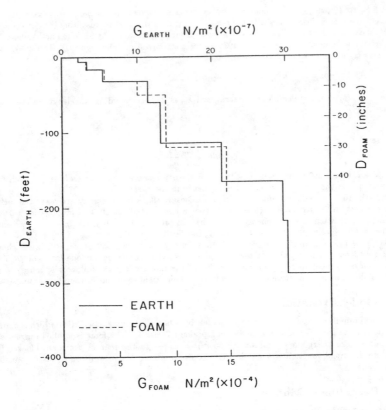

Figure 1: A plot of shear modulus versus depth in the Earth (solid line, after SW/AA), and in the foam (dashed line).

Mechanical Drive

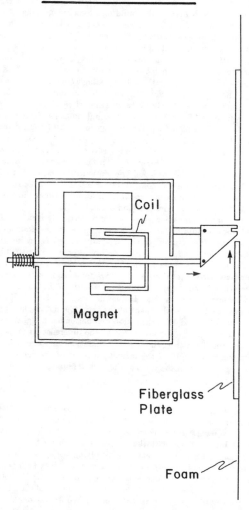

Figure 2: Illustration of electromagnetic driver

3 Summary of Results

3.1 Vertically Incident SH-Wave

Shannon and Wilson, Inc. and Agbabian Associates used $SHAKE/FLUSH$ codes (1980) to compute the free field and building response for vertically incident shear wave in their 2-D numerical model of the El Centro Terminal Substation Building. Results were obtained for SH-wave polarized in an east-west direction (perpendicular to the building's long axis), using as input motion the N-S component of the October 31, 1935 Helena, Montana earthquake. Their results predicted that for frequencies above 1.5 Hz, there is considerable filtering of high frequency energy, ranging from 20% to over 100% (of the response in the building), and they attributed this to the significant mass of the building's foundation (our results, described below, indicate that much of the filtering is due to simple scattering). In the foam rubber model of the El Centro TSB, excitation was both perpendicular (x-axis) and parallel (y-axis) to the building's *long* axis, with an impulse as the input motion (*Figure 3.a*). Then the convolution with the N-S component of the *Helena* 1935 earthquake was carried out for comparison with SW/AA's 2-D model results (*Figure 3.b*). *Figure 4* shows the response spectra of free field and the TSB, for input motions along x and y axes. There is a significant drop (up to 200%) in high frequency energy in the building for frequencies above 4 Hz. Below 4 Hz there is little filtering.

In the 2-D numerical model (SW/AA), filtering starts at about 1.5 Hz, while in the 3-D foam rubber model, it starts at about 4 Hz (*Figure 5*). This discrepancy could be a direct result of the difference between 2-D and 3-D models. Soil-Foundation Interaction is expected to be significant when the wavelength of incident wave is comparable or smaller than the size of the structure. Near the surface, the wavelength corresponding to 1.5 Hz is about 5 times the length of the foundation. At 4 Hz the wavelength is about 2 times the foundation length.

Figure 6 shows a plot of foundation response for a vertically incident SH-wave (polarized in x-direction). In addition , results are shown for a case with the superstructure removed from the foundation. The fact that the removal of the superstructure has an insignificant effect at high frequencies, suggests that the foundation alone is mainly responsible for the high frequency filtering.

Figure 7 shows the response spectra for the foundation alone and for a hollow, rigid foundation which represents a nearly zero mass foundation (TSB foundation with most of the mass removed and then braced to retain its rigidity). for frequencies above 7 Hz there is little difference in the response curves. This indicates that at high frequencies most of the reduced response amplitude is due to the energy being scattered by the shape of the rigid foundation, and not by its inertial mass. For frequencies between 1.5 and 4 Hz, the hollow foundation has a higher response amplitude than the solid foundation. This difference is attributed to the difference in mass. The solid foundation has a higher response amplitude than the hollow foundation for frequencies between 4 and 7 Hz. At these frequencies the wavelength of the incident wave is of the same order as the size of the foundation and could easily deform the walls of the hollow foundation, reducing its response.

3.1.1 El Centro May 18, 1940 Earthquake

We use results of the vertically incident SH-wave to estimate the *free field* response at the El Centro Terminal Substation during the May 18, 1940 El Centro earthquake. To do so we have to deconvolve the El Centro record using the transfer function for the TSB to obtain the input motion and then convolve the input motion with the transfer function for the free field to obtain the free field ground acceleration. This is equivalent to multiplying the spectrum of El Centro earthquake (N-S component) by the ratio of the two spectral transfer functions (T_{ff}/T_{TSB}). *Figure 8.a* shows the Fourier spectra of the two transfer functions. At frequencies above $10 Hz$ the signal to noise ratio of the TSB transfer function is very low. This makes the deconvolution process unstable at high frequencies. However, we can estimate reasonable cutoff frequencies based on free field records for 1979 Imperial Valley earthquake, which recorded little energy beyond $10 Hz$ in frequency. Thus we will estimate the free field response during 1940 earthquake up to frequencies near $10 Hz$. *Figure 8.b* shows the peak acceleration of about $.6g$ for cutoff frequency of $10 Hz$. The deconvolution process could overestimate high frequency content of the free field if those frequencies were not present at

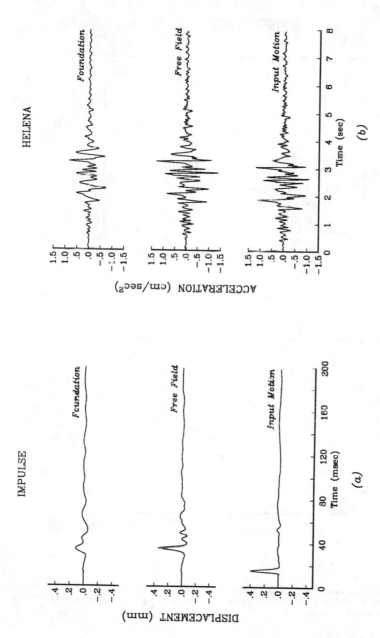

Figure 3: Free field response and *TSB* foundation response for vertically incident *SH*-wave polarized perpendicular to the building's long axis, for (*a*) an impulse, and (*b*) the N-S component of the Helena, Montana, October 31, 1935 earthquake as the input motion.

INCIDENT ANGLE = 0°

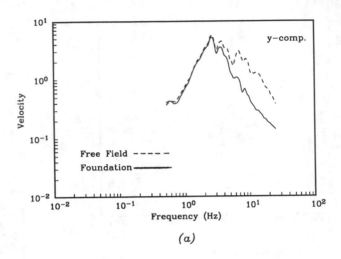

(a)

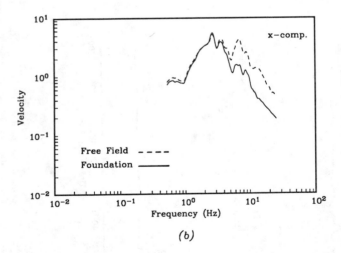

(b)

Figure 4: Comparisons of free field and foundation response spectra for vertically incident SH-wave, (a) polarized parallel, and (b) perpendicular to the building's long axis. N-S component of Helena earthquake is used as input.

INCIDENT ANGLE = 0°

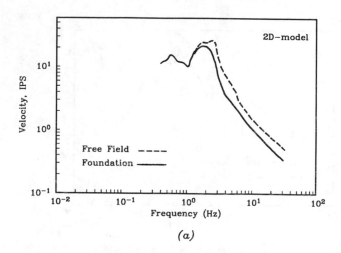

(a)

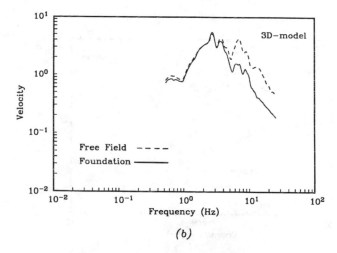

(b)

Figure 5: Comparisons of *2-D* numerical model results obtained by SW/AA (1980) using SHAKE/FLUSH codes (a) with experimental results obtained from *3-D* foam rubber modeling (b).

INCIDENT ANGLE = 0°

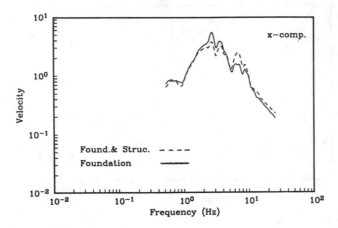

Figure 6: Comparison of foundation response in the presence (dashed curve) and absence (solid-curve) of the superstructure.

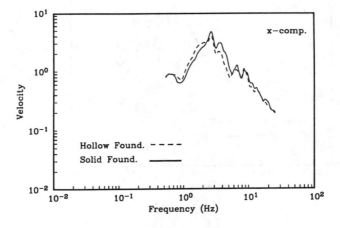

Figure 7: Comparison of the response of solid (solid curve) and hollow (dashed curve) foundations.

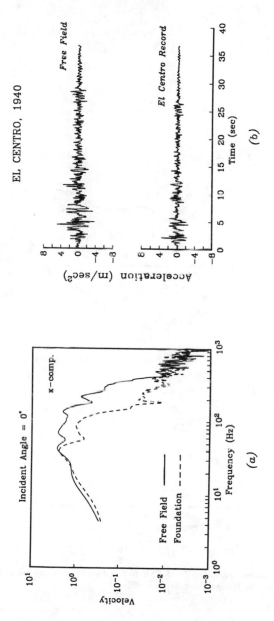

Figure 8: (a) Fourier spectra for free field and foundation transfer functions, (b) El Centro free field accelerogram obtained by deconvolution with cutoff frequency of 10 Hz. The lower figure is the N-S component of may 18, 1940 El Centro earthquake.

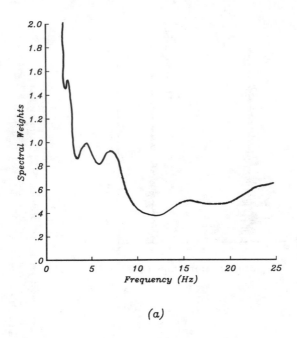

(a)

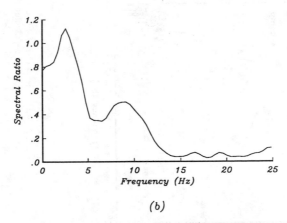

(b)

Figure 9: Comparison of the spectral ratio of *TSB* to free field in the foam model (b), with the spectral weights (a), used by Munguia and Brune (1980).

INCIDENT ANGLE = 50°

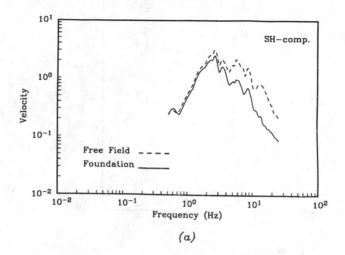

(a)

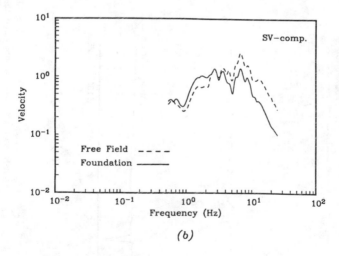

(b)

Figure 10: Comparisons of free field response (dashed curve) and foundation response (solid curve) for both SH-wave and SV-wave cases. Incident angle is 50°.

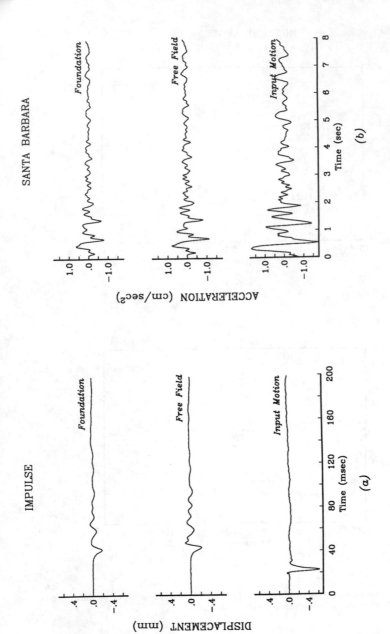

Figure 11: Free field response and *TSB* foundation response for horizontally incident wave for (*a*) an impulse, and (*b*) the S45°E component of the Santa Barbara, California, June 30, 1940 earthquake as the input motion.

INCIDENT ANGLE = 90°

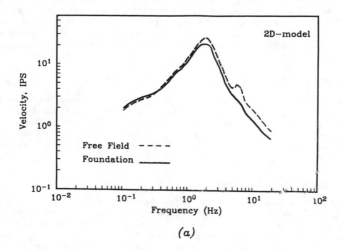

(a)

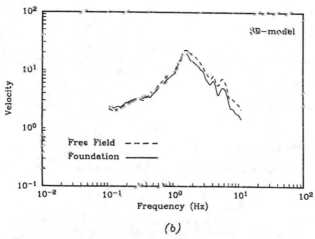

(b)

Figure 12: Comparisons of *2-D* numerical model results obtained by SW/AA Using TRI/SAC codes (*a*) with the experimental results obtained from *3-D* foam rubber modeling (*b*). S45°E component of Santa Barbara earthquake is used as input.

ne earthquake source in the first place *i.e.* are a result of noise, either from the field data or the model data. *Figure 9.b* shows the spectral ratio of the *TSB* to free field in the foam model which was used to estimate the free field ground acceleration during 1940 El Centro earthquake. *Figure 9.a* shows the weighting function used by Munguia and Brune (1984) to make the synthetic El Centro 1940 accelerogram agree with the recorded accelerogram. The shape of the both graphs is very similar, suggesting that most of the filtering observed by them was due to Soil-Structure Interaction, and not as they suggested, due to the source.

3.2 Obliquely Incident S-Wave

Results for S-wave with the incident angle of about $50°$ are shown in *Figure 10*. For the SH-wave (polarized along y-axis) results are very similar to the corresponding results obtained for vertical incidence, except for the overall lower response amplitude due to the longer travel path (attenuation) for obliquely incident waves. For the SV component (polarized perpendicular to the long axis of the TSB), besides the substantial reduction at higher frequencies, there are also some differences at lower frequencies.

3.3 Horizontally Incident Wave

The response of the TSB foundation and the free field for a horizontally incident wave is given for both an impulse (*Figure 11.a*), and the S45°E component of the Santa Barbra, California June 30, 1941 earthquake (*Figure 11.b*), as input motion. *Figure 12* shows the response of the TSB foundation and free field in the foam model (*b*). The Santa Barbra earthquake was used as input motion following SW/AA. The results indicate filtering at almost all frequencies. The filtering at higher frequencies is less than the case of the vertically incident wave. This is due to the fact that free field record does not have much high frequency content. Our results are in good agreement with SW/AA results in this case (*a*).

Conclusion

Our foam rubber model of El Centro Terminal Substation Building indicates that during an earthquake the massive foundation of the building—the site of accelerographs—tend to filter out high frequencies, in approximate agreement with the *2-D* numerical results. Our results also showed that at high frequencies much of the filtering is due to the scattering by the shape of the foundation. Experimental results for different angles of incidence also indicate reduction of high frequncy in the building. With foam rubber modeling, very complex, *3-D*, Soil-Structure Interaction problems can be studied.

Acknowledgments

This study was done under a grant from the Electric Power Research Institute (EPRI rp 2556-2).

References

[1] Munguia, L. & Brune, J. N.,1984. Simulations of strong ground motion for earthquakes in the Mexicali-Imperial Valley region, Geophys. J. R. aster. Soc. (1984) 79,747-771.

[2] Shannon and Wilson, Inc. and Agbabian Associates (1980). Site-dependant response at El Centro, California accelerograph station including soil/structure interaction effects, NUREG/CR-1642,U.S. Nuclear Regulatory Commission, Washington, D.C., 150pp.

SHEAR WAVE VELOCITY IN GRAVELS

By A.J.L. Bolognesi*, M. ASCE, O.A. Vardé**, and F.L. Giuliani***, M. ASCE

ABSTRACT

Shear wave velocity at small strains related to a reference confining pressure, V_{s1}, is a basic dynamic soil property suitable for layer characterization and usually the most important field measurement to introduce in the evaluation of the dynamic response of layered soils to earthquake motion. For the specific case of gravels this paper presents the known range of its values and it tries to evaluate which factors influence it. It presents its relationship with the dimensionless maximum shear modulus coefficient, $(X)_{max}$, results of field determinations showing the dispersion of values and of attempts to use it as an indirect measurement of other properties.

INTRODUCTION

In the past decade geotechnical engineers have recognized the usefulness of shear wave velocities determinations to solve soil dynamic problems. This paper treats in detail that subjet for gravels.

Only shear wave velocities, V_s, at small strains, of the order of 10^{-4} or smaller, such as those corresponding to the maximum shear modulus, G_{max}, or the cross-hole techniques, are considered. Also, because the paper is biased toward practical engineering problems which must take into account the dispersal of field determinations, approximate relationships are used. As reaffirmed by Seed et al (1984), for most practical purposes, the shear moduli vary with the 0.5 power of effective mean principal stress. Consequently, shear wave velocities vary with the 0.25 power. This approximate relationship is used to relate shear wave velocities to a reference confining pressure, which in this paper is equal to one atmosphere, P_a. This shear wave velocity, V_{s1}, is a property which characterizes each gravel.

Seed and Idriss(1970) introduced the shear modulus parameter, K_2, $(K_2)_{max}$ for G_{max}, valid only for the U.S. customary system of units. Hardin(1978) introduced the dimensionless parameter $G_{max}/(P_a \times \sigma'_0)^{0.5}$, which in this paper is designated $(X)_{max}$ valid for any system of units. Relationships can be established between V_{s1} and $(X)_{max}$ which makes it possible to obtain both values either from field or laboratory determina-

---

* Partner Bolognesi-Moretto. Ingenieros Consultores, Luis Saenz Peña 250, Buenos Aires(1110), Argentina.
** President Vardé y Asociados, Av. Pres. Quintana 585,Buenos Aires(1129), Argentina.
*** Chief Civil Engineering Department, Hidronor S.A.,Pte.Irigoyen 378, Cipolletti(8324), Argentina.

ions.

The paper is organized, besides the Introduction and Conclusions, in four subtitles: a) Shear Wave Velocity in Ground Motion Evaluation, which places it within the main purpose of the Conference. b) Approximate Relationships, which contains the equations used and appropriate comments. c) Available Data, where meaningful data is gathered and compared and d) Measurements of Shear Wave Velocities Along the Limay River, which shows the dispersion of actual values in river alluviums and embankment gravelly shells, layer characterization by means of dynamic soil properties and the results of attempts to use V_{s1} as an indirect measurement of other properties.

SHEAR WAVE VELOCITY IN GROUND MOTION EVALUATION

Martin and Seed(1982)presented a study of three different approaches available to determine the dynamic response of horizontally layered soils deposits to earthquake motion: 1) the equivalent linear method; 2) the method of characteristics and 3) the finite element method. Of the three the first is the most used because it is the base of the SHAKE (Schnabel et al 1972)program distributed by NISEE(reference 11)and now available in microcomputer version(1985). In all three methods, as well as in most of the computer programs for earthquake engineering and soil dynamics, the dynamic soil properties are summarized by: (1) The shear modulus at low strain, G_{max}; (2) the modulus attenuation curve with strain; and (3) the damping ratio. Seed et al(1984)have indicated that the modulus attenuation curve with strain amplitude for most cohesionless soils is about the same and that damping ratios for sands and gravels are very similar, from which they concluded that the shear moduli for any given soil are generally characterized by the modulus coefficient measured at low strain for that soil.

When dealing with a practical engineering problem the sequence of determinations is: a)By means of shear wave measurements, preferably by cross-hole techniques for gravels, characterize the different layers of the natural deposit by means of V_{s1}. In this paper two examples are given of this layer characterization; b) For the average V_{s1} of each of the layers thus determined compute the values of $(X)_{max}$, or $(K_2)_{max}$, and G_{max}. This paper contains the appropriate equations for that purpose; c) From published results select the modulus attenuation and damping ratio versus strain curves. For gravels Seed et al(1984)have presented values from laboratory tests and,only for the attenuation curve, Ortigosa et al (1985)from cyclic plate tests; d)Select the earthquake base motion; e) Perform the ground response analysis using the selected computer program.

The above sequence of determinations shows how important site dynamic characterization is for the resolution of practical problems. The authors consider that for that purpose shear wave velocity related to a reference confining pressure of one atmosphere, V_{s1}, is a convenient basic property, independent of confining pressures which blurr layer characterization. This paper presents,for the specific case of gravels,the probable range of values of V_{s1} and how it is affected by the many factors, physical properties, geologic and earthquake history, man originated modifications, etc, that must be taken into consideration, and examples of field dispersion of results and layer characterization.

APPROXIMATE RELATIONSHIPS

SHEAR WAVE VELOCITY

Equations used in this paper

$$\sigma'_o = (\sigma'_1 + \sigma'_2 + \sigma'_3) / 3 \tag{1}$$

$$G_{max} / P_a = (X)_{max} (\sigma'_o / P_a)^{0.5} \tag{2}$$

$$(X)_{max} = G_{max} / (P_a \times \sigma'_o)^{0.5} \tag{3}$$

in which G_{max} = shear modulus at small strains, P_a = the atmospheric pressure in the system of units used for G_{max} and σ'_o and $(X)_{max}$ = the dimensionless maximum shear modulus parameter.

$$G = V_s^2 \ \gamma \ /g \tag{4}$$

in which G = shear modulus, V_s = shear wave velocity, γ = total unit weight and g = acceleration due to gravity.

$$V_{s1} = V_s / (\sigma'_o / P_a)^{0.25} \tag{5}$$

in which V_{s1} = shear wave velocity when the confining pressure, σ'_o, is equal to P_a.
At small strains

$$V_s = [(X)_{max} (P_a \times \sigma'_o)^{0.5} \times g / \gamma]^{0.5} \tag{6}$$

Which is obtained by substituting G in equation(4)with the value of G_{max} from equation(3). By definition $\sigma'_o \sim P_a$ for V_{s1}, from which it follows

$$V_{s1} = (X)_{max}^{0.5} \times (P_a \times g / \gamma)^{0.5} \tag{7}$$

$$(X)_{max} = (V_{s1})^2 \times [\gamma / (P_a \times g)] \tag{8}$$

which are the relationships permitting the interchangeable use of either V_{s1} or $(X)_{max}$ to define dynamic soil properties.

Influence of confining stress conditions on shear wave velocity

Hardin and Richart(1963)published the first paper indicating that for small strain amplitudes the shear modulus varies with the 0.5 power of effective mean principal stress. Seed and Idriss(1970)proposed the equation

$$G = 1000 \ K_2 \ (\sigma'_m)^{0.5} \ psf \tag{9}$$

in which $\sigma'_m = \sigma'_o$ (psf) and K_2 is a shear modulus parameter. Seed et al(1984)reaffirmed that "for most practical purpose, the dynamic shear moduli of granular soils (sands and gravels) can be conveniently expressed" by equation(9).

K_2 is valid only when G and σ'_o are measured in psf. If in equation (2), P_a is made equal to 2116.22 psf, atmospheric pressure in psf, and the value of G_{max} thus obtained is introduced in equation(9)for $G = G_{max}$ it results:

$$(X)_{max} = 21.738 \ (K_2)_{max} \tag{10}$$

which gives the relationship between the numbers expressing $(X)_{max}$ and $(K_2)_{max}$.

In this paper, tables and figures present the dimensionless parameter $(X)_{max}$, the values of which can be readily transformed in those corresponding to $(K_2)_{max}$ by equation (10).

Yu and Richart (1984) and Stokoe et al (1985) made studies employing sands and concluded separately that shear wave velocities and, hence, shear moduli depend about equally on the principal effective stresses in the direction of wave propagation and particle motion and are essentially independent of the third principal stress. These results confirm that equations based on G'_o are simply approximations but for the ordinary stress conditions encountered in situ differences between computed and correct values seem to be relatively unimportant.

Reference confining pressure of shear wave velocities

For the same material, V_S increases as the confining pressure increases.

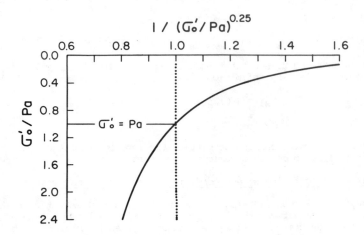

FIG.1—CHART FOR COMPUTING V_{s1}.

From the application of equation 5 it follows that all velocities, V_S, with confining pressures below P_a are multiplied by a number larger than 1 to obtain V_{s1} and, correspondingly, for a number smaller than 1 when $G'_o > P_a$. Fig.1 shows the magnitude of these corrections. It also serves as a graphical method to obtain V_{s1}.

Approximately, for a submerged gravel deposit, $V_S = V_{s1}$ at a depth of 42.5 feet (13.0 m) when the reference confining pressure is 1 atmosphere. For dry deposits such approximate depth is 25 feet (7.5 m).

Void ratios and relative densities

Hardin(1978)pointed out that e is the appropriate parameter for correlation of shear wave velocities, but this cannot be shown from tests
on a single sand where D_r, relative density, is proportional to e, void
ratio. Iwasaki and Tatsuoka(1977)had arrived at the same conclusion
as Hardin.

The figures in this paper present correlations between $(X)_{max}$ and e,
and between V_{s1} and e when it is pertinent. D_r scales, which are different for different materials are also incorporated.

AVAILABLE DATA

In their presentation of a comprehensive series of tests concerning
the shear modulus-strain relationship for gravelly soils Seed et al(1984)
pointed out the meager data available on the subject. These tests were
performed on 12-inch diameter isotropically consolidated samples under
undrained cyclic loading conditions. The modeled sample of Oroville and
the sample from the natural deposit of Livermore have been selected from
these series. The particles of the Oroville gravel were well-rounded
and hard. Those of Livermore were well-rounded to rounded and relatively hard.

The values of $(K_2)_{max}$ in Fig.11, 14, 15 and 16 of Seed et al(1984)were
transformed into $(X)_{max}$ by means of equation 10 and the corresponding
values of e computed from the data on Table 3 of the same paper. With Y
values obtained from the above mentioned Table 3, the values of V_{s1} were
computed from the previously determined $(X)_{max}$ by means of equation 7.
Diagrams correlating $(X)_{max}$ and V_{s1} with e and D_r were drawn, from
which the values presented in Table 2 and figures 3 and 4 were taken.

With the purpose of trying to estimate the influence on V_{s1} of: a)the

TABLE 1

GRAIN SIZE AND PLASTICITY OF THE GRAVELS IN TABLE 2

Identifi-cation.	Refer-ence N°	D_{max} mm	% > than 75 mm	than 50 mm	D_{50} mm	D_{10} mm	% < than 0.075 mm	I_p
K and E[*]	S	75	0	5	12	1	1	NP
Limay 2	**	150	10	20	20	0.5	1	NP
Livermore	17	50	0	0	10	0.6	8	HP
Oroville	17	50	0	0	12	0.4	4	NP
	2	150	20	32	21	1.2	2	NP
	19	610	30	35	25	0.6	6	NP
Santiago	12	250	25	40	36	1	3	5-20
Limay 1	**	150	30	40	25	1	2-3	15

*Kokusho and Esashi.
**This paper.

TABLE 2

e, D_r, $(X)_{max}$ AND V_{s1} OF THE GRAVELS IN TABLE 1

Identification	Reference N°	Place of determination	e	D_r %	$(X)_{max}$	V_{s1} m/s	ft/s*
K and E	8	Laborat.	0.315	—	1530[2]	254[3]	830[3]
Limay 2	**	Field	0.206	90	1845[4]	285[4]	930[4]
Livermore	{17	Laborat.	0.206	86	1870[5]	282[6]	920[6]
	17	Laborat.	0.23	78	1600[5]	260[6]	850[6]
Oroville	{17	Laborat.	0.23	81	2400[5]	305[6]	1000[6]
	2-16	Field	0.23[1]	86[1]	3804[7]	389[8]	1280[8]
Santiago	12-13	Field	0.255	—	6521[9]	540[10]	1770[10]
Limay 1	**	Field		90	6403[12]	520[11]	1700[11]

*To the nearest 10 feet.
**This paper.
NOTES: (1)From reference 2. (2)Average of values computed from Fig.5, reference 8. (3)Computed by means of equation 7 for $\gamma = 2.4\,\gamma_w$. (4) From this paper Limay 2 test embankment. (5)Computed from reference 17. (6)Computed from reference 17. (7)Computed by equation 10 for $(K_2)_{max} = 175$. (8)Computed by equation 7 for $\gamma_d = 2.355\,\gamma_w$ and $\gamma_s = 2.90\,\gamma_w$, references 2 and 19. (9)Computed by equation 10 for $(K_2)_{max} = 300$. (10) Computed by equation 7 for $\gamma = 2.26\,\gamma_w$. (11)From Fig.2. Approximate average value for the layer of gravel with voids partially filled with clay. (12)Computed by equation 8 for $\gamma = 2.4\,\gamma_w$.

$\gamma_w = 9.81$ kN/m^3 = 1 tn/m^3 = 62.428 pcf.

characteristics of each gravel (grading, plasticity index, shape and hardness of particles, etc), b)the percent of particles larger than the maximum gravel size, 75 mm, c)the maximum size of cobbles, d)the geological history of natural deposit and e)the effect of aging and seismic events since the end of construction on earth structures, Table 1 and 2 were prepared, including both laboratory and field determinations. They include results of both the Oroville and Livermore laboratory tests at values of e which permit comparisons plus other determinations which are briefly described as follows.

The laboratory results mentioned above are supplemented for Oroville with those obtained from measurements of the dam response to earthquakes. Seed(1980)reported than the true value of $(K_2)_{max}$ for the soil in the shells was computed to be about 170. The grain size of said shells is given in Table 1 from two sources, (references 2 and 19).

Kokusho and Esashi(1981)made undrained cyclic loading tests on 12" (300 mm) diameter, isotropically consolidated, fully saturated specimens.

SHEAR WAVE VELOCITY

Ortigosa et al(1981,1985)defined the attenuation curve, obtained from cyclic plate tests, of the Santiago(Chile)gravel, for which the value of $(K_2)_{max}$ from refraction surveys is, as reported by Acevedo et al(1973), of the order of 300.

Results of in situ shear wave velocity measurements by means of cross-hole seismic tests made at three places along the Limay River(Argentina) are included in this paper. In Tables 1 and 2 are values from a test embankment, Limay 2, and a gravel with cobbles and plasticity index, the layer of gravel in the alluvium with voids partially filled with clay at Limay 1, which are convenient for the comparisons that follow.

Laboratory results from Livermore gravel and field cross-hole results from the Limay 2 test embankment give the lowest values of V_{s1} or $(X)_{max}$ published so far for dense gravels (e of the order of 0.21, D_r in the 85 to 90% range, V_{s1} of the order of 900 ft/s).

Even under laboratory conditions two gravels with practically the same grain size and plasticity, with particles which do not differ much in shape and hardness and with the same void ratio (e = 0.23)have different values of V_{s1}, as shown by the tests of Livermore ($V_{s1} \sim 850$ ft/s) and Oroville (V_{s1} = 1000 ft/s) gravels.

For the same void ratio (e = 0.23) the differences between Oroville laboratory (V_{s1} = 1000 ft/s) and field ($V_{s1} \simeq 1280$ ft/s) can be attributed in undetermined proportions to two factors. One factor is the presence in the dam shells of a fraction of the order of 30 to 35% of particles larger than 2"(50 mm), which are not present in the modeled sample. The other is the effect of aging and seismic events since the dam was built.

If the values of V_{s1} from Oroville field, 1280 ft/s, are compared with those of Santiago natural deposit, 1770 ft/s, it is found that these two gravels similar in grain size and percentage of cobbles differ in the plasticity index, from which the differences in V_{s1} can be assigned to the influence of the cohesion, though the effects of the intense seismic activity in Santiago cannot be disregarded. If the velocities were corrected in order to reduce both to the same value of e, their differences would be somewhat larger.

The approximate average value of V_{s1} = 1700 ft/s, for the layer of gravel with voids partially filled with clay at Limay 1, seems to confirm that dense gravels with about 30% of cobbles and plasticity are at the uppermost range of shear wave velocities.

In the above comparisons V_{s1} was measured only in ft/s because all matching values in m/s and ft/s are given in Table 2.

MEASUREMENTS OF SHEAR WAVE VELOCITIES ALONG THE LIMAY RIVER

Two major dams have already been built and one is under construction along the Limay River located in Argentina on the eastern side of the Andes and further developments are contemplated. Extensive measurements of shear wave velocities, mostly by the cross-hole seismic method have been made during the last five years and publications on the subject, a list of which is given by Bolognesi et al(1987), have appeared during this period.

This paper has been prepared essentially from actual test data furnished by the coauthors Giuliani, who was in charge of general supervi-

sion and Vardé who made the laboratory tests at Limay 1 and 2, while at
Limay 3 they were made by the Argentina firm Geotecnia S.A.. Cross-hole
seismic testing and data analysis were performed by K.H. Stokoe II, from
Austin, Texas, U.S.A. at Limay 1 and by the Argentine firm PROINGEO S.A.
at Limay 2 and 3. Both firms used similar techniques though it must be
pointed out that at Limay 1 a digital oscilloscope was used as recording
equipment and cement grouting fixed the PVC casing while at Limay 2 and
3 an enhancement seismograph registered the signals while the annular
space between PVC casing and borehole was filled with pea gravel.

The local name of the sites designated in this paper as Limay 1,Limay
2 and Limay 3 are Alicura, Pichi Picún Leufú and Michihuao,respectively.

Shear wave velocities made at Limay 1 and Limay 3 are of interest because they cover the whole range of known possible values. Limay 1 is
located approximately 120 miles(190km)upstream from Limay 3. Al Limay 1
the width of the river-valley is of the order of 1150 ft (350m) and the
slope of the riverbed of the order of 2.5‰, while at Limay 3 the respective values are 8200 ft (2500m) and 1‰.

Characterization of the dynamic soil properties of layers of river alluvium at Limay 1 and Limay 3.

Fig.2 shows the values of V_{S1} from cross-hole seismic tests at three
different locations at Limay 1 and at four different locations at Limay
3. V_{S1} measurements were made with the water table practically at the

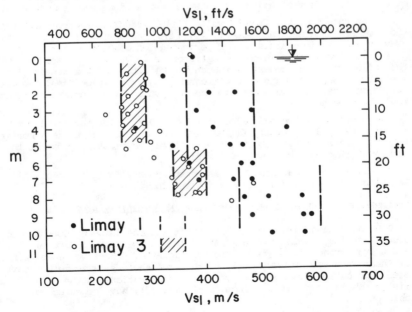

FIG.2—VALUES OF V_{S1} IN RIVER ALLUVIUM AT LIMAY 1 AND LIMAY 3.

surface in the alluvium deposit. In both cases reliable samples, and, consequently, good correlations were obtained between shear wave velocity, grading, physical properties and relative density of the gravels because: 1) at Limay 1 the shear wave velocity measurements were made at sites where later on the trench for the impervious core was excavated and; 2) at Limay 3 the water table was later on lowered 16.5 ft (5 m) inside a 7.5 ft (2.3 m) diameter shaft and eleven approximately 3.5 to 4ft (1 to 1.2 m) diameter case borings drilled down to bedrock with rotating buckets allowed good samples to be obtained for grain size and plasticity determinations in the full depth.

Fig.2 contains only the values useful for comparisons. Where determinations were made bedrock was at an average depth of approximately 50 ft (15 m) at Limay 1 and 35 ft (11 m) at Limay 3. It is known that the proximity of rock introduces distortions in the values of V_{s1} and these values cannot be assigned to the gravels at the bottom, which justifies their exclusion for the comparisons made in this paper. At Limay 1 were also excluded the cemented gravels layers and those containing a large amount of big boulders, which give very high values of V_{s1}.

Data and comments pertaining to Limay 1.

The matrix of the alluvium is made of gravel and sand with cobbles. Its properties can be defined by the following values:

D_{max} : 150 mm   % > than 75 mm : 30   % > than 50 mm : 40

D_{50} : 25 mm D_{10} : 1 mm   % < than 0.075 mm : 2-3

The matrix is nonplastic and the gravel and coarse sand well rounded, while the rest of the sand is angular and sometimes with flat particles. The grains are made of tuff, basalt and andesite.

The material described occupies the upper 20 ft (6 m) of the alluvium and is characterized by values of V_{s1} ranging approximately from 1200 to 1600 ft/s (365 to 490 m/s). D_r is of the order of 95 - 100%. From Tables 1 and 2 it can be seen that this upper layer has properties similar to those of Oroville shells. Approximately from 20 to 33 ft (6 to 10m) the voids of the matrix are partially filled with clay, the I_p is of the order of 15 and D_r of the order of 90%. This layer is characterized by values of V_{s1} ranging approximately between 1500 and 2000 ft/s (460 to 610 m/s). These properties are similar to those of Santiago gravel as presented in Tables 1 and 2. In the following layer down to approximately 40 ft (12m) the matrix is cemented and large boulders are present. Below 40 ft (12m) values are unreliable because of the distortions introduced by the layer above and the proximity of rock at the bottom.

Data and comments pertaining to Limay 3.

The average grain size for approximately 2 to 30 ft (0.6 to 9.1 m), with the only exception of two somewhat finer layers around 2 and 20 ft (.6 and 6 m), can be represented by the following values:

D_{max} : 150 mm   % > than 75 mm : 9   % > than 50 mm : 20

D_{50} : 17 mm D_{10} : 0.4 mm   % < than 0.075 mm : 1 a 3

The gravel is nonplastic with an important portion of flat particles.

The grains are made of tuff, basalt, andesite and granite.
Relative density determinations in the top 16.5 ft (5 m) layer show
that they vary between approximately 40 and 100%.

From Fig.2 it can be seen that V_{s1} between the surface and 16.5 ft (5m)
varies between approximately 800 and 950 ft/s (244 and 290 m/s), velocities
of the order of those of the Livermore gravel. In the following layer
down to 26 ft (8 m) between 1115 and 1310 ft (340 and 400 m). Below this
depth results are unreliable because of the already mentioned distortions
created by the proximity of bed rock.

Shear wave velocities as an indirect measurement of other properties.

Attempts at Limay 2.

Since shear wave velocities can be measured in situ with considerable
accuracy and confidence their use as an indirect method for knowing
other properties has the potential advantage that they can be determined
on hard-to-sample soils such as gravels(reference 10). The most sought
after correlation is between shear wave velocity and relative density.
It has been already pointed out that seismic velocity as an index test
for liquefaction resistance is not yet a proven technique(reference 10).
Potential liquefaction behaviour is associated in gravels with low relative densities and, consequently, it is part of the determination of
relationships between shear wave velocities and relative densities.

Table 2 shows the wide range of values of V_{s1} for gravels and Fig. 2
the dispersion of results found in natural deposits. The first conclusion is that the shear wave velocity by itself can not be used to determine relative densities.

If the deposits or the layer to be studied is moderately homogeneous
a calibration between relative density and shear seismic waves can be attempted by means of a test embankment having the same particles,grading
and plasticity as the natural deposit with different intensities of compaction. As an example, results of such an attempt made at Limay 2, a
site located about 31 miles (50km) upstream of Limay 3, are represented
in Fig.3 together with those of laboratory tests of other sands and
gravels for comparisons. As has been already indicated, equation 7 and
8 permit the interchangeable use of either V_{s1} or $(X)_{max}$, to define
dynamic soil properties. When in a figure it is convenient to accentuate
contrasts $(X)_{max}$, which is a function of $(V_{s1})^2$, can be used. Furthermore, Fig.3 reproduces the part of Hardin's(1978)diagram concerning
sands and gravels, including the coefficients of elastic stiffness, S,
in his equation:

$$G_{max} = \frac{S}{2(1 + \nu) F(e)} \; P_a^{(1-n)} \; \sigma_o'^n \qquad (11)$$

in which $F(e) = 0.3 + 0.7 \, e^2$.

Fig.3 shows very clearly why e must be used for comparisons between
different materials and why either D_r or e can be used for a specific
case. To the Hardin's(1978)diagram have been added the results: a)from
Chung et al(1984)laboratory tests and proposed equations for Monterrey
N°0 sand, which fall well within the Hardin's limits for sands. b)from
Seed et al(1984)laboratory tests of gravels from the natural deposit at
Livermore. The values shown in Fig.3 were obtained from Fig.14 of ref-

erence 17. It shows that the extrapolations to obtain $(K_2)_{max}$, from which $(X)_{max}$ was computed probably introduced imprecisions. The line linking the values of $(X)_{max}$ with e seems to be too steep. c) 23 correlations between $(X)_{max}$ and e, obtained from be test embankment are represented by small circles in Hardin's (1978) diagram. Due to the dispersion of results and the limited range of values of e, or D_r, the data does not fit linear regression or any curve fit with an acceptable correlation coefficient. Consequently, there is not a reliable relationship showing the variation of $(X)_{max}$, or $(K_2)_{max}$, or V_{s1}, with e or D_r. On the contrary, with the majority of pairs of values in the 80 to 100%

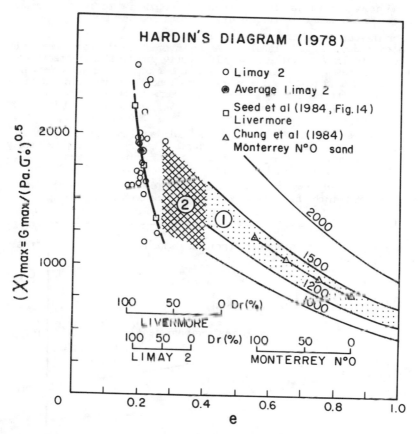

FIG.3—CORRELATIONS BETWEEN e AND $(X)_{max}$.

NOTE: 1 Clean sands. Laboratory tests.
        2 Dense well-graded gravel-sand with some fines. Laboratory tests.

        The S curves (1000, 1200, 1500, 2000) represent Hardin's elastic
        stiffness coefficient for $\nu = 0.12$ and $n = 0.5$

D_r range reliable averages were obtained as follows:

e = 0.206 D_r = 90% V_{s1} = 935ft/s(285m/s) $(X)_{max}$ = 1845

with standard deviations of 0.024, 14%, 85 ft/s (26 m/s), and 332 respectively, which apply to the test embankment material, which has the following characteristics:

D_{max} : 150 mm   % > than 75 mm : 10   % > than 50 mm : 20

D_{50} : 20 mm D_{10} : 0.5 mm   % < than 0.075 mm : 1

The gravel is nonplastic with an important portion of flat particles. The grains are made of tuff, basalt, andesite and granite.

For the purpose of the dynamic analysis of dam embankments the test embankment gave valuable results. An improvement in the test embankment procedure would be to prepare the differently compacted sections in such a way as to have an approximately equal number of results at $D_r \simeq$ 0%, $D_r \simeq$ 50% and $D_r \simeq$ 100%. Extrapolating the concept of Hardin's elastic stiffness coefficient, S, to gravel sizes seems to open a promising procedure to define a $(X)_{max}$ versus e, or V_{s1} versus e, curve in spite of the dispersion of results, as can be seen in Fig.3.

It must be pointed out that Fig.3 is a window showing the smallest values of $(X)_{max}$, which according to Table 2 can be up to approximately 3.5 times larger than those presented in this figure. The whole picture can be better represented in the V_{s1} - e diagram shown in Fig.4.

For the record there follows some brief information on the test embankment from which the values shown above were obtained. The upper

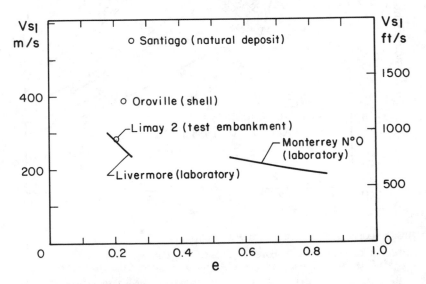

FIG.4—CORRELATIONS BETWEEN e AND V_{s1}.

platform was approximately 12500 ft^2 (1160 m^2) and the height approximately 23 ft(7 m). Between the approximate depths of 3 to 22 ft(0.9 to 6.7 m) 10 correlations between measured shear wave velocity and density were made for each of three different sections compacted with 2, 6 and 10 passes of vibratory roller. For the 30 points, e, D_r, V_{s1} and $(X)_{max}$ were determined. 7 values of V_{s1}, too high or too low were discarded. The remaining 23 correlations are the ones represented in Fig.3. The values of e were computed from density determinations made during construction and trenches excavated later on.

CONCLUSIONS

Shear wave velocity at small strain, V_s, is usually the most important field measurement to introduce in the evaluation of the dynamic response of layered soils to earthquake motion. The most reliable results are obtained by applying the cross-hole method.

For granular materials, when related to a reference confining pressure it is a basic dynamic soil property. The confining pressure of reference proposed and used in this paper is one atmosphere, P_a.

It is assumed that shear wave velocities vary with the 0.25 power of effective mean principal stress, σ'_o. This is only an approximation sufficient for most practical purposes considering the normal dispersion of results obtained from natural gravel deposits and even man-made gravelly structures.

The shear wave velocity related to a confining pressure of one atmosphere is represented by the symbol V_{s1}, the value of which is $V_{s1} = V_s / (\sigma'_o/P_a)^{0.25}$. Values of V_{s1} for gravels with predominantly small percentages of particles passing the 200 sieve (mostly $<$ 5%) with or without cobbles, obtained from laboratory and field tests, are given and commented on in this paper.

Depending on the grading, plasticity index, shape and hardness of particles of each gravel, on the percent of particles larger than maximum gravel size, on the maximum size of cobbles, the geological history of natural deposits and the effect of aging and seismic events since the end of construction on earth structures, V_{s1} has very different values. As an example, from the information given in this paper, for a value of e of the order of 0.20, V_{s1} can vary between a minimum of the order of 800 ft/s(260 m/s), e.g. Limay 3 and Livermore to a maximum of the order of 1700 ft/s(520 m/s), e.g. Limay 1 and Santiago.

V_{s1} is a property very appropriate for layer dynamic-characterization, not affected by the blurring effects of confining pressure.

Considering the wide range of values of V_{s1} for gravels, the normal dispersion of results in natural deposits and the relatively poor sensitivily of V_{s1} to variations of relative density, D_r, shear wave velocity by itself alone cannot be used to determine relative density. If the deposit or layer under study is moderately homogeneous a test embankment may furnish sufficient points relating D_r to V_{s1} as to obtain a linear or curve fit with an acceptable correlation coefficient. Results of an attempt to use such technique and comments and recommendations pertaining to it are given in this paper. Obviously, engineering judgement plays a fundamental part when information of this type is applied to the solution of practical problems.

APPENDIX—REFERENCES

1. Acevedo, P., Avendaño, M.S., Araneda, M., "Propiedades Dinámicas de las Gravas de Santiago", Revista del IDIEM, Volumen 12, N°3. December 1973.
2. Banerjee, N.G., Seed, H.B., Chan, C.K., "Cyclic Behaviour of Dense Coarse-Grained Materials in Relation to the Seismic Stability of Dams", Report.No.UCB/EERC-79/13, Earthquake Engineering Research Center, University of California, Berkeley, Calif., Jun. 1979.
3. Eolognesi, A.J.L., Vardé, O.A., Giuliani, F.L., "Velocidad de Ondas de Corte de Gravas", VIII Panamerican Conference on Soils Mechanics and Foundation Engineering, Cartagena, Colombia, Aug. 1987, Vol. 3, pp. 533-544.
4. Chung, R.M., Yokel, F.Y., Drnevich, V.O., "Evaluation of Dynamic Properties of Sands by Resonant Column Testing", Geotechnical Testing Journal, ASTM, June 1984, Vol. 7, No. 2, pp. 60-69.
5. Hardin, B.O., and Richart, F.E., Jr., "Elastic Wave Velocities in Granular Soils", Journal of the Soil Mechanics and Foundations Division, ASCE, Vol. 89, No. SM1, Feb. 1963, pp. 33-65.
6. Hardin, B.O., "The Nature of stress-strain Behaviour for Soils", ASCE Specialty Conference. Earthquake Engineering and Soil Dynamics, Pasadena, CA, Jun. 1978, Vol. 1, pp. 3-90.
7. Iwasaki, T., Tatsouka, F., "Effects of Grain Size and Grading on Dynamic Shear Moduli of Sands", Soils and Foundations, Sep. 1977, Vol. 17, No. 3, pp. 19-35.
8. Kokusho, T., Esashi, Y., "Cyclic Triaxial Test on Sands and Coarse Materials", Tenth International Conference on Soil Mechanics and Foundation Engineering, Stockholm, 1981, Vol. 1, pp. 673-676.
9. Martin, P.P., and Seed, H.B., "One-Dimensional Dynamic Ground Response Analysis", ASCE, Vol. 108, No. GT7, Jul. 1982, pp. 935-952.
10. National Research Council. Commission on Engineering and Technical Systems. "Liquefaction of Soils During Earthquakes",Washington, DC, Nov. 1985.
11. NISEE. National Information Service for Earthquake Engineering. Earthquake Engineering Research Center. University of California, Berkeley.
12. Ortigosa, P., Musante, H., Kort, I., "Mechanical Properties of the Gravel of Santiago", Tenth International Conference on Soil Mechanics and Foundation Engineering, Stockholm, 1981, Vol.2, pp.545-548.
13. Ortigosa, P., Musante, H., Retamal, E., "Cyclic Plate Tests on Granular Soils", Eleventh International Conference on Soil Mechanics and Foundation Engineering, San Francisco, 1985, Vol. 2, pp. 917-920.
14. Schnabel, B., Lysmer, J., Seed, H.B., "SHAKE, A Computer Program for Earthquake Response Analysis of Horizontally Layered Sites". Report No. EERC 72-12, Dec. 1972, Earthquake Engineering Research Center. University of California, Berkeley.
15. Seed, H.B., Idriss, I.M., "Soil Moduli and Damping Factors for Dynamic Response Analyses", Report No. EERC 70/10, Earthquake Engineering Research Center, University of California, Berkeley, Dec. 1970.
16. Seed, H.B., "Lessons from the Perfomance of Earth Dams During

Earthquakes", Conference Held at the Institution of Civil Engineers London, Oct. 1980, pp. 97-104.

17. Seed, H.B., Wong, R.T., Idriss, I.M., Tokimatsu, K., "Moduli and Damping Factors for Dynamic Analysis of Cohesionless Soils", Report No.UCB/EERC-84/14, Earthquake Engineering Research Center, University of California, Berkeley, Sep. 1984.

18. Stokoe, K.M., Lee, S.H.H., Knox, D.P., "Shear Moduli Measurements under True Triaxial Stresses", ASCE Session "Advances in the Art of Testing Soils under Cyclic Conditions", Detroit, Michigan, Oct. 1985, pp. 166-185.

19. The August 1, 1975 Oroville Earthquake Investigations, Department of Water Resources, State of California, Bulletin 203-78, Feb.1979.

20. Yu, P., Richart, F.E., "Stress Ratio Effects on Shear Modulus of Dry Sands", Journal of Geotechnical Engineering, ASCE, Mar. 1984, Vol. 110, No. 3, pp. 331-345.

SHEAR WAVE VELOCITY MEASUREMENTS
IN EMBANKMENT DAMS

Phil C. Sirles*

ABSTRACT

The U.S. Bureau of Reclamation (USBR) is presently
emphasizing the importance of site-specific shear-wave
velocity measurements for application in dynamic analyses
of embankment dams. Results of approximately 50 USBR
crosshole-seismic investigations show that significant
variations in shear wave velocity exist with depth, and
that these measured variations qualitatively correlate
with material type and structural loading conditions;
however, they are much more prominent than assumed,
calculated or predicted velocity trends. Four case
studies are presented where measured shear-wave velocity
profiles are compared with standard penetration test data,
material type and computed peak accelerations. The
results illustrate that the near surface shear-wave
velocity distribution, and consequently the accelerations
expected to occur, are indeed complex and site dependent.

INTRODUCTION

Dynamic analyses of embankment dams for liquefaction
potential or seismic-stability requires the knowledge of
subsurface material properties in order to accurately
predict ground accelerations expected to occur during
earthquake loading. Elastic properties of soil deposits
overlying bedrock are among the most important input
parameters for dynamic studies. Not only are the values
of these elastic properties important, but perhaps more
significant is their distribution with depth. Subsurface
modeling of these site-specific elastic properties through
theoretical predictions or correlation principles (e.g.,
depth, SPT-N value or material type) can result in
substantial differences in computed strong ground motions
(Aubeny, 1984). However, the severity of ground shaking
can accurately be evaluated using one, two, or three
dimensional analysis with a high degree of confidence when
utilizing well defined (measured) subsurface material
elastic properties.

---

* Geophysicist, U.S. Bureau of Reclamation, P.O. Box
25007 - MC-D-1631, Denver, CO 80225.

Currently, the USBR is emphasizing the importance of site-specific, shear-wave velocity measurements to better understand how the observed variations in shear wave velocity with depth affect ground accelerations. This paper is intended to: 1) Illustrate significant variations in measured shear-wave velocity with depth obtained at four dams located in the western United States; 2) Qualitatively compare these observations with material type and corrected SPT N-values; and, 3) Demonstrate how these variations in shear wave velocity affect peak ground accelerations. For the purpose of first-cut approximations at the response of a soil column to earthquake loading expected to occur at a particular site, the equivalent-linear method as programmed in SHAKE (Schnabel et al., 1972) was used for these analyses. Nonlinear, strain dependent properties for modulus and damping were incorporated in the analyses. The case histories reported herein represent only a fraction of site-specific investigations conducted for dynamic analysis of embankment dams by the USBR. Figure 1 shows the location of dams and other structures where crosshole shear-wave velocity measurements have been performed.

FIGURE 4.   USBR crosshole-seismic investigations.

TEST METHOD

Equipment and analysis procedures used for crosshole investigations have been developed to a point where routine measurements are performed under the USBR Safety of Dams program. During the past decade the USBR has developed a sophisticated geophysical data acquisition system designed specifically for conducting crosshole seismic investigations. State-of-the-art equipment is utilized for the surveys such that downhole compressional- and shear-wave sources are used in conjunction with downhole hydrophones and vertically oriented geophones, respectively. This field set-up allows optimum generation and recovery of the respective wave energy. Wireline winches are used to transmit downhole signals to a high-resolution, digital recording system, which provides means for fast efficient data acquisition. Data acquired in the field are stored on floppy diskettes allowing analysis to be performed with desktop computer software.

The preparation of boreholes, deviation surveys and the analysis procedures are all conducted in accordance with ASTM (1984) standard test procedures for crosshole seismic investigations. USBR crosshole investigations are typically performed at a minimum of two sites; one on the crest or at a predetermined "critical" mid-slope location, and one near the downstream toe of the embankment. There are several advantages to utilizing two crosshole test sites:

1) Obtain measurements within the embankment
2) Obtain measurements within the foundation materials-
   - while under static load beneath the structure
   - while in a free-field condition at the toe
3) Verify lateral continuity of elastic properties
4) Allow dynamic response analyses to be performed with in situ velocities at the toe and through the embankment

CASE HISTORIES

In the presentation of the following case studies in situ velocity profiles obtained for each investigation are plotted and compared with: Peak accelerations for the foundation materials (base motion for the embankment), computed by the equivalent-linear method; standard penetration test data (where available), corrected for overburden and energy delivered to the drill rod (SPT $N_{1(60)}$); and laboratory soil types presented in the Unified Soil Classification system (USC). These case studies were selected to illustrate the aforementioned advantages to performing crosshole investigations.

COLD SPRINGS DAM - Cold Springs Dam, located in north-
central Oregon and completed in 1908, is a zoned
earthfill embankment. It has a structural height of 115
ft (35 m), hydraulic height of 85 ft (26 m) and a crest
length of 3,450 ft (1050.5 m) at elevation 629.8 ft (192
m). Foundation materials consists of two alluvial soil
units: An upper unit composed of loose, clean-sands and
silts; and, a lower unit which is composed of dense to
very dense gravels with interbedded sandy silt layers.
The alluvium is approximately 40 ft (12.2 m) thick and
overlies bedrock which is fractured basalt.

Crosshole velocity measurements were performed at the
downstream toe and a mid-slope location on the embankment
and the results are presented in Figures 2 and 3,
respectively (Sirles, 1987). Shear wave velocities
measured in the embankment steadily increase with depth
from 544 to 865 fps (166-264 mps). The shear wave
velocities obtained in the foundation materials delineate
the different stiffnesses of the two alluvial units with
distinct ranges in velocity: Within the upper unit they
range from 361 to 737 feet/second (fps) (110-225
meters/second (mps) at the toe of the structure, and from
526 to 890 fps (160-271 mps) beneath the embankment; and,
within the lower unit they range from 929 to 1,670 fps
(283-509 mps), which includes the measurements at the toe
and beneath the embankment. These data show the effect of
structural loading which has increased the velocity in the
upper alluvial soils by approximately 55 percent. The
figures show an irregular velocity/depth distribution
which corresponds to variable materials and SPT-N values.
The basaltic bedrock at the sites has an average velocity
of 2,953 fps (900 mps) with a trend of slightly increasing
velocity with depth.

To approximate the accelerations expected to occur at Cold
Springs Dam the Pacoima-Tart II modified accelerogram was
selected to model the base excitation. The record was
scaled to .61g to adjust the maximum acceleration of the
record (0.75g) to an appropriate amplitude for the site.
As shown in Figure 2 there is considerable amplification
of the input base motion as it propagates to the ground
surface. The strong base excitation beneath the
embankment is amplified in the lower alluvial unit, and
then damped out in the upper alluvial unit (Figure 3).

RYE PATCH DAM - Rye Patch Dam, located in central Nevada
and completed in 1936, is a rolled earthfill embankment.
It has a structural height of 78 ft (23.8 m), hydraulic
height of 70 ft (21.3 m) and crest length of 1,074 ft
(327.4 m). The embankment is composed of a (variable)
firm to hard clay, sand, and gravel mixture. Foundation
materials are alluvial deposits consisting of very loose
to very dense poorly graded sand, silty-sand, silt and

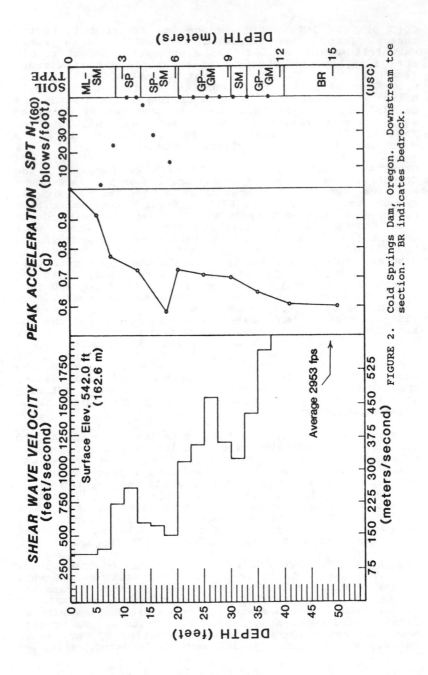

FIGURE 2.   Cold Springs Dam, Oregon. Downstream toe
            section. BR indicates bedrock.

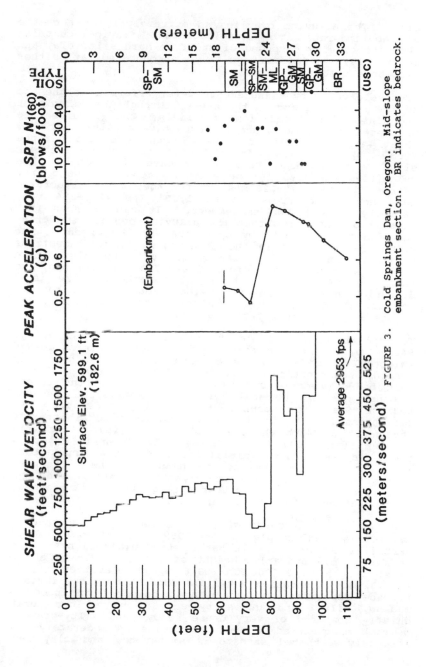

FIGURE 3. Cold Springs Dam, Oregon. Mid-slope embankment section. BR indicates bedrock.

clay. The alluvium overlies an extremely thick sequence
of non-indurated to semi-indurated, fluvio-lacustrine
deposits chiefly composed of stiff to hard silt, clay and
sandy clay. Bedrock was not encountered at either
crosshole test site.

Crosshole velocity measurements were performed at the toe
of the structure and on the crest, through the embankment
and into the foundation materials; the results are
presented in Figures 4 and 5, respectively (Sirles, 1986).
Shear wave velocities obtained in the embankment range
from 1,023 to 1,359 fps (312-414 mps), and although
slightly varied they tend to increase with depth. The
velocities measured in the alluvial deposits range from
586 to 740 fps (179-226 mps) at the toe of the structure;
and they range from 949 to 1,005 fps (289-306 mps)
directly beneath the embankment. It can be seen from
these two test sites that the alluvium beneath the crest
has a higher velocity by approximately 33 percent. The
velocities obtained in the thin-bedded fluvio-lacustrine
sediments typify the varied nature of materials. Here
velocities range from 1,040 to 1,840 fps (317-561 mps) and
there is no consistent trend, but rather, there is a
dramatic variation in the velocity/depth distribution.
The variation does, however, correlate well with SPT-N
values and changing material types.

To approximate the accelerations expected to occur at Rye
Patch Dam a synthetic near-field bedrock accelerogram was
selected to model the base excitation. The synthetic
record, representing a magnitude 7.5 earthquake, has a
peak bedrock acceleration of .63g, therefore the record
did not need to be scaled for this analysis. Because
bedrock was not encountered at either crosshole site
elevations for the input base excitation were arbitrarily
chosen at a depth where the shear wave velocities were
greater than 1,500 fps (457 mps). As shown in Figure 4
there is respectable amplification of the input base
motion as it propagates to the ground surface. The base
excitation beneath the embankment is attenuated and
amplified as it is transmitted through the thick sequence
of foundation sediments (Figure 5).

CASITAS DAM - Casitas Dam, located in south-west
California and completed in 1959, is a rolled-zoned
earthfill embankment. It has a structural height of 334
ft (101.8 m), hydraulic height of 261 ft (79.5 m) and a
crest length of 2,000 ft (609.6 m) at elevation 585 ft
(178.3 m). Foundation materials consist of two alluvial-
terrace deposits: An upper unit composed of loose to
medium, silt, sand and silty sands; and a lower unit
chiefly composed of very dense sandy- and silty-gravel.
Bedrock encountered at each test site is moderately- to
intensely-weathered sandstone, siltstone, and claystone

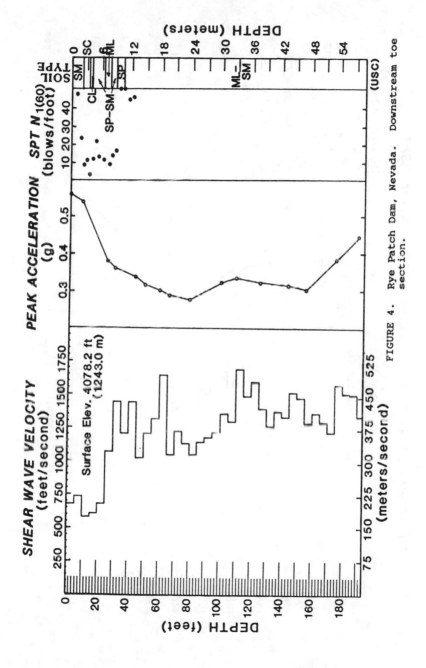

FIGURE 4. Rye Patch Dam, Nevada. Downstream toe section.

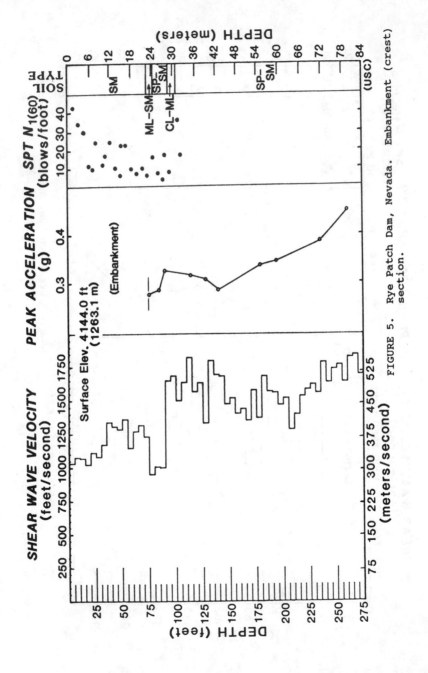

FIGURE 5. Rye Patch Dam, Nevada. Embankment (crest) section.

strata of the Tertiary Sespe Formation.

Crosshole velocity measurements were performed at the downstream toe and a mid-slope location on the embankment, and the results are presented in Figures 6 and 7, respectively (Sirles, 1986). Shear wave velocities delineate the different stiffnesses of the two alluvial units with fairly distinct ranges in velocity: Within the upper unit they range from 657 to 989 fps (200-301 mps) at the toe, and from 1,008 to 1,286 fps (307-392 mps) beneath the embankment; and, within the lower unit they range from 1,041 to 1,478 fps (317-450 mps) at the toe, and from 1,498 to 1,942 fps (457-592 mps) beneath the dam. The effect of the structural load has increased the velocity in the upper unit by approximately 25-35 percent. The figures show a irregular velocity/depth distribution which correspond to variable materials and SPT-N values. The bedrock has an average velocity of 2,150 fps (655 mps) beneath the structure and slightly lower than that at the toe.

To approximate the accelerations expected to occur at Casitas Dam the Pacoima-Taft II modified accelerogram was selected to model the bedrock excitation. The record was scaled to .68g to adjust the maximum acceleration of the record (0.75g) to an appropriate amplitude for the site. As shown in Figure 6 there is considerable amplification of the input base motion as it propagates through the lower alluvial soil deposit, then as it passes through the upper soils it is attenuated and amplified to the ground surface. The base excitation input beneath the embankment (Figure 7) exhibits a similar character to that observed in the soil column at the toe.

BULL LAKE DAM - Bull Lake Dam, located in central Wyoming and completed in 1938, is a modified homogeneous earthfill embankment. It has a structural height of 80 ft (24.4 m), a hydraulic height of 68 ft (20.7 m) and a crest length of 3,446 ft (1050 m) at elevation 5813.0 ft (1771.8 m). The foundation materials are glacial related deposits which can generally be separated into two units: Fluvio-lacustrine sediments composed of medium to stiff, sandy fat-clay; and, glacial till deposits composed of very dense, thick, heterogeneous boulder-gravel-sand-silt-clay mixtures. Bedrock was not encountered by the boreholes at either crosshole test site.

Crosshole velocity measurements were performed at the toe of the structure and on the crest, through the embankment and into the foundation materials, and the results are presented in Figures 8 and 9, respectively (Sirles, 1987). Shear wave velocities obtained in the embankment range from 811 to 1,710 fps (247-521 mps), although there appear to be thin high and low velocity layers, overall

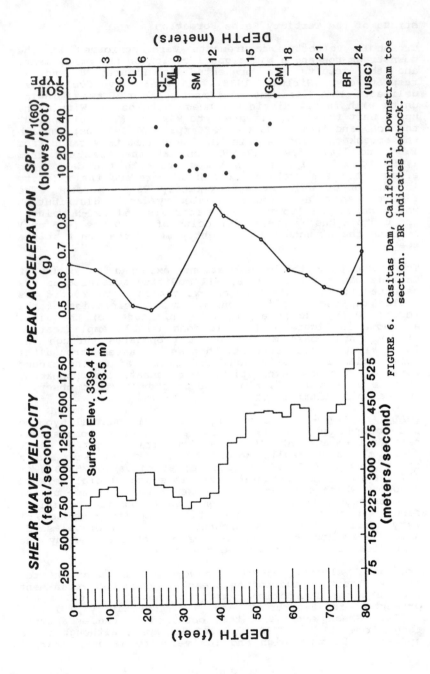

FIGURE 6.   Casitas Dam, California. Downstream toe
section. BR indicates bedrock.

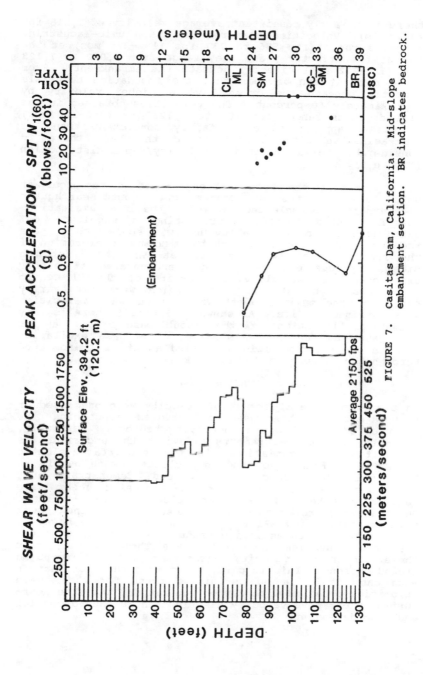

FIGURE 7. Casitas Dam, California. Mid-slope embankment section. BR indicates bedrock.

there is a fairly consistent average velocity of 1,218 fps
(371 mps). Velocities measured in the fluvio-lacustrine
deposits range from 596 to 737 fps (182-225 mps) at the
toe of the structure; and they range from 1,062 to 1,123
fps (324-342 mps) directly beneath the embankment. From
these data it can be seen that the foundation materials
directly beneath the crest have a higher velocity by
approximately 40 percent. The velocities obtained in the
till deposits range from 788 to 1,123 fps (240-342 mps)
and there appears to be a fairly consistent trend of
increasing velocity, except at depth where there is
considerable variation in the velocity/depth distribution
(Figure 8).

To approximate the accelerations expected to occur at
Bull Lake Dam the El Centro 1940 South earthquake
accelerogram was selected to model the base excitation.
The El Centro record, representing a magnitude 6.7
earthquake recorded on alluvium, was scaled to a peak
acceleration of .30g to adjust the maximum acceleration of
the record (.35g) to an appropriate amplitude for the
site. Because bedrock was not encountered at either
crosshole site the alluvial-recording of the El Centro
event was input at elevations which were arbitrarily
chosen as the maximum depth where shear wave velocities
were obtained in situ. As shown in Figure 8 there is only
slight amplification of the input base motion as it
propagates to the ground surface. The base excitation
beneath the embankment is amplified as it is transmitted
through the foundation deposits (Figure 9).

CONCLUSIONS

In the past, general trends of velocity were predicted or
calculated and then used in studies concerning the
dynamics of a soil column overlying bedrock. Currently,
reliable in situ measurements alleviate the uncertainties
associated with variations in the subsurface elastic
properties. Four case studies, selected among nearly 50
USBR crosshole-seismic investigations, demonstrate just
how substantial the variation of in situ shear wave
velocity may be with depth and also from site-to-site.
Qualitative correlation exists between the obtained shear
wave velocities and SPT $N_{1(60)}$ or material type, however,
it is difficult to establish unique trends to these data.
The case histories also illustrate the influence that
these subsurface velocity variations have on computed
ground motions. This in part, stems from the fact that in
each case there is no definitive trend for shear wave
velocities obtained with depth; but rather, that the near-
surface shear wave velocity distribution is indeed complex
and site dependent.

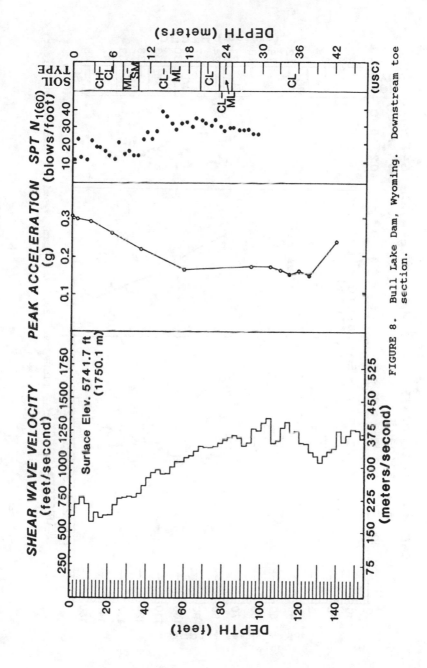

FIGURE 8. Bull Lake Dam, Wyoming. Downstream toe section.

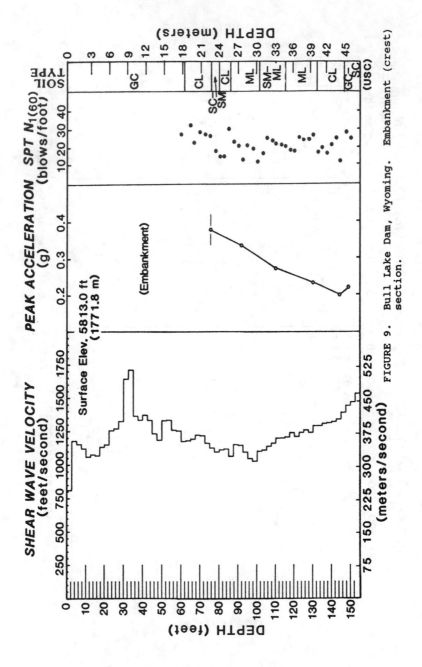

FIGURE 9. Bull Lake Dam, Wyoming. Embankment (crest) section.

ACKNOWLEDGEMENTS

The author greatly appreciates the help and guidance in preparing this document by Andy Viksne and Dave Route in the USBR Geophysics Section. Insight and support for the study were provided by Messrs. R.L. Dewey, J. L. Nettleton, and L.J. Von Thun of the Embankment Dams Branch, USBR, and thanks are extended to them.

REFERENCES

American Society for Testing Materials (1984). "Standard Test Methods for CROSSHOLE SEISMIC TESTING." ASTM D-4428 M-84.

Aubeny, C.P. (1984). "The Amplification of Seismic Shear Waves Propagating Through Horizontally Stratified Soil Deposits." M. S. Thesis, The University of Colorado, Boulder, Colorado.

Schnabel, P. B., Lymer, J., and Seed H.B. (1972). "SHAKE: A Computer Program for Earthquake Response Analysis of Horizontally Layered Sites." Report No. EERC 72-12, University of California, Berkeley, California.

Sirles, P.C. (1986). "In Situ Compressional and Shear Wave Velocity Investigation - Casitas Dam, Ventura Project, California." USBR Report, Engineering and Research Center, Denver, CO.

_____ (1986). "In Situ Compressional and Shear Wave Velocity Investigation - Rye Patch Dam, Humboldt Project, Nevada." USBR Report, Engineering and Research Center, Denver, CO.

_____ (1987). "In Situ Compressional and Shear Wave Velocity Investigation - Bull Lake Dam, Riverton Project, Wyoming." USBR Report, Engineering and Research Center, Denver, CO.

_____ (1987). "In Situ Compressional and Shear Wave Velocity Investigation - Cold Springs Dam, Umatilla Project, Oregon." USBR Report, Engineering and Research Center, Denver, CO.

IN SITU SEISMIC TESTING OF HARD-TO-SAMPLE SOILS BY SURFACE WAVE METHOD

Kenneth H. Stokoe, II[1], M. ASCE, Soheil Nazarian[2], M. ASCE, Glenn J. Rix[3], Ignacio
Sanchez-Salinero[4], Jiun-Chyuan Sheu[5], and Young-Jin Mok[2]

ABSTRACT

The Spectral-Analysis-of-Surface-Waves (SASW) method is an in situ seismic method for determining the shear wave velocity and shear modulus profiles of geotechnical sites. Field testing involves the generation and measurement of surface waves which permits all testing to be performed on the ground surface. Measurements are made at strains below 0.001 percent where elastic parameters of geotechnical materials are essentially independent of strain amplitude. Because boreholes are not required, the method is especially well suited for in situ testing of hard-to-sample soils like gravelly materials and debris slides. An overview of the SASW method is presented, and three case histories are discussed. A means of inferring the densities of hard-to-sample materials from in situ shear wave velocities is also presented. Estimations of in situ densities in terms of loose, medium dense or dense states are presented for each case history.

INTRODUCTION

Shear moduli of engineering materials such as soil, rock, concrete, and asphalt are important parameters in characterizing the mechanical behavior of these materials under many different types of loading. In geotechnical engineering, low-amplitude shear moduli (measured at strains less than about 0.001 percent) are employed in designing facilities such as vibrating machine foundations, as reference levels for evaluating dynamic soil performance and liquefaction potential during earthquake shaking, and for in situ evaluation of hard-to-sample deposits like gravels and cobbles (National Research Council, 1985). Because shear wave velocity is directly related to the stiffness of the material skeleton through which shear waves propagate, it is possible to measure shear wave velocities and then derive material parameters, such as shear modulus, or infer material parameters, such as in situ density, from measured wave velocities. This relationship forms the basic idea behind the use of seismic methods to assess in situ material parameters.

Seismic methods often used to profile near-surface soils include the crosshole, downhole, surface refraction, steady-state Rayleigh-wave method and Spectral-Analysis-of-Surface-Waves (SASW) method. Of these techniques the crosshole and the downhole methods (body wave methods) are the most widely used today. However, these methods require boreholes and thus can be difficult to use in hard-to-sample soils. The SASW method, a variation of the steady-state Rayleigh-wave method, is a promising new method that has all of the advantages of the steady-state method without having its disadvantages. In the SASW method, both the source and receivers are placed on the ground surface. Surface waves are generated by applying a vertical impulse to the ground surface. The propagation of these waves along the surface is then monitored. From the velocities of

[1]Brunswick-Abernathy Regents Professor, University of Texas, Austin, Texas 78712
[2]Research Associate, University of Texas, Austin, Texas 78712
[3]Graduate Research Assistant, University of Texas, Austin, Texas 78712
[4]Research Engineer, GEOCISA, Madrid, Spain
[5]Assistant Engineer, Dames & Moore, San Francisco, California 94105

propagation, the stiffness profile of the site is calculated through an inversion process. Because both the source and receivers are located on the ground surface, the method is especially well suited for testing hard-to-sample soils.

An overview of the SASW method is presented herein. Three case histories illustrating the use of the method in evaluating hard-to-sample soils are then presented. One case history involves testing gravelly materials which have previously liquefied. Another case history involves in situ evaluation of the effectiveness of dynamic compaction. The third case history involves evaluation of the compactness of debris slides at Mount St. Helens. In each case history, in situ densities of the materials are inferred by comparing in situ shear wave velocities with those predicted empirically for sands and gravels of various densities.

BACKGROUND INFORMATION

Two key points in SASW testing are the generation of primarily first-mode surface wave energy and the measurement of the surface waves (Rayleigh waves) at reasonable distances from the source. In layered media, the velocity of propagation of a surface wave depends on the frequency (or wavelength) of the wave. This variation of velocity with frequency is called dispersion and arises because waves of different wavelengths sample different parts of the layered medium. For example, high-frequency (short wavelength) waves propagate only in near-surface layers. Lower-frequency waves with longer wavelengths propagate through the near-surface layers as well as deeper layers. Therefore, by using surface waves over a wide range of frequencies, one can effectively sample different portions of the material profile.

In the original steady-state Rayleigh-wave method, a steady-state vibrator acting vertically on the surface of the soil or pavement produced vibrations of a known frequency that propagated along the surface (Jones, 1962 and Ballard, 1964). A vertically oriented sensor (velocity transducer or accelerometer) was moved progressively away from the vibrator and successive positions were found at which the vertical surface motions were in phase with the vibrator (Richart et al, 1970). The distance between any two of these successive positions was assumed to correspond to one wavelength (L_R) of the propagating wave. Since the frequency (f) of vibration was known, the velocity (V_R) of the wave propagating at that frequency could be calculated using:

$$V_R = L_R \cdot f. \tag{1}$$

A plot of surface wave phase velocity (V_R) versus frequency or wavelength is obtained by repeating this process for different excitation frequencies. Such a plot is called a dispersion curve. This curve can be considered the "raw" field data.

The steady-state technique is easily understood and performed. Field testing is, however, very time consuming. With the development of digital electronic equipment of the 1970's, this shortcoming has been overcome. Instead of using a steady-state vibrator at fixed frequencies, an impulsive or random-noise load is applied at the surface of the soil deposit. The general testing configuration is shown in Fig. 1. Two vertical receivers located on the surface are used to monitor the wave train generated by the source as it passes by them. The signals produced by the receivers are digitized and recorded by a dynamic signal analyzer. Each recorded time signal is transformed to the frequency domain

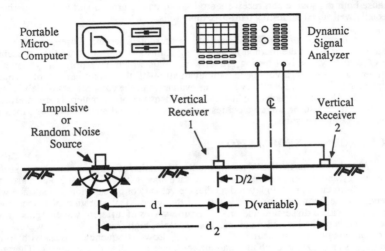

Fig. 1 - General Configuration of Source and Receivers Used in SASW Testing.

domain using a Fast Fourier Transform algorithm, and the phase difference (ϕ) between the two signals is calculated for each frequency. A travel time (t) between receivers can then be obtained for each frequency by:

$$t = \phi/2\pi f \tag{2}$$

where the phase difference (ϕ) is in radians and the frequency (f) is in cycles per second. Since the distance between receivers (d_2-d_1) is known, wave velocity is calculated using:

$$V_R = (d_2-d_1)/t \tag{3}$$

The corresponding surface wave wavelength is determined using:

$$L_R = V_R/f. \tag{4}$$

The calculation steps outlined in Eqs. 2 through 4 can be performed for each frequency, and the results plotted in the form of a dispersion curve.

In addition to the use of digital electronics and impulsive or random-noise sources, another significant advance incorporated in the SASW method with respect to the original steady-state technique involves the manner in which the shear moduli of the layers in the soil profile are back-calculated from the experimental dispersion curve. This process of determining layer moduli from an experimental dispersion curve is given the generic term "inversion" (Nazarian, 1984). The first attempts at surface wave inversion used in the original steady-state method were crude by today's standards. It was assumed that the bulk of the surface wave energy travelled through a zone about one wavelength deep. By assuming that the surface wave phase velocity obtained at a particular frequency was representative of the parameters at a depth equal to one-half of the wavelength, a plot of surface wave phase velocity versus depth (rather than wavelength) was obtained (Richart et al, 1970). (This hypothesis is only an approximation since the parameters of the materials above and below a depth of one-half of a wavelength indeed affect the propagation velocity of that frequency.) To complete this method of inversion, shear wave velocity was obtained from V_R by assuming Poisson's ratio. The ratio of surface wave velocity to shear wave velocity ranges from 0.874 to 0.955 for Poisson's ratio ranging from 0.0 to 0.5. For many practical applications, the ratio of surface wave velocity to shear wave velocity can be considered equal to 0.92.

The inversion process currently used is an iterative procedure based on forward modeling described by Nazarian (1984). In this procedure a theoretical dispersion curve is matched to the experimental curve obtained in the field. To calculate the theoretical dispersion curve, the actual site is modeled as a layered half-space with uniform, elastic layers of infinite horizontal extent. The shear wave velocities and thicknesses of the layers in the profile are assigned initial values. (Values for Poisson's ratio and mass density are also assigned to each layer.) For this assumed layering, a theoretical dispersion curve is calculated using a modified version of the Haskell-Thomson matrix solution (Thomson, 1950; Haskell, 1953; and Nazarian, 1984). The theoretical dispersion curve is then compared with the experimental dispersion curve obtained in the field. The assumed shear wave velocities and thicknesses of the layers in the model are adjusted until satisfactory agreement between the theoretical and experimental dispersion curves is obtained. At that point it is assumed that the shear wave velocities and thicknesses of the layers in the model accurately represent the actual stiffnesses and layering of the site. (Variations in values of Poisson's ratio and mass density have less than about a 10 percent effect on the final shear wave velocities for reasonable choices (Ewing et al, 1957) and, therefore, are generally not varied after the initial choice is made.) This inversion procedure has increased the accuracy of the material profiles determined with the SASW method and has significantly expanded the variety of site conditions under which the method can be successfully applied.

FIELD PROCEDURE

The general configuration of the source, receivers and recording equipment is shown in Fig. 1. Vertical velocity transducers (Mark Products Model L-4C) with a natural frequency of 1 Hz have been found to perform very well as receivers at soil sites. The arrangement of receivers which has been adopted for the SASW method is called the common receiver midpoint geometry. With this arrangement the two receivers are located equidistant from an imaginary centerline which is kept fixed. Although in theory it should be possible to use one receiver spacing for the entire test, practical considerations such as attenuation dictate that several different receiver spacings be used. Therefore, testing is performed with various receiver spacings always keeping the centerline midway between the receivers. The distance between the source and first receiver (d_1 in Fig. 1) is varied but an attempt is made to keep the distance approximately equal to the distance between the two receivers (d_2-d_1 in Fig. 1). In addition, the location of the source is reversed so that testing is performed for both forward and reverse profiles. Typically, distances between receivers of 4, 8, 16, 32, 64, and 128 ft (1.2, 2.5, 5, 10, 20, and 40 m) are used if soil is to be evaluated to depths of about 60 ft (18 m).

Several types of sources are used to generate energy over the required frequency ranges. At close receiver spacings, small, hand-held hammers are used. At spacings ranging from 8 to 16 ft (2.5 to 5 m), sledge hammers or large dropped weights weighing from 50 to 100 lbs (220 to 440 N) are employed. For receiver spacings greater than 16 ft (5 m), a variety of sources have been used including dropped weights ranging from 150 to 2000 lbs (670 to 8900 N), bulldozers, and a very large weight used for dynamic compaction (32 tons, 290 kN). The large, dynamic compaction weight is an excellent generator of low frequencies, permitting wavelengths as long as 500 ft (150 m) to be measured at sites where it has been used. (However, use of this source cannot be considered nondestructive testing.) Bulldozers have proven to be a valuable source of surface waves for those situations in which impact sources such as dropped weights were not capable of generating sufficient energy over a wide range of frequencies or where these sources were not available.

The recording device used is a Hewlett-Packard 3562A Dynamic Signal Analyzer. A dynamic signal analyzer is a digital oscilloscope that by means of a microprocessor has the ability to perform calculations in either the time or frequency domains. Other instrumentation certainly can be used. However, it is strongly recommended that any instrumentation be capable of performing frequency-domain calculations in real time during field testing.

A typical set of SASW records collected in the field at one receiver spacing is presented in Fig. 2. The most important of these four records is the phase of the cross power spectrum (Fig. 2a) from which the phase difference between receivers as a function of frequency is obtained. The phase difference is used to calculate the dispersion curve as described in Eqs. 2 through 4. The coherence function (Fig. 2b) indicates the quality of the signals being recorded. For two or more averages, the coherence will take on a value between 0 and 1. A value near one corresponds to a very high signal-to-noise ratio. Similarly, a value near zero indicates a poor signal-to-noise ratio. Using this information, data collected in the field can be checked and, if necessary, modifications to the test setup can be made to improve the data. In addition, the coherence function can be used during data reduction to remove undesirable data from consideration. Finally, the auto power spectra of the two receivers (Figs. 2c and 2d) are recorded and stored to provide an indication of the source characteristics. The cross power spectrum shown in Fig. 2a is combined with other spectra measured at other receiver spacings to form the composite dispersion curve like the one presented in Fig. 3.

CASE HISTORIES

Debris Slides at Mount St. Helens, Washington

Mount St. Helens is located in the Cascade range of Southern Washington State. A plan view of the general location of Mount St. Helens and its vicinity is presented in Fig. 4. Mount St. Helens is about 100 miles (160 km) to the southeast of Seattle. The mountain became volcanically active in March, 1980. The eruption process started with a magnitude 4 earthquake on March 20, 1980 and ended with a catastrophic landslide and a major eruption on May 18, 1980. The eruption resulted in a 560-cubic-mile (2300 km3) rockslide-avalanche (Christiansen and Peterson, 1982). The avalanche along with the pyroclastic flows and blast deposits created three debris blockages which impounded Spirit, Coldwater and Castle Creek Lakes.

The Spirit Lake blockage consists of about 1 ft (0.3 m) of fine, sand-size ash underlain by debris consisting of a heterogeneous mixture of silt, sand, pebbles, cobbles, and boulders which is well-graded and loosely packed (Youd et al, 1982). The Coldwater Lake and Castle Creek Lake blockages consist of mostly rounded clasts and wood in a brown muddy and sandy matrix (Voight et al, 1982). The thickness of this debris varies from about 3 to 300 ft (1 to 90 m). In all three cases, significant portions of the blockages are composed of materials which cannot be tested in situ by conventional geotechnical means. Therefore, SASW tests were performed. A total of five sites was tested: two at both the Spirit Lake and Castle Creek Lake blockages and one at the Coldwater Lake blockage. The reason for this testing program was to measure the shear wave velocity profiles so that preliminary stability estimations of the three blockages could be made.

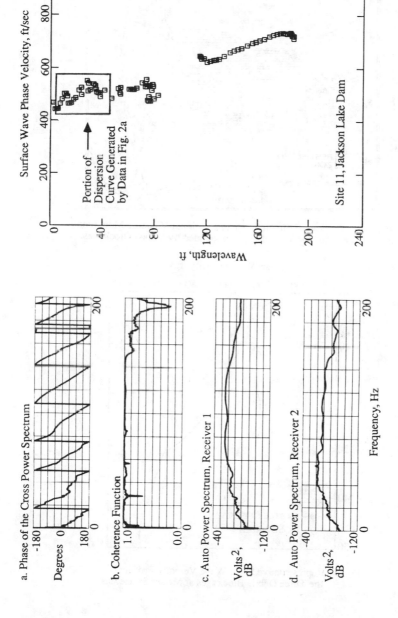

Fig. 3 - Composite Dispersion Curve Associated with Phase of the Cross Power Spectrum Given in Fig. 2.

Fig. 2 - Typical Record Set Obtained During SASW Testing (Receiver Spacing = 16 ft).

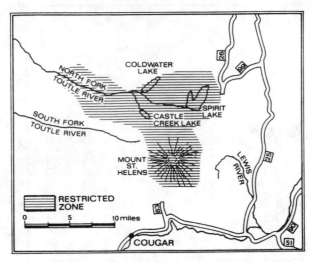

Fig. 4 - Plan View of Mount St. Helens National Volcanic Monument
Showing the Three Lakes Impounded by Debris Blockages.

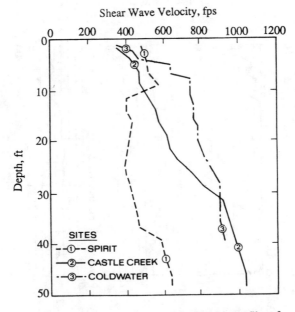

Fig. 5 - Representative Shear Wave Velocity Profiles of
the Three Debris Blockages at Mount St. Helens.

Representative shear wave velocity profiles determined from the inversion process are shown in Fig. 5. At each site, a total of 20 layers was used in the inversion process. The SASW method was found to work well in this hard-to-sample material. Based on the velocity profiles, the Coldwater Lake blockage is quite stiff while the Spirit Lake blockage is quite soft. In fact, the Spirit Lake blockage exhibits shear wave velocities similar to those of very loose sands which liquefied during the 1979 and 1981 earthquakes in Imperial Valley, California (Stokoe and Nazarian, 1985). (Of course, the profiles are representative of only those areas tested which constitute a small fraction of each blockage). In situ densities of the debris blockages are inferred from the V_S values as discussed in the next section.

Gravelly Soils Subjected to the 1983 Borah Peak, Idaho Earthquake

During the 1983 Borah Peak, Idaho earthquake (M_s = 7.3), liquefaction of near-surface soils resulted in extensive damage at several sites. Many of these soils had a significant gravel content and were previously believed to be too stiff and/or too well drained to be susceptible to liquefaction. The purpose of this study was to make in situ measurements of the shear wave velocities of these materials at a site located 5 miles (8 km) from the epicenter of the earthquake. At this particular site, ground shaking caused a house to move several inches off its foundation and lateral spreading induced by liquefaction of gravelly soils resulted in damage to farm roadways. Because of the difficulties encountered when boring and sampling gravelly soils, it was decided to use the SASW method to test these materials. Stiffness profiles were evaluated to a depth of approximately 30 ft (9 m) at five locations in July, 1985.

A typical range in grain-size distribution curves for the gravelly soils which experienced liquefaction is shown in Fig. 6. More than 50 percent by weight of these materials may be classified as gravels. An example of the shear wave velocity profiles determined using the SASW method is presented in Fig. 7 along with the material profile at this location. The results indicate that the sandy gravels which liquefied range in depth from 3 to 16 ft (1 to 5 m) and have shear wave velocities which vary between 300 and 400 ft/sec (90 and 120 m/sec). Soils with shear wave velocities in this range are so loose that liquefaction would be expected for the levels of peak ground surface acceleration ($\cong 0.50$ g) estimated near this location. In addition, the generation of large porewater pressures at this site is greatly enhanced by the relatively impermeable cap on top of the gravelly soils.

This case study indicates that gravelly soils which were previously believed to be too well drained to liquefy may experience liquefaction if the gravels are very loose, large ground accelerations are present and a relatively impermeable cap covers the gravelly material. It is also interesting to note that the range in wave velocities and associated depths for the gravelly soils in Idaho are essentially the same as those found for sandy soils which liquefied in the Imperial Valley of California (Stokoe and Nazarian, 1985). This study has further demonstrates that the SASW method can be successfully applied to sites where other in situ tests may be inappropriate or difficult to use because of the problems encountered during drilling and sampling.

Dynamic Compaction of Foundation Soils at Jackson Lake Dam, Wyoming

In July, 1985 the Bureau of Reclamation began a four-year effort to strengthen and modify Jackson Lake Dam, located in the Grand Teton National Park, near Jackson, Wyoming. Detailed analyses had revealed that the dam would likely experience substantial damage during a large earthquake due to liquefaction of the foundation soils. In the summer of 1986, the existing earth embankment was removed, and dynamic compaction was used to densify and strengthen the foundation soils. As part of a program to monitor the dynamic compaction, The University of Texas was invited to use the SASW method at

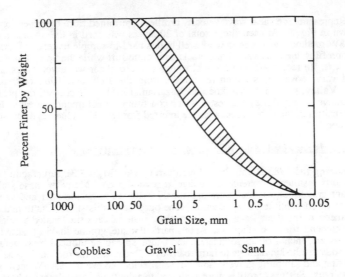

Fig. 6 - Typical Gradation Curves of Gravelly Soils that Liquefied
During the 1983 Borah Peak, Idaho Earthquake.

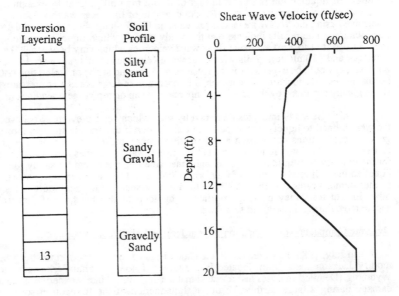

Fig. 7 - Typical Composite Profile of Liquefaction Site near Borah Peak, Idaho.

13 sites in Sectors, A, B and C of the embankment to assess changes in the stiffness of the foundation soils before and after dynamic compaction.

The first series of SASW tests took place in July, 1986 prior to dynamic compaction. A second series of post-compaction tests was conducted in May, 1987 at the same locations used in the first series of tests. Unfortunately, the second series of tests could not be performed immediately after dynamic compaction was completed because of construction delays and the onset of winter. The time lag between completion of dynamic compaction and the second series of tests introduced additional variables in the comparison of pre- and post-compaction results, i.e. the effects of time, a freeze-thaw cycle, etc. However, these additional variables are felt to have little effect on the comparisons.

The two shear wave velocity profiles obtained before and after dynamic compaction at Site 6 are shown in Fig. 8. (Sixteen layers were used in inverting each profile.) The results indicate that near-surface soils less than 4 ft (1.2 m) in depth were less stiff after compaction than before compaction. This decrease in velocity (stiffness) resulted from the fact that soil dumped in the holes created by dynamic compaction had not yet been compacted. Materials in the range of depths from 4 to 28 ft (1.2 to 8.5 m) increased significantly in stiffness after compaction. The key zone identified for compaction consisted of silty and sandy materials in the depth range of about 15 to 30 ft (4.5 to 9 m). Materials in this depth range significantly increased in stiffness as shown in Fig. 8. (This stiffness increase is indicative of a density increase as discussed in the next section.) Finally, the stiffness of soils from 28 to 60 ft (8.5 to 18.3 m) remained approximately the same. This general pattern was exhibited by 4 of the 13 sites tested.

The other general pattern in stiffness change resulting from dynamic compaction is shown in Fig. 9 for Site 13. In this case, the zone identified for compaction between 15 and 30 ft (4.5 and 9 m) exhibited a stiffness increase, and hence, some density increase. However, a loosening in the zone from about 27 to 40 ft (8.1 to 12.2 m) at Site 13 seemed to occur. This apparent decrease is an interesting occurrence and deserves further study. However, such a study was not part of this work. The general pattern exhibited at Site 13 was exhibited by 5 of the 13 sites tested. The remaining profiles (4 out of 13) showed little or no change (two cases) or the results were inconclusive (two cases).

One of the benefits of the SASW method is that sites can be reoccupied at future times so that changes in stiffness profiles can be evaluated. The study at Jackson Lake Dam clearly demonstrates this benefit. In addition, a layer of gravelly soil existed over the zone of interest at the dam. This gravelly layer made drilling difficult. However, it was penetrated (and sampled) easily and effectively by the SASW method.

INFERENCE OF DENSITY FROM IN SITU V_S

Basic Approach

One reason for performing seismic tests in hard-to-sample soils is to infer in situ densities from measured shear wave velocities. This can be done for sands and gravels by comparing measured values of V_s with values determined from empirical relationships like the ones developed by Seed et al (1986). Seed et al have expressed the small-strain value of shear modulus, G_{max}, as:

$$G_{max} = 1000 \ K_2 \ (\bar{\sigma}_m)^{0.5} \tag{5}$$

where K_2 is an empirical constant that takes into account the density (void ratio) of the material, $\bar{\sigma}_m$ is the mean effective principal stress expressed in pounds per square foot

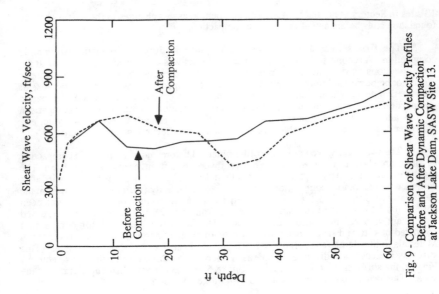

Fig. 9 - Comparison of Shear Wave Velocity Profiles
Before and After Dynamic Compaction
at Jackson Lake Dam, SASW Site 13.

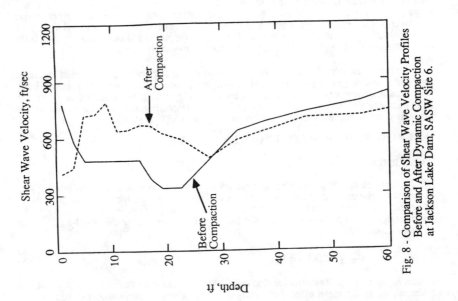

Fig. 8 - Comparison of Shear Wave Velocity Profiles
Before and After Dynamic Compaction
at Jackson Lake Dam, SASW Site 6.

(psf), and G_{max} is expressed in psf. Shear wave velocity and shear modulus are related by:

$$G = (\gamma/g) \bullet V_s^2 \tag{6}$$

where γ is total unit weight and g is gravitational acceleration. By combining Eqs. 5 and 6, shear wave velocity can be written as:

$$V_s = [(1000)(g/\gamma)(K_2)]^{0.5} \bullet (\bar{\sigma}_m)^{0.25} \tag{7}$$

With Eq. 7, the variation of V_s with density and depth at a site can be calculated. This is done by assuming different values of K_2 which reflect density and by calculating values of $\bar{\sigma}_m$ for each depth. The value of $\bar{\sigma}_m$ is typically calculated from:

$$\bar{\sigma}_m = \bar{\sigma}_v (1 + 2K_o)/3 \tag{8}$$

in which $\bar{\sigma}_v$ is the vertical effective stress at the depth of interest and K_o is the effective coefficient of earth pressure at rest. In the use of Eq. 8, the following assumptions are generally implicitly made: (1) level ground, (2) principal stresses are oriented in the vertical and horizontal directions, (3) the intermediate and minor principal stresses are equal ($\bar{\sigma}_2 = \bar{\sigma}_3$, and (4) the age of the deposit can be neglected. In addition, an important assumption is made about the natural cementation of the granular material; that is, little or no cementation exists. The assumption of no cementation occurs from the fact that empirical equations, like Eq. 5, have generally been developed using reconstituted specimens in the laboratory. If one chooses to try to account for cementation, then other equations have to be developed specifically including the effect of varying degrees of cementation.

Case Histories

The approach outlined above was used to compare in situ shear wave velocities with shear wave velocities of sands or gravels of varying densities to obtain a qualitative sense of the densities. It is important to stress that only a qualitative sense for the densities is determined. (Precise values of density or void ratio are not assumed in these comparisons.) To perform the calculations, values of K_2 of 30, 50, and 70 were taken as representative of loose, medium dense, and very dense sands, respectively. Values of K_2 equal to 40, 80 and 120 were similarly used for gravels.

Shear wave velocities determined by the SASW tests and the empirical method are compared in Fig. 10 so that in situ densities can be inferred. As a starting point, a total unit weight of 120 pcf (18.9 kN/m3) was used in Eq. 7 and a K_o of 0.5 was used in Eq. 8 for all cases. Other values could be used if more information were known about the sites. However, variations in the values have only a small effect because of the fractional values of the exponents.

Several interesting points are evident in Fig. 10. First, at Mount St. Helens (Fig. 10a), the debris slide at Spirit Lake is very loose in the depth range from 10 to 40 ft (3 to 12.2 m) while the debris slide at Coldwater Lake is quite dense in the same depth range. Second, the gravelly soils which liquefied in the 1983 Idaho earthquake (Fig. 10b) are very loose in the depth range from 3 to 12 ft (1 to 3.7 m). As already noted, the values of V_s in this depth range are very similar to values of V_s which have been measured in the liquefiable sands in Imperial Valley (Stokoe and Nazarian, 1985) even though one deposit is basically a sand while the second is basically a gravel. Finally, dynamic compaction of the silty materials at Jackson Lake Dam increased the stiffness of the materials as desired.

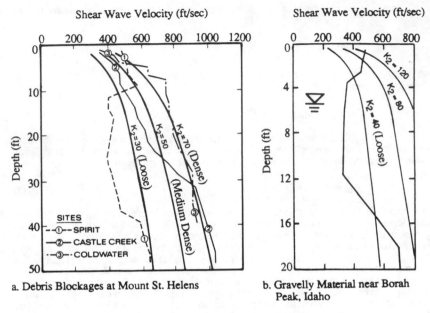

a. Debris Blockages at Mount St. Helens

b. Gravelly Material near Borah Peak, Idaho

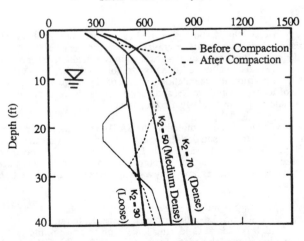

c. Silts and Silty Sands at Jackson Lake Dam, Wyoming

Fig. 10 - Estimation of a Qualitative Sense of In Situ Densities of Granular Soils by Comparing In Situ and Empirical Shear Wave Velocities.

This stiffness increase occurred, however, because of a combination of density and stress changes. An increase in horizontal stress by a factor of two would only increase the value of V_S by about 20 percent, and therefore the significant velocity increase in the depth range from 15 to 25 ft (4.6 to 7.6 m) shows a significant change in density.

SUMMARY AND CONCLUSIONS

In situ testing of geotechnical materials can be performed by many different methods. Selection of a method or combination of methods depends on the purpose of the tests and the advantages and/or limitations of the methods. If deformation characteristics in the small-strain range are required for either direct or indirect use, then in situ seismic methods offer the most direct measure of these characteristics. One seismic method, the Spectral-Analysis-of-Surface-Waves (SASW) method, is discussed. The SASW method represents an improved version of the steady-state Rayleigh-wave method of in situ seismic testing. The method is well-suited for determining shear wave velocity and shear modulus profiles of geotechnical sites. One distinct advantage of the SASW method is that all testing is performed on the ground surface. Hence, no boreholes are required. This advantage makes the SASW method very useful for profiling hard-to-sample soils such as gravelly materials and debris slides.

In SASW testing, two important elements are the generation and measurement of surface waves (Rayeleigh waves). In addition, real-time computation of spectral functions is necessary in the field. Using the distance between measurement points and the phase shift for each surface wave frequency, phase velocity and associated wavelength can be calculated from which a dispersion curve, a plot of phase velocity versus wavelength, can be constructed. By applying an inversion process, an analytical technique for reconstructing the shear wave velocity profile from the dispersion curve, layering and the shear wave velocity and shear modulus of each layer can be readily obtained. The most important step in SASW testing is the inversion process. Through inversion, one is able to accutately evaluate soft material located beneath stiff material without drilling boreholes. The theoretically-based inversion method used in this study results in more accurate profiles than was possible with the empirically-based methods in use previously.

Three case studies are presented to illustrate the utility of the SASW method. Evaluation of the stiffness of debris slides at Mount St. Helens and of previously liquefied gravelly materials near Borah Peak, Idaho show the advantages of using this method in hard-to-sample soils. Use of the method to monitor the magnitude and extent of dynamic compaction at Jackson Lake Dam, Wyoming further demonstrates the attributes of the SASW method.

Finally, the use of measured shear wave velocities to infer, in a qualitative sense, in situ densities of hard-to-sample soils is presented. This indirect evaluation of density is performed by comparing velocities measured in situ with those predicted by empirical means for sands and gravels of various densities. Consideration must be given in applying this approach to stress-state, site age and degree of natural cementation. The results of the three case histories clearly demonstrate the value of this approach. For instance the gravelly materials that liquefied in the 1983 Borah Peak earthquake were found to be very loose (as expected) while the debris slide at Coldwater Lake which was created by the eruption at Mount St. Helens was found to be quite dense. In each of these cases, intact sampling of the materials was impossible.

ACKNOWLEDGEMENTS

This work was supported by the Texas State Department of Highways and Public Transportation, the United States Geological Survey, the United States Bureau of Reclamation and the United States Air Force Office of Scientific Research. The authors would like to thank these organizations for their support. The assistance of Dr. T.L. Youd and Mr. P. Sirles in the case histories is also appreciated.

REFERENCES

1. Ballard, R.F., Jr., (1964), "Determination of Soil Shear Moduli at Depth by In Situ Vibratory Techniques," Miscellaneous Paper No. 4-691, U.S. Army Engineer Waterways Experiment Station, Vicksburg, Mississippi, 9 pp.
2. Christiansen, R.L., and Peterson, D.W., (1982), "Chronology of the 1980 Eruptive Activity," U.S. Geological Survey Professional Paper 1250, The 1980 Eruption of Mount St. Helens, Washington, D.C., pp. 17-30.
3. Ewing, W.M., Jardetzky, W.S., and Press, F., (1957), Elastic Waves in Layered Media, McGraw-Hill Book Company, Inc., New York, 380 pp.
4. Haskell, N.A., (1953), "The Dispersion of Surface Waves in Multilayered Media," Bulletin of the Seismological Society of America, Vol. 43, pp. 17-34.
5. Jones, R., (1962), "Surface Wave Technique for Measuring the Elastic Properties and Thickness of Roads: Theoretical Development," British Journal of Applied Physics, Volume 13, pp. 21-29.
6. National Research Council, (1985), "Liquefaction of Soils During Earthquakes," Proceedings, Workshop organized by R. Dobry and R.V. Whitman, National Academy Press, 240 pp.
7. Nazarian, S., (1984), "In Situ Determination of Elastic Moduli of Soil Deposits and Pavement Systems by Spectral-Analysis-of-Surface-Waves Method," Ph.D. Dissertation, The University of Texas at Austin, 458 pp.
8. Richart, F.E., Hall, J.R., and Woods, R.D., (1970), Vibrations of Soils and Foundations, Prentice-Hall, Inc., New Jersey, 414 pp.
9. Seed, H.B., Wong, R.T., Idriss, I.M., and Tokimatsu, K., (1986), "Moduli and Damping Factors for Dynamic Analyses of Cohesionless Soils," Journal of the Geotechnical Engineering Division, Vol. 112, No. 11, ASCE, 17 pp.
10. Stokoe, K.H., II, and Nazarian, S., (1985), "Use of Rayleigh Waves in Liquefaction Studies," Proceedings, Measurement and Use of Shear Wave Velocity for Evaluating Dynamic Soil Properties, Geotechnical Engineering Division, ASCE, pp. 1-14.
11. Thomson, W.T., (1950), "Transmission of Elastic Waves Through a Stratified Solid," Journal of Applied Physics, Vol. 21, pp. 89-93.
12. Voight, B., Glicken, H., Janda, R.J. and Douglass, P.M., (1982), "Catastropic Rockslide Avalanche of May 18," U.S. Geological Survey Professional Paper 1250, The 1980 Eruption of Mount St. Helens, Washington, D.C., pp. 347-377.
13. Youd, T.L., Wilson, R.C., and Schuster, R.L., (1982), "Stability of Blockage in North Fork Toutle River," U.S. Geological Survey Professional Paper 1250, The 1980 Eruption of Mount St. Helens, Washington, D.C., pp. 821-828.

SASW and Crosshole Test Results Compared

Dennis R. Hiltunen, S.M.ASCE[1] and Richard D. Woods, M.ASCE[2]

ABSTRACT

Spectral-analysis-of-surface-waves (SASW) tests and crosshole tests were conducted at the same site to compare the resulting shear wave velocity profiles. Previous research has suggested that, while the profiles obtained from the two methods usually compare favorably, the velocities from the crosshole test are often slightly larger. One plausible explanation for this discrepancy is that wave path curvature is frequently neglected when reducing the data from crosshole tests. It was found that this phenomenon partially explained the differences noted between SASW and crosshole results, accounting for approximately 30 percent of the discrepancy.

INTRODUCTION

The need for accurate in situ shear wave velocity profiles to evaluate the dynamic response of soil or structures supported on soil during earthquake loading, machine loading, or other types of dynamic loading has long been recognized (Hoar and Stokoe [1978], Woods [1978], Woods and Stokoe [1985], and Woods [1986]). More recently shear wave velocity data has been found useful in the evaluation of liquefaction potential as well (Dobry et al. [1980], Dobry et al. [1981], De Alba et al. [1984], Dobry et al. [1984], and Stokoe and Nazarian [1985]). Geophysical techniques such as crosshole and downhole are usually employed to determine low-amplitude in situ shear wave velocities.

The spectral-analysis-of-surface-waves (SASW) method is a new testing technique for determining in situ shear wave velocity profiles. The method has essentially been developed since the first earthquake engineering and soil dynamics conference held in Pasadena, California in 1978. The SASW method is based upon the generation and detection of Rayleigh waves at the surface of a soil system and hence has an advantage over crosshole or downhole techniques in that it does not require boreholes. The present paper compares the shear wave velocity profiles from SASW and crosshole tests conducted at the same site.

THE SASW METHOD

The spectral-analysis-of-surface-waves (SASW) method is a testing procedure for determining shear wave velocity profiles of soil systems in situ. The test is performed from the ground surface and thus requires no boreholes. Measurements are made at

[1] Research Investigator, Dept. of Civil Engrg., Univ. of Michigan, Ann Arbor, MI 48109-2125
[2] Professor, Dept. of Civil Engrg., Univ. of Michigan, Ann Arbor, MI 48109-2125

strain levels below 0.001 percent, where elastic properties of soil are considered independent of strain amplitude. The key elements in SASW testing are the generation and measurement of Rayleigh waves.

A number of publications in recent years have described in detail the SASW method (Nazarian [1984], Nazarian and Stokoe [1984], Stokoe and Nazarian [1985], Woods and Stokoe [1985], and Woods [1986]). A schematic of the experimental arrangement for SASW tests is presented in figure 1. Current practice calls for locating two vertical receivers on the ground surface a known distance apart and a transient wave containing a large range of frequencies is generated in the soil by means of a hammer. The surface waves are detected by the receivers and are recorded using a Fourier spectrum analyzer. The analyzer is used to transform the waveforms from the time to the frequency domain and then to perform spectral analyses on them. The spectral analysis functions of interest here are the phase information of the cross power spectrum and the coherence function. Knowing the distance and the relative phase shift between the receivers for each frequency, the phase velocity and wavelength associated with that frequency are calculated. The final step is application of an inversion process that constructs the shear wave velocity profile from the phase velocity-wavelength information (dispersion curve).

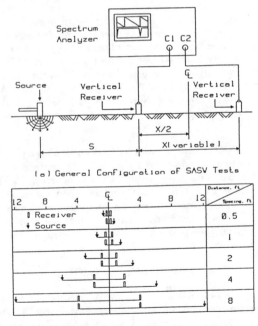

(a) General Configuration of SASW Tests

(b) Common Receivers Midpoint Geometry

Figure 1 — Schematic of Experimental Arrangement for SASW Tests
(after Nazarian [1984])

THE CROSSHOLE TEST

The crosshole test is a well established technique for determining in situ shear wave velocity profiles (Stokoe and Woods [1972], Woods [1978], Hoar [1982], Woods and Stokoe [1985], and Woods [1986]). The test consists of first establishing a series of boreholes (usually 2 or 3) along a common line a known distance apart. In one borehole a source is inserted to create a seismic wave. Receivers are placed in the remaining hole(s) to measure the arrival of the seismic wave. By measuring the travel time of the seismic wave between the boreholes, the wave velocity is determined by simply dividing the known travel distance by the measured travel time. The velocity of both the compression and shear wave can be determined in this manner.

PREVIOUS COMPARISONS

Nazarian and Stokoe (1984) presented two case studies in which SASW and crosshole shear wave velocity profiles were compared. In both cases the maximum difference in shear wave velocities was less than 10 percent.

Nazarian (1984) conducted numerous case studies to compare SASW results with the results from both crosshole and downhole tests. He found that the results from SASW tests were usually within 20 percent of those from crosshole and downhole. He also found that the shear wave velocities obtained from SASW tests were usually less than those from the crosshole test. No explanation for this observation was provided. One possible explanation is that ignoring wave path curvature in the crosshole test may yield shear wave velocities which are too high. The wave path followed by seismic waves in a medium that varies in stiffness with depth is curved and not a straight line. It is common practice in the crosshole test with small borehole spacings to ignore the effects of wave path curvature. This assumption was checked in the present case study.

TEST RESULTS AND DISCUSSION

SASW tests were conducted at a soil site on the University of Michigan campus hereafter referred to as the Beal St. Field Site. The site has been used for many years for both research and teaching purposes and the soil profile is relatively well known. The soil is a glacially-overconsolidated silty sand to a depth of at least 26.2 ft (8 m) (Hryciw and Woods [1988]). In addition to the SASW tests crosshole tests were performed as well. The crosshole test is a well established method for determining shear wave velocity profiles as discussed previously, and thus has served to validate (or invalidate) the SASW test and data analysis procedures.

The SASW tests were conducted following the CRMP geometry (figure 1). Data was collected at receiver spacings of 4, 8. and 16 ft (1.22, 2.44, and 4.88 m). Two low-natural-frequency (1 Hz) velocity transducers were used as receivers. An 8 lb (35.6 N) sledge hammer source was used for the 4- and 8-ft (1.22- and 2.44-m) spacings, while a standard penetration test (SPT) hammer was used for the 16-ft (4.88-m) spacing. The crosshole tests were performed using a 3-borehole array (see figure 6[b] in Woods [1986]). The average dispersion curve from the SASW test is shown in figure 2. The program "INVERT" (Nazarian [1984] and Stokoe and Nazarian [1985]), was then used to obtain the shear wave velocity profile from the experimental dispersion curve (inversion process). The soil profile was divided into 2-ft (0.61-m) layers of constant shear wave velocity, Poisson's ratio, and unit weight to a depth of 16 feet (4.88 m) (equal to one-third the maximum wavelength measured). A homogeneous half-space was assumed below a depth of 16 feet (4.88 m). Constant values of Poisson's ratio (0.33) and unit weight (110 pcf [17.28 kN/m^3]) were assumed for the entire profile as Nazarian (1984) has shown that the effect of these parameters is very small in comparison to the shear wave velocity. Further, it is only the ratio of unit weights of the individual layers that enter into the calculations, and thus the actual value chosen is not significant. Of course, the actual unit weight of the soil is required to calculate the shear modulus from the shear wave velocity. A shear wave velocity was then assigned to each of the layers and the theoretical dispersion curve was calculated and compared with the experimental. The assumed shear wave velocity profile was then systematically adjusted until the theoretical and experimental dispersion curves matched to within a reasonable tolerance. The resulting match between the two dispersion curves is shown in figures 3–7. Figure 3 demonstrates the two curves for all wavelengths, while figures 4–7 show selected portions at a larger scale to further demonstrate the match. It is observed that the agreement between the theoretical and experimental curves is excellent for all wavelengths.

The assumed shear wave velocity profile that results in the desired match between the theoretical and experimental dispersion curves is taken as the shear wave velocity profile representing the site. The shear wave velocity profile used to produce the theoretical dispersion curve shown in figures 3–7 is shown in figure 8. Also shown in figure 8 is the shear wave velocity profile obtained from the crosshole test. The crosshole test results shown in this figure have been designated as "uncorrected". They were obtained by dividing the horizontal distance between the boreholes by the travel time measured in the field. The travel path is thus assumed to be a horizontal straight line. No correction for possible refraction or wave path curvature effects was made. It is observed that the shear wave velocity profiles from the two methods compare very favorably. In all but one case the results differ by less than 10 percent. The difference between the results at a depth of approximately 6.5 ft (1.98 m) is about 15 percent. It is also observed that all but one of the crosshole velocities are larger than the SASW results. Both of these observations follow very closely the results from previous investigations.

Figure 9 compares the same SASW results as in figure 8 with "corrected" crosshole results. The crosshole results were corrected for possible wave path curvature by an approximate method described in Hoar (1982). The correction procedure assumes that the shear wave velocity varies linearly with depth, which is a reasonable assumption for the profile under study. The correction was found to be small, as all correction factors were between 0.97 and 1, and thus suggests that wave path curvature did not significantly influence the crosshole test results in this case study. The crosshole test results were reduced, but they are still larger than the velocities from the SASW test. Averaged over the range of available data, wave path curvature accounted for approximately 30 percent of the discrepancy between SASW and

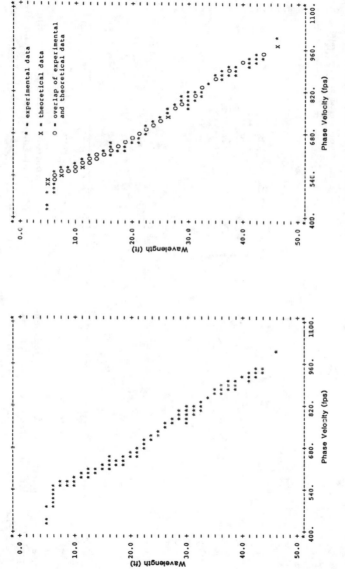

Figure 2 — Average Experimental Dispersion Curve for Beal St. Field Site

Figure 3 — Comparison of Average Experimental and Theoretical Dispersion Curves for Beal St. Field Site (all wavelengths)

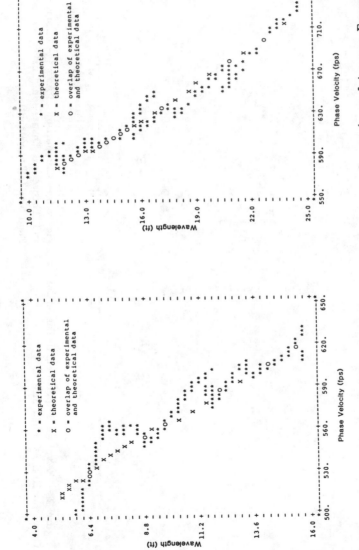

Figure 4 — Comparison of Average Experimental and Theoretical Dispersion Curves for Beal St. Field Site (4- to 16-ft [1.22- to 4.88-m] wavelengths)

Figure 5 — Comparison of Average Experimental and Theoretical Dispersion Curves for Beal St. Field Site (10- to 25-ft [3.05- to 7.62-m] wavelengths)

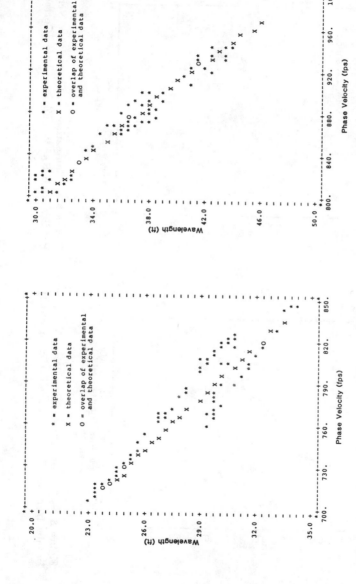

Figure 6 — Comparison of Average Experimental and Theoretical Dispersion Curves for Beal St. Field Site (20- to 35-ft [6.10- to 10.67-m] wavelengths)

Figure 7 — Comparison of Average Experimental and Theoretical Dispersion Curves for Beal St. Field Site (30- to 50-ft [9.14- to 15.24-m] wavelengths)

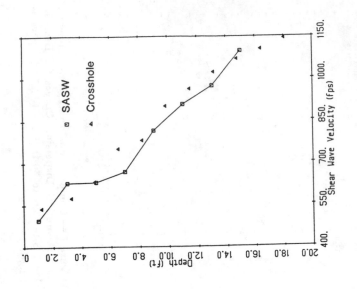

Figure 8 — Comparison of Shear Wave Velocity Profiles from SASW and Crosshole (Uncorrected) Tests

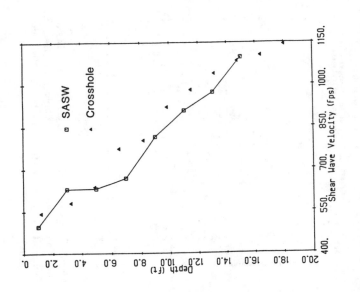

Figure 9 — Comparison of Shear Wave Velocity Profiles from SASW and Crosshole (Corrected) Tests

crosshole velocities. Thus, wave path curvature in the crosshole test, as corrected for following the method of Hoar (1982), cannot fully explain the discrepancies usually observed between SASW and crosshole results. Another possible explanation for the differences noted is that the two methods rely upon different physical behaviors to interpret the shear wave velocity. The crosshole method is based on identifying the initial disturbance propagating through the material, while the SASW method looks at the behavior much later in the time history. Furthermore, it is likely that shear waves and Rayleigh waves respond differently to the in situ stress conditions since the Rayleigh wave contains both dilatational and shear components. Differences in strain amplitude could also cause the discrepancy, though this influence is likely very small. All merit further investigation. It also should be remembered that the velocities from the SASW test are obtained by fitting a theoretical model to experimental data that contains scatter. This scatter alone may account for all of the observed differences. In the end, however, the noted differences between the two methods are small and for most practical applications of shear wave velocity data are likely insignificant.

CONCLUSIONS

The results from crosshole tests, an established method for determining the shear wave velocity profile of soil systems, were compared with the results from SASW tests conducted at the same site. The results were found to compare very favorably and they followed closely the trends observed in previous investigations. In particular, it was noted that the velocities from the crosshole test were slightly larger than the velocities obtained from SASW. Wave path curvature in the crosshole test was found to partially explain these discrepancies.

ACKNOWLEDGEMENTS

The authors are grateful to Drs K. H. Stokoe, II and Soheil Nazarian of the University of Texas at Austin for providing the program "INVERT" for use in SASW research at the University of Michigan. The continued support of SASW research at the University of Michigan by the Geotechnical Laboratory of the U. S. Army Engineer Waterways Experiment Station, Vicksburg, Mississippi is sincerely appreciated.

APPENDIX—REFERENCES

[1] De Alba, P., Baldwin, K., Janoo, V., Roe, G., and Celikkol, B. (1984), "Elastic-Wave Velocities and Liquefaction Potential," Geotechnical Testing Journal, ASTM, Vol. 7, No. 2, June, pp. 77–87.

[2] Dobry, R., Mohamad, R., Dakoulas, P., and Gazetas, G. (1984), "Liquefaction Evaluation of Earth Dams—A New Approach," Proceedings of the Eighth World Conference on Earthquake Engineering, San Francisco, California, Volume III, July 21–28, pp. 333–340.

[3] Dobry, R., Powell, D. J., Yokel, F. Y., and Ladd, R. S. (1980), "Liquefaction Potential of Saturated Sand—The Stiffness Method," Proceedings of the 7th World Conference on Earthquake Engineering, Istanbul, Turkey, Vol. 3, September, pp. 25–32.

[4] Dobry, R., Stokoe, K. H. II, Ladd, R. S., and Youd, T. L. (1981), "Liquefaction Susceptibility from S-Wave Velocity," Insitu Testing to Evaluate Liquefaction Susceptibility, Proceedings of Geotechnical Engineering Division Session at ASCE Convention, St. Louis, Missouri, October 27.

[5] Hoar, R. J. (1982), "Field Measurement of Seismic Wave Velocity and Attenuation for Dynamic Analyses," Ph.D. Dissertation, The University of Texas at Austin, May, 478 pp.

[6] Hoar, R. J. and Stokoe, K. H. II (1978), "Generation and Measurement of Shear Waves In Situ," Dynamic Geotechnical Testing, ASTM STP 654, American Society for Testing and Materials, pp. 3–29.

[7] Hryciw, R. D. and Woods, R. D. (1988), "DMT-Cross Hole Shear Correlations," Accepted for publication in Proceedings of the First International Symposium on Penetration Testing, Orlando, Florida, March 20–24.

[8] Nazarian, S. (1984), "In Situ Determination of Elastic Moduli of Soil Deposits and Pavement Systems by Spectral-Analysis-of-Surface-Waves Method," Ph.D. Dissertation, The University of Texas at Austin, 453 pp.

[9] Nazarian, S. and Stokoe, K. H. II (1984), "In Situ Shear Wave Velocities From Spectral Analysis of Surface Waves," Proceedings of the Eighth World Conf. on Earthquake Engineering, San Francisco, California, Vol. III, July 21–28, pp. 31–38.

[10] Stokoe, K. H. II and Nazarian, S. (1985), "Use of Rayleigh Waves in Liquefaction Studies," Measurement and Use of Shear Wave Velocity for Evaluating Dynamic Soil Properties, Proceedings of a Geotechnical Engineering Division Session at ASCE Convention, Denver, Colorado, May 1, pp. 1–17.

[11] Stokoe, K. H. II and Woods, R. D. (1972), "In Situ Shear Wave Velocity by Cross-Hole Method," Journal of the Soil Mechanics and Foundations Division, ASCE, Vol. 98, No. SM5, May, pp. 443–460.

[12] Woods, R. D. (1978), "Measurement of Dynamic Soil Properties," Proceedings of the Earthquake Engineering and Soil Dynamics Conference, ASCE, Pasadena, California, Vol. I, June 19–21, pp. 91–178.

[13] Woods, R. D. (1986), "In Situ Tests for Foundation Vibrations," Use of In Situ Tests in Geotechnical Engineering, Geotechnical Special Publication No. 6, Proceedings of an ASCE Geotechnical Engineering Division Specialty Conference, Virginia Tech, Blacksburg, Virginia, June 23–25, pp. 336–375.

[14] Woods, R. D. and Stokoe, K. H. II, (1985), "Shallow Seismic Exploration in Soil Dynamics," Richart Commemorative Lectures, Proceedings of a Geotechnical Engineering Division Session at ASCE Convention, Detroit, Michigan, October 23, pp. 120–156.

IN-SITU SEISMIC VELOCITY AND ITS RELATIONSHIP WITH SOIL CHARACTERISTICS IN OSAKA BAY

Yoshinori Tomozawa Iwasaki*

ABSTRACT

In-situ PS-wave velocities were measured at two bore holes with depth of about 140 meters in south Osaka Bay. The subsurface geology in the area consists of quaternary age alternating clay and sand layers down to a few hundred meters. However, the most part of the formation of two borings is clay layers. P-waves are found within the range of 1,400m/sec in very soft surface layer to 2,000m/sec at dense gravel layer. The P-wave velocities are found to depend mainly upon void ratio. Shear wave velocity varies from 30m/sec in very soft clay to 500m/sec in dense gravel layer. S-wave velocities are evaluated by rigidities of soils(G), which have been historically expressed by an experimental function of void ratio(e) and confining pressure(P) in the form of

$$G = A\frac{(B-e)^2}{1+e}P^n$$

where A, B and n are constant parameters from laboratory test results.

Taking into account general consolidation relation between e and $logP$, the above expression is modified into a linear relationship between e and log G with the liquid limit parameter of the clay.

In-situ measured rigidities of clays, calculated by shear wave velocity, are also found to have linear relationship between e and $logG$ if void ratios are plotted against G in semi-log scale. The relation between G and e is expressed by the following relation.

$$e = e_0 - Cr \times \log\frac{G}{Gr}$$

where G_r is a reference rigidity at void ratio e_0, and, Cr termed the rigidity index was found to be a function of LL such that: $Cr = -0.0027*(LL+155)$. The above expression is a more general formula to express rigidity of clayey soils which is easily understandable with the relation between consolidation characteristics.

Introduction

As is well known, shear wave velocity plays an important role to consider ground motion characteristics under earthquake condition. In soil dynamics, the shear wave characteristics are mainly studied through laboratory tests of various methods. Ishimoto and Iida(1937)[1] were the first who developed resonant column testing apparatus and studied soil characteristics of several soils. Hardin and Richart(1963)[2] proposed the relation among the rigidity, void ratio and consolidation pressure based upon resonant column study for sand. Seed and Idriss(1970)[3] made clarified the usefulness of the equivalent linear moduli to treat the soils under earthquake motions and formalized the strain dependent characteristics of clayey and sandy soils. Since then various soils have been tested and the results are summarized recently by Kokusho(1987)[4].

----

*Director, Geo-Research Institute, Osaka Soil Test Lab.

1-8-4, Utsubo-Honmachi, Nishi-ku, Osaka 550, Japan

290

IN SITU SEISMIC VELOCITY

The practical calculation of earthquake response analysis of soil layer based upon the equivalent linear moduli has been based upon the rigidities at small strain(G_o) as well as the rigidity reduction and the strain dependent damping ratio. Eventhough soils show strain dependent characteristics under earthquake condition, the rigidities at small strain, hence shear wave velocity, has been widely realized to be one of the basic factor to give a reference value of the strain dependent characteristics.

In-situ measurement of S wave velocities has been widely recognized as to give the initial rigidity of soils to be used in the response calculation. PS logging had been applied in the field of seismic design purpose in various geological conditions by refraction as well as down hole method. The shear wave velocities have been given as average values to the layers and the relationship among Vs, N-values and overburden pressures for different soil types have been discussed by several researchers. These averaged velocities are not considered accurate enough to discuss the relationship between rigidities and soil characteristics in detail.

Recently, Kitsunezaki(1980)[5] developed a new technology, called as the suspension PS-logging system, which consists of shot source and receivers in the bore hole without fixing the receivers against wall. The wave velocity is calculated as the arrival time difference at two receivers with distance of one meter, which have enabled to measure the S-wave velocity for every one meter along the bore hole. This paper describes the results of suspension PS-logging applied to measure PS-velocities along two bore holes in the south of Osaka Bay and made the comparisons of the dynamic properties and the soil mechanical characteristics.

PS-logging System

PS-logging is to measure P and S waves along a bore hole wall. For shear wave measurement, the down hole method of a single receiver fixed against the borehole wall with the source on the surface of the ground has been widely accepted to evaluate the velocities of various geological conditions from very soft alluvium to hard rock formations. In Osaka area, the result of PS logging for approximately 100 borings have been obtained by this down-hole method. The relationship among shear wave velocity(Vs), over burden effective pressure (P) and standard penetration test N-value has been studied for different soils. Assuming the relationship given by equation (1), regression analysis was made to obtain the most suitable values of parameters to relate S-wave velocity and other factors as follows:

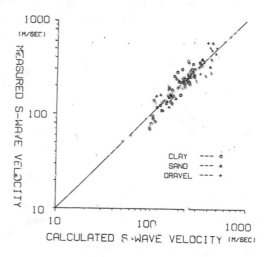

Fig.1 Calculated vs. Measured Shear Wave Velocity

$$Vs = \alpha \times F_i P^\beta (N+1)^\gamma \qquad (1)$$

where

Vs; shear wave velocity(m/sec)

α, β, γ ; constant

$\alpha = 38.8, \beta = 0.175, \gamma = 0.244$

F_i ;formation factor

for clay Fc = 1.000

sand   Fs = 0.922

gravel Fg = 0.851

P: effective overburden stress

N: SPT value

F_i termed formation factor is to express the eefect of different soil type normalizing by the clayey formation factor as $F_c = 1.0$. The measurement results in thin layers(less than 3m of thickness) were avoided to base reliable data sets. Among the constants, SPT-N value is found as the most significant factor for shear wave velocity. Fig.1 shows the comparison between the calculated and observed velocities of shear waves in Osaka area. The log scaled standard diviation of the error was $\sigma = 0.079$. Shear wave velocity obtained by down-hole method decreases its accuracy if the interval distance decreases and any attempt to study the characteristics of rigidity with detailed soil properties could not be made.

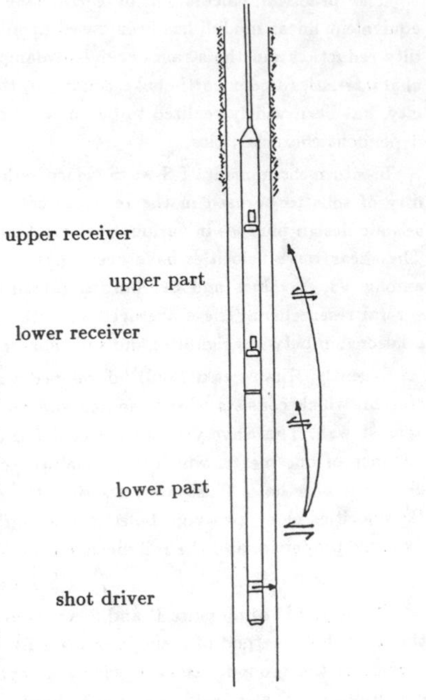

upper receiver

upper part

lower receiver

lower part

shot driver

Fig.2 Suspension Type PS-logging

boring No.56-2

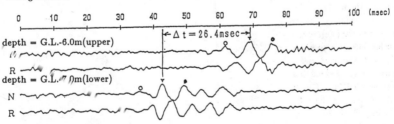

N:normal shot direction / R:reverse shot direction

(symbols at peaks of the wave corresponds to the same phase motin of upper and lower receivers)

Fig.3 An Example of Record of S-wave

Recent development of the suspension type PS-logging system by Kitsunezaki(1980)[5] is shown in Fig.2. The upper part contains twin receivers each of which consists of vertical and horizontal geophones. The lower part is equipped with a magnetic hammer which strikes the piston driver inducing horizontal impulse pressure onto bore hole wall through mud fluid. The shear wave radiated from the source induces horizontal particle motion of soils due to the passage of the shear waves. The borehole itself is forced to move horizontally with the surrounding soil.

IN SITU SEISMIC VELOCITY

The floating receivers in the mud liquid follows the same movement as the borehole since the receivers are designed to have enough flexibility and nearly equal bulk density as the mud fluid. Fig.3 shows an example of the recorded S waves at the depth of − 6 and − 7m from the sea bottom ground at borehole No.56- 2. At each depth, a pair of observed waveforms denoted as N and R are shown.

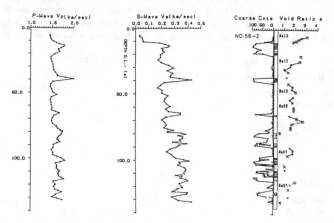

Fig.4 PS-Logging Result for Boring No.56-2

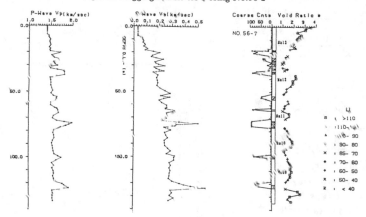

Fig.5 PS-Logging Result for Boring No.56-7

These pairs are different only in the direction of the impact shot(normal or reverse). It is expected that comparison of shear waves from the normal and reverse shots helps identify the "S-wave". The travel time difference between two records of GL-6.0 and GL-7.0 is shown as $dt = 26.4msec$ for one meter distance and the resulting S-wave velocity is 38m/sec.

Fig.4 and Fig.5 show the results of P and S wave velocities obtained in two borings denoted No.56-2 and No.56-7.

EARTHQUAKE ENGINEERING

Osaka Bay is a tectonically sinking basin and sedimentary layers are results of combination effects due to eustatic sea level change and the sinking basin. Alternation of sand-gravel and clay layers are found in the area. The change of the void ratio with depth, liquid limit as well as coarse material contents are also shown in the same Figs.4 and 5. The void ratio(e) decreases with depth from $e = 3.0$ at the surface to about $e = 1.0$ at the deeper clay layer. 1. The S-wave velocity is extremely low on the order of 30 to $60m/sec$. in the alluvial clay denoted Ma13 and increases with depth. P-wave velocities are rather constant with depth compared with S wave velocity. Both P and S-waves increase abruptly at the depth where the coarse contents increase.

P-S Wave Velocities and Soil Characteristics

Theoretical treatments of wave propagation through soils as porous media for P-wave velocity were done by several researchers. According to Ishihara(1976)[6], P-wave velocity is expressed by the following equation,

$$V_p^2 = \frac{(\lambda+\frac{4}{3}G)}{\rho} = \frac{K}{\rho} \quad (2)$$

where

$\lambda = \alpha b + 2\beta c + Kc$

K:bulk modulus(kgf/cm^2)

G:rigidity(kgf/cm^2)

ρ:unit mass of soil (g/cm^3)

$\alpha_b = \dfrac{((n^2-1)C_p+(n-1)C_s+C_w)}{\beta}$

$\alpha_c = \dfrac{-n((n-1)C_p+C_s)}{\beta}$

$K_c = \dfrac{nC_b}{\beta}$

$\beta = C_b(C_w-C_s)+C_pC_s$

$n = \dfrac{e}{1+e}$

C_s :compressibility of
soil particle(cm^2/kgf)

C_w :compressibility of
water(cm^2/kgf)

$(C_w = 4.8\times10^{-5}cm^2/kgf)$

C_b :compressibility of soil skelton

C_p :compressibility of
void($C_p = \dfrac{(C_b-C_s)}{n}$)

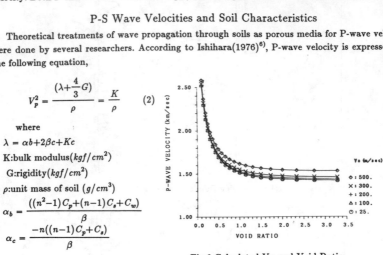

Fig.6 Calculated Vp and Void Ratio

Fig.7 Measured Vp and Void Ratio

Based upon equation (2), Vp may be expressed as a function of void ratio as a parameter of Vs and plotted in Fig.6. The measured P-wave velocities are plotted in Fig.7. The P-wave velocity itself is rather constant in the range of the void ratio of the clay layers.

IN SITU SEISMIC VELOCITY

Eventhough scattered, the tendency of the increase of the P-wave with S-wave velocity found in the measured velocity of Fig.7 as well as computed relationship in Fig.6.

Shear wave velocity and Characteristics of Soils

As well known, the rigidity is expressed by shear wave velocity Vs and unit mass of the soil ρ as follows,

$$G = \rho \, V_s^2 \tag{3}$$

Hardin and Richart(1963)[2] and Hardin-Black(1968)[7] proposed the following relationship among the rigidity G, void ratio e and confining pressure (P_c) for sand and clay,

$$G = AF(e)(P_c)^n \tag{4}$$

where A: constant($= 330 - 700$)

F(e): function of void ratio

$$F(e) = \frac{(B-e)^2}{1+e}$$

where $B = 2.17-2.97$ according to soils studied

n: constant($= 0.5$)

P_c: confining pressure

Table-1   Parameters in eq-4 obtained by various researchers

researcher	A	B	n	soil	remarks
Hardin-Richart(1963)	700	2.17	0.5	sand	round Ottawa sand
Hardin-Richart(1963)	330	2.97	0.5	sand	angular crushed quartz
Shibata-Soelarno(1975)	4200	*	0.5	sand	F(e) (see below)
Iwasaki-Tatsuoka et al.(1978)	900	2.17	0.38	sand	eleven kinds clean sand
Kokusho(1980)	840	2.17	0.5	sand	Toyoura sand
Yu-Richart(1984)	700	2.17	0.5	sand	three kinds clean sand
Hardin-Black (1968)	330	2.97	0.5	clay	kaolinite, etc.
Marcuson-Wahls(1972)	450	2.97	0.5	clay	kaolinite, Ip=35
Marcuson-Wahls(1972)	45	4.40	0.5	clay	Bentonite, Ip=60
Zen-Umehara(1984)	200-400	2.97	0.5	clay	Remolded clay, Ip=0-50
Kokusho-Yoshida et al.(1982)	14.1	7.32	0.6	clay	Undisturbed, Ip=40-85

$$* \quad F(e) = 0.67 - (\frac{e}{1+e})$$

The same relationship has been confirmed throuth laboratory tests by other researchers (Shibata and Soelarno(1975)[8], Iwasaki, Tatsuoka and Takagi(1978)[9], Kokusho(1984)[10], Yu and Richart(1984)[11], Marcuson and Wahls(1972)[12], Zen and Umehara(1978)[13] and Kokusho,Yoshida and Esashi(1982)[14]) and those parameters obtained by various researchers are summarized by Kokusho(1987)[4] and listed in Table-1.

For sand and gravel soils, the range of void ratio is rather small and the pressure effects is the most important factor to rigidity of sand. For clays, void ratio changes with much wider range caused by consolidation.

The equation (4) is found that the rigidity is a function of void ratio and consolidation pressure. Since the void ratio is related with the consolidation pressure through compression index(C_c), the rigidity (G) may be expressed by only void ratio or consolidation pressure with compression index.

The tested ranges of the void ratio to evaluate parameters by each researcher are listed in Table-2. The function $F(e)$ in equation (4) was calculated for different proposed equations and shown in Fig.8. In Fig.8 the solid lines correspond to the tested ranges and the functions are found to have linear relationship with void ratio within the tested ranges.

Table-2 B-values and tested range of void ratio

No	reseacher	A	B	void ratio range
1	Hardin-Black(1968)	330	2.97	0.6 - 1.5
2	Marcuson-Wahls(1972)	45	4.40	1.5 - 2.5
3	Zen,Umehara-Hamada(1978)	200-400	2.97	0.4 - 1.7
4	Kokusho,Yoshida-Esashi(1982)	14.1	7.32	1.5 - 4.0

$$logF(e) = C_1 - C_2 e \qquad (5)$$

Taking logarithm of each side of the equation(4) results in the following equation.

$$logG = logA + logF(e) + logP_c^n \qquad (6)$$

For normally consolidated clays, the void ratio decreases with increase of the consolidation pressure as follows;

$$e = e_o - C_c log(\frac{P_v}{P_r}) \qquad (7)$$

where C_c: compression index
P_v: effective overburden pressure
P_r: reference overburden pressure
e_o: reference void ratio at P_r

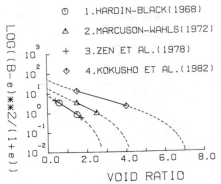

① 1.HARDIN-BLACK(1968)
△ 2.MARCUSON-WAHLS(1972)
+ 3.ZEN ET AL.(1978)
◇ 4.KOKUSHO ET AL.(1982)

Fig.8 Function $F(e)$ Value and Void Ratio

Assuming $P_c = \dfrac{1+2K_o}{3}P_v$ and $P_{cr} = \dfrac{1+2K_o}{3}P_r$, where K_o is the coefficient of lateral pressure, the equation (7) becomes as follows,

$$logP_c = \frac{e_o - e}{C_c} + log(P_{cr}) \qquad (8)$$

Putting the equations(5) and (8) into (6), log(G) can be expressed by the function of only void ratio(e).

$$log(G) = D - C \times e$$

or

$$e = e_o - C_r log(\frac{G}{G_r}) \qquad (10$$

where

$$D = logA + nlog(\frac{3}{2}P_r) + C_1 + \frac{n}{C_c}$$

$$C = \frac{n}{C_c} + C_2$$

$$C_r = \frac{C_c}{n + C_c C_2} \quad : \text{Rigidity Index}$$

e_o : reference void ratio

G_r : reference rigidity at e_o

The relation between rigidity(G) and void ratio(e) can be easily calculated for normally consolidated clays based upon equation (10). The rigidity index (C_r) defined as the decrease of void ratio to increase the rigidity for ten times or one cycle of log scale and is obtained if the compression index (C_c) and a constant (C_2) are known.

Table-3 Typical relation among LL, C_c and e_o

I.L(%)	30	50	80	120
e_o	1.2	1.8	3.0	4.4
C_c	0.18	0.36	0.63	0.99
$C_r(B = 7.32)$	0.27	0.51	0.79	1.09

(eo at $P_r = 0.01 kgf/cm^2$)

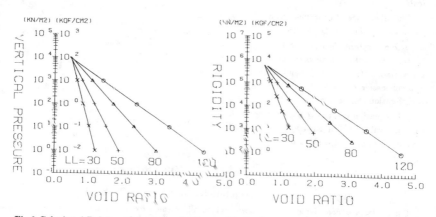

Fig.9 Calculated Relation between $e - logP$ Fig.10 Calculated Relation between $e - logG$

e compression index may be expressed with liquid limit(LL) as follows,

$$Cc = 0.009(LL-10) \tag{11}$$

Typical values of several variables and constants for clayey soils are read from the e-P curves given by Lambe and Whitman(1969)[15] in their textbook and listed in Table-3.

Another constasnt C_2 is a tangent of $logF(e)$ against e and changes with the parameter of B.

As a typical example, the equation proposed by Kokusho and Yoshida et al.(1982)[14], which covers the widest range of void ratio and applied to various clayey soils from Ip=40 to 100, was chosen to see the relation between the rigidity(G) and void ratio(e). Based upon these relationships with equating $P_c = \frac{2}{3}P_v(K_0$ is assumed as 0.5),

$$G = \frac{14.1(7.32-e)^2}{(1+e)} \frac{2}{3}P_v^{\ n} \tag{12}$$

the relation of e and $logG$ as well as e and $logP_v$ are plotted in Figs.9 and 10. Fig. 9 shows the relation of $e - \log P_v$ for clay with characteristics shown in Table-3. Fig.10 shows the relation of e and G expressed by the equation (12) in $e-\log G$ axis. C_2 is about 0.32 for this case of B = 7.32. The correspoinding C_r is also listed in Table-3. It is interesting to recognize the linear relationship between e and log G similar to the compression curve. This linear relationship may be also confirmed for other proposed values of the constants in Table-1.

It is also realized that the amount of the void ratio change to increase the rigidity is different for clays with different liquid limit.

Fig.11 shows the change of the measured rigidities for the wide range of void ratio for a clay with liquid limit of about LL=100 by laboratory test(Nakagawa and Okuda(1978)[16]). This also shows linear relatioship between e and $logG$.

The advantage of equations (9) or (10) compared to the equation(4) are as follows,

1) The constants in equation (4) are difficult to correlate with the physical relation among the variables. On the other hand, equation (10) is much easier understood by comparison with consolidation characteristics.

2) In equation (4), the void ratio must be less than the value B. The equation has the limitation of the applicable void ratio.

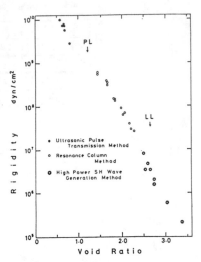

Fig.11 Rigidity Change with Void Ratio
(after Nakagawa and Okuda(1978)[16])

Rigidity and Soil Characteristics of Clays in Osaka Bay

Osaka Basin is considered to have been tectonically formed by the subsidence of the Osaka basin block with inclining northwards. The Osaka basin block was separated by the surrounding mountains with active faults. Geological study of 900m depth boring in 1970 have revealed that the settlement rate of the basin rock was confirmed as about 500m for the last one million years

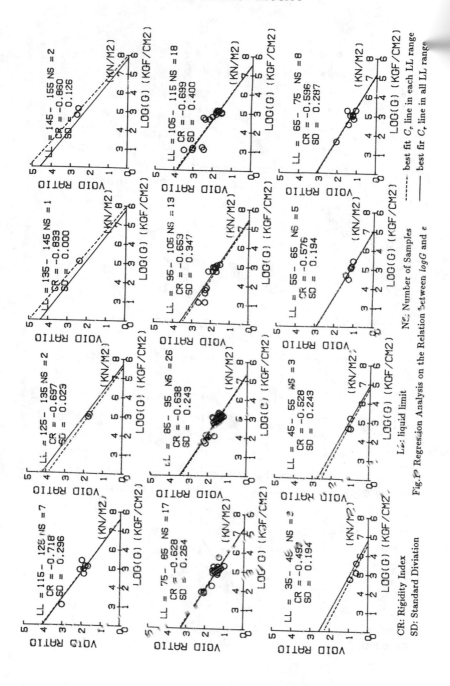

LL = 145 − 155 NS = 2
CR = −0.860
SD = 0.126

LL = 135 − 145 NS = 1
CR = −0.833
SD = 0.000

LL = 125 − 135 NS = 2
CR = −0.697
SD = 0.023

LL = 115 − 125 NS = 7
CR = −0.718
SD = 0.296

LL = 105 − 115 NS = 18
CR = −0.699
SD = 0.400

LL = 95 − 105 NS = 13
CR = −0.653
SD = 0.347

LL = 85 − 95 NS = 26
CR = −0.638
SD = 0.243

LL = 75 − 85 NS = 17
CR = −0.628
SD = 0.264

LL = 65 − 75 NS = 8
CR = −0.596
SD = 0.287

LL = 55 − 65 NS = 5
CR = −0.576
SD = 0.194

LL = 45 − 55 NS = 3
CR = −0.528
SD = 0.243

LL = 35 − 45 NS = 3
CR = −0.492
SD = 0.194

------ best fit C_r line in each LL range

—— best fir C_r line in all LL range

NC: Number of Samples

LL: liquid limit

CR: Rigidity Index

SD: Standard Diviation

Fig. 10 Regression Analysis on the Relation between $\log G$ and e

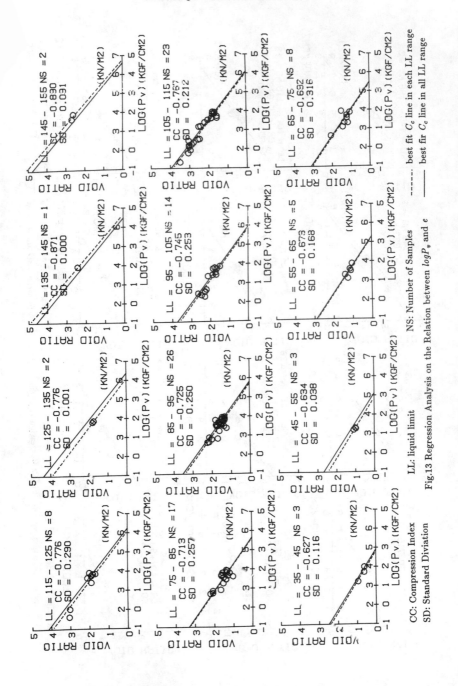

Fig.13 Regression Analysis on the Relation between $logP_v$ and e

CC: Compression Index
SD: Standard Diviation

LL: liquid limit
NS: Number of Samples

------- best fit C_c line in each LL range
——— best fir C_c line in all LL range

of quarternary age(Huzita and Maeda(1985)[16]). Two borings to which PS-logging was ap_ in the south Osaka Bay were among the many borings made for the soil investigation of Kan. airport project. The southern Osaka Bay area is considered to have been basically the same sed. mentary environments as the northern Bay area(Nakaseko et al.(1984)[17]). To study the relationship between the rigidity and soil characteristics, the rigidities are calculated based upon the shear wave velocities and unit weight of soil samples at the depth where the velocities were measured. Fig.12 shows the relation between $\log G$ and e for various range of liquid limit. Eventhough scattered, e is found to have some linear relationship with $\log G$.

The compression index C_c is also obtained by plotting void ratio and effective overburden stress Pv as in Fig.13 .

Assuming the equation (9) between e and $\log G$, the constants of e_o and C_r were searched to satisfy the best fit the observed data set by linear regression analysis(method of least squares).

The best fit rigidity index(C_r) and standard diviation for each range of liquid limit is shown in the Figures noted as CR and SD. Fig.14 shows the relationship between the rigidity index and the middle value of the liquid limit range analysed. The radiuses of the plotted circle correspond to the number of the samples for each range of the liquid limit analysed. These obtained rigidity indices are further analysed by the method of least squares to get the relationship between the rigidity index and the liquid limit.

The weight was given to

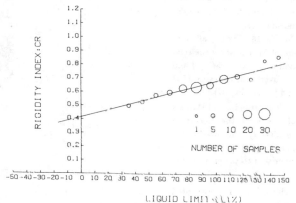

Fig.14 Rigidity Index and Liquid Limit

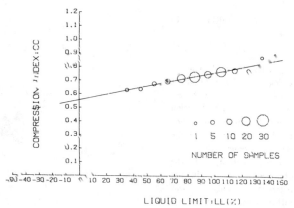

Fig.15 Compression Index and Liquid Limit

each rigidity index(C_r) as proportional to the number of the samples in the corresponding range of liquid limit. The analysis resulted in the following relation.

$$C_r = -0.0027(LL+155)$$ (13)

The compression indices(C_c) are also plotted against liquid limit and the following relation is obtained.

$$C_c = -0.0020(LL+280) \quad (14)$$

Table-4 shows the analytical results of rigidity index and compression index with liquid limit for Osaka marine clay.

In Fig.16 Fig.17 , all data are plotted to show the relation of the void ratio(e) and the overburden pressure(Pv) and the rigidity(G).

The standard diviations of errors of the calculated rigidity(G) and overburden pressure(Pv) in log scale compared to the measured

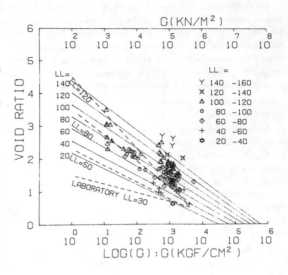

Fig.16 In-situ Void Ratio and Rigidity of Clay in Osaka Bay

rigidity or the estimated pressure calculated by sample density are obtained as follows,

$$\sigma_{logG} = 0.43 \quad (15)$$

$$\sigma_{logP_v} = 0.27 \quad (16)$$

These standard diviation in log-scale of σ being about 0.3 - 0.4 is equivalent to say that the standard diviation range is within from one third to three times of the estimated value.

The sensitivity of the obtained constants in the regression analysis was rather weak because of the limited number of samples in

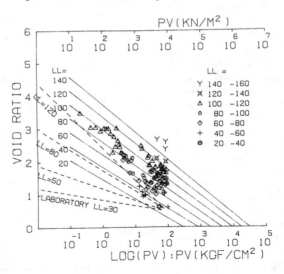

Fig.17 In-situ Void Ratio and Oververden Pressure of Clay in Osaka Bay

several range of the liquid limit.

The above results may be changed if the adequate number of samples are provided.
The comparison between Tables-3 and 4 and Figs.16 and 17 indicates an interesting differenc
on the laboratory and field behaviour of normally consolidated clays.
The laboratory results of $e - logP$ and $e - logG$ lines are replotted as dotted lines in Figs.16
and 17. The in-situ void ratio at the same pressure is larger than laboratory value. The in-situ
rigidity at the same void ratio is greater than laboratory value. These difference can be explained
mainly by aging effects on clays in-situ. The sedimeted clay decreases void ratio as it continues
to consolidate by additional overburden pressure caused by newly sedimented formation. At the
same time, it gains bonding strength between clay paticles which increases rigidity at the same
void ratio and remains larger void ratio at the same overburded pressure.

Table-4. C_c and C_r based upon in-situ marine clay in Osaka Bay

Liquid Limit	40	50	60	70	80	90	100	110	120	130	140
Compression Index C_c	0.64	0.66	0.68	0.70	0.72	0.74	0.74	0.76	0.78	0.80	0.82
Rigidity Index C_r	0.52	0.55	0.58	0.61	0.63	0.66	0.69	0.72	0.74	0.77	0.77

Conclusion

Based upon seismic wave velocities obtained by PS- logging in two boreholes at south Osaka
Bay, and comparison of the velocities with soil characteristics resulted in the following conclusions concerning the dynamic properties of clayey soils in Osaka Bay:

1. P-wave velocity is mainly a function of void ratio and corresponds fairly well to theoretical
calculation by Ishihara.

2. Rigidity(G) has been expressed by a function of void ratio(e) and consolidation pressure(P_c)
for soils.

This relationship can be also expressed with a linear relationship between log G and e using
the constant (C_r) as a parameter for Liquid Limit as follows,

$$e = e_0 - C_r \log_{10}(\frac{G}{G_r})$$

The in-situ measured rigidity was confirmed to have the same relationship as shown above for
the wide ranges of void ratios, from 0.5 to 3.5, and of rigidity, from 10^1 to $10^4 kgf/cm^2$.
Rigidity index C_r is found to be a function of liquid limit as follows,

$C_r = 0.0020(LL+155)$.

3. The physical meaning of the increase of rigidity by consolidation may be easily understood
by introducing the same formula as compression curve of $e-\log P$ In-situ measured rigidity was
found to be several times higher than those obtained through lab testing consolidation of an
undisturbed sample having the same void ratio and LL. The increase of rigidity under natural
state is probably due to a combination of such effect as secondary consolidation and aging.

Acknowledgments

The author is grateful to Profs. A.Kitsunezaki and K.Nakagawa who made strong support to
carry out PS-logging and useful discussions on shear waves. The author also expresses his grati-
tude to Profs.Huzita and Nakaseko for their stimulating discussion on sedimentary characteristics
of Osaka Basin.

EARTHQUAKE ENGINEERING

References

1) Ishimoto,M.and Iida,K.(1936),(1937),"Determination of Elastic Constants of Soils by Means of Vibration Method", Part 1. and 2., BERI.,Univ. of Tokyo, Vol.14,15,pp.632-656, pp67- 86.

2) Hardin,B.O. and Richart,F.E.(1963),"Elastic Wave Velocities in Granular Soils", Proc. of ASCE,Vol89,SM1,pp.33-65. Hardin,B.O. and Black,W.L.(1968),"Vibration modulus of normally consolidated clay",Proc. of ASCE,Vol.94 SM2,pp353-369.

3) Seed,H.B.and Idriss,I.M.(1970),"Soil Moduli and Damping Factors for Dynamic Response Analysis", Report No.EERC 70-10, Univ.of California.

4) Kokusho,T.(1987),"In-situ Dynamic Soil Properties and Their evaluations", Preprint, Theme Lecture,The 8th Asian Regional Conference on SMFE pp.721-728.

5) Kitsunezaki,C.(1980),"A New Method for Shear-wave Logging", Geo-physics,45,1489-1506.

6) Ishihara,K.(1967),"Propagation of Compressional Waves in a Satu-rated Soil",Proc. of International Symposium on Wave Propagation and Dynamic Properties of Earth Material,Univ. of New Mexico Press.

7) Hardin,B.O. and Black,W.L.(1968),"Vibration modulus of normally consolidated clay", Proc.,ASCE,Vol.94,SM2,pp.353-369.

8) Shibata,T. and Soelarno,D.S.(1975),"Stress-strain characteristics of sands under cyclic loading",Proc. Japan Society of Civil Engineerings,No.239,pp.57-65(in Japanese).

9) Iwasaki,T.,Tatsuoka,F. and Takagi,Y.(1978),"Shear moduli of sands under cyclic torsional shear loading",Soils and Foundations,Vol.18,No.1,pp.39-56.

10) Kokusho,T.(1980),"Cyclic triaxial test of dynamic soil properties for wide strain range",Soils and Foundations,Vol.20,No.2,pp.45-60.

11) Yu,P. and Richart,F.E.(1984),"Stress ratio effects on shear modulus of dry sands",Proc.ASCE,Vol.110,GT3,pp.331-345.

12) Marcuson,W.F. and Wahls,H.E.(1972),"Effects of time on dynamic shear modulus of clays",Proc.ASCE,Vol.98,SM12,pp.1359-1373.

13) Zen,K.,Umehara,Y. and Hamada,K.(1978),"Laboratory tests and in-situ seismic survey on vibratory shear modulus of clayey soils with various plasticities",Proc. 5th Japan Earthquake Engineering Symposium,pp.721-728.

14) Kokusho,T. and Yoshida,Y. and Esashi,Y.(1982),"Dynamic soil properties of soft clay for wide strain range", Soils and Foundations,Vol.22,No.4,pp.1-18

15) Lambe,T.W. and Whitman,R.V.(1969),"Soil Mechanics",Wiley,New York, p.320.

16) Nakagawa,K.and Okuda,T.(1978),"Experimental Study on Elastic Moduli of Clayey Sediments under Undrained State", Journal of Geoscience, Osaka City Univ.,Vol.21, Art.1,pp.1-16.

17) Huzita,K. and Maeda,Y.(1985),"Geology of the OSAKA-SEINAMBU (West-south Osaka) DISTRICT",Quadrangle Series, Kyoto(11)No.62 Geological Survey of Japan(in Japanese).

18) Nakaseko,K. et al.(1984),"Geological Survey of the Submarime Strata at the Kansai International Airport in Osaka Bay, Central Japan",Report of the Calamity Science Institute,Osaka(in Japanese).

IN SITU DAMPING MEASUREMENTS
BY CROSSHOLE SEISMIC METHOD

Young Jin Mok[1], Ignacio Sanchez-Salinero[2], Kenneth H. Stokoe, II[3], M. ASCE
and Jose M. Roesset[4], F. ASCE

ABSTRACT

In situ seismic measurements in soil are generally used to evaluate wave velocities and, hence, to characterize elastic moduli. However, material damping characteristics of soil are also important but are rarely measured in situ. The crosshole seismic method is shown to be a very useful method for performing attenuation (damping) measurements at small strains. The theoretical framework and field implementation of such measurements are discussed. Important variables are: 1. the frequency content of the waveform, 2. the ratio of the wavelength to the source-receiver spacing and 3. the ability to track accurately wave particle motions. The use of windowing time-domain records to minimize interference from reflected and refracted waves and the calculation of spectral ratios in the frequency domain are necessary approaches to data processing in these tests. Two case studies are presented which illustrate these points.

INTRODUCTION

If soils were perfectly elastic materials, the static and dynamic parameters could be characterized by seismic wave velocities (or elastic moduli) alone. However, soils are not perfectly elastic materials. Even at small strains on the order of 10^{-3} percent or less, soils exhibit some energy absorbing capacity upon cyclic loading. This capacity, commonly called damping, is generally quite small at small strains and, hence, has been ignored in the application of most in situ seismic tests. The primary use of these tests has been to characterize the elastic components of moduli. However, a more complete characterization of soil requires the measurement of damping characteristics in addition to seismic wave velocities.

There are two major motivations for developing field methods to measure the damping (attenuation) properties of soils. First, computer codes used to model dynamic soil-structure interaction problems require the attenuation properties of the foundation materials underlying the structure as well as the dynamic stiffness properties of these materials. Second, attenuation characteristics combined with stiffness characteristics have significant potential to reveal considerably more information about most soils than stiffness characteristics alone. Information such as soil type, soil structure, natural cementing or fissuring, and inclusion of inhomogeneities may be determinable by combining both measurements in situ.

The amplitudes of seismic waves decrease as the waves propagate through soil. This attenuation of wave amplitude is caused by two mechanisms: 1) spreading of wave energy from a source, generally called geometrical or radiational damping, and 2) dissipation of energy due to mainly frictional losses (particle sliding) in the soil itself,

1 Research Associate, University of Texas, Austin, Texas 78712
2 Research Engineer, GEOCISA, Madrid, Spain
3 Brunswick-Abernathy Regents Professor, Univ. of Texas, Austin, TX 78712
4 Paul D. and Betty Robertson Meek Centennial Prof., Univ. of Texas, Austin, TX 78712

.monly known as attenuation, material or internal damping. The phenomenon of _enuation is much more complex than the elastic aspects of seismic wave propagation. Both laboratory and field measurements of material damping have proven to be difficult to make. One reason is that the exact nature of material damping is poorly understood and may occur from several phenomena whose magnitudes vary with strain amplitude.

Laboratory measurements of attenuation in soil and soft rock samples under varying pressures and strain amplitudes can be performed. The resonant column test is frequently used in geotechnical engineering. However, reliance on this method assumes that the specimen is undisturbed, that its natural environment has been recreated in the laboratory, and that the model used to analyze the results represents the phenomenon correctly. These can be highly questionable assumption. Moreover, the small specimens used in these types of tests cannot reflect the macroscopic features of a large site. This problem is especially important in highly fractured rock and structured clays where the results of laboratory tests are felt to be biased toward lower values of material damping.

Due to the above-mentioned shortcomings inherent in laboratory tests, field seismic measurement of material damping (attenuation) is very desirable. The basic approach to the field measurement is the generation of seismic energy in the form of compression or shear waves at one point and the monitoring of the amplitude of the energy at other points. For the crosshole seismic method presented herein, this process consists of installing three or more boreholes at a site and then conducting measurements with a source and receivers placed at similar depths in the boreholes. Care is necessary in performing amplitude measurements under these conditions, and consideration must be given to the following points and/or possible problems: 1) precise calibration of the receivers is necessary; 2) interference from reflected and refracted waves adversely affects the measurements; 3) electrical or mechanical noise decreases signal quality; 4) intimate coupling between the boreholes and receivers is necessary; 5) the boreholes should not alter the free-field motions; 6) the three-dimensional orientation of the receivers must be accurately controlled; 7) the radiation pattern of the source must be understood; and 8) coupling between additional near- and far-field components has to be considered. If these points are properly taken into account, field measurement could be an accurate means of representing the damping characteristics of the site, at least at small strains.

The main objectives of this paper are to investigate variables affecting in situ attenuation measurements and to initiate development of a practical method of measuring in situ attenuation characteristics. As with velocities, attenuation in shear is of more concern than in compression because of the ability to measure shearing characteristics below the water table and because of the importance of shear behavior in geotechnical engineering analyses. Therefore, particular attention is paid to measurements with shear waves. However, compression waves are also considered and one example is presented. The theoretical framework for these measurements is first discussed. Implementation of the work in crosshole testing is then presented followed by two case studies.

THEORETICAL FRAMEWORK

Definitions and Terminology

The elastic properties of soils and rocks are uniquely defined by elastic moduli and/or compression (P) and shear (S) wave velocities. However, the phenomenon of attenuation is complex, and attenuation properties can be specified by a wide range of measures.

The common·measure of attenuation used by geophysicists and seismologists is t quality factor, Q, and its inverse, Q-1, the dissipation factor. The intrinsic definition of Q originally from electrical circuit theory, may be expressed as:

$$\frac{1}{Q} = \frac{\Delta E}{2\pi E} \tag{1}$$

where ΔE = the amount of energy dissipated per cycle of harmonic excitation in a certain volume, E = the peak elastic energy stored in the same volume, and π = 3.1459.... . The units of energy are force-length/unit volume, and Q and Q-1 are dimensionless.

According to dissipative processes, there are several measures of attenuation. These are defined and related to one another as follows (Knopoff, 1965; and O'Connell and Budiansky, 1978):

$$\frac{1}{Q} = \frac{\Delta E}{2\pi E} = \frac{\alpha V}{\pi f} = \frac{\alpha \lambda}{\pi} = \frac{\delta}{\pi} = \frac{\Delta f}{f_r} \tag{2}$$

where α = attenuation coefficient (units of length-1), δ = logarithmic decrement (dimensionless), λ = wavelength (units of length), Δf = resonance width (Hertz), V = propagation velocity (length/time), f = frequency (Hertz), and f_r = resonant frequency (Hertz).

In geotechnical engineering, the frictional energy loss associated with a system of equations for a dynamic problem is often used as a measure of attenuation. This energy loss can be defined by the following relationship:

$$D = \frac{\Delta E}{4\pi E} \tag{3}$$

where D is called damping ratio and is commonly expressed as a percent. D is generally assumed to be independent of the frequency and amplitude of particle motion at small strains (less than 10-3 percent) and is often called hysteretic damping.

Most data suggest that, for dry rocks and dry soils, Q is indeed independent of frequency (Pandit and Salvage, 1973; Toksoz et al, 1979; Tittmann et al, 1981; and Stoll, 1985), and intergrain friction (grain sliding) is the dominant cause of energy loss. On the other hand, water-saturated rocks, sands and silts show a definite dependency of Q on frequency (Stoll, 1985). Saturated clays are the least understood, but the dependency of Q on frequency seems to be relatively small. Therefore, in the work reported herein which deals with dry sand and saturated clay, the approximate frequency independence of Q-1 is assumed.

In terms of relating Q and D, it seems more appropriate to relate Q and D using the intrinsic definition rather than to relate Q and fraction of critical damping (Hoar, 1982; and Redpath et al, 1982). (It should be noted that fraction of critical damping is based upon a Kelvin-Voigt model which shows frequency dependent attenuation while damping ratio is frequency independent. Damping ratio and fraction of critical damping measured in the laboratory are matched at resonance in an attempt to eliminate the frequency dependency of the latter.) The relationship between Q and D is:

$$D = \frac{1}{2Q} \tag{4}$$

Since both velocity and attenuation are associated with a particular mode of wave propagation, one experimental method may yield a P-wave velocity and/or a dissipation factor for the P-wave, Q_p^{-1}, while another may determine the S-wave velocity and/or Q_s^{-1}. Finally, it should be mentioned that an appropriate account must be taken of geometrical spreading, reflections and refractions in the analysis of spatially travelling waves in a layered half-space. This point is addressed in more detail in the following sections.

Spectral Ratios

The amplitudes of seismic waves propagating through a medium are attenuated by two separate mechanisms: 1. geometrical spreading of a fixed amount of energy due to wave front expansion, and 2. dissipation of energy due to attenuation mechanisms such as frictional losses at grain boundaries where relative movement or sliding occurs and/or viscosity and flow of pore fluids. These two mechanisms are often known as geometrical or radiational damping and material or internal damping, respectively.

For a harmonic wave propagating in an infinite homogeneous medium, the change in body wave amplitude with distance from the source due to both geometrical and material damping can be described by:

$$A_R = S_0 \, \frac{1}{R} \, \exp^{(-\alpha R)} \tag{5}$$

where A_R = the body wave amplitude at a distance R from the source, S_0 = the amplitude of the harmonic source, α = the attenuation coefficient, and exp = the base of the natural logarithm. The term $1/R$ accounts for geometrical damping, and the exponential term accounts for material damping. It should be noted that this equation is only approximate because any additional near-field terms are ignored. Hence, it is often referred to as a far-field expression.

In practice, amplitudes of waves at two different distances from the source R_1 and R_2 are measured. The ratio of two amplitudes is obtained from Eq. 5 as:

$$\frac{A_2}{A_1} = \frac{R_1}{R_2} \, \exp^{(-\alpha \, [R_2 - R_1])} \tag{6}$$

in which A_1 = body wave amplitude at distance R_1 from the source, and A_2 = body wave amplitude at distance R_2 from the source. The term A_2/A_1 is designated as a spectral ratio. Again, the term R_1/R_2 accounts for geometrical spreading and the exponential term accounts for material damping.

As mentioned earlier, one measure of material damping is the attenuation coefficient α which can be expressed as:

$$\alpha = \frac{\ln \, [A_1 R_1 / A_2 R_2]}{(R_2 - R_1)} \tag{7}$$

However, in geotechnical engineering, the common way of expressing material damping is in terms of a damping ratio. By combining Eqs. 2, 4 and 7, it can be shown that material damping ratio is simply:

$$D = \frac{\ln [A_1R_1/A_2R_2]}{2\pi t_I f}$$

in which t_I = the interval travel time of the wave between distances R_1 and R_2, and f = the frequency of the wave. To use Eq. 8 to calculate material damping, the following assumptions are made: 1. material damping is independent of frequency and strain amplitude for small strains (less than 10^{-3} percent), 2. measured wave amplitudes are not affected by reflected or refracted waves, 3. any additional near-field effects on amplitude are negligible, and 4. particle motions (A_1 and A_2) are accurately tracked at both measurement points.

Equations 5 through 8 are based on measurement of a harmonic wave. To utilize these equations for waves generated from a point source, frequency domain analysis should be adopted. Two time records at distances R_1 and R_2 from the point source need to be converted to amplitude spectra using Fourier transforms. At each frequency, the spectral ratio is calculated by dividing the Fourier amplitude of the signal at R_1 by that of the signal at R_2. Damping ratio for each frequency is then calculated using Eq. 8.

IMPLEMENTATION

Field Measurements

Measurement of wave amplitudes (particle motions) in the field is performed herein by collecting time-domain records using the crosshole method. Simultaneous monitoring of two receivers in the receiver boreholes for the same source impulse is necessary. Therefore, three boreholes are required as illustrated in Fig. 1. As with wave velocity measurements, the equipment needed for damping measurements consists of a source, receivers and recording equipment.

Source - The in-hole source (wedge-type mechanical source) can be used for damping measurements as well as wave velocity measurements. The source impulse is rich in shear motion and therefore is very good for S-wave attenuation measurements. (It is not a good source for P-wave attenuation measurements). The radiation pattern of shear wave energy for this source gives the maximum energy in the horizontal direction.

The auto-power spectra of the source impulse should be determined at each site. One can then determine the frequency range over which energy is present at both receivers, and this frequency range can be used as one criterion to decide over which frequency bandwidth reasonable damping ratios can be calculated. The predominant frequency range of shear energy in most soils is typically from 10 Hz to about 400 Hz.

Receivers - Usually three-component velocity transducers are used as receivers with pneumatic packers (coupling system). Vertical receivers are used to monitor S-waves from the in-hole source. Radial receivers are used to monitor P-waves from explosive-type or sparker-type sources. However, S-wave measurements are stressed herein.

Because the change in wave amplitudes due to material damping is rather small compared with changes from other undesirable causes, extra precautions must be taken to assure that wave amplitudes are accurately monitored. First, receivers should be properly calibrated, and the difference, if any, between frequency responses of the receivers should be taken into account. Second, the coupling of each receiver in the test should be the same. In addition, any effect of coupling on the frequency response of the receiver must be investigated. However, investigation of coupling effects is a difficult task. In this study,

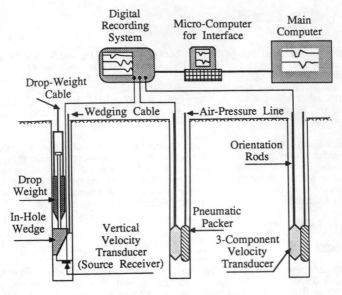

Figure 1 - Schematic Diagram of Data Aquisition and Anaysis System
for Damping Measurements

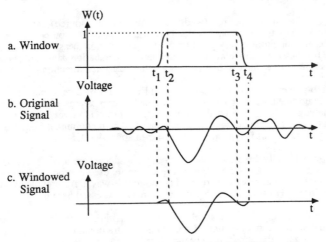

Figure 2 - Use of an Extended-Cosine-Bell Window
to Gate Original Time-Domain Window

the frequency response of the receivers is assumed to be the same when the receivers are coupled with pneumatic packers pressurized to the same air pressure. (This assumption deserves future study.)

Finally the verticality of the receivers should also be taken into account. However, with the present receivers, it is impossible to keep tracking of verticality. Therefore, borehole verticality was always measured. In the tests presented herein, the boreholes were essentially vertical, and any verticality effect was thus neglected.

Recording Equipment - A digital oscilloscope (Nicolet Explorer III or Hewlett-Packard 5423A Analyzer) should be used to monitor time-domain signals. Records can then be stored on magnetic diskettes for later analyses. In these tests, the records were transferred to the main computer (Dual Cyber 170/750) using a micro-computer (Hewlett-Packard, Model 9836) for the interface between the digital oscilloscope and the main computer (see Fig. 1). This recording and analysis system was preferred to a digital spectral analyzer which has built-in spectral analysis functions due to the flexibility of using the main computer in these studies.

Data Analyses

Time-Domain Window - As already mentioned, Eq. 8 is valid for direct waves in an infinite homogeneous medium. Unfortunately, real signals almost always include reflected and/or refracted waves, as well as the direct wave. Thus the effect of reflected and/or refracted waves on the amplitude spectra must be taken into account. In this study, a time-domain window was adopted to reduce the effect from wave interference. The direct wave was gated from the waveform with an extended-cosine-bell window, W(t), as illustrated in Fig. 2 (Ramirez, 1985). In this figure, t_2 is the first arrival point of the direct wave of concern, t_1 is the cross-over point before the direct wave, t_3 is the cross over point after the first trough and first peak of the direct wave, and t_4 is the following cross-over point. This method of gating avoids abruptly terminating the seismic trace at some value away from zero, which could cause fictitious frequency components to appear in the spectral analysis. It should be noted that this window is only valid under the assumption that the first cycle of the direct wave is not overlapped by reflected and/or refracted waves.

Selection of Frequency Bandwidth - To decide the frequency bandwidth for computing damping ratios, two criteria have been adopted. First, the predominant frequency range of the source impulse is considered. The shear wave impulse generated by the wedge-type mechanical source has the most shear energy typically between frequencies of 10 Hz and 400 Hz, with most energy below 300 Hz (Mok, 1987). The upper-bound of the bandwidth was thus taken as 300 Hz.

Second, the wavelength associated with the source-receiver spacing must be considered. Sanchez-Salinero (1987) backcalculated damping ratios from two signals monitored at distances R_1 and R_2 from theoretically generated source impulses (see insert in Fig. 3) in a given material defined by wave velocities and a damping ratio. Dispersion in the apparent damping ratio occurs because of the combination of additional near-field and far-field terms at distances close to the source. Backcalculated damping ratios start to become equal (and stable) to the actual damping ratio in shear at a value of $R_1/\lambda_S = 2$ for $R_2/R_1 = 2$, where λ_S is the shear wavelength and R_2/R_1 of 2 is the case of the crosshole setup used herein. Typical results of this effect are shown in Fig. 3a. For compression wave motion, the actual damping ratio becomes stable at a value of $R_1/\lambda_S = 4$ for $R_2/R_1 = 2$, where λ_S is still the shear wavelength. Typical results for P-motion are shown in Fig. 3b.

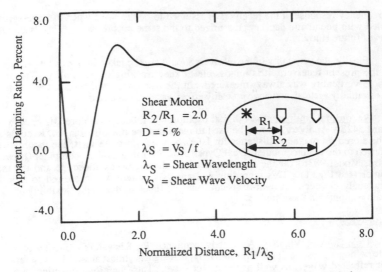

a. - Apparent Damping Ratio for Three-Dimensional Shear Motion

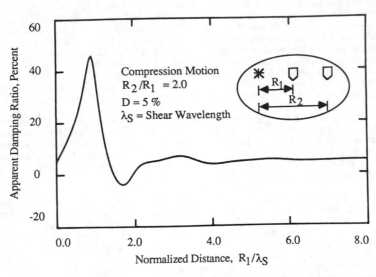

b. - Apparent Damping Ratio for Three-Dimensional Compression Motion

Figure 3 - Apparent Damping Ratios for Three-Dimensional Seismic Motions
in a Medium with 5 Percent Damping and Poisson's Ratio of 0.25

IN SITU DAMPING MEASUREMENTS

In general terms, wavelength, frequency and wave velocity are related by:

$$V = f\lambda \tag{9}$$

For the shear wavelength to be less than $R_1/2$, the frequency must be higher than $2V/R_1$. Therefore, the lower-bound of the frequency bandwidth, f_{lower}, is given by:

$$f_{lower} \geq \frac{2V}{R_1} \tag{10}$$

The complete frequency bandwidth for computing damping ratios for shear waves in the crosshole tests used herein can then be expressed as:

$$\frac{2V_S}{R_1} \leq f \leq 300 \tag{11}$$

where R_1 = the distance of the first receiver from the source.

In the same manner, the bandwidth for calculation of damping ratios for compression waves would be:

$$\frac{4V_S}{R_1} \leq f \leq f_{upper} \tag{12}$$

where, V_S = shear wave (not compression wave) velocity, and f_{upper} = the upper-bound frequency from the auto-power spectrum generated by the compression wave source.

FIELD CASE STUDY

Material Damping from S-Waves in Clay

Crosshole records from shear wave velocity measurements in a clay layer at a depth of 25 ft (7.5 m) at O'Neill Forebay Dam are shown in Fig. 4. The in-hole source was used to generate these shear waves, and the predominant frequency range was found to be about 10 Hz to 300 Hz. The distances of the first and second receivers from the source were 9.0 ft and 18.0 ft (2.7 m and 5.5 m), respectively. Measured shear wave velocity based on initial arrivals in time-domain records, was about 770 fps (235 m/sec).

The windowed time-domain signals are also presented in Fig 4. The same window shown in Fig. 2 was applied to the original signals. Elimination of energy not transmitted by the direct shear wave is clearly shown by comparing the windowed and non-windowed signals.

Apparent phase velocities before and after windowing are shown in Fig. 5. The term "apparent" in apparent phase velocity is used to denote a velocity calculated simply by dividing distance (d) by travel time (t). The term "phase" in apparent phase velocity is used to denote that the velocity of a particular frequency (f) is being calculated. Therefore, apparent phase velocity is d/t(f), where t(f) is the travel time of the frequency of interest (Sanchez-Salinero, 1987). Apparent phase velocities from the windowed signals are more stable than those from the non-windowed signals and are around 770 fps (235 m/sec) over a frequency range of about 100 Hz to 360 Hz. The 100 Hz frequency corresponds to a wavelength of 7.7 ft (2.3 m) which is about equal to R_1. Without filtering, a value of λ_S equal to 0.5 R_1 is necessary before a stable value of velocity is reached. This behavior

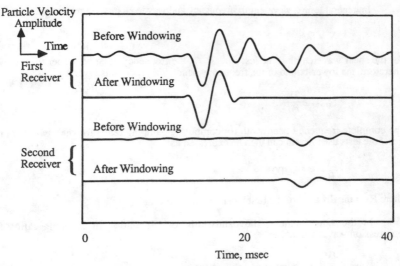

Figure 4 - Shear Wave Records from a Depth of 25 ft in a Clay Layer
at O'Neill Forebay Dam, Califonia

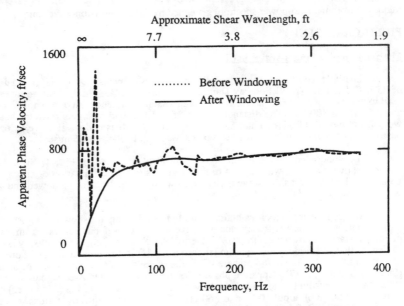

Figure 5 - Dispersion Curves of Shear Wave Velocities of Clay
at O'Neill Forebay Dam, California

shows one of the beneficial effects of the window. The variation in apparent phase velocities at low frequencies (long wavelengths) is caused by coupling between the additional near-field and far-field terms. Also the slight increase in apparent phase velocity with increasing frequency above 100 Hz is possibly due to dispersion created by material damping, but more research is necessary to understand this behavior.

Calculated damping ratios are shown in Fig. 6. Apparent damping ratios determined without windowing fluctuate significantly with frequency. On the other hand, the window reduces the fluctuations. Based on Eq. 11, the frequency bandwidth over which the average damping ratio should be selected is from 170 Hz to 300 Hz. The average damping ratio over this bandwidth is about 5 percent, with a variation from about 4 to 7 percent.

Unfortunately no laboratory samples were obtained. Therefore, no independent damping measurements could be made. However, much other laboratory data confirm that this value of material damping is quite reasonable (Hardin and Drnevich, 1972). In addition, to verify further the damping ratio of 5 percent, the synthetic signal at the second receiver was generated by applying geometrical spreading and the calculated damping ratio of 5 percent to the first observed signal. These signals are shown in Fig. 7. The overall agreement between the observed and synthetic signals is very good which strongly supports the 5 percent value of material damping.

LABORATORY CASE STUDY

Material Damping from P-Waves in Sand

Figure 8 shows two P-wave records obtained with accelerometers buried in a dry sand sample confined in the large-scale triaxial device. This device is a cubical device with dimensions of approximately 7 ft (2.1 m) on a side and was built to study the effects of stress state on compression and shear wave velocities. Wave velocity studies have been performed for the past six years and are presently continuing. The sand was placed by the pluviation method, and the accelerometers were placed in the sand by hand at the proper locations by stopping the pluviation process at specified filling depths (Lee, 1985). Thus, these accelerometers should exhibit less problems with coupling than receivers in typical crosshole tests. The frequency response of each accelerometer is flat over a frequency range of 5 Hz to 6 kHz. The distances between the source and two receivers are 2 ft and 4 ft (0.6 m and 1.2 m), respectively. The predominant frequency range of the P-wave source is 600 to 3500 Hz (Lee, 1985). The measured P-wave velocity in this material based on initial arrivals in the time-domain records was about 1700 fps (518 m/sec), and the measured shear wave velocity was about 1100 fps (336 m/sec).

As shown in Fig. 8, the direct P-wave pulse seems to be followed by reflected waves from the cube boundaries and/or the pulse due to the additional near-field term (White, 1983; and Sanchez-Salinero, 1987). In this analysis, the assumption is made that the direct wave is not overlapped by following reflected waves. Windowing was thus performed on the records to eliminate any reflections in other parts of the record. The signals gated by the windows based on this assumption are also shown in Fig. 8. To investgate how the window worked, apparent phase velocities were calculated from both the original and windowed signals. As shown in Fig. 9, the windowed signals yield more stable phase velocities around 1700 fps (518 m/sec) over the frequency range from 800 Hz to 3000 Hz.

Damping ratios based on windowed and unwindowed records are presented in Fig. 10. Again, the window improves the results. The frequency bandwidth for computing

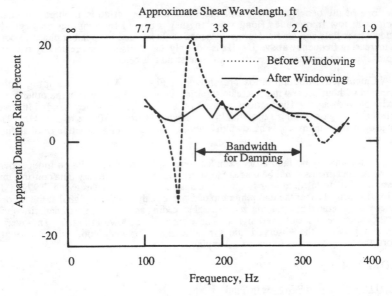

Figure 6 - Apparent Damping Ratios from Shear Wave Measurements Shown in Figure 4

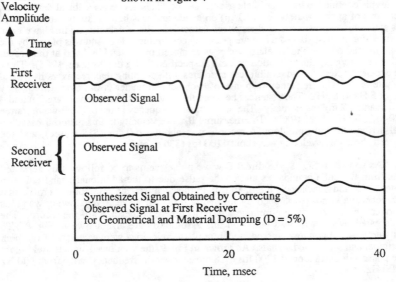

Figure 7 - Observed and Synthesized Signals of Shear Wave at Second Receiver in Clay Layer

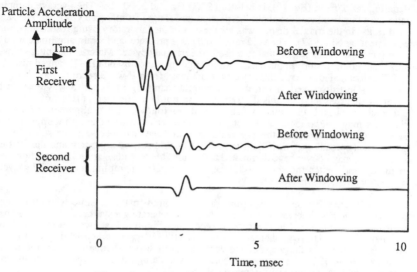

Figure 8 - Compression Wave Records from Two Accelerometers
Buried in Sand in the Large-Scale Triaxial Device

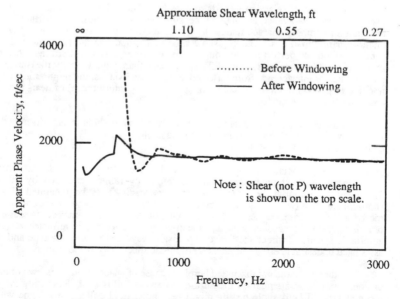

Figure 9 - Dispersion Curves of Compression Wave Velocities
in Sand in the Large-Scale Triaxial Device

damping ratio, from (Eq. 12), is between 2200 Hz and 3500 Hz. The average damping ratio over this bandwidth is about 2.5 percent and ranges from 2 to 3 percent. The same sand tested in the triaxial device was also tested in the laboratory using resonant column equipment by Ni (1987). The void ratio, confining pressure, and strain amplitude used in testing the specimens was similar to those in the triaxial device. Material damping ratios in shear (not compression) in the small-strain range measured by Ni (1987) ranged from 0.6 to 0.7 percent. However, damping ratio calculated from P-wave records was about 2.5 percent as just discussed. Even though material damping in P- and S-waves could be different, the reason(s) for this discrepancy can not be reconciled at this time. Several inferences about this discrepancy can be made as follows. First, alignment of the two receivers may not be identical which could cause calculation of a high damping ratio. Second, distances between the P-wave source and the two receivers are so close that geometrical spreading of near-field terms (R-2) could still be important and the far-field term (R-1) may not yet have dominated the result. More studies are necessary, however, to understand better the relationship between material damping measured by cyclic laboratory tests and seismic tests.

To verify further the damping ratio measured in the seismic tests, the signal observed at the first receiver is used as the basis for generating a synthetic pulse to compare with the pulse observed at the second receiver. Figure 11 shows the observed and synthetic signals. The overall agreement between the observed and synthetic pulses is very good, particularly in the view of the complexity of wave attenuation. Therefore, the measured value of 2.5 percent seems correct, even though it is unexpectedly high when compared with the results from resonant column tests.

CONCLUSIONS

Most uses of seismic methods to date have been directed towards evaluation of wave velocities. However, attenuation measurements are becoming important because of the need for damping characteristics along with elastic parameters in characterizing soils under dynamic loadings. The crosshole seismic method has the potential to be a very useful method for performing attenuation measurements at small strains. Attenuation measurements are, however, more difficult to make than wave velocity measurements for reasons such as interference from reflected and refracted waves, the requirement of precise tracking of particle motion, and coupling (and resulting interference) of near- and far-field components of wave motion.

Attenuation or material damping as measured in crosshole tests is discussed herein. It is shown that such measurements are possible if carefully performed. The major conclusions of this initial study are as follows: 1. Spectral analysis is necessary in attenuation measurements so that spectral ratios can be used to calculate apparent phase velocities and associated damping ratios, 2. The frequency bandwidth for computing damping ratios must be based upon the frequency range generated by the source. Wavelengths appropriate with the source-receiver spacing must be generated. The source should be at least two wavelengths from the first receiver for S-wave measurements and more for P-wave measurements. 3. Windowing of the time-domain records seems to be a promising method for reducing the adverse effect of reflected and/or refracted waves on the amplitude spectra of direct shear or compression waves. The window type and possible associated undesirable effects, if any, on amplitude spectra needs more study.

Although it was not discussed herein because of space limitations, two other points of consideration in attenuation measurements are as follows: 1. Reversing the source-receiver array in crosshole testing might be a good field collection scheme which can eliminate many of the variables which might adversely affect the amplitude spectra.

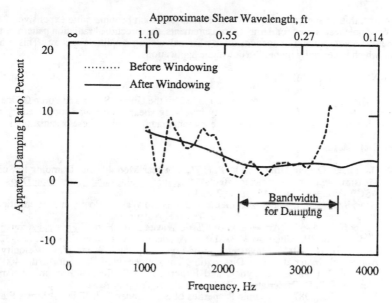

Figure 10 - Apparent Damping Ratios from Compression Wave Measurements Shown in Figure 8

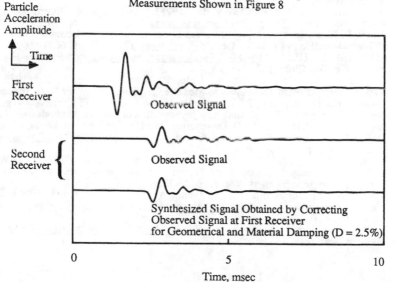

Figure 11 - Observed and Synthesized Signals of Compression Wave at Second Receiver in Sand

However, this arrangement requires four boreholes which become quite expensive. 2. If the downhole test is used for damping measurements, the directional radiation pattern of the impulsive source should be accounted for when calculating amplitude spectra. This is one point which is generally ignored in downhole applications.

ACKNOWLEDGEMENTS

The majority of this work was supported by the United States Air Force Office of Scientific Research. The one case study involving shear waves was supported by the United States Bureau of Reclamation. The authors gratefully thank these organizations.

REFERENCES

1.   Hardin, B. O., and Drnevich, V. P. (1972), "Shear Modulus and Damping in Soils: Measurements and Parameter Effects," J. Soil Mechanics and Foundation Division, ASCE, Vol. 98, No. SM6, pp. 603-624.

2.   Hoar, R. J. (1982), "Field Measurement of Seismic Wave Velocity and Attenuation," Ph.D. Dissertation, The University of Texas at Austin, 522 p.

3.   Knopoff, L. (1965), "Attenuation of Elastic Waves in the Earth," Physical Acoustics, Chpt. 7, Vol. III, edited by W.P. Mason, Academic Press, New York, pp. 287-324.

4.   Lee, S. H. H. (1985), "Investigation of Low-Amplitude Shear Wave Velocity in Anisotropic Material," Ph.D. Dissertation, The University of Texas at Austin, 400 p.

5.   Mok, Y. J. (1987), "Analytical and Experimental Studies of Borehole Seismic Methods," Ph.D. Dissertation, The University of Texas at Austin, 272 p.

6.   Ni, S. H. T. (1987) "Dynamic Properties of Sand under True Triaxial Stress States from Resonant Column/Torsional Shear Tests," Ph.D. Dissertation, The University of Texas at Austin, 421 p.

7.   O'Connell, R. J., and Budiansky, B. (1978), "Measures of Dissipation in Viscoelastic Media," Geophysical Research Letters, Vol. 5, No. 1, pp. 5-8.

8.   Pandit, B. I., and Savage, J. C. (1973), "An Experimental Test of Lomnitz's Theory of Internal Friction in Rocks," J. Geophysical Research, Vol. 78, pp. 6097-6099.

9.   Ramirez, R. W. (1985), "The FFT, Fundamentals and Concepts," Prentice-Hall, Inc., Englewood Cliffs, N.J., 178 p.

10. Redpath, B. B., Edwards, R. B., Hale, R. J.,and Kintzer, F. Z. (1982), "Development of Field Techniques to Measure Damping Values for Near-Surface Rocks and Soils," Report - URS/John A. Blume and Assoc., San Francisco, 120 p.

11. Sanchez-Salinero, I. (1987), "Analytical Investigation of Seismic Methods used for Engineering Applications," Ph.D. Dissertation, The Univ. of Texas at Austin, 401 p.

12. Stoll, R.D., (1985), "Marine Sediment Acoustics," J. Acoustical Society of America, Vol. 77, No. 5, pp. 1789-1799.

13. Tittmann, B. R., Nadler, H., Clark, V. A., Ahlberg, L. A., and Spencer, T. W. (1981), "Frequency Dependence of Seismic Dissipation in Saturated Rocks," Geophysical Research Letters, Vol. 8, pp. 36-38.

14. Toksoz, M. N., Johnston, D. H., and Timur, A. (1979), "Attenuation of Seismic Waves in Dry and Saturated Rocks: I. Laboratory Measurements," Geophysics, Vol. 44, pp. 681-690.

15. White, J. E., (1983), "Underground Sound, Application of Seismic Waves," Elsevier, New York, 253 p.

Dynamic Strain in Silt and Effect on Ground Motions
Sukhmander Singh* Member, ASCE
Robert Y. Chew**

Abstract

Cyclic load testing and collection of data on the cyclic strength deformation characteristics of silty soils were made. Data from the testing of undisturbed samples of Valdez silt, Yukon silty soils, soils from tailings dam in Chile, and results of cyclic load testing of loess, were examined. A summary of the results is presented in this paper. In almost all the silt samples tested, a gradual softening of the samples was observed from the start of the tests and large strains were reached without the development of significant pore pressure in a silt sample. Sands with less than 20 percent silt will usually behave similar to clean sands when subjected to cyclic loading. Increasing silt content modifies the behavior. When about 60 percent silt is present the behavior is controlled by the silt. The differences in permeability and fabric of soil containing various amounts of silts lead to various different pore pressure generation characteristics. The data points out the difficulty in establishing dynamic strength deformation characteristics for silts in terms of joint pore pressure and deformation characteristics. Unlike sands, the strains develop before a significant pore pressure increase is recorded. In cases where deformations reach a significant magnitude well before the pore pressures reach 100 percent, it is suggested that the evaluation of the transmission of ground motions from the bed rock to the ground surface through deposits of silty soils should take into account the dynamic strength loss and liquefaction characteristics of these soils by defining the characteristics only in terms of the percent of the strain irrespective of the magnitude of the pore pressures. For loose silts, where 100 percent pore pressure ratio was reached and accompanied by large strains, these criterias can be similar to the criteria presently used for sands.

Introduction

One of the principal research aims of the Earthquake Hazard Reduction Program has been to improve the capability to predict the character and effects of damaging ground motions. In the past 20 years the subject of the dynamic strength deformation characteristics of sand deposits has been the area of intensive research. As a result, significant improvement has occurred in the ability to measure and predict the probable performance of a wide

*Associate Professor of Civil Engineering, Santa Clara Univesity, Santa Clara, California, 95053
**Project Engineer, Trans Pacific Geotechnical Consultants, Inc., Concord, California, 94524

321

range of sandy soils during an earthquake.   A noticeable gap exists
when the soils subjected to shaking are silts and sandy silts.   The
main objective of this study was to ascertain whether there were
sufficient data available to develop an understanding of the dynamic
strength deformation characteristics of silty soils and their
influence on ground motion transmission through these soils.   The
plan to achieve this was to collect and evaluate existing test data
on silts in an attempt to understand the cyclic-strength-deformation
characteristics of these soils.   Much of the data was obtained from
Dames & Moore projects involving an extensive testing of Valdez
silt, Yukon silty soils, and soils from tailings dam projects
(References 1, 2 and 3).   In addition, the results of cyclic load
testing of silts carried out by Prakash and Puri (1982), Tokimatsu
et al (1981) and Zhou (1981) were also examined.   Works of Youd and
Perkins (1978), Youd (1980) and Seed (1986) on the liquefaction
induced deformations of soils were also studied.   Experimental
studies of the behavior of sands containing various amounts of silts
conducted by Chang, et al (1982) were also evaluated.   Most of the
Dames & Moore samples were undisturbed piston samples and all were
tested under undrained cyclic load triaxial testing.   The following
sections summarize the author's observations on the behavior of
silty soils when subjected to cyclic loading and possible influence
on ground response.

Test Results

Figure 1 shows the results of a cyclic triaxial test on a silt
sample from Valdez.   The upper portion of the plotted curves shows
the development of strain with increasing numbers of cycles and the
lower portion of the curves shows the rate of pore water pressure
increase.   It is readily apparent that the increase in both strain
and pore water pressures takes place gradually.   This is quite
different from the behavior of sands (Seed and Lee, 1966; Singh,
Donovan and Park, 1980; and Finn, 1981) where strains which had
previously been small start to increase rapidly as the pore water
pressure approaches about 60 percent or more of the confining
pressure.

Figure 2 shows the peak to peak axial strain in percent plotted
against the pore water pressure increase, also expressed as a
percentage (pore pressure ratio equals 100 percent when the pore
water pressure equals the confining pressure) for a wide range of
samples.   For most of the samples whose results are shown on Figure
2 large strains are reached without the development of significant
pore pressures.   The exception appears to occur for relatively loose
silts (sample 6) which approximate the curves for sand shown on
Figure 2.   The envelope curves for the grain size of the soils whose
test results are given in Figure 2 are shown in Figure 3.

Figure 4 shows a comparison of pore pressure development in a
typical silt sample representative of samples 1 through 5 and 7 and
a typical loose sand sample.   For most of the silt samples tested,
the trend of pore pressure development was different than for sands.
It takes a large number of cycles of loading to reach the 100
percent pore pressure ratio which is similar to the build up of pore

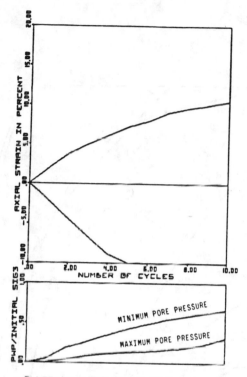

ALYESKA VALDEZ BRG 3 SAMPLE 2

STRESS-CONTROLLED,CYCLIC TRIAXIAL TEST

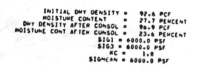

INITIAL DRY DENSITY   =   92.6 PCF
MOISTURE CONTENT   =   27.7 PERCENT
DRY DENSITY AFTER CONSOL =   96.9 PCF
MOISTURE CONT AFTER CONSOL = 23.6 PERCENT
SIG1 = 6000.0 PSF
SIG3 = 6000.0 PSF
KC = 1.0
SIGMEAN = 6000.0 PSF

FIGURE 1

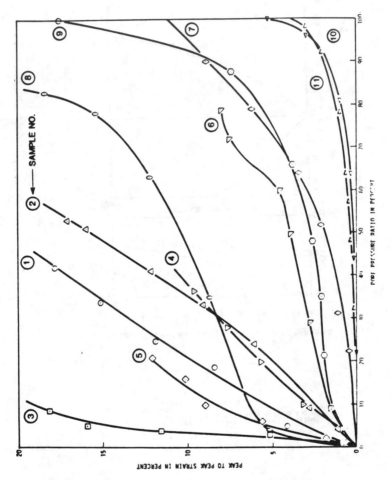

FIGURE 2

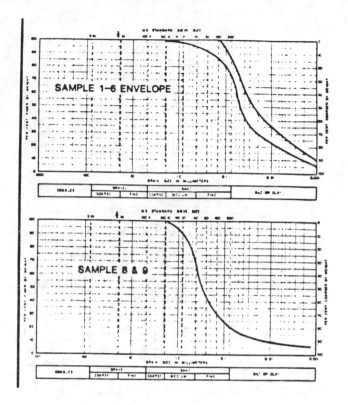

SAMPLE NO.	DESCRIPTION	INITIAL DRY DENSITY p.c.f.	MOISTURE CONTENT %	DRY DENSITY AFTER CONS... p.c.f.
1	Gray silt	92.6	27.7	98.9
2	Gray clayey silt	89.4	29.6	98.2
3	Gray silt	93.39	32.3	98.70
4	Gray silt	102.0	23.3	104.0
5	Gray clayey silt	99.69	26.3	103.44
6	Gray silty clay to clayey silt	77.04	54.0	81.52
7	Phase II Gray clay silt	91.69	32.2	95.7
8	Layered silt and medium fine silty sand	104.3	24.1	109.3
9	Gray silty medium fine sand	116.7	7.0	120.8
10	Silty fine sand	88.0	31.2	92.0
11	Gray clayey silty fine sand	90.2	30.2	91.5

Figure 3a

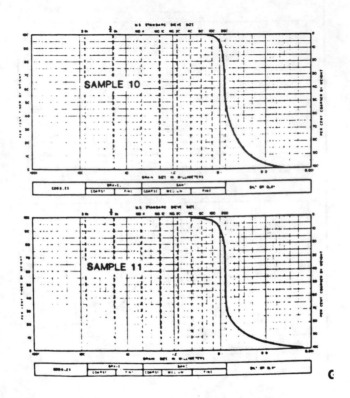

SAMPLE NO.	DESCRIPTION	INITIAL DRY DENSITY p.c.f.	MOISTURE CONTENT %	DRY DENSITY AFTER CONSOL. p.c.f.
1	Gray silt	92.6	27.7	98.9
2	Gray clayey silt	89.4	29.6	98.2
3	Gray silt	93.39	32.3	98.76
4	Gray silt	102.0	23.3	109.0
5	Gray clayey silt	99.69	26.3	107.44
6	Gray silty clay to clayey silt	77.04	54.0	81.52
7	Phase II Gray clay silt	91.69	32.2	95.7
8	Layered silt and medium fine silty sand	104.3	24.1	109.3
9	Gray silty medium fine sand	116.7	7.0	120.8
10	Silty fine sand	88.0	31.2	92.0
11	Gray clayey silty fine sand	90.2	30.2	91.5

Fig.3b

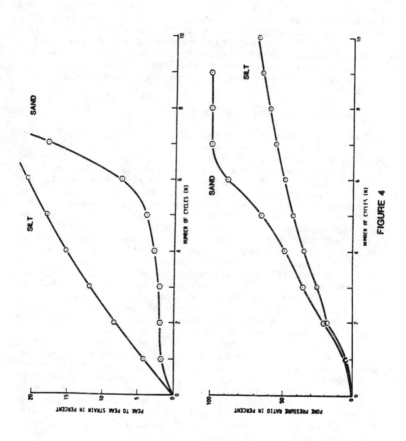

FIGURE 4

pressures in a dense sand sample. In some of the silt samples a
pore pressure ratio of 100 percent was never reached. It is
uncertain whether this is a true representation of the soil behavior
or a consequence of the difficulty of measuring transient pore
pressure changes at the base of a relatively impervious sample.

The presence of sand can offset the slow build up of pore
pressure in a silt sample. Chang, et al (1982) pointed out that the
difference in permeabilities and fabric of soils containing various
amounts of fines lead to vastly different pore pressure generation
characteristics. According to Chang, et al, a noticeable change in
the development of pore pressure in a sand sample takes place with
an addition of 30 percent fines (i.e., silt). When the silt content
reaches 60 percent or more, the soil fabric becomes one of sand
grains embedded in silt with practically no sand grain to sand grain
contact so the specimen behavior is totally determined by the silty
fines. Conversely, it may be suggested that a silt sample
containing about 30 percent or more of sand content will have its
pore pressure generation characteristics approaching that of sand.
However, additional data is needed to draw definitive conclusions on
this aspect.

Cyclic Strength Deformation of Silts

Although pore pressure and deformation characteristics under
cyclic loading are closely related, there is a problem in
establishing a dynamic strength loss criteria for silts in terms of
joint pore pressure and deformation characteristics. In almost all
the silt samples tested, a gradual softening of the sample was
observed from the start of the test. In cases where deformations
reach a significant magnitude well before the pore pressure reaches
100 percent, the dynamic strength loss characteristics may be
defined only in terms of the percent of strain irrespective of the
magnitude of the pore pressure. This can be an important
consideration in the ground response analysis of a deposit
containing a significant layer of silt. For example an early
deformation of the silt during the initial stages of the ground
shaking when only insufficient pore pressures have developed to
affect the stiffness characteristics of the surrounding sandy
layers, can influence the overall response of the deposit as well as
the nature of the ground motions propagating through it. Although
more work is needed, but it is likely that in the case of the
evaluation of surface ground motions and ground deformations during
large earthquakes where deformations reach a significant magnitude,
the ground response analysis for silty deposits may underestimate
the ground motions if the dynamic strength loss criteria in terms of
joint pore pressure and deformation characteristics as presently
used for sands is used for silts. For loose silts where the 100
percent pore pressure ratio was reached and accompanied by large
strains, the liquefaction criteria can be described by a 100 percent
pore pressure ratio with 10 percent of strain potential. This is
similar to the criteria presently used for sands.

Conclusions

The dynamic strength deformation characteristics of silts appears to be different from that of sands and it is important to recognize their influence on ground motion transmission. The following conclusions can be made:

1) Sands with less the 20 percent silt will usually behave similar to clean sands when subjected to cyclic loading. Increasing silt content modifies the behavior. When about 60 percent silt is present the behavior is controlled by the silt.

2) There is a problem in establishing criteria for cyclic strength deformations for silts in terms of joint consideration of pore pressure and deformation.

3) For sands it is possible to use pore pressure criteria to describe terms such as initial liquefaction and estimated strain potential. For most silts this is not possible because the 100 percent pore pressure increase is not reached during testing. Unlike sands, the strains develop before a significant pore pressure increase is recorded.

4) It is suggested that the evaluation of the transmission of ground motions from the bed rock to the ground surface through deposits of silty soils should take into account the dynamic strength loss and liquefaction characteristics of these soils by defining the characteristics only in terms of the percent of the strain irrespective of the magnitude of the pore pressures. For loose silts, where 100 percent pore pressure ratio was reached and accompanied by large strains, these criterias can be similar to the criteria presently used for sands.

References

Chang, N.Y., S.I. Yeh, and L.P. Kaufman (1982), "Liquefaction Potential of Clean and Silty Sands," Third International Earthquake Microzonation Conference, Proceedings, Seattle, Washington, Vol. II, pp. 1017-1032.

Chang, N.Y., N.P. Hsieh, D.L. Samuelson and M. Horita (1982), "Static and Cyclic Undrained Behavior of Monterey No. 0 Sand," Proceedings, Third International Earthquake Microzonation Conference, Seattle, Washington, Vol. II, pp. 929-944.

Dames & Moore (1975), Outfall Diffuser and Dike Stability Studies, Valdez Terminal, D&M Project 8354-059-20.

Dames & Moore (1980), Laboratory Dynamic Soil Testing, Prez Caldera No. 1 Tailings Dam, Chile, D&M Project 10438-003-03.

Dames & Moore (1982), Geotechnical Studies, Kluane Lake Crossing, Robinson-Dames & Moore.

De Alba, Pedro (1975), "Determination of Soil Liquefaction Characteristics by Large Scale Laboratory Tests," Ph.D. Dissertation, University of California, Berkeley.

Donovan, N.C. and S. Singh (1978), "Liquefaction Criteria for TransAlaska Pipe Pipeline," Journal of the Geotechnical Engineering Division, Vol. 104, No. GT4 (April), pp. 447-462.

Finn, W.D., Liam (1982), "Liquefaction Potential: Development Since 1976," International Conference on Recent Advances in Geotechnical Earthquake Engineering and Soil Dynamics, St. Louis, Missouri.

Prakash, Shamsher and V.K. Puri (1982), "Liquefaction of Loessial Soils," Third International Earthquake Microzonation Conference, Proceedings, Seattle, Washington, Vol. II, pp. 1101-1107.

Seed, H.B. and K.L. Lee (1966), "Liquefaction of Saturated Sands During Cyclic Loading," Journal of Soil Mechanics and Foundations, ASCE, Vol. 92, No. SM6, Proc. Paper 4972.

Seed, H.B., "Design Problems in Soil Liquefaction", Report No. UCB/EERC-86/02 College of Engineering, University of California Berkeley, February, 1986.

Singh, S., N.C. Donovan and T. Park (1980), "A Re-examination of the Effect of Prior Loading on Liquefaction of Sands," Proceedings, Seventh World Conference on Earthquake Engineering.

Youd, T.L. and D.M. Perkins (1978), "Mapping of Liquefaction Induced Ground Failure Potential," Journal: Geotechnical Engineering Division, ASCE, Vol. 104, No. GT4.

Youd, T.L. (1980), "Ground Failure Displacement and Earthquake Damage to Buildings," Proceedings of the Second Conference on Civil Engineering and Nuclear Power, ASCE, Vol. 2.

SMALL-STRAIN SHEAR MODULUS OF CLAY

William A. Weiler, Jr.*

ABSTRACT

The results of an evaluation of small-strain shear modulus of Boston Blue Clay using the resonant column test are presented, and compared to similar evaluations on five other naturally-deposited, inorganic clays. Relationships describing small-strain shear modulus (G_0) of the six clays in terms of confining pressure and overconsolidation ratio (OCR) are presented. Normalizing the small-strain shear modulus by dividing by undrained shear strength (S_u) discloses that G_0/S_u is highly sensitive to plasticity index and type of strength test, moderately sensitive to OCR, and slightly to moderately sensitive to confining pressure.

INTRODUCTION

The small-strain shear modulus of naturally-deposited clays has been much studied over the past 20 years (Anderson, 1974; Anderson and Stokoe, 1977; Arango et al, 1978), primarily because of its importance in assessing site amplification effects for earthquake-resistant design of structures. One of the primary tools used to evaluate the small-strain modulus (G_0, determined at shear strains less than 10^{-3} percent) is the resonant column test, performed in the laboratory on "undisturbed" samples recovered from test borings. The purpose of this paper is to describe and discuss the use of the SHANSEP (Stress History and Normalized Soil Engineering Parameters) procedure for determining small-strain modulus of saturated clay. SHANSEP was developed by Ladd and Foott (1974) for mitigating the effects of sample disturbance on the laboratory measurement of undrained shear strength (S_u) of clay. The results of an evaluation of small-strain modulus of Boston Blue Clay are described in this paper, and presented in comparison to similar evaluations of other naturally-deposited clays. Also, the use of a G_0/S_u ratio for evaluating the small-strain modulus of clay is discussed, based on the SHANSEP methodology.

TESTING PROCEDURE

It has long been recognized that even the best quality thin-wall tube samples of clay are not truly undisturbed. The effects of sample disturbance have been discussed in detail by others (Baligh et al, 1987; Ladd et al, 1977). Ladd et al (1977) noted that the recompression index, virgin compression index, and the preconsolidation (maximum past) pressure of clay as observed in

---

* Senior Engineer, Haley & Aldrich, Inc., 58 Charles Street, Cambridge, Massachusetts 02141

laboratory consolidation tests are all significantly affected by
sample disturbance. Baligh et al (1987) used a new technique to
simulate some tube sampling disturbance effects on the undrained shear
strength of normally-consolidated, resedimented Boston Blue clay, as
measured in a triaxial cell. Their results indicate that simply
reconsolidating a disturbed sample to its in-situ stress state
(Recompression method) leads to an overestimate of S_u compared to
the undisturbed test result, while using SHANSEP testing procedures
and normalizing with respect to consolidation pressure results in more
precise measurements of S_u.

Based on the aforementioned studies of disturbance effects on soil
compressibility and strength characteristics, it is considered likely
that the small-strain shear modulus of clay is significantly affected
by sample disturbance, especially at confining pressures less than or
equal to the preconsolidation pressure of the clay sample. To attempt
to reduce the effects of sample disturbance on the test results, the
SHANSEP testing method was chosen for this evaluation.

The SHANSEP procedure was developed for use in measuring undrained
shear strength, but it has also been useful for other soil engineering
properties, such as shear or Young's modulus. In fact, a SHANSEP-type
program is particularly easy to perform for small-strain shear modulus
of clay, because of the non-destructive nature of the resonant column
test. The procedure consists of the following steps:

1. Consolidate the sample using a 1.5:1 to 2:1 load increment ratio
    to a confining pressure at least 1.5 times the preconsolidation
    pressure of the sample.

2. Determine the equation of the shear modulus vs. confining
    pressure relationship in the normally consolidated range. It
    should be noted that the SHANSEP procedure for undrained shear
    requires that S_u vary linearly with confining pressure. For
    small-strain shear modulus, SHANSEP must be extended such that
    G_0 vs. confining pressure is linear on a log-log plot, yielding
    the following equation:

$$G_{0-nc} = K\sigma_c^{\ n} \hspace{3cm} (1)$$

    in which σ_c is the effective confining pressure and K and n are
    constants herein referred to as the modulus multiplier and
    modulus exponent.

    All the resonant column testing described herein was performed on
    isotropically consolidated samples. The octahedral (mean) normal
    stress should be substituted for σ_c for anisotropic stress
    conditions.

3. Unload to overconsolidation ratios (OCR) of 2, 4, and 6 or 8, and
    determine G_0 at each OCR. For each OCR, calculate G_{0-nc} at
    that confining pressure using equation (1), and plot G_0/G_{0-nc}
    versus OCR on a log-log plot. The slope of this line is the

parameter m for the general equation:

$$G_0 = K\sigma_c{}^n OCR^m \qquad (2)$$

in which m is a constant herein referred to as the OCR exponent.

It should be noted that a SHANSEP testing program requires testing at significantly higher confining pressures than commonly used for resonant column testing, to define the normally-consolidated soil behavior range. A distinct break in the shear modulus vs. confining pressure plot should be noted when the preconsolidation pressure of the soil is reached. Apparently, the highest confining pressure typically used for resonant column testing is 100 psi (Anderson, 1974; Isenhower, 1979; Lodde, 1980; Trudeau, 1973). A confining pressure of 100 psi is not high enough to define the normally consolidated range for all but very soft to soft clays.

The usual cautions with respect to the use of the SHANSEP approach are in order here, also. Some soils (such as quick clays or naturally-cemented soils) may have a structure which is broken down when the preconsolidation pressure is exceeded. The use of SHANSEP for those soils may lead to an underestimate of S_u or G_0. The Recompression method should be used for laboratory resonant column testing of such soils.

Ladd and Foott (1974) recommend that for undrained shear strength testing, the duration of consolidation at the final confining pressure be long enough to complete primary consolidation plus one log cycle of secondary compression. This is consistent with the standard time of 1000 minutes for reporting G_0 data only if the primary consolidation time is about 100 minutes. For the typical solid cylindrical resonant column test specimen with a 1.5 to 2 in. diameter, 3 to 4 in. height, and filter paper strips to allow radial drainage, the time to completion of primary consolidation for most clays is 100 to 200 minutes (Anderson, 1974; Isenhower, 1979; Lodde, 1980; Trudeau, 1973). For that test configuration, the value of G_0 at 1000 minutes does include about one log cycle of secondary compression. For test configurations with larger specimens or without radial drains, the time to complete primary consolidation alone may exceed 1000 minutes (Koutsoufas and Fischer, 1979). The importance of time effects in the measurement of G_0 has been reported by others (Anderson, 1974; Anderson and Stokoe, 1977; Isenhower, 1979; Trudeau, 1973) and will not be discussed here.

It should also be noted that the use of a SHANSEP-type program requires a good knowledge of the in-situ preconsolidation pressures. For stiff to hard clays, the apparent preconsolidation pressure may be significantly lower than the actual because it is usually necessary to drive tubes or core these soils to obtain samples for testing. The disturbance caused by sampling results in a rounded, poorly-defined break in the void ratio (or vertical strain) vs. log pressure plot, which in turn leads to an underestimate of the preconsolidation

pressure. For soft to medium stiff clays, thin-wall tube samplers can
be pushed, but large diameter (greater than 3 in.), thin wall, fixed
piston samplers and careful drilling techniques are required.

BOSTON BLUE CLAY TEST RESULTS

The testing program for this evaluation consisted of two series of
resonant column/torsional shear tests performed at the University of
Texas at Austin. The clay samples were obtained with a three-inch
diameter, fixed piston sampler from a site in Medford, Massachusetts.
Sample UP1 was obtained at a depth of 39-41 feet, and consisted of
uniform Boston Blue Clay. Sample UP2 was taken at 94-96 ft., and
consisted of Boston Blue Clay with fine sand pockets and partings.
The effective overburden pressure and estimated preconsolidation
pressure for each sample are listed below:

Sample No.	Depth (ft.)	In-Situ Overburden Pressure (TSF)	Estimated Preconsolidation Pressure (TSF)	Approx. OCR
UP1	39-41	1.3	4.0	3
UP2	94-96	2.8	3.0	1.1

The preconsolidation pressures have been estimated based on oedometer
tests performed on clay samples obtained at different depths on the
same site. The liquid limit of the clay at the site ranges from 40 to
45, and the plasticity index ranges from 20 to 23.

The results of the resonant column tests performed in the normally
consolidated range are summarized on Figure 1. Also shown on Figure 1
are eight data points from a previous study of small-strain shear
modulus of Boston Blue Clay performed at MIT (Trudeau, 1973). These
eight points (from 5 different tests) were extracted from that study
because the confining pressures for those points appear to be above
the sample preconsolidation pressure. A (log-log) linear regression
analysis of all these data (12 points) yields the equation

$$G_{o \cdot nc} \text{ (TSF)} = 250 \, \sigma_c^{0.86} \qquad (3)$$

with a correlation coefficient of 0.84.

The effects of overconsolidation ratio on the small-strain shear
modulus of Boston Blue Clay are illustrated on Figure 2, which depicts
the increase of G_o with increasing OCR, using normalization to
compensate for confining pressure differences. The two tests for this
study, along with tests from the previous MIT study, show remarkable

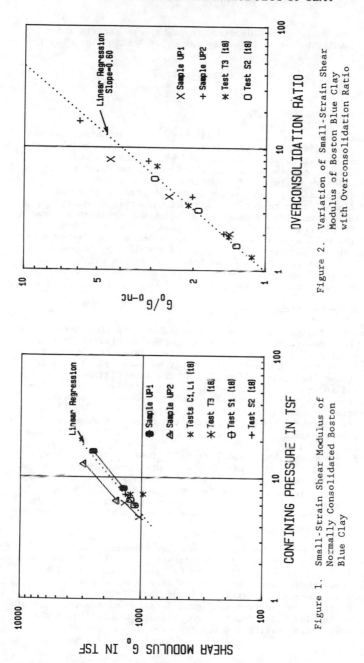

Figure 2.   Variation of Small-Strain Shear
Modulus of Boston Blue Clay
with Overconsolidation Ratio

Figure 1.   Small-Strain Shear Modulus of
Normally Consolidated Boston
Blue Clay

agreement. The slope of a line drawn through these data is about
0.60. The SHANSEP-based relationship for G_0 of Boston Blue Clay is
therefore

$$G_0 \ (\text{TSF}) \ = \ 250 \ \sigma_c{}^{0.86} \text{OCR}^{0.60} \tag{4}$$

COMPARISON WITH PUBLISHED DATA

Other investigators (Koutsoufas and Fischer, 1979; Singh and Gardner,
1979) have used SHANSEP-type procedures for small-strain modulus of
clay. Others (Anderson, 1974; Isenhower, 1979) gathered sufficient
data from resonant column tests that the normally-consolidated range
of those particular soils could be defined. Table 1 provides a
summary of the data which were found in the literature on shear
modulus-confining pressure relationships of naturally-deposited clay
in the normally-consolidated range. The equations in Table 1 were
determined by the writer by linear regression of the published data,
except for the Gulf of Alaska relationship, which was published by the
authors (Singh and Gardner, 1979). Most of the data by Anderson
(1974) could not be used because the preconsolidation pressures of
most of the clays tested were apparently not exceeded at the highest
confining pressures used (80-100 psi).

Confining Pressure Effects

One interesting feature about the data in Table 1 is that the modulus
multiplier (K) appears to decrease with increasing plasticity index,
while the modulus exponent (n) appears to increase. Both these trends
are logical, however. A plastic clay would be expected to be less
stiff (lower modulus) than a lean clay at the same OCR and confining
pressure, and so would have a lower modulus multiplier. Also, since
the void ratio of a more plastic clay would normally decrease faster
with increasing confinement than that of a lean clay, it would be
expected that the shear modulus increase (that is, the modulus
exponent) would be greater.

It should be noted that the relationships in Table 1 are based on a
consolidation time of 1000 minutes for each pressure increment, except
for the Gulf of Alaska clays, for which the times were not described.
However, the AGS clay data (Koutsoufas and Fischer, 1979) do not
include the "standard" log cycle of secondary compression since the
authors indicate that shear modulus was measured at the end of primary
consolidation.

The apparent relationship between the modulus exponent and plasticity
index is illustrated on Figure 3. The data for this figure are those
from Table 1, supplemented by some data by Egan and Ebeling (1985).
This figure does indicate a general trend of modulus exponent
increasing with increasing plasticity index, but there is a large
amount of scatter. The scatter in Figure 3 is indicative that factors
not reflected in plasticity index (such as clay structure) and the
details of sample preparation and testing procedures, are also
significant.

TABLE 1

Small-Strain Shear Modulus and Undrained Shear Strength of Six
Naturally Deposited Clays

Clay	Liquid Limit	Plasticity Index	G_o (TSF)	$S_{u\text{-}dss}/\sigma_{vc}$	$G_o/S_{u\text{-}dss}$
Gulf of Alaska Clays (Ref.15, 24 points)	34-35	14-15	$375\sigma_c\ 0.85 OCR^{0.59}$	$0.230 CR^{0.75}$	$1630\sigma_c^{-0.15} CR^{-0.16}$
AGS CL Clay (Ref. 9, 1 test)	32-39	16-22	$440\sigma_c\ 0.84 OCR^{0.27}$	$0.230 CR^{0.8}$	$1910\sigma_c^{-0.16} CR^{-0.53}$
Boston Blue Clay (Ref. 18, 7 tests)	40-46	19-23	$250\sigma_c\ 0.86 OCR^{0.60}$	$0.200 CR^{0.8}$	$1250\sigma_c^{-0.14} OCR^{-0.20}$
San Francisco Bay Mud (Ref. 7, 13, 4 tests)	88	43	$165\sigma_c\ 0.95 OCR^{0.51}$	$0.250 CR^{0.8}$	$660\sigma_c^{-0.05} CR^{-0.29}$
AGS CH Clay (Ref. 9, 1 test)	63-64	32-38	$125\sigma_c\ 1.18 OCR^{0.69}$	$0.250 CR^{0.8}$	$500\sigma_c\ 0.18 OCR^{-0.18}$
Leda Clay (Ref. 1, 2 tests)	67-69	37-40	$97\sigma_c\ 1.08$	*	*

* No data found for small-strain shear modulus of Leda clay in overconsolidated state

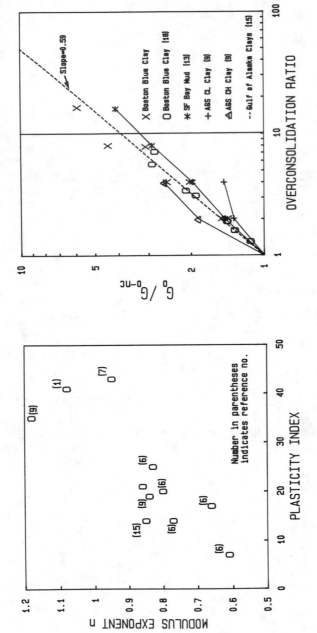

Figure 4.   Variation of Small-Strain Shear
Modulus of Five Naturally-
Deposited Clays with Over-
consolidation Ratio

Figure 3.   Variation of Modulus Exponent
with Plasticity Index for
Naturally-Deposited Clay

Overconsolidation Ratio Effects

Data on small-strain shear modulus during unloading from a normally consolidated state are available for 5 of the 6 clays listed in Table 1 (no such data were available for Leda Clay). The OCR responses of the Gulf of Alaska clays were illustrated by Singh and Gardner (1979), but could not be analyzed because the pressures from which the samples were unloaded were not identifiable. An OCR exponent (m) of 0.59 for the Gulf of Alaska clays was indicated by Singh and Gardner based on their analysis of the data. However, it appears that the shear modulus becomes relatively constant during unloading to high OCR (greater than about 8), which means that the OCR exponent increases significantly at high OCR.

The influence of OCR on small-strain shear modulus for 5 of the clays from Table 1 is illustrated on Figure 4. The slope of this plot is the OCR exponent, m. Nearly all the data fall within a range of m from 0.3 to 0.7, with an average of about 0.6. These values are consistent with the results from 3 clays listed by Egan and Ebeling (1985), which were reported to have had OCR exponents of 0.40, 0.43 and 0.67 with corresponding plasticity indexes of 13-21, 25 and 5-7.

There does not appear to be any trend of variation of m with clay plasticity, but it does appear that m may not be a constant. The OCR exponent appears to increase significantly with increasing OCR for Boston Blue Clay and the Gulf of Alaska clays, while it appears to decrease significantly for the AGS CL and CH clays.

G_o/S_u RATIO

The usefulness of a defined G_o/S_u ratio for estimating the small-strain shear modulus G_o based on undrained shear strength measurements is apparent. The fact that a unique G_o/S_u ratio does not exist for all clays under all conditions has been identified in several studies (Arango et al, 1978; Egan and Ebeling, 1985). The reasons for this variation can be evaluated using the SHANSEP approach.

One of the primary problems in the use of a G_o/S_u ratio is the definition of the undrained shear strength. Arango et al (1978) noted that for clays they studied G_o/S_u ranged from 800 for shear strengths obtained using consolidated-undrained triaxial compression tests (CIU-TC) to 1800 for shear strengths measured in unconfined compression (UC) tests, using G_o values from in-situ measurements. The variability of shear strength with numerous factors was the reason for the development of SHANSEP. The factors which have been identified to cause variation in the measured undrained shear strength of clay are:

1. Disturbance of sample.
2. Anisotropy of soil (inherent and induced).
3. Boundary conditions of test method.
4. Rate of strain in test.

For example, higher undrained shear strengths will typically be
measured for CIU or CK_0U triaxial compression tests, which are run
on consolidated samples at relatively low strain rates, compared to UC
tests run on unconfined samples at high strain rates. Also, at a
given confining pressure, an unconsolidated-undrained (UU) test may
indicate a higher strength than a CIU-TC test if the sample is of high
quality, but a disturbed sample will probably indicate a UU strength
significantly lower than the CIU-TC result. The undrained shear
strength of the clay sample therefore depends on how the above factors
combine. The variability of shear strength, and therefore the
G_0/S_u ratio, can be reduced if the SHANSEP method is used and
factors 2, 3 and 4 above are standardized by use of a specific
strength test and sample orientation. Since a major use of
small-strain shear modulus is for analysis of site response to
vertically-propagating shear waves, the undrained shear strength on
horizontal planes is most appropriate for use. The direct simple
shear (CK_0UDSS) test would therefore appear to provide the best
representation of shear strength for such analyses. Although this
test is not widely used, Ladd and Edgers (1972) have shown that
S_{u-dss}/σ_{vc} for a wide variety of normally-consolidated clays is
0.23 ± 0.03 and the OCR exponent for undrained shear strength is about
0.8, so that S_{u-dss} can easily be estimated.

For 5 of the clays listed in Table 1, CK_0UDSS tests have been run
and reported in the literature. It is instructive to use the G_0
relationships from Table 1 and divide them by the CK_0UDSS undrained
shear strength (S_{u-dss}) relationships reported in the literature.
The S_{u-dss} relationships and the calculated G_0/S_{u-dss} ratios are
indicated in Table 1. For undrained shear strength, an OCR exponent
of 0.8 was assumed for clays for which this exponent was not available
(Ladd et al, 1977).

It is recognized that calculation of G_0/S_{u-dss} in this manner does
not account for differences in the particular samples used to
determine G_0 and S_{u-dss}. This limitation is partially overcome by
using average relationships based on a number of tests. Since data
from only one resonant column test were available for the two AGS
clays, those data must be viewed cautiously.

The calculated relationships indicate the following apparent trends
for G_0/S_{u-dss}:

1. It is highly dependent on clay plasticity, with higher ratios for
    less plastic clays.

2. It always decreases with increasing OCR.

3. It decreases with increasing confining pressure for low to medium
    plasticity clays, and increases with increasing confining
    pressure for highly plastic clays.

These three trends are the results of observations on five inorganic,
low to medium plasticity clays, and are not intended to be

representative of all clays. There are no data included in Table 1 for clays with plasticity indexes less than 10 or greater than 50. Also, the G_0 relationships in Table 1 (and therefore also $G_0/S_{u \cdot dss}$ in Table 1) become less accurate with increasing OCR, as discussed previously, and their accuracy at high confining pressures is unknown. The use of these relations to estimate G_0 in heavily overconsolidated clays or shales, or in very deep clay deposits, should be approached with caution.

The range of values of $G_0/S_{u \cdot dss}$ for typical OCR of 1 to 5 and confining pressures from 1 to 5 TSF can easily be calculated from the equations in Table 1. Based on such calculations, the following G_0/S_u ratios may be used to estimate G_0 if undrained shear strength is based on CK_0UDSS data:

$$G_0/S_{u \cdot dss}$$

Plasticity Index	OCR=1	OCR=2	OCR=5
15-20	1500	1250	1000
20-25	1100	950	800
35-45	600	520	450

Observation of the above table leads to the conclusion that even when the definition of undrained shear strength is standardized, a single value of G_0/S_u is not appropriate for all clays. The use of the above values should result in a significant improvement over the constant G_0/S_u values of 2200 or 1000 proposed elsewhere (Seed and Idriss, 1970; BSSC, 1985).

It is important to note that the above ratios would be decreased significantly if the undrained shear strength was based on triaxial compression tests. For the clays listed in Table 1, the undrained shear strength in triaxial compression ($S_{u \cdot tc}$, as measured using the SHANSEP procedure) is typically 40 to 50% higher than $S_{u \cdot dss}$. If consolidated-undrained triaxial tests are used to determine S_u, the following $G_0/S_{u \cdot tc}$ ratios may be used to estimate G_u:

$$G_0/S_{u \cdot tc}$$

Plasticity Index	OCR=1	OCR=2	OCR=5
15-20	1100	900	600
20-25	700	600	500
35-45	450	380	300

Effect of Time of Confinement on G_0/S_u

The effect of time of confinement on small-strain shear modulus is well-known (Anderson, 1974; Anderson and Stokoe, 1977; Trudeau, 1973). It is interesting to compare the increase in G_0 with time to

that of S_u. Ladd and Edgers (1972) provide data from Taylor (1952)
that indicate the increase in S_u of remolded Boston Blue Clay with
time of confinement. The test method used by Taylor was a direct
shear test modified to maintain a constant volume of specimen. Taylor
performed a number of tests, and the average results quoted by Ladd
and Edgers (1972) indicate that S_u increases by about 10 to 15
percent per log cycle of time during secondary compression. This gain
in strength is comparable to the 15 to 20 percent increase in
small-strain shear modulus per log cycle of time during secondary
compression observed in the tests performed on Boston Blue Clay by
Trudeau (1973). It is interesting to note that the rate of gain in
S_u with time is close to the rate of gain in G_o. When using a
G_o/S_u ratio and undrained shear strength data to estimate G_o,
time effects can therefore be incorporated by using S_u data measured
after one day on confinement, and correcting the estimated G_o for
in-situ duration of pressure using published data (Anderson and
Stokoe, 1977).

Strain Rate Effects on G_o/S_u

The importance of rate of strain in the measurement of undrained shear
strength of clay has long been recognized (Taylor, 1948). Strain rate
effects have also been found to be significant in the measurement of
small-strain modulus of clay. Isenhower (1979) performed cyclic
torsional simple shear and resonant column tests on samples of San
Francisco Bay Mud and noted that small-strain shear modulus varied
with cyclic strain rate.

The resonant column tests, performed at strain rates of 700 to 6,000
percent/hour (frequencies of 28 to 36 Hz) indicated higher values of
G_o than cyclic torsional simple shear tests performed at strain
rates of 1/2 to 150 percent per hour (frequencies of 0.03 to 1 Hz).
The gain in shear modulus was about 5% per log cycle of strain rate.

Predominant frequencies of earthquake shaking may vary from less than
1 Hz to as much as 15 Hz for different earthquakes. As a result of
the strain rate effect, resonant column tests will result in higher
values of shear modulus (although by only 5 to 10 percent) than would
be obtained by testing at the actual earthquake frequency.

It is interesting to note that the data by Taylor (1948) on strain
rate effects on undrained shear strength (unconfined compression
tests) on remolded Boston Blue Clay indicate an increase in strength
of about 5 percent per log cycle of strain rate. This is the same as
the increase in small-strain shear modulus of San Francisco Bay Mud
seen in the data by Isenhower (1979).

When using laboratory resonant column tests or undrained shear
strength tests it is important to be cognizant of the effects of
strain rate.

SUMMARY AND CONCLUSIONS

The following conclusions can be made based on the evaluations discussed herein:

1. The SHANSEP laboratory testing procedures for undrained shear strength can be adapted to small-strain shear modulus (G_o) as measured in the resonant column test. The benefit of using this procedure is that some of the effects of sample disturbance can be mitigated. An empirical relationship of the form

$$G_o = K\sigma_c{}^n OCR^m$$

can be developed to describe small-strain shear modulus for a particular clay.

2. The modulus exponent (n) appears to vary from less than 1 for low plasticity clays to in excess of 1 for highly plastic clays, as illustrated in Figure 3. The overconsolidation ratio exponent (m) appears to be about 0.6 ± 0.2, based on available data. The OCR exponent may vary at high OCR, however.

3. The use of a G_o/S_u relationship to estimate small-strain shear modulus based on undrained shear strength measurements is greatly restricted by the variability of S_u with different testing methods. If the shear strength is defined by means of a particular test, the effect of shear strength variability on the G_o/S_u ratio can be minimized. Even so, the $G_o/S_{u\cdot dss}$ ratio is found to be sensitive to clay plasticity and OCR.

4. Both small-strain shear modulus and undrained shear strength increase with increasing time of confinement and increasing strain rate. Since site response analyses require knowledge of both G_o and S_u, the effects of time of confinement and rate of strain on each of these parameters should be considered when performing such analyses.

Acknowledgements

The Haley & Aldrich, Inc , Professional Development Program provided funding for final preparation of the text and figures. Prof. Ken Stokoe and Dr. Tony Ni of the University of Texas at Austin coordinated and performed the resonant column tests on Boston Blue Clay. Discussions with Dr. Robert Pyke were invaluable and his contribution to this paper is appreciated. Michael Christian, Star Poole, and Wayne Wilson typed the text. Acey Welch assisted in final preparation of the figures.

Appendix I.- References

1. Anderson, D. G., "Dynamic Modulus of Cohesive Soils," Dissertation submitted in partial fulfillment of the requirements for the degree of Doctor of Philosophy, University of Michigan, 1974.

2. Anderson, D. G., and Stokoe, K. H., "Shear Modulus: A Time-Dependent Soil Property," Dynamic Geotechnical Testing, ASTM STP 654, 28 June 1977, pp. 66-90.

3. Arango, I., Moriwaki, Y., and Brown, F., "In-situ and Laboratory Shear Velocity and Modulus," Proc. ASCE Specialty Conf. on Earthquake Engineering and Soil Dynamics, June 1978, Vol. I, pp. 198-212.

4. Baligh, M. M., Azzouz, A. S. and Chin, C.-T., "Disturbances Due to 'Ideal' Tube Sampling," ASCE Journal of Geotechnical Engineering, Vol. 113, No. 7, July 1987, pp. 739-757.

5. Building Seismic Safety Council, NEHRP Recommended Provisions for the Development of Seismic Regulations for New Buildings, Part 2, Commentary, 1985.

6. Egan, J. A. and Ebeling, R. M., "Variation of Small-Strain Shear Modulus with Undrained Shear Strength of Clay," Soil Dynamics and Earthquake Engineering, Proc. 2nd International Conference, June-July 1985, pp. 2-27 to 2-36.

7. Isenhower, W. M., "Torsional Simple Shear/Resonant Column Properties of San Francisco Bay Mud," Geotechnical Engineering thesis GT80-1, University of Texas at Austin, December 1979.

8. Koutsoufas, D., and Fischer, J. A., "In-Situ Undrained Shear Strength of Two Marine Clays," ASCE Journal of the Geotechnical Engineering Division, Vol. 102, No. GT9, September 1976, pp. 489-1005.

9. Koutsoufas, D. C. and Fischer, J. A., "Stress History Effects on the Dynamic Properties of Two Marine Clays," Proc. ASCE Session on Soil Dynamics in the Marine Environment, Boston, MA, 1979, 24 pp.

10. Ladd, C. C. and Edgers, L., "Consolidated-Undrained Direct Simple Shear Tests on Saturated Clays," Research in Earth Physics Phase Report No. 16, Massachusetts Institute of Technology, July 1972.

11. Ladd, C. C. and Foott, R., "New Design Procedure for Stability of
    Soft Clays," ASCE Journal of the Geotechnical Engineering
    Division, Vol. 100, No. GT7, July 1974, pp. 763-786.

12. Ladd, C. C., Foott, R., Ishihara, K., Schlosser, F., and Poulos,
    H. G., "Stress-Deformation and Strength Characteristics," Proc.
    9th ICSMFE, 1977, Vol. 2, pp. 421-494.

13. Lodde, P. F., "Dynamic Response of San Francisco Bay Mud,"
    Geotechnical Engineering thesis GT82-2, University of Texas at
    Austin, October 1980.

14. Seed, H. B., and Idriss, I. M., "Soil Moduli and Damping Factors
    for Dynamic Response Analyses," Earthquake Engineering Research
    Center Report No. EERC 70-10, December 1970.

15. Singh, R. D. and Gardner, W. S., "Characterization of Dynamic
    Properties of Gulf of Alaska Clays," Proc. ASCE Session on Soil
    Dynamics in the Marine Environment, Boston, MA, 1979, 24 pp.

16. Taylor, D. W., Fundamentals of Soil Mechanics, J. Wiley & Sons,
    1948, p. 378.

17. Taylor, D. W., "A Direct Shear Test with Drainage Control,"
    Direct Shear Testing of Soils, ASTM STP131, 1952, pp. 63-74.

18. Trudeau, P. J., "The Shear Wave Velocity of Boston Blue Clay,"
    Massachusetts Institute of Technology Soils Publication No. 317,
    February 1973, 62 pp.

Appendix II.- Units

1 inch (in.)	= 25.4 millimeters (mm)
1 foot (ft.)	= 0.3048 meters (m)
1 pound force (lb.)	= 4.45 Newtons (N)
1 pound per sq. in. (psi)	= 6.9 kiloPascals (kPa)
1 ton per sq. ft. (TSF)	= 95.8 kiloPascals (kPa)

IMPULSE AND RANDOM TESTING OF SOILS

By

M.S. Aggour[I], M. ASCE, K.S. Tawfiq[II] and M.R. Taha[III]

ABSTRACT: Cohesive soils at different confining pressures were tested in a resonant column device using both impulse and random torsional excitation in addition to conventional sinusoidal excitation. The damping and shear modulus from both impulse and random loadings were determined by both the power spectral density function and the transfer function methods. Evaluation of the dynamic soil properties for sinusoidal vibration followed conventional procedures. The results indicated that during impulse and random loadings the damping values were higher and the shear moduli lower than the values obtained from sinusoidal loading at the same root mean square strain. Thus in soil structure interaction problems and in the evaluation of ground motion at a site, soil properties from routine sinusoidal tests should be adjusted to obtain realistic values that are representative of actual field conditions. Additional testing covering a wide range of variables may be needed to develop correction factors.

INTRODUCTION

The random vibration theory has recently been utilized in soil structure interaction analyses and has contributed to recent advances in analytical methods for evaluating ground motion. For these analyses the geotechnical engineer needs to determine soil properties under random loading conditions.

For the determination of the dynamic properties of soils, almost all laboratory testing techniques use sinusoidal loading as the type of excitation force. In field testing, waves are generated by either an impact force or detonation of small charges. Such generating systems transmit energies to the soil; however, the energy transmitted in the soil does not have the same frequency content as either laboratory or earthquake loadings. In order to determine the dynamic soil properties that can be used in ground motion evaluation and in soil structure interaction problems, nonperiodic loadings should be utilized in laboratory testing.

Under random (white noise) loading several methods of data analysis could be used, including autocorrelation analysis, power spectral density analysis, the random decrement technique, and the

---

I. Professor, Department of Civil Engineering, University of Maryland, College Park, MD 20742
II. Research Associate, Department of Civil Engineering, University of Maryland, College Park, MD
III. Graduate Student, Department of Civil Engineering, University of Maryland, College Park, MD

maximum entropy method. Soil properties under random loading were
determined in the laboratory by the autocorrelation function, the
power spectral density function (Amini et al., 1986), and the random
decrement technique (Aggour et al., 1982a, 1982b; Al-Sanad et al.,
1983). In general, the main problem with the techniques used in the
analysis of random loading testing is that only the output (measured
response) could be used. For an accurate determination of dynamic
soil properties, an input-output relationship should be developed and
utilized. Another problem encountered in random vibration tests is
the ability to predict the displacement and strain under random
loading tests. For the sinusoidal vibration test, the strain
amplitude can be obtained using the deterministic approach; in random
vibration tests, however, the uncertainty associated with the
probabilistic nature of the random response signal should be
considered.

Recently, Amini et al. (1988) presented a comprehensive study of
the effect of random loading on the dynamic properties of cohesionless
soils. In their study the transfer function method, a technique that
utilizes the input/output relationship for the determination of soil
properties under random excitation conditions, was used. In addition,
both the peak shearing strain and the root mean square strain were
introduced and utilized.

The first objective of this study was to simulate both the random
and impulse types of loading that are encountered in the field in
laboratory testing, and then to compare the properties of clayey soils
from random and impulse loading tests obtained by the existing
technique that utilizes only the response measurements, such as the
power spectral density function, with those obtained using the
transfer function method. The second objective was to compare the
soil properties determined from conventional sinusoidal loading tests
with those determined from impulse and random loading tests at
different strain levels; these comparisons will provide an understanding of the relationship between conventional laboratory testing
and the different types of field testing.

RANDOM VIBRATION ANALYSIS

Virtually all testing techniques that are currently used to
extract modal parameters, in this case the damping and natural
frequencies, are intended for linear systems. Most types of
nonlinearities are amplitude dependent (such as for soils); therefore,
if the frequency response function and modal parameters are
determined, keeping the displacement response at a constant level,
then the system could be approximated as a linear system (Ewins,
1984). Certainly the computed parameters, i.e., damping and natural
frequencies, would be valid only for the particular vibration test and
at that specific strain level.

Two sets of functions can be used to describe random processes:
one based in the time domain (i.e., correlation function) and the
other in the frequency domain (i.e., power spectral density function).
The autocorrelation function is exactly proportional to the free
vibration decay due to an initial displacement only for an ideal white
noise input. The power spectral density function, as contrasted with
the autocorrelation function, describes the general frequency
composition of the data. The autocorrelation function and the power

spectral density function are related through a Fourier transform. When the excitation is an ideal white noise, the damping ratio of a system can be obtained from the power spectral density of the response by the half-power bandwidth method (Bendat and Piersol, 1980).

A more reliable method of evaluating the damping and natural frequencies is based on the transfer function method, which uses both excitation and response. The transfer function of a system can be obtained from the relationship between the input spectral density and the output spectral density functions (Crandall and Mark, 1963). A distinct advantage of the transfer function method is that any type of excitation input can be used since the measurement being made is a response signal divided by the input causing it. Thus, by measuring and analyzing the excitation, the true response is obtained. In the transfer function method the damping and natural frequencies can be determined by a number of methods, including the magnitude of the transfer function (or peak-amplitude method), the real part of the transfer function, the imaginary part of the transfer function, and the Nyquist plot. These methods were utilized in the testing of different cohesionless soils by Amini (1986) and were found to give the same damping and natural frequency for all sands tested.

SOIL TESTED AND SAMPLE PREPARATION

The required specimens of cohesive soil for this study were prepared from kaolinite clay known as Edger Plastic Kaolin, EPK. This type of clay is a pure commercial kaolinite with a liquid limit of 52%, plastic limit of 31%, and specific gravity of 2.67. It was necessary to have available a large number of specimens for this study. The requirements were for specimens of clay with as high a degree of saturation as possible and with the clay structure duplicated as closely as possible, including the void ratio, degree of saturation, particle orientation or fabric, mineralogy, and composition of both the double layer and the pore water. Such duplication for large numbers of specimens could only be hoped for in remolded specimens extruded from compacted soil. To obtain compacted specimens with as high a degree of saturation as possible, the relationship between the moisture content and dry density for the EPK clay was first established using the modified compaction test. Then the optimum moisture content and curves of the degree of saturation were determined. Once the moisture-density-saturation relationship was established the required samples were then prepared at a specified moisture content and degree of saturation. The soil specimens for the dynamic properties measurements were then obtained from the compacted sample using a tube with an inner diameter of 3.81 cm and a height of 20.32 cm. The tube was first lubricated inside and outside to eliminate friction with the soil. The tube was pushed vertically into the compacted soil in the mold, so that all of the obtained specimens would have the same soil structure configuration with regard to the preferred particle orientation (Tawfiq, 1987). The properties of the remolded specimens have a void ratio of 1.3 and water content of 50% at the beginning of the testing program.

TEST EQUIPMENT AND EXPERIMENTAL PROCEDURE

For the purpose of this investigation the main testing equipment was the Drnevich-type resonant column apparatus. Methods, procedures, and apparatus descriptions of resonant column testing are presented in Drnevich et al. (1978) and in the ASTM Standard D4015 for resonant testing. However, some modifications were introduced in the method of testing and analysis to accommodate random vibration testing. Provisions were also made for the purpose of testing cohesive soil specimens.

In the conventional sinusoidal torsional loading test, the excitation signals were generated by a variable frequency sine-wave generator. The shear moduli were calculated from the resonant frequencies, and the damping ratios were determined using the magnification factor method, which ultilizes the input current and output acceleration. For comparison, the logarithmic decrement method, which describes the decay of the free vibration of a single-degree of freedom system, was also used to measure the damping characteristics of the soil specimens.

For the random loading test, the excitation was provided by a white noise generator (random wave generator). The output of the white noise generator was passed through a band-pass filter before it was connected to the torsional driving coils of the apparatus via a power amplifier. Both the input and output signals were connected to a Fast Fourier Transform (FFT) Analyzer (Rockland 5050B) for analysis. The purpose of the FFT analyzer in random testing was to transfer the time history records of vibration signals from the time domain (magnitude vs. time) to the frequency domain (magnitude vs. frequency). This transformation of the random signals facilitated the extraction of the following vibration parameters that were needed to determine the dynamic soil properties: the signal intensity (amplitude), the system resonance, and the decay rate. For the impulse loading test, the excitation was provided by built-in signal sources in the FFT analyzer. A schematic of the equipment that was used in this testing program is shown in Figure 1.

The FFT analyzer was first calibrated in accordance with the characteristics of the signals and the system damping ratios. In addition, four important factors had to be defined before the analyses could start. First, the number of averages had to be selected. It was found that a number of averages from 130 to 250 were needed to obtain a smooth frequency response function; therefore, the number of averages was set at 250 to be consistent and to reduce the effect of the number of averages (Tawfiq, 1987). Second, the frequency resolution had to be established. It was found that the resonance frequency was not sensitive to the range of the resolution chosen, whereas the damping values were sensitive to the frequency resolution. The effects of resolution on damping values were very small for the range of resolution from 0.1 to 0.5 Hz. Using a resolution of 0.1 Hz required a long testing time and might not be practical for testing some soil types, whereas the use of a resolution of about 0.5 Hz provided the required accuracy in a much shorter time. In this study, all tests were compared at a frequency resolution of 0.50 Hz (Tawfiq, 1987). Third, the full scale sensitivity, which refers to the maximum signal level the analyzer can accept without overloading, had to be set. Fourth, the proper weighting function to reduce the leakage had to be selected. Once these four factors were set in the FFT analyzer, the random and impulse testing in the resonant column device could be conducted. A microcomputer was used for the purpose of analyzing the

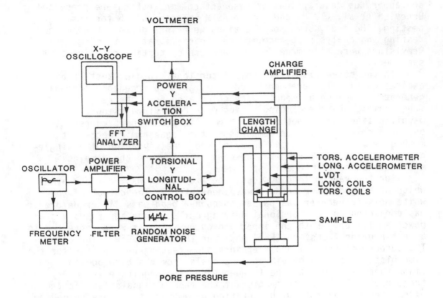

Fig. 1. Schematic of the equipment used

outputs and to evaluate the root mean squares of the power spectral functions of the output signals.

Three series of tests were performed to determine the dynamic properties of the soil specimens. The first series of tests was performed on the soil specimens using sinusoidal excitations, while the second series used random excitations and the third series used impulse loading. The clay specimens were prepared as described before and mounted in the resonant column device using the procedures described by Drnevich et al. (1978). Three confining pressures were used in the testing program, 34.45, 68.9, and 206.7 kPa.

In the sinusoidal testing the input, output, and resonant frequency of the system were recorded for different excitation levels. In the random or impulse testing, the excitations were applied to the soil specimens and the intensity of the signals were varied from low to high. The input and output were then analyzed by the FFT analyzer. For each test at a different strain level, the FFT analyzer provided the excitation and response time histories, excitation and response power spectral density function, and the transfer function.

STRAIN CALCULATION

To provide accurate comparisons of the values of both the shear moduli and damping values from random loading tests with those obtained from sinusoidal vibration tests, both tests had to be at the same strain level. For the sinusoidal vibration test, the strain

amplitude was obtained using the deterministic approach, see Drnevich
et al. (1978). In the random vibration test, the calculation of
strain was complicated by the uncertainty that is associated with the
probabilistic nature of the random response signal. Two strain
definitions, the peak shearing strain and root mean square (rms)
strain, are discussed.

The peak shearing strain in the case of sinusoidal loading
corresponded to the maximum amplitude of the constant sine-wave. In
the case of random loading, the response had an irregular pattern, and
thus, calculation of the strain was complicated by the uncertainty
associated with the random response signal. The task was equivalent
to computing the probability that a random process will ever exceed
some given level during a given interval of time. Analytical
techniques are available for the determination of the extreme value of
a random response signal (Vanmarcke, 1983) and these techniques in
conjunction with rms strain were used in the testing of cohesionless
soils (Amini et al., 1988).

It has been shown that the rms strain for the case of random
vibration represented a strain with peaks having the highest
probability of occurrence. For the determination of the rms strain,
the FFT analyzer was used to display the power spectral density of the
acceleration from which the power spectral density of the displacement
was obtained. Once obtained, the area below the curve represented the
mean-square value of the displacement. The root mean square of the
displacement was evaluated by taking the square root of the area under
the displacement power spectrum, which is defined as the displacement
rms. Utilizing the specimen dimension and equipment calibration
factors, the rms strain from the displacement rms was then determined.

It should be noted that while the peak shearing strain for
sinusoidal loading represented a steady peak, the maximum strain for
the case of random vibration was only an indication of an
instantaneous peak. In addition, the peak shearing strain for
sinusoidal vibration, which is deterministic, was obtained by
multiplying the rms strain by $\sqrt{2}$; whereas in the case of random
vibration, the peak strain was determined by multiplying the rms
strain by a mean peak factor that was not a constant but primarily a
function of the duration of the event. In this paper, data obtained
from sinusoidal, random, and impulse loading testing were compared at
the same rms strain.

RESULTS

Three different types of excitations were used in this study:
sinusoidal, random, and impulse. The excitation and response time
domain functions for a typical random vibration is shown in Figure 2.
Figure 3 shows the input and response time domain functions of an
impulse generated by the built-in signal sources in the FFT analyzer.
The pulse is a sine wave burst at a frequency equal to the center of
the span. Figure 4 shows the input and response of the generated
impulse in the frequency domain. As shown in Figure 4, the frequency
spectrum of the impulse signal is nearly flat over a wide frequency
range, which is similar to that of a random excitation.

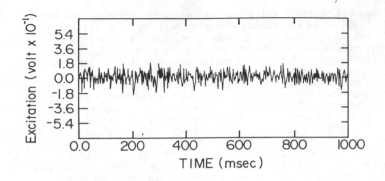

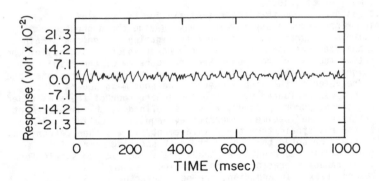

Fig. 2. Excitation and response time domain function
for a typical random loading

Comparison Between Different Techniques for Evaluating Soil Properties

In field testing for the evaluation of soil properties, it is
difficult to delineate a confined, homogeneous system that would be
needed for transfer function measurements, hence necessitating the use
of the output response as the only data that can be used for
determining properties in the field. However, this problem is
overcome in the laboratory where both input and output signals can be
readily and accurately utilized in the use of the transfer function
method.

In order to assess the field values based on output measurements
only, a comparison of the results from the transfer function method
with other methods, such as the power spectral density function, is
necessary. Figure 5 shows a typical example of the power spectral
density function of the output, and the transfer function of the same
output for both a random and an impulse vibration test. As shown in
the figure, both are almost identical. Thus, it can be concluded that

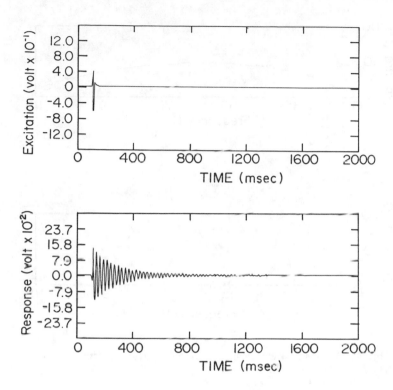

Fig. 3. Typical excitation and response of an impulse in the time domain

both the damping values and the shear moduli for a clayey soil under either random or impulse excitation will be the same if determined from either the power spectral density function or from the transfer function. This is true within the range of strains used in this study and should be checked for other strain levels.

The accuracy of the transfer function estimation is normally determined by a single function called the coherence function. For an ideal measurement the coherence should be unity, and we must look for this condition to ensure that the measurements have been accurately made (Ewins, 1984). It was found that for the cohesive soils tested the coherence function is very close to unity, as shown in Figure 6.

Comparison of Properties Obtained from Random and Sinusoidal Vibrations

The second objective was to compare the dynamic properties of the

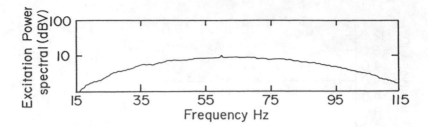

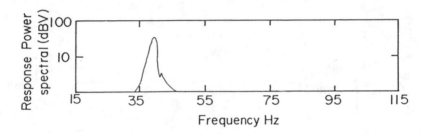

Fig. 4.   Excitation and response of an impulse loading in the
frequency domain

soils under both random loading and impulse loading with those
properties obtained from conventional sinusoidal tests. To study the
effect of the loading type on the damping ratio, Figure 7 is provided
as an example of the data obtained. The damping values from both
random and impulse loading were compared with those values obtained by
conventional sinusoidal tests. From this figure it is concluded that
the damping values from both random and impulse loading were higher
than the ones obtained by the sinusoidal vibration at the same rms
strain. As the rms strain decreased, the differences in the damping
values decreased. This can be explained by the fact that the damping
of soil is highly strain-dependent. Since the random signal has
higher peaks than the sinusoidal signal at the same displacement rms,
damping from random vibration should be higher than that corresponding
to sinusoidal loading.
     It should be mentioned here that the difference shown between the
damping for impulse, random, and sinusoidal loading as a function of
rms strain will be different if the damping is plotted as a function
of the peak shearing strain. In sinusoidal vibration, the peak
shearing strain is deterministic and is obtained by multiplying the
rms strain by $\sqrt{2}$. In random vibrations the peak shearing strain is
only an indication of an instantaneous peak and is obtained by

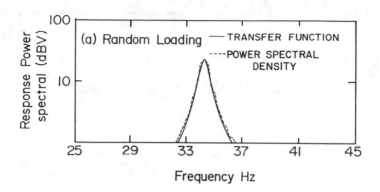

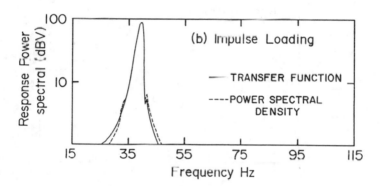

Fig. 5. Power spectral density and transfer function
        for (a) Random loading; (b) Impulse loading

multiplying the rms strain by a mean peak factor that is not a
constant, but is primarily a function of the duration of the random
event. Thus, for different durations of events, different damping
plots can be generated for impulse and random loadings as a function
of the peak strain.

Similarly, the shear moduli from impulse, random, and sinusoidal
vibrations were compared as shown in Figure 8. The shear moduli of
impulse and random loadings were lower than the shear moduli of
sinusoidal loading at the same rms strain. At a low rms strain, the
differences between both types of testing were small, and for higher
rms strains the differences were larger. Again if the shear modulus is
to be plotted against the peak shear strain, such differences will
change with the change of the duration of the random event.

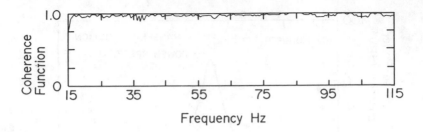

Fig. 6. Example of the coherence function for an impulse test

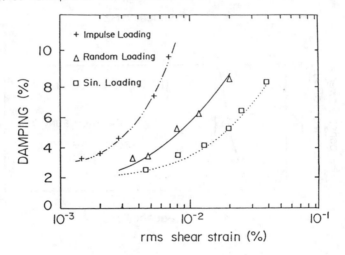

rms shear strain (%)

Fig. 7. Effect of type of loading on damping values
         as a function of rms strain
         (Confining pressure = 34.45 kPa)

SUMMARY AND CONCLUSIONS

In this study impulse and random loading tests, as well as
conventional sinusoidal vibration tests, were performed on a clayey
soil in a resonant column device. The testing program utilized the
transfer function and the power spectral density function. In
calculating the soil strain, the concept of the rms strain was used.

The results of this limited investigation showed that within the
range of variables studied, the shear modulus and damping of clayey
soils determined by either the transfer function or the power spectral
density function were found to be the same. Thus the existing
technique of using only the measured response in determining soil
properties is valid.

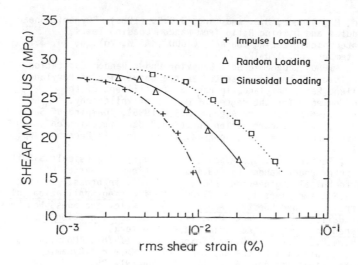

Fig. 8. Effect of type of loading on shear modulus
as a function of rms strain
(Confining pressure = 34.45 kPa)

In impulse and random loading testing, it was found that damping
values were higher and the shear moduli lower than the values obtained
by sinusoidal loading at the same rms strain. The differences
increased as the strain amplitude increased. Thus in soil structure
interaction problems and in the evaluation of ground motion at a site,
soil properties from routine sinusoidal tests should be adjusted to
obtain realistic values that are representative of actual field
conditions. Additional testing that covers a wide range of variables
may be needed to develop correction factors.

ACKNOWLEDGMENTS

The research presented herein was sponsored by the Maryland State
Highway Administration. The authors are grateful for the support
provided.

APPENDIX I. REFERENCES

Aggour, M.S., Yang, J.C.S. and Al-Sanad, H. (1982a), "Application of
the Random Decrement Technique in the Determination of Damping of
Soils, "Proceedings, Seventh European Conference on Earthquake
Engineering, Athens, Greece, Vol. 2, September, pp. 337-344.
Aggour, M.S., Yang, J.C.S., Chen, J., Amer, M. and Al-Sanad, H.
(1982b), "In-Situ Determination of Damping of Soils,"
Proceedings, Seventh Symposium on Earthquake Engineering,
Roorkee, India, Vol. 1, November, pp. 365-370.

Al-Sanad, H., Aggour, M.S. and Yang, J.C.S. (1983), "Dynamic Shear
    Modulus and Damping Ratio from Random Loading Tests,"
    Geotechnical Testing Journal, GTJODJ, ASTM, Vol. 6, No. 3,
    September, pp. 120-127.
Amini, F. (1986), "Dynamic Soil Behavior Under Random Excitation
    Conditions," thesis presented to the University of Maryland,
    College Park, Maryland, in partial fulfillment of the
    requirements for the degree of Doctor of Philosophy.
Amini, F., Tawfiq, K.S. and Aggour, M.S. (1986), "Damping of Sandy
    Soils Using Autocorrelation Function," 8th Symposium on
    Earthquake Engineering, Roorkee, India, Vol. I, December, pp.
    181-188.
Amini, F., Tawfiq, K.S. and Aggour, M.S. (1988), "Cohesionless Soil
    Behavior Under Random Excitation Conditions," Journal of
    Geotechnical Engineering, ASCE, 114 (GT8), in press.
Bendat, J.S., and Piersol, A.G. (1980), Engineering Applications of
    Correlation and Spectral Analysis, John Wiley and Sons, New York.
Crandall, S.H. and Mark, W.D. (1963), Random Vibration in
    Mechanical Systems, Academic Press, New York.
Drnevich, V.P., Hardin, B.O. and Shippy, D.J. (1978), "Modulus and
    Damping of Soils by the Resonant-Column Method," Dynamic
    Geotechnical Testing, ASTM, STP 654, pp. 91-125.
Ewins, D.J. (1984), Modal Testing: Theory and Practice, John Wiley and
    Sons, New York.
Tawfiq, K.S. (1987), "Effect of Time and Anisotropy on Dynamic
    Properties of Cohesive Soils," thesis presented to the University
    of Maryland, College Park, Maryland, in partial fulfillment of
    the requirements for the degree of Doctor of Philosophy.
Vanmarcke, E. (1983), Random Fields, The MIT Press, Cambridge,
    Massachusetts.

INTERNAL DAMPING OF COMPOSITE
CEMENTITIOUS MATERIAL: CEMENTED SAND

BY Tzyy-Shiou Chang,[1] A. M. ASCE and Richard D. Woods,[2] M. ASCE

ABSTRACT

Recent research and practical experience have shown that an small amount of cementing material partially or totally filling the void space between soil particles would strongly affect the dynamic behavior of the cemented soil.(1,2,3,4,5,6,7,14) This investigation was conducted to obtain the internal damping characteristics of cemented sand as a function of degree of cementation, curing time and confining pressure. The tests were conducted using resonant column device at strain levels less than 10^{-3}%. The free-vibration decay method was used in this research to determine the damping ratio of sand cemented with sodium silicate.

The 3-parameter visco-elastic model was used successfully to characterize the internal damping of cemented sand. This model also successfully predicted observed test results for internal damping.

INTRODUCTION

Internal damping is an important property of soil in many practical applications such as earthquake-resistant design, machine foundation design, and seismic wave propagation surveys. Attenuation of ground surface wave propagation is partially controlled by internal damping of the soil deposit.

It is well known that a small amount of cement injected into a soil mass can significantly increase the stiffness and reduce the liquefaction potential of the soil mass under seismic loading.(2,3,4) In soil deposits overlying bedrock changes in stiffness and internal damping due to cementation of soil particles will strongly affect the upward propagating ground motion during earthquake shaking.

----

[1] Staff Engineer, Soil and Materials Engineers, Inc.,
Ann Arbor, Michigan, and formerly, Research Associate,
Dept. of Civil Engineering, University of Michigan, Ann Arbor.

[2] Professor, Department. of Civil Engineering, University of
Michigan, Ann Arbor, Michigan.

In this research, several types of sand and sodium silicate were used because of the well known properties of sand and the popularity of the sodium silicate for soil improvement. Tests were conducted at low strain levels to observe the behavior of cemented sand with the parameters of degree of cementation, curing time, confining pressure and basic index properties of sand. The basic index properties of sand, D_{10} and C_u, were the most important factors affecting the number of contact points per volume and groutability of sand. The number of contact points between soil particles at which chemical bonds can develop is one of the major concerns in studying the behavior of cemented sands. The index properties of sand, C_u and D_{10}, also strongly affect the groutability of sand mass which is also one of the major concerns in grouting project. (1,2,3,4)

For this study, cemented sand was considered as a composite cementitious material. The visco-elastic model, known as 3-parameter solid,(16) was employed to analyze the complex modulus of the cemented sand. Using this model, the damping ratio was related to the viscosity of the composite and modulus of the cemented soil skeleton.

Damping ratios of cemented sand specimens were obtained in this study using the free-vibration decay method. The measured damping ratios were used to extract values for the 3-parameter solid. Furthermore, the equations derived based on 3-parameter solid, as shown in Eq. 11 and Eq. 11a, are also useful in studying the internal damping of any composite cementitious materials which can be reasonably modeled by a 3-parameter solid.

BASIC PROPERTIES OF SANDS AND CEMENTS USED

Soils

Three fine-grained and one medium-grained sands were tested in this investigation. The major index properties of interest were their effective grain diameter, D_{10}, and coefficient of uniformity, C_u. The soils are described as:

1. Ottawa 20–30: rounded, poorly graded, $C_u = 1.06$,
   $D_{10} = 0.66$ mm.
2. Muskegon sand: angular to subrounded, poorly graded,
   $C_u = 1.50$, $D_{10} = 0.28$ mm.
3. Mortar sand: angular, well graded, $C_u = 4.42$,
   $D_{10} = 0.24$ mm.
4. Crushed medium sand: angular, poorly graded,
   $C_u = 1.06$, $D_{10} = 1.45$ mm.

Cement

The selection of a cementing material for this investigation was based on the following considerations:(12,13)

1. low cost,

2. high effectiveness,
3. ready availability, and
4. non-toxicity.

Using the above considerations, sodium silicate, which is one of the most popular chemical grouts in the United States, was selected. Three cementing mixtures made up of sodium silicate were used as follows:
Sodium silicate/H_2O/Ethylacetate & foramide:

Type (SS60): 60:30:10.
Type (SS50): 50:42:8,
Type (SS40): 40:50:10.

PREPARATION OF SPECIMENS AND TEST APPARATUS

Preparation of Soil Specimens

The injection method as shown in Fig.1 was employed to prepare specimens with sodium silicate because the sodium silicate grout used in this study had a viscosity less than 10 cP. At this viscosity the soils selected ($k = 10^{-1} - 10^{-5}$ cm/sec, and k:permeability) were rated as groutable. The target degrees of cementation (C) for this investigation were , 10%, 20%, 30%, 45%, 60%, 75% and 90%, where degree of cementation is defined as C = (volume of grout / void volume) x 100%.

Test Apparatus

The Hall-type resonant column devices was employed in this investigation to determine the damping ratio of cemented sand by the free vibration decay method.(8,9,10,11,15)

EXPERIMENTAL PROCEDURE

All the tests were conducted at six pressures: 3 psi, 6 psi, 10 psi, 15 psi, 20 psi and 30 psi (20.7 KN/m^2 − 207 KN/m^2) with the shearing strain level less than 10^{-5}. The confining pressures used in this research approximate the in-situ depths of 3 feet to 30 feet within which most practical grouting projects take place. In addition, this zone is also the major critical zone for studying the attenuation and screening of surface waves except for those with very low-frequency. Long term test as long as 120 days were conducted on the specimens. The specimens were cured in between tests in desiccators at high relative humidity. The specimens could be retested many times as required with only very minor specimen disturbance.

MODELING OF COMPOSITE CEMENTITIOUS MATERIAL

Cemented sand can be considered as one type of composite cementitious material for which the internal damping can be represented by a visco-elastic model, known as 3-parameter solid or standard linear solid, "SLS".(16) This model contains a Maxwell body plus a spring in parallel as shown in Fig.2.

In this model:

$\bar{E}_s(t)$ = elastic modulus of soil skeleton
after cementation,

$E_g(t)$ = elastic modulus contribution of the
cement occupying part of the voids
between soil particles, and

μ = viscosity of the composite.

It should be pointed out that when the "E_g" part of the model is removed, the model becomes a "Kelvin-Voigt" model which is a common model for clean, uncemented sand. The internal damping of clean sand has been studied by Hardin using a Kelvin-Voigt model.(9,10,15) The result of his analysis reveals that damping ratio, D, which is related to logarithmic decrement, can be expressed as:

$$\delta = 2\pi D/(1-D^2)^{1/2}$$
$$= \pi\tan(\delta_L)$$
$$= \pi(\mu\omega/G) \qquad\qquad --\text{Eq.1}$$

in which :

δ = Logarithmic decrement,
D = Damping ratio,
μ = viscosity,
ω = frequency of oscillation,
G = shear modulus of clean sand, and
δ_L = loss angle of complex modulus.

The internal damping of a cemented sand can be represented by the damping in a the 3-parameter solid. The stress-strain relationship can be developed using the free-body diagram as shown in Fig.3.

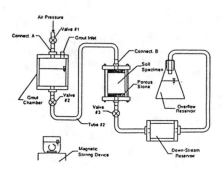

Fig.1   Schematic of Injection Apparatus

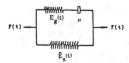

Fig.2   3-Parameter Solid Represents The
Soil Skeleton of Cemented Sand

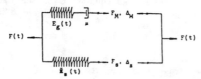

Fig.3 Free Body Diagram of 3-Parameter Solid

Let "M" denote the Maxwell model, i.e. spring $E_g(t)$ and viscous dashpot, μ, are in series, the basic equations used for the 3-parameter model are as follows:

a. $F(t) = F_M + F_s$, force balance

b. $\Delta = \Delta_M = \Delta_s$, geometry

c. $F_s = \bar{E}_s \Delta_s$, F-Δ for spring

d. $(D/E_g + 1/\mu)F_M = D\Delta_M$, F-$\Delta$ for Maxwell model

Where F = stress, Δ = strain and
    D = first derivative with respect to time.

Based on the above F-Δ-Time relationship,

we get: $(D/E_g + 1/\mu)(F_s + F_M) = \bar{E}_s(D/E_g + 1/\mu)\Delta_s + D\Delta_M$

rearrange,

$$(D/E_g + 1/\mu)F = \{D[(\bar{E}_s/E_g) + 1] + \bar{E}_s/\mu\}\Delta$$

or, $(1/E_g)DF + (1/\mu)F = [1 + (\bar{E}_s/E_g)]D\Delta + (\bar{E}_s/\mu)\Delta$

which implies the stress-strain relationship as follows:

$$p_0 \sigma + p_1 \dot{\sigma} = q_0 \gamma + q_1 \dot{\gamma}$$

$---$ Eq.2

where: $p_0 = 1/\mu$, $p_1 = 1/E_g$, $q_0 = \bar{E}_s/\mu$, $q_1 = 1 + (\bar{E}_s/E_g)$.
    σ = stress of 3-parameter solid,

    $\dot{\sigma}$ = first derivative of stress with respect

to time,

γ = strain of 3-parameter solid, and

$\dot{\gamma}$ = first derivative of strain with respect to time.

In an attempt to solve the differential equation Eq.2, the Laplace Transform method and the initial condition shown in Eq.3 below were employed to find the solution:

$$p_1 \sigma(0^+) = q_1 \gamma(0^+)$$ $---$Eq.3

The relaxation response (stress), $G_r(t)$, required to maintain a constant strain, $\gamma(t)$ as shown in Eq.4, can be derived to be as Eq.5 and shown in Fig.4:

$$\gamma(t) = \gamma_0 1(t)$$ $---$Eq.4

$$G_r(t) = \sigma(t)/\gamma(t)$$
$$= G_r(\infty) + [G_r(0) - G_r(\infty)]e^{-\lambda t}$$
$$= G_r(\infty) + \Delta G_r(t)$$ $---$Eq.5

where: $\lambda = p_0/p_1 = E_g/\mu$, $\Delta G_r(t) = [G_r(0) - G_r(\infty)]e^{-\lambda t}$, and

$$G_r(\infty) = q_0/p_0 = \bar{E}_s$$ $---$Eq.5a

$$G_r(o) = q_1/p_1 = E_g + \bar{E}_s$$ $---$Eq.5b

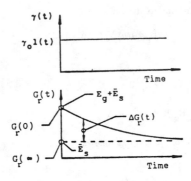

Fig.4   Stress-Strain Relationship of 3-Parameter Solid

Now, we consider the 3-parameter solid undergoing a sinusoidal oscillation as described by:

$$\gamma(t) = \gamma_0 \mathrm{Sin}\omega t$$

$$- - - \mathrm{Eq.6a}$$

The sinusoidally varying strain can be mathematically considered as the imaginary part of the complex strain history as shown below:

$$\gamma(t) = \gamma_0 e^{-i\omega t}$$
$$= \gamma_0 (\mathrm{Cos}\omega t + i\mathrm{Sin}\omega t)$$

$$- - - \mathrm{Eq.6b}$$

The complex modulus of a 3-parameter solid under such a sinusoidal strain history can be described also in terms of shear stiffness as:

$$\sigma(t) = G^*(\omega)\gamma_0 e^{-i\omega t}$$
$$= G^*(\omega)\gamma(t)$$

$$- - - \mathrm{Eq.7}$$

where $G^*(\omega) = $ complex shear modulus, and

$$G^*(\omega) = G_1(\omega) + iG_2(\omega)$$

$$- - - \mathrm{Eq.8}$$

where:

$G_1(\omega) = $ elastic component of complex modulus, and

$$G_1(\omega) = [\lambda^2 G_r(\infty) + \omega^2 G_r(0)]/(\lambda^2 + \omega^2)$$

$$- - - \mathrm{Eq.8a}$$

$G_2(\omega) = $ viscous component of complex modulus, and

$$G_2(\omega) = \{\omega\lambda[G_r(0) - G_r(\infty)]\}/(\lambda^2 + \omega^2)$$

$$- - - \mathrm{Eq.8b}$$

The internal damping of cemented sand is governed by the loss angle of complex modulus where defined as:

$$\delta_L = \mathrm{Tan}^{-1}(G_2(\omega)/G_1(\omega))$$

$$- - - \mathrm{Eq.9}$$

By substituting Eq.5a, 5b, Eq.8a and 8b into Eq.9, we get:

$$\delta_L = \mathrm{Tan}^{-1}\{\mu\omega/[\bar{E}_s + (\mu\omega/E_g)^2(\bar{E}_s + E_g)]\}$$

$$- - \mathrm{Eq.10}$$

By substituting Eq.10 into Eq.1, and substituting shear deformation in place of dilatational deformation, we get:

$$\delta = 2\pi D/(1 - D^2)^{1/2}$$
$$= \pi\mu\omega/[\bar{G}_s + f_D(g)]$$

$$= \pi\mu\omega/G_D$$

$$- - - Eq.11$$

where:

$G_D = \bar{G}_s + f_D(g)$, defined as equilibrium damping stiffness,

$$f_D(g) = (\mu\omega/G_g)^2(\bar{G}_s + G_g)$$

$$- - - Eq.11a$$

$\bar{G}_s$ = shear modulus of soil skeleton of cemented sand,

G_g = shear modulus representing the effect of

　　partial filling of voids with cement,

D = damping ratio of soil-grout mixture,

ω = frequency of oscillation, and

μ = viscosity of soil-grout mixture.

As shown in Eq.11 and Eq. 11a, $f_D(g)$ represents the contribution of cementing material to the internal damping of cemented soil. For an uncemented, clean sand, $f_D(g)$ is equal to zero, and Eq.11 becomes exactly the same as Eq.1 which was derived by Hardin to represent the internal damping of clean sand and G_D is equal to G as shown in Eq. 1. Eq.11 also reveals that, in terms of log decrement, internal damping can be functionally expressed as:

$$\delta = 2\pi D/(1-D^2)^{1/2}$$

$$= f(\omega, \bar{G}_s, G_g, \mu)$$

$$- - - Eq.12$$

From Eq.11, it can be seen that the internal damping of cemented soil is much more complicated than that of uncemented, clean sand. An increase in G?-$_s$ reduces the damping ratio which implies that an increase in confining pressure reduces the damping ratio. The parameters μ and G?-$_s$ also increase with an increasing degree of cementation, C, thus making the situation more complicated because:

1. Viscosity μ increases with an increasing C, thus increasing the damping ratio. On the other hand,

2. Increase of μ results in increase of $f_D(g)$, thus reducing the damping ratio, and

3. $\bar{G}_s$ increases with C, thus reducing the damping ratio.

It is not well understood if G_g increases or remains constant as C increases. However it is reasonable to assume at a certain curing time, Gg increases with increasing C because more space within soil particles is occupied by cementing material. In addition, it appears at a certain C, Gg increases with time as the cement hardens.

It is rather difficult to conclude whether the increase of C% increases or reduces the damping ratio of cemented sand. It depends upon how the degree of cementation, C%, affects the parameters μ, $\bar{G}_s$, G_g and $f_D(g)$ respectively. Thus, experiments are required to find a critical value of C% at which the damping ratio reaches the relative maximum or minimum value or approaches a constant value.

DAMPING RATIO OF CEMENTED SAND AT AN EARLY AGE

The first measurement of the damping ratio was made at 1,440 minutes when the shear modulus of the cemented sand was about 40% of its ultimate value which suggests that the process of bonding soil particles had not yet been completed. The test results reveal that the damping ratio of cemented sand at an early age is relatively high, up to 4-5% as shown in Figs.5-6, compared with 1-2% of uncemented, clean sand, as shown for C=0 in Fig.9-10. As shown in Fig. 5-6, it is clearly seen that the damping ratio increases only slightly with an increasing degree of cementation, C. This indicates that, at an early age, the increase of viscosity, μ, of cemented sand is much more predominant than the increase of $\bar{G}_s$ and the change of $f_D(g)$ as shown in Eq.11. More grout injected into a soil mass results in higher viscosity, thus increasing the damping ratio of the cemented sand.

Furthermore, as expected, the damping ratio decreases with an increasing confining pressure. This is because of the increase of $\bar{G}_s$ and $f_D(g)$ due to the increasing confining pressure while viscosity, μ, remains constant.

There is no evidence showing that the damping ratio at an early age has any functional relationship to the grain size distribution of cemented soil.

TIME EFFECT ON DAMPING RATIO OF CEMENTED SAND

The damping ratio of cemented sand is also time-dependent but only due to the time-dependent chemical reaction of cement. A typical figure representing the damping ratio vs time relationship of cemented sand is shown in Fig.7. The damping ratio decreases as the specimen hardens over time. This is reasonable because, as expressed in Eq.11, $\bar{G}_s$ and $f_D(g)$ increase over time, thus reducing the damping ratio of the cemented sand.

A characteristic figure representing the relationship between damping ratio, D, shear modulus, $\bar{G}_s$, and time for all the test results is shown in Fig.8. It reveals that the damping ratio is the exact inverse of the shear modulus of cemented sand as described below:

1. Damping ratio decreases over time while shear modulus increases.

2. Damping ratio decreases with an increasing confining pressure while shear modulus increases.

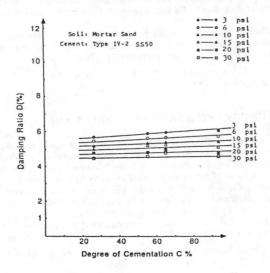

Fig.5 D vs C for Mortar Sand with Cement Type SS50 at 1,440 Minutes

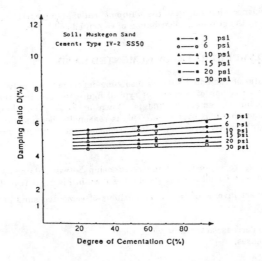

Fig.6 D vs C for Muskegon Sand with Cement Type SS50 at 1,440 Minutes

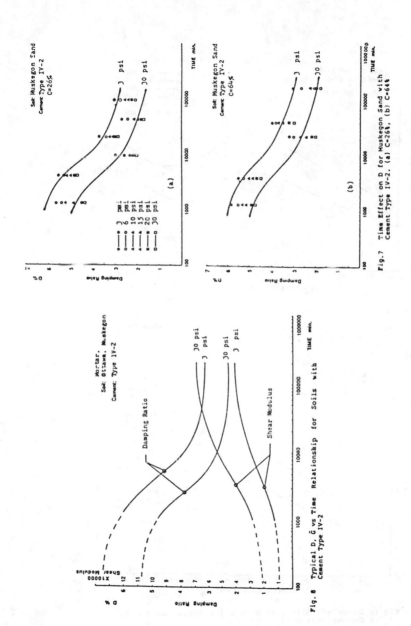

Fig.7 Time Effect on D for Muskegon Sand with Cement Type IV-2. (a) C=26%, (b) C=64%

Fig. 8 Typical D, Ḡ vs Time Relationship for Soils with Cement Type IV-2

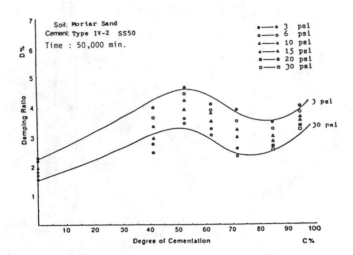

Fig.9 Long-Term Effect of D vs C for Mortar Sand with
       Cement Type SS50

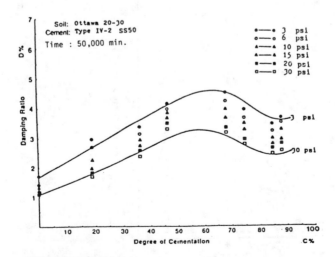

Fig.10 Long-Term Effect of D vs C for Ottawa Sand with
        Cement Type SS50

It should be pointed out that, for sodium silicate grout, both shear modulus and damping ratio reach the ultimate value at about 50,000 minutes (35 days) after the specimen is prepared. This suggests the complete curing of the chemical bonds between soil particles of cemented sand.

EFFECT OF C ON DAMPING RATIO OF CEMENTED SAND

As shown in Figs 5, 6, 9 and 10, both at a early age and after the curing process has completed, all the specimens showed higher damping ratios than those of uncemented clean sand (in the range of 3-6 % for cemented sands compared to 1-2 % for uncemented clean sands). This implies even a small amount of viscous material partially occupying the void space of soil mass would significantly change the internal damping of this composite material.

The damping ratio taken at 50,000 minutes, which approaches the ultimate value, was used to investigate the long-term effect of the degree of cementation, C, on damping ratio, D. Two typical results are shown in Figs.9-10. For all the soils used, damping ratio increases with an increasing C and reaches the maximum value at C=55%-65%, and then drops slightly. It is important to point out that the test results show a relative maximum value of the damping ratio at a degree of cementation which implies the complete curing of cement between soil particles.

These results can be well explained by Eq.11 and 11a. The effect of the degree of cementation on damping ratio of cemented sand is very complicated because all the parameters shown in Eq.11 and 11a are affected by it. Viscosity, μ, is increased by increasing C, and the shear modulus of soil skeleton, $\bar{G}_s$, is increased by increasing C as well. But $f_D(g)$ is tricky because it also depends on μ, G_g and $\bar{G}_s$ as indicated in Eq. 11 and 11a., and $f_D(g)$ can be either increased or decreased by increasing C depending on how C affects μ, G_g and $\bar{G}_s$. As assumed early, G_g increases with increasing C thus reducing $f_D(g)$ and hence increasing the damping ratio. On the other hand, increase in μ and $\bar{G}_s$ due to an increase in C increases $f_D(g)$ and reduces the damping ratio. Based on the above, the turning point of D is expected at a particular degree of cementation where the damping ratio reaches the relative maximum or minimum value. This is confirmed by the test results.

CONCLUSION

Based on a 3-parameter solid, the internal damping of cemented soil, which is one type of composite cementitious material, can be expressed as the following:

$$\delta = \pi\mu\omega/[\bar{G}_s + f_D(g)] \qquad\qquad ----- \text{Eq.}11$$

Test results revealed even a small amount of viscous cementing material like sodium silicate used in this research injected into soil mass would result in significant change of internal damping both at a early age and after curing process has completed.

Whether an increase in degree of cementation increases or reduces the damping ratio of cemented soil depends on how the degree of cementation affects μ, $\bar{G}_s$, G_g, and

$f_D(g)$ respectively for each individual soil. In this investigation, the test results show that the damping ratio of cemented sand is relatively high, up to 4-6% compared with 1-2 % for uncemented, clean sand, at an early age. This is a reasonable finding because the cement is still relatively viscous at this time. Later, the damping ratio decreases as the soil skeleton stiffens over time and reaches a constant value approximately at the time when the shear modulus remains constant. This represents the complete curing of the bonds between soil particles and both the shear modulus and damping ratio remain stable. Test results reveal that D increases with an increase of C%, starting at C=0%, and reaches a relative maximum value of damping ratio when C is about 60%. For C greater than 60%, an increase of C results in a slight decrease of the damping ratio.

All the conclusions made in this investigation regarding internal damping of cemented sand can be very well interpreted by the Eq.11 and 11a. It should be emphasized that Eq.11 also can be used to study the internal damping of any types of composite cementitious material which can be reasonably modeled by a 3-parameter solid.

Based on this research, internal damping of cemented sand at low strain and high frequency is well understood. However further research on high-strain and low frequency is necessary especially for earthquake and ground motion due to blasting.

Acknowledgement

The encouragement to perform these specific studies by Wallace Hayward Baker and Joseph Welsh of the Hayward Baker Company is gratefully acknowledged as are the supplies of grout and reactant which made testing possible. Professor F. E. Richart also served as advisor and provided inspiration for continuing studies of dynamic properties of soils.

References

1) Chang, T. S., (1986), "Dynamic Behavior of Cemented Sand", Ph.D. Dissertation, The University of Michigan, Ann Arbor, Michigan.

2) Chang, T.S. and Woods, R.D., "Effect of Confining Pressure on Shear Modulus of Cemented Sand", Development in Geotechnical Engineering Vol. 43, The 3rd International Conference on Soil Dynamics and Earthquake Engineering, Princeton University, New Jersey, U.S.A., June 22-24, 1987, pp193-208.

3) Clough, G. W., Kuck, W. M. and Kasali, G., (1979) "Silicate-Stabilized Sand," Journal of Geotech. Engng Div., Proc. ACSE, Jan, No. GT1.

4) Clough, G. W. and Rad, N. S., (1982) "The Influence of Cementation on The Static and Dynamic Behavior of Sand," December, Final Report for The United States Geological Survey, Dept. of The Interior, Office of Earthquake Studies.

5) Chae, Y. S., Au, W. C. and Chiang, Y. C., (1981) "Determination of Dynamic Shear Modulus of Soil From Static Strength," Oct., Proc. of International Conference on Recent Advance in Geotech. Earthquake Engng. and Soil Dynamics, St. Louis, Vol. 1 pp33-38.

6) Clough, G. W., Sitar, N. and Bachus, R. C., (1981) "Cemented Sand Under Static Loading," June, J. of Geotech Engng Div., ASCE 107, GT6, pp799–817.

7) Chiang, Y. C. and Chae, Y. S., (1972) "Dynamic Properties of Cement Treated Soils," Highway Research Record 379, pp39–51.

8) Hall, J.R., Jr., and Richart, F.E., Jr., (1963) "Dissipation of Elastic Wave Energy in Granular Soils," J. of Soil Mech. and Found. Div., Proc. ASCE, Vol. 89, No. SM 6, Nov., pp27–56.

9) Hardin., B.O., and Richart, F.E., Jr. (1963), "Elastic Wave Velocities in Granular Soils," J. Soil. Mech. and Found. Div., Proc. ASCE, V.89, No. SM1, Feb.,pp.33–65.

10) Hardin, B.O. (1961),"Study of Elastic Wave Propagation and Damping in Granular Materials," Ph.D. Dissertation, Univ. of Florida, Aug., 207 pp.

11) Hardin, B. O. (1965), "The Nature of Damping in Sands," J. Soil Mech. and Found. Div., Proc. ASCE, Vol.91, No. SM1, Jan., pp 63–97.

12) Karol, R.H., (1982) "Chemical Grouts and Their Properties," Proc. of The Conference on Grouting in Geotechnical Engng, ASCE. New Orleans, Louisiana, Feb. 10–12, pp433–440.

13) Karol, R.H., (1983) Chemical Grouting, Marcel Dekker, Inc., New York and Basel.

14) Partos, A., Woods, R.D. and Welsh, J.P., (1982) "Soil Modification for Relocation of Die Forging Operations," Proc. of The Conference on Grouting in Geotechnical Engng, ASCE, New Orleans, Louisiana, Feb., pp938–958.

15) Richart, F.E., Jr., Hall, J.R. and Woods, R.D., (1970) Vibration of Soils and Foundations, Englewood Cliffs, N. J., Prentice-Hall Inc..

16) Ward, I. M., (1979),"Mechanical Properties of Solid Polymers", Second Edition, John Wiley & Sons, 475pp.

Using the CPT for Dynamic Site Response Characterization

By Richard S. Olsen[*], M. ASCE

ABSTRACT

The Cone Penetrometer Test (CPT) can provide important information about site stratigraphy variation and provides estimates of many geotechnical properties which are required for a dynamic site response analysis. This paper will describe the techniques and methodology for using the CPT to estimate the following geotechnical parameters: soil type, SPT blow count, SPT silt corrected blow count, maximum shear modulus and high strain shear stress ratio (i.e. cyclic shear stress ratio to achieve at least 5% shear strain potential).

INTRODUCTION

A dynamic site response evaluation generally requires input parameters taken from the interpretation of borings and field/laboratory tests. Dynamic response studies have in the past concentrated on determining the most accurate values using choice data rather than consideration of the site variability. Significant amounts of in-situ variability at a site may not be apparent when the field investigation is performed with borings. Also, while laboratory testing and field shear wave velocity measurements provide point location information, site variability of important dynamic response inputs can be the real problem. The Cone Penetrometer Test (CPT) has the potential for providing better answers concerning site variability and estimates of geotechnical properties.

The CPT can provide estimates of several important parameters required for dynamic site response analysis, which this paper will address: 1) small-strain shear modulus G_{max}, 2) soil layering at any CPT probe location, 3) soil layer variation across a site, 4) soil types 5) site, areal or regional geotechnical property variation with elevation interval, 6) index of cyclic shear stress ratio to achieve at least 5% shear strain, and 7) water table depth. The CPT cannot provide estimates of the following: 1) soil unit weights, 2) modulus degradation with strain level, 3) damping or 4) permeability constants (unless a CPT piezocone is used).

---

[*]Research Civil Engineer, U.S. Army Waterways Experiment Station, Earthquake Engineering & Geophysics Division, Geotechnical Laboratory, P.O. Box 631, Attention GH, Vicksburg, MS 39180-0631

SITE CHARACTERIZATION USING THE CPT

The areal variation of soil properties within elevation intervals may be very important for soil characterization and the ultimate determination of the dynamic site response. The transmission of the earthquake motion can occur through stiffer soil pockets laterally to weaker deposits and over-strain these weaker soil pockets. The result at a non homogenous site may be too much apparent transmission through the weak pockets to the upper soil layers because the earthquake motion has been transmitted through the adjacent stiff zones.

Using borings to determine the site or areal variation can be expensive compared to CPT probing. While borings can provide soil samples for elaborate laboratory tests, in-situ variability of subsurface soils can overshadow the sophistication of the tests. Standard Penetration Test (SPT) measurements address this problem by sampling more soil to determine the site-based blow count. Unfortunately, the SPT can only provide rough indexes of any geotechnical property. Research toward using the CPT to predict the SPT blow count has also shown that the same SPT blow count, in sand, can be achieved by varying in-situ conditions such as overconsolidation, sensitivity, relative density and grain size distribution (Schmertmann, 1979, Olsen and Malone, 1988). Therefore, in-situ stratigraphy and factors controlling geotechnical properties are more complex than can be determined by simple in-situ tests such as the SPT.

The CPT is an excellent tool for stratigraphy evaluation. The CPT provides two strength measurement indexes, with depth, which can be used to estimate continuous profiles of geotechnical properties. The CPT also has a per-foot unit cost approximately 4 to 5 time less than the total cost for borings when soil sampling, laboratory index testing, and drafting of the boring logs are included (Olsen & Farr, 1986). The CPT's main virtue continues to be site characterization. CPT probes can and should be performed in strings across the site and at a close spacing. The CPT can then be analyzed and the results plotted. These CPT plots can be arranged in cross sections for stratigraphy evaluation and to study the geotechnical property variation across a site. Only then should boring locations be planned (and located directly next to CPTs), and soil sample depths assigned (Olsen and Farr, 1986).

METHODOLOGY FOR DETERMINING CPT BASED ESTIMATES

The approach used in this paper for developing the CPT based estimates are based on the use of : 1) site-specific correlations and 2) established empirical correlations. Site-specific correlations may be used, for example, for CPT prediction of: SPT blow count based on correlations of SPT measurement with nearby CPTs, percent passing the #200 sieve from correlations of CPTs with nearby soil samples, etc. The use of established empirical correlations refers to relationships such as SPT blow count to liquefaction potential, relative density to shear modulus relationships, relative density from the CPT q_c based on laboratory large diameter chamber tests, etc.

The techniques presented in this paper for estimating the cyclic shear stress to achieve 5% shear strain and the maximum shear modulus are based on the following methodologies;

1) Use of the CPT Soil Characterization chart which relates CPT corrected cone resistance, q_c, and corrected friction ratio, FR_1, to predict soil types and geotechnical properties.

2) Estimation of the SPT blow count, N, using the CPT q_{c1} and FR_1 correlated to field measurement observation (Olsen and Malone, 1988).

3) Estimation of an equivalent clean sand (i.e. silt-corrected) SPT blow count, N_{1c}, based on CPT predicted SPT blow count, N, and on the CPT determined soil characterization type (i.e. soil classification).

4) Use of the SPT liquefaction technique (Seed, et.al., 1985) to convert the CPT estimated N_{1c} to a corrected cyclic shear stress ratio to achieve at least 5% strain potential $((\tau_{av}/\bar{\sigma}_v)_{5\%})_1$

5) Use of correlations between corrected CPT measurements and normalized laboratory cyclic tests from soil samples taken from nearby borings (Olsen, 1984) to determine the $((\tau_{av}/ \bar{\sigma}_v)_{5\%})_1$ for material with sand through clay classifications.

6) Use of site-specific and established empirical correlations to develop trends of corrected maximum shear modulus, $(G_{max})_1$.

THE CPT TEST

The typical electric CPT test in the U.S. consists of a 1.4 inch (3.57 cm) diameter electrical CPT probe pushed into the earth at 2 cm/second (3.9 feet/minute) using 1 meter (3.28 feet) segment rods against the reaction of a 20-ton truck. Generally, two measurements are recorded: cone resistance, q_c, which is an end bearing stress, and the sleeve friction resistance, f_s, which is a localized, large-strain index of shear strength. Both measurements are usually reported in terms of tons per square foot (tsf) (one tsf equals 0.977 kg/cm^2).

Strength and stratigraphy evaluation require quality CPT data, which can only be achieved with: the correct electrical cone design type, electrical equipment quality assurance, up-to-date calibrations, ability to measure electronic zero drift, and experienced field technicians. Depending on the calibration quality, different cone designs can yield variations in the accuracy of friction sleeve resistance measurement. For sites with sensitive clay and/or metastable sand potential, CPTs should be performed using either a tension or compression cone design rather than a subtracting cone design (Olsen and Farr, 1986).

CPT DATA NORMALIZATION

In order to use the CPT for determining the variability of geotechnical properties within soil layers, CPT data must be initially normalized (i.e. corrected) with respect to vertical stress before being converted to indices of soil classification and strength. The most widely used technique for normalization is to convert the CPT data to an equivalent value at a standard vertical effective stress of

1 tsf (0.977 kg/cm^2). The cone resistance, q_c, can be converted to a corrected value, q_{c1}, for representation at a vertical effective stress of 1 tsf (0.977 kg/cm^2) using either a C_p factor or an exponent technique. The C_p factor is similar in concept to the C_n factor used to correct the Standard Penetration Test (SPT) blow count data (see Marcuson and Bieganousky, 1977). The C_p factor was originally developed based on large scale laboratory chamber tests using fine sands (Schmertmann, 1978). The C_p factor, which can be calculated as shown by equation (1) (Olsen, 1984), was developed for all soil types and is based primarily on large-scale laboratory chamber tests (see Olsen and Malone, 1988).

$$q_{c1} = q_c \, C_p = q_c \, \frac{1}{(\bar{\sigma}_v)^n} \tag{1}$$

$\bar{\sigma}_v$ = vertical effective stress in tsf (1 tsf = 0.977 kg/cm^2)

For normally consolidated conditions, the exponent n value ranges from approximately 0.56 for coarse sands to slightly under 1 for clays and 0.75 to 0.9 for soil mixtures and silts. In sand, n accounts for the effective bearing pressure bulb in front of the cone which increases volumetrically at less than a linear rate with increasing vertical effective stress.

The CPT sleeve friction, f_s, is an index of medium- to high-strain shear strength after the soil has been remolded during passage around the cone tip. For clean sands, void ratio will generally decrease during the soil passage around the cone tip to the sleeve, creating a denser configuration with a resulting measured f_s which is higher than the in-situ drained strength.

If a soil mixture has a sensitive sand/silt grain structure and a significant clayey fraction, the clay matrix may be underconsolidated because initial static overburden stresses are transferred through the silt/sand grains. During the passage around the cone tip, the soil is remolded with little pore pressure dissipation because the matrix is clay. The total void ratio change will be negligible because it is basically undrained. The partially remolded strength of the clay matrix will dominate, with the resulting f_s much lower than the in-situ undrained grain to grain strength. The f_s can be an index of the near remolded high-strain undrained strength because the grain contacts are broken during the cone passage and the clay matrix dominates the strength measurement.

Water table depth determination with the CPT can best be done using the piezo-cone in sand layers. But, the low strength profile between a desiccated zone and a normally consolidated layer can be identified using the profile of CPT f_s and will generally be lower than the present water table because it represents the lowest water table-affected strength.

Sleeve friction, f_s should be converted to a normalized value, f_{s1} as shown in equation 2. For normally consolidated, saturated clay, the f_s will range between 1/4 and 1/2 the difference between the

consolidated undrained remolded (S_{ur}) and peak (S_u) shear strengths
(Douglas, et al, 1985) as measured by the consolidated undrained
triaxial test. But if a clay is partially saturated, of normal
sensitivity, and medium stiff to stiff, f_s should be close to the in-
situ undrained peak strength (S_u) .

$$f_{s1} = \frac{f_s}{\bar{\sigma}_v} \tag{2}$$

The CPT friction ratio, FR, as shown by equation (3) has been used
as a CPT normalizing parameter for the last 20 years because q_c and f_s
are both to some degree dependent on vertical effective stress.
However, the in-situ vertical and horizontal effective stresses affect
q_c and f_s differently because the mechanism of failure different. As
a result, the friction ratio is only useful for normally consolidated,
relatively non-sensitive soils near a vertical effective stress of 1
tsf (0.977 kg/cm^2). A better approach is to use a corrected friction
ratio, FR_1, based on corrected CPT measurements as shown in equation
(4), based on q_{c1} and f_{s1} from equations (1) and (2).

$$FR = \frac{f_s}{q_c} \, 100 \tag{3}$$

$$FR_1 = \frac{f_{s1}}{q_{c1}} \, 100 = \frac{f_s}{q_c} \, \frac{1}{(\bar{\sigma}_v)^{(1-n)}} \, 100 \tag{4}$$

CPT SOIL CLASSIFICATION

The most common soil classification system in the U.S. is the
Unified Soil Classification System (USCS). The USCS classifies soil
into distinct groups (i.e. SM, CL, ML, et.al.,) based on remolded soil
conditions rather than in-situ conditions. Slight variation of soil
indices can change the USCS group (Douglas and Olsen, 1981). A CPT
soil classification system is based on strength behavior because it is
determined from in-situ strength measurements. The CPT can predict a
continuous range of strength based soil classifications from clay to
sand/gravel.

The author has previously published CPT soil characterization
(i.e. soil classification) charts which were based on q_{c1} and f_{s1}
rather than the standardized q_c and friction ratio (FR). The FR was
not utilized because it is only valid for a standard vertical
effective stress of 1 tsf (0.977 kg/cm^2) and useful only for soil
classification, yet, it has become an established parameter in the CPT
industry and cannot be ignored. Therefore, the CPT Soil
Characterization chart was replotted as shown in figure 1 (Olsen and
Malone, 1988), and is now based on q_{c1} and corrected friction ratio,
FR_1. Note that the horizontal corrected friction ratio axis, FR_1, is
plotted as a log scale. This places greater emphasis on corrected
friction ratios below 2 percent. The chart can be used to identify
unstable soil structures and over-consolidation. The steepness of the

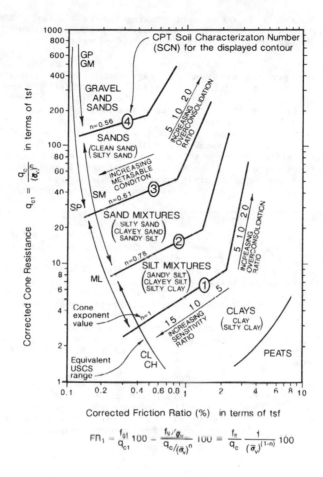

Corrected Friction Ratio (%) in terms of tsf

$$FR_1 = \frac{f_{s1}}{q_{c1}} 100 = \frac{f_s/\bar{\sigma}_v}{q_c/(\bar{\sigma}_v)^n} 100 = \frac{f_s}{q_c} \frac{1}{(\bar{\sigma}_v)^{(1-n)}} 100$$

FIG. 1.— CPT Soil Characterization Chart

CPT soil characterization lines in the over-consolidation zone results from using a normally consolidated mean stress to determine the corrected cone resistance.

The major CPT soil characterization lines (i.e. Soil Classification Numbers of 1, 2, 3 and 4) represent strength based estimates of soil classification. The Soil Classification Number (SCN) 3 is between the sand (e.g. SP to SM) and sand mixture (e.g. SM to ML) zones and has an equivalent fines content (i.e. percent passing the #200 sieve) of 0 to 10 percent. SCN 2 is between the sand mixture (e.g. SM to ML) and silt mixture (e.g. ML to CL) zones and has an equivalent fines content of 40 to 60 percent. The USCS boundary between a ML and SM is 50 percent passing the #200 sieve. SCN 1 is between silt mixtures (ML to CL) and clay (e.g. CL to CH) and has an equivalent content of approximately 80 to 100 percent.

Determining q_{c1}, FR_1 and soil type (e.g. SCN) with this chart is an iterative process because the exponent n value required to determine q_{c1} is a function of the soil characterization type. The f_{s1} is not dependent on the exponent n value. Initially, the lowest exponent n value (e.g. 0.56) is assumed and used to calculate q_{c1} and FR_1, which can be correlated to a Soil Classification Number (SCN) using figure 1. The exponent n value associated with that SCN will be higher than the assumed exponent n value used to calculate q_{c1} and FR_1. The normalization computations are repeated using a new exponent n value between the assumed and chart determined SCN exponent n value until it converges satisfactorily, say to within a tolerance of 0.02. Generally, fewer than 5 iterations are required for convergence.

CPT PREDICTION OF SPT BLOW COUNT

Accurate prediction of the SPT blow count (N) using CPT data is a requirement toward formulation of the techniques for 1) estimating the cyclic shear stress ratio to achieve at least 5% shear strain potential and 2) maximum shear modulus. The SPT sampler is resisted during penetration partially by side friction and inertia forces but primarily by end bearing stress. The relative contribution of these three components are dependent on soil type, over-consolidation effects, and strength consistency. The CPT is an ideal tool for predicting the SPT blow count because the CPT measures forces which are similar to those that resist the SPT penetration. The SPT parameter is a common geotechnical parameter that all geotechnical engineers have used.

The corrected SPT blow count, N_1, as shown in equation 5, can be predicted using corrected CPT parameters as shown in Figure 2 (see Olsen and Malone, 1988). It will be assumed for this paper that SPT blow count is for 60% energy efficiency and the conventional 60 subscript will not be shown to simplify the descriptors.

$$N_1 = N\, C_n = N\, \frac{1}{(\bar{\sigma}_v)^m} \tag{5}$$

with m = 0.55   loose sands (relative densities less than 55%)
     0.45   dense sands (relative densities greater than 70%)

Historically, q_c/N ratios of 4 to 5 were used to predict the SPT blow count. However, by excluding the SPT side friction effects (i.e. CPT sleeve friction), the SPT blow count predictions are only approximate. Ratios of q_{c1}/N_1 were calculated based on the predicted N_1 and the q_{c1} vertical axis from figure 2 as shown for illustration also in Figure 2. The q_{c1}/N_1 was used in lieu of q_c/N because the two parameters can be more meaningfully compared if both are properly normalized. Because the q_{c1}/N_1 contours are not parallel to the SCN contours in figure 2, using a relationship of q_c/N ratio to grain size index (e.g. q_c/N to D_{50}) may not be not valid. Also, note that for any q_{c1} level, the predicted N_1 and q_{c1}/N_1 ratio can vary over a wide

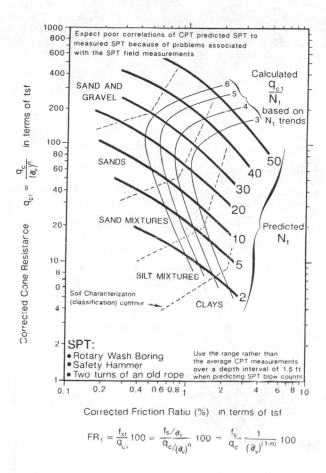

FIG. 2.— CPT procedure for estimating corrected SPT blow count (N_1)

range. For sands having a typical q_{c1} of 50 to 120 tsf (48 to 117 kg/cm^2), the q_{c1}/N_1 ratio from Figure 2 can range between 2 and 8. But, if the sand is medium grain size, generally uniform, medium dense, normally consolidated, and non-metastable the q_{c1}/N_1 ratio range decreases to between 4 and 5 which is also the range from the simplified q_c/N to D_{50} relationship.

CPT PREDICTION OF EQUIVALENT CLEAN SAND SPT BLOW COUNT

The prediction of an equivalent clean sand (e.g. silt co
SPT blow count is important because it is required for ind
the SPT liquefaction technique to determine the corrected

stress to achieve at least 5% shear strain potential $((\tau_{av}/\bar{\sigma}_v)_{5\%})_1$. The equivalent clean sand blow count (N_{1c}) can be determined as shown by equation 6 using the silt-correction blow count, ΔN_{Liq} determined from figure 3, and based on the fines content (i.e. percent passing the #200 sieve).

$$N_{1c} = N_1 + \Delta N_{Liq} \tag{6}$$

The c subscript designates equivalent clean sand. The SPT silt-corrected blow count (ΔN_{Liq}) shown in figure 3 was developed directly from the SPT liquefaction chart (Seed, et.al., 1985). While the SPT silt-correction blow count, ΔN_{Liq}, is based on a mechanical grain size index such as the percent passing the #200 sieve, a better alternative would be a strength based soil classification index such as the CPT Soil Characterization Number (SCN). The CPT SCN contours from the CPT soil characterization chart (figure 1) can be assigned equivalent fines contents as previously discussed with corresponding

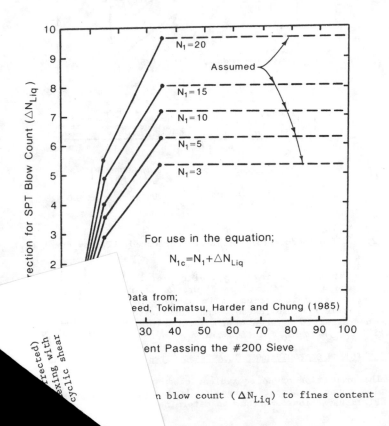

For use in the equation;

$N_{1c} = N_1 + \Delta N_{Liq}$

Data from;
Seed, Tokimatsu, Harder and Chung (1985)

ent Passing the #200 Sieve

n blow count (ΔN_{Liq}) to fines content

silt-correction SPT blow counts (ΔN_{Liq}) also assigned (from figure 3) as shown in figure 4. An equivalent clean sand SPT blow count, N_{1c}, of 18 will be used as an illustrative example for developing the CPT procedure for determining N_{1c} trends. This N_{1c} of 18 also relates to a silt-correction blow count, ΔN_{Liq}, of 5 for 15% fines content and 9 for 35% fines content (from figure 3). Therefore, a N_{1c} of 18 will correspond at 15 percent fines, to a N_1 of 13 plus a ΔN_{Liq} of 5 and at 35 percent fines, to a N_1 of 9 plus a ΔN_{Liq} of 9 as shown in figure 4. As also shown in figure 4, a N_{1c} of 18 will start to bend away from the N_1 contour of 18 at a SCN contour of 3.2 which has an estimated fines content of 5%, then cross the N_1 contour of 13 at the SCN contour of 2.6 which has an estimated fines content of 15% and then finally cross the N_1 contour of 9 at the SCN contour of 2.3 which has an estimated fines content of 35%.

Developing the CPT procedure for estimating trends of cyclic shear stress ratio to achieve at least 5% shear strain, $((\tau_{av}/\bar{\sigma}_v)_{5\%})_1$

The technique for determining the relationship of corrected cyclic shear stress ratio to achieve at least 5% shear strain potential, $((\tau_{av}/\bar{\sigma}_v)_{5\%})_1$, to CPT data is based on two components: 1) using the SPT liquefaction technique together with the CPT predicted N_{1c} blow count for sands and 2) correlations of cyclic laboratory tests correlated to nearby CPT probes for all soil classifications. For loose to medium dense, clean sand, a corrected cyclic shear stress ratio to achieve at least 5% shear strain potential, $((\tau_{av}/\bar{\sigma}_v)_{5\%})_1$, will generally correspond to the classic definition of liquefaction (Seed, et.al., 1985). While a 5% shear strain level is an arbitrary value, it is a means of assigning a normalized, high strain shear stress to all soil types. The site and depth specific ($\tau_{av}/\bar{\sigma}_v)_{5\%}$ can be calculated from the corrected $((\tau_{av}/\bar{\sigma}_v)_{5\%})_1$ which is in terms of a vertical effective stress of one tsf (0.977 kg/cm^2) and 15 equivalent shear stress cycles using the factors described by Seed et.al. (1985).

Cyclic laboratory tests (triaxial and simple shear) performed on soil samples taken from borings were correlated to results from nearby CPT probes. Corrected laboratory cyclic shear stress ratio values, $((\tau_{av}/\bar{\sigma}_v)_{5\%})_1$, were calculated using cyclic laboratory data based on the following criterion: use of standard Mohr envelope concepts, cyclic laboratory factors such as c_{rb} using an equivalent effective vertical stress of 1 tsf (0.977 kg/cm^2) acting on the failure plane, equivalent 15 shear stress cycles and using a failure criterion of 5% peak to peak axial strain. The elevation interval of the soil samples used for each set of laboratory cyclic tests were compared to the nearby CPT probe to determine the corresponding range of corrected CPT measurements. This range of q_{c1} and FR_1 parameters were then drawn as circular zones on the CPT soil characterization chart with the corresponding $((\tau_{av}/ \bar{\sigma}_v)_{5\%})_1$ displayed inside each circle as shown in figure 4.

The previously used example N_{1c} of 18 corresponds to a $((\tau_{av}/\bar{\sigma}_v)_{5\%})_1$ of 0.2 (Seed, et. at., 1985) and can now be extended down from the fines content of 35% to the higher than normal soil sensitivity clay zone based on the cyclic laboratory data as shown in

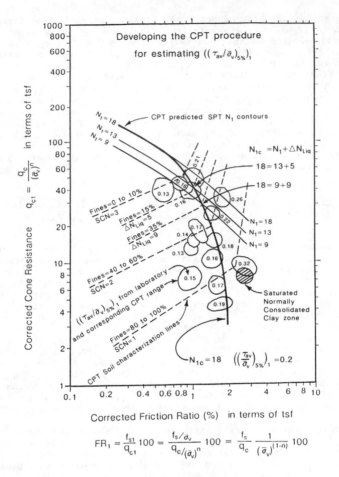

$$FR_1 = \frac{f_{s1}}{q_{c1}} 100 = \frac{f_s/\bar{\sigma}_v}{q_c/(\bar{\sigma}_v)^n} 100 = \frac{f_s}{q_c} \frac{1}{(\bar{\sigma}_v)^{(1-n)}} 100$$

FIG. 4.— Developing the CPT procedure for estimating $((\tau_{av}/\bar{\sigma}_v)_{5\%})_1$

figure 4. Note the general agreement of both techniques for a $((\tau_{av}/\bar{\sigma}_v)_{5\%})_1$ of 0.2 within the soil classifications of sand to silty sand.

The final $((\tau_{av}/\bar{\sigma}_v)_{5\%})_1$ trends are shown in figure 5 for values between 0.1 and 0.3. If a simplified or equivalent linear site response analysis indicates that earthquake induced shear stresses are exceeding this threshold high strain cyclic shear stress, a more complex analytical model is probably required.

The point on the Soil Characterization chart where normally consolidated clay is generally found (see figures 1 and 4) corresponds

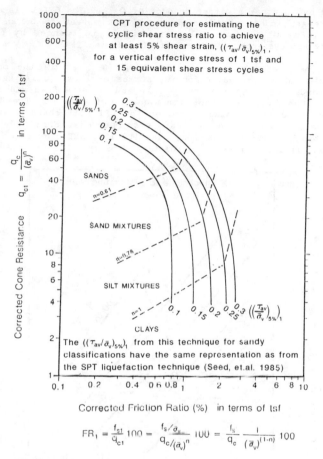

$$FR_1 = \frac{f_{s1}}{q_{c1}} 100 = \frac{f_s/\bar{\sigma}_v}{q_c/(\bar{\sigma}_v)^n} 100 = \frac{f_s}{q_c} \frac{1}{(\bar{\sigma}_v)^{(1-n)}} 100$$

FIG. 5.— Final CPT procedure for estimating $((\tau_{av}/\bar{\sigma}_v)_{5\%})_1$

to a cyclic stress ratio range of 0.25 to 0.3 (figure 5). This level of cyclic induced shear stress corresponds to approximately 80 to 100 percent of the static undrained peak strength for clay. Therefore, when the induced cyclic shear stresses are near the static peak strength for normally consolidated clays, this technique is indicating that large strains (i.e. 5% peak strain) may be possible. Theirs and Seed (1968) indicate the same general conclusions.

CPT ESTIMATION OF THE MAXIMUM SHEAR MODULUS, $(G_{max})_1$

The CPT procedure for estimation of the maximum shear modulus, G_{max}, is a difficult process because modulus is a low strain parameter. The CPT q_c is basically a high strain measurement which is influenced by, but not a direct product of G_{max}. The CPT f_s is a high strain strength index and therefore would not appear to be an index for estimating G_{max}, but, G_{max} is highly influenced by fabric sensitivity and over-consolidation effects, as is f_s.

The normalized maximum shear modulus, $(G_{max})_1$, is the maximum shear modulus at a vertical effective stress of 1 tsf (0.977 kg/cm^2). The CPT trends for prediction of $(G_{max})_1$ and K_2 (see Seed and Idriss (1970) for the K_2 definition) are shown in figure 6. Trends for $(G_{max})_1$ and K_2 can be characterized on the CPT soil characterization chart because the following can also be characterized: soil classification, overconsolidation, sensitivity, SPT blow count, the normal consolidation zone, and relative density. The actual CPT

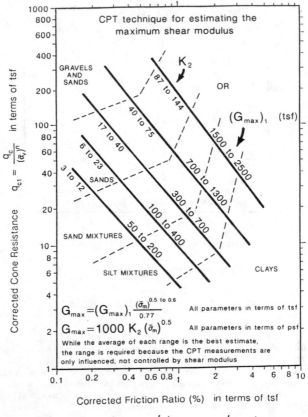

$$FR_1 = \frac{f_{s1}}{q_{c1}} \, 100 = \frac{f_s / \bar{\sigma}_v}{q_c / (\bar{\sigma}_v)^n} \, 100 = \frac{f_s}{q_c} \, \frac{1}{(\bar{\sigma}_v)^{(1-n)}} \, 100$$

FIG. 6.— CPT procedure for estimating maximum shear modulus, $(G_{max})_1$

estimated trends of $(G_{max})_1$ and K_2 in figure 6 were developed using the following relationships: 1) correlations of G_{max} to SPT blow count for alluvial and diluvial deposits (Imai and Tonouchi, 1982), 2) correlations of relative density to $(G_{max})_1$ (Seed and Idriss, 1970), 3) CPT prediction of relative density (Schmertmann, 1979 and Tokimatsu and Seed, 1987), 4) CPT prediction of SPT blow count and overconsolidation ratio (Olsen and Malone, 1988) and 5) observational shear wave velocity data to CPT data (Douglas and Strutynsky, 1984 and Douglas et al, 1985). The range, rather than the average, of the estimates $(G_{max})_1$ or K_2 from figure 6 should be used because as previously stated, the CPT measurements are only influenced, not controlled by shear modulus.

The $(G_{max})_1$ trends on the CPT soil characterization chart (figure 6) indicates that G_{max} is more proportional to f_{s1} than q_{c1}. The f_{s1} contours can be constructed on the CPT soil characterization chart and would be at a one to one slope extending toward the bottom right. The CPT f_{s1} is primarily an index of sensitivity and over-consolidation effects. Also, for any level of over-consolidation or sensitivity in figure 6, the sand $(G_{max})_1$ is approximately 1.5 to 4 times higher than clay. Another interesting observation is that the $(G_{max})_1$ from figure 6 for over-consolidated clay is approximately the same as for medium dense normally consolidated non-sensitive sand. That confirms the general usage of the 75% relative density sand modulus relationship for clays. The CPT estimation of $(G_{max})_1$ appears to be dependent in order of importance, on; overconsolidation, sensitivity, strength, and soil classification.

Shear modulus degradation can be calculated from the maximum shear modulus (figure 6) to the calculated modulus at 5% strain (figure 5) and termed percent modulus degradation. This calculated percent modulus degradation for sand from these figures appears to ranges between 0.4 and 0.6 percent. For normally consolidated clay, the percent modulus degradation is just over one percent, for higher than normal clay sensitivity it increases to 2 or 3 percent and for highly over-consolidated clay, it ranges between 0.6 and 0.8 percent. These percent modulus degradations are approximately the same level as from Seed and Idriss (1970).

CONCLUSIONS

This paper has described the components and methodology for CPT estimation of cyclic shear stress to achieve 5% shear strain, and maximum shear modulus. The CPT techniques for estimating: 1) soil classification is shown in figure 1, 2) SPT corrected blow count, N_1, is shown in figure 2, 3) corrected cyclic shear stress to achieve at least 5% shear strain potential, $((\tau_{av}/\bar{\sigma}_v)5\%)_1$, is shown in figure 5, and 4) maximum shear modulus, $(G_{max})_1$, is shown in figure 6. Because these techniques are relatively new with limited field correlations, some level of effort should be expended toward SPT blow count determination, undisturbed soil sampling for laboratory testing and cross/down hole shear wave velocity measurements. The proper planning of a CPT/boring field program is a critical ingredient toward minimizing cost and maximizing the return of the CPT to boring correlations. See the paper by Olsen and Farr (1986) concerning planning a CPT/boring field program. The main virtue of the CPT is still stratigraphy evaluation.

ACKNOWLEDGMENTS

I gratefully acknowledge the project support of the Vicksburg and Nashville Districts of the U. S. Army Corps of Engineers and the reviewer contributions of John Potter, Irving E. Olsen, Paul Bluhm and Paul Hadala. Permission was granted by the Chief of Engineers to publish this information. The views of the author to not purport to reflect the view of the Corps of Engineers or the Department of Defense.

APPENDIX — REFERENCES

1. Douglas, B. J., and Strutynsky, A.I., Mahar, L.J. and Weaver, J. (1985) "Soil strength determination from the Cone Penetrometer Test", Arctic 85, ASCE, San Francisco, March 1985
2. Douglas, B. J., and Strutynsky, A.I. (1984) "Cone Penetrometer Test, pore pressure measurements and SPT hammer energy calibration for liquefaction hazard assessment", report to USGS, Aug. 1984
3. Douglas, B. J., and Olsen, R. S. (1981) "Soil Classification Using the Electric Cone Penetrometer", Proc. Cone Penetration Testing and Experience, ASCE, St. Louis, MI
4. Imai, T., and Tonouchi, K. (1982) "Correlation of N value with S-wave velocity", Proceedings of the Second European Symposium on Penetration Testing, Amsterdam, May 1982.
5. Marcuson, W. F., and Bieganousky, W. A. (1977) "SPT and Relative Density in Coarse Sands", Journal of Geotechnical Engineering, ASCE, Vol.103, No. GT11
6. Olsen, R. S. and Malone, P. G. (1988) "Soil Classification and Site Characterization using the CPT", First International Symposium on Penetration Testing (ISOPT-1), Florida, March 1988.
7. Olsen, R. S. and Farr, J. V. (1986) "Site Characterization Using the Cone Penetrometer Test", INSITU 86, ASCE, Blacksburg, VA.
8. Olsen, R. S. (1984) "Liquefaction Analysis using the Cone Penetrometer Test (CPT)", Eighth World Conference on Earthquake Engineering, San Francisco, California.
9. Schmertmann, J. H. (1979) "Statics of the SPT", Journal of Geotechnical Engineering, ASCE, Vol.105, No. GT5.
10. Schmertmann, J. H. (1978) "Study of Feasibility of Using Wissa-Type Piezometer Probe to Identify liquefaction Potential of Saturated Fine Sands", U.S. Army Waterways Experiment Station, Technical report S-76-2.
11. Seed, H.B., Tokimatsu, K., Harder, L.F., and Chung, R. M. (1985) "Influence of SPT procedures in Soil Liquefaction Resistance Evaluation," Journal of Geotechnical Engineering, ASCE, Dec. 1985.
12. Seed, H.B. and Idriss, I.M. (1970) "Soil moduli and damping factors for dynamic response analysis", Earthquake Engineering Research Center, Report EERC 70-10, Dec. 1970.
13. Tokimatsu, K. and Seed, H.B. (1987) "Evaluation of settlement in sands due to earthquake shaking", Journal of Geotechnical Engineering, ASCE, Vol. 113, No. 8, Aug. 1987
14. Thiers, G. R. and Seed, H. B. (1969) "Strength and stress-strain characteristics of clays subjected to seismic loading conditions, in vibration effects of earthquakes on soils and foundations", ASTM 71st annual meeting, Publication 450, San Francisco, 1968.

CORRELATIONS BETWEEN DYNAMIC SHEAR RESISTANCE
AND STANDARD PENETRATION RESISTANCE IN SOILS

by

David W. Sykora*, Associate Member, ASCE and
Joseph P. Koester*, Associate Member, ASCE

ABSTRACT: Dynamic shear resistance, as indicated by either shear modulus or shear wave velocity, is a prerequisite parameter for the dynamic
analysis of earthen structures, foundations for superstructures, and
free-field seismic response. Dynamic shear resistance parameters are
expensive to determine in situ and in the laboratory.

Numerous researchers and practitioners have examined the viability of
correlations between dynamic soil properties and basic, more common
engineering parameters. These correlations appear to have evolved
because of the expense of active measurement to augment (in some cases,
replace) designated testing. Later studies seem to capitalize on a
rapidly expanding data base of measured values that was nonexistent
even a decade ago.

This study presents, discusses, and compares correlations involving
shear modulus or shear wave velocity and Standard Penetration
Test (SPT) N-value to date in the United States and Japan. The objective of this presentation is to provide the reader with a comprehensive understanding of correlations in that they may appreciate their evolution and use the technology appropriately. Specific recommendations
are made to assist the reader in applying correlations to geotechnical
engineering problems.

INTRODUCTION

The dynamic response of a soil mass subjected to excitation is the
focus of much attention among engineers both in research studies and in
application of state-of-the-art technology to practical problems. A
key property necessary to properly evaluate dynamic response of soil
both quantitatively and qualitatively, is dynamic shear modulus (modulus of rigidity), G.
    Values of G are determined either by measurement in the labora-
tory on "undisturbed" soil samples or by calculations involving assumptions of linear elasticity using shear wave velocity, V_s, measured
in situ, and the mass density of the soil. Mass density, ρ, may be
determined using "undisturbed" soil samples or in situ density tests.
Shear modulus measured at small shear strain (less than 0.001 percent),

---

*   Research Civil Engineer, U.S. Army Engineer Waterways Experiment
    Station, P.O. Box 631, Vicksburg, MS 39180.

referred to as G_{max} , ultimately is the desired initial design parameter.

Investigators have been attempting to develop correlations between G_{max} and V_s and various geotechnical parameters for at least two decades. These correlations have evolved from measurements made in both the field and laboratory, although the accuracy and applicability of such correlations developed in these two environments differ considerably. Under controlled laboratory conditions, precise and detailed analyses of factors affecting G_{max} and V_s have been performed. Laboratory studies have been very useful in determining soil properties and test conditions upon which G_{max} and V_s are most dependent. However, laboratory-prepared samples which offer consistency to the investigator cannot be conditioned completely to simulate age and cementation effects which occur over tens of thousands of years in situ. Conversely, field correlations involving V_s are crude due to considerable scatter of the data and the limited availability of the exact parameters known to affect V_s . Field correlations to date have proven to be functional only to a limited extent in geotechnical engineering practice.

Myriad correlations presently exist in the literature and are available to the practitioner. Unfortunately, review of one or even a couple of articles may not adequately reflect the variability and individuality of each particular study. Furthermore, many studies were written and exist in Japanese and must be translated into English.

The intention of this review is to communicate important findings that have surfaced throughout the past 25 years in a manner that allows the reader to feel comfortable in implementing correlations to estimate V_s or G_{max} . Specifically, correlations used to estimate V_s or G_{max} from values of Standard Penetration Test (SPT) N-value are considered. The primary bases for comparison are best-fit relations and correlation coefficients reported in the studies. Specific recommendations for effective use of correlations are provided at the conclusion of the report. An extensive examination of existing correlations to predict V_s and G_{max} , including laboratory studies, is reported by Sykora (1987).

INITIAL STUDIES

The first few studies performed to determine methods of estimating V_s appear to be well conceived but indirect. Most initial studies conclude with a relationship between SPT N-value and V_s that was derived from theory or laboratory measurements as opposed to a field-derived data base. Consideration of these studies is deemed important to understand and appreciate the evolution of V_s correlations.

Correlative relations proposed by some initial studies are compiled in Table 1. These studies, in general, represent rather crude analyses involving very few, if any, data. No details were provided with regard to SPT equipment or procedures used. For this reason, these studies should only be considered useful to appreciate the

Table 1.

Study	Application	Proposed Relation(s)
Kanai (1966)		$V_s = 62\ N^{0.6}$ (fps*)
Yoshikawa (date unknown)		$V_s = 127(N + 1)^{0.5}$ to $178(N + 3)^{0.5}$
Sakai (1968)		$V_s = 49\ N^{0.5}$ to $110\ N^{0.5}$ (fps)
Shibata (1970)	Sands	$V_s = 104\ N^{0.5}$ (fps)
Ohba and Toriuma (1970)	Alluvial soils	$V_s = 280\ N^{0.31}$ (fps)

* SI units conversions are given in Appendix I.

evolution process of shear wave velocity correlations and the potential range of values produced from relations.

RECENT STUDIES

Numerous correlative studies have been conducted to directly examine a relationship between SPT N-value and V_s . Most of these studies were performed in the 1970's in Japan. Since then, a few similar studies have been reported in the United States.

It is apparent that empirical equations resulting from the various studies were not intended to replace in situ measurements. These correlations would fall considerably short of the accuracy and consistency produced by in situ seismic measurements. Rather, these correlative studies were conducted with the hope that, in time, equations useful in supplementing in situ measurements could be developed.

Uncorrected N-value

Ohsaki and Iwasaki (1973) performed simple statistical regression analysis on over 200 sets of data accumulated from seismic explorations (using predominantly downhole techniques) throughout Japan. The authors were primarily concerned with determining a basic correlation between G and N , but they did analyze the effects of geologic age and soil type by considering subsets of data collected. Best-fit relations correlation coefficients were developed and comparisons were made by dividing the data into groups according to soil type and geologic age divisions. Regression analysis was used to fit a relation for all data at:

$$G = 124\ N^{0.78} \text{ (tsf)} \tag{1}$$

This best-fit relation corresponds to a reported correlation coefficient of 0.886.

Ohsaki and Iwasaki (1973) did not propose a relationship to predict V_s using N-value. Such a relation may be inferred, however, by assuming a representative value or range of unit weight for soil, γ . Typical values of γ in Japan were examined by Ohsaki (1962). This

study appears to be accepted in the Japanese literature. Average values of γ for sand, silt, and clay proposed by Ohsaki (1962) are
112.4, 100.9, and 93.2 pcf, respectively. The value of 112.4 pcf was
used for this study to produce an approximate relation between V_s and
N-value.

Ohta and Goto (1978a,b) used detailed statistical analyses on
nearly 300 sets of data from soils in Japan. These studies supersede
studies by Ohta et al. (1970) and Ohta and Goto (1976). Each data set
consisted of values of V_s, SPT N-value, depth, geologic age, and soil
type. The result of the analyses was the evolution of 15 different
equations, with varying correlation coefficients for predicting V_s.

In using this approach, variables and combinations of variables were
examined to determine their effect on V_s predictions and also to
determine which combinations of variables produced the most accurate
results (highest correlation coefficients). Initially, six divisions
of soil type were used (Ohta and Goto 1978a). Later, Ohta and Goto
(1978b) narrowed the soil divisions to three groups—clays, sands, and
gravels. This simplification produced only slightly lower correlation
coefficients for correlations involving soil-type divisions. Ohta and
Goto (1978a,b) also reported the only known studies which examined V_s
correlations with both N-value and depth in the same equation.

The best-fit relation proposed by Ohta and Goto corresponding to
all data is:

$$V_s = 280 \ N^{0.348} \ \text{(fps)} \tag{2}$$

This best-fit relation corresponds to a correlation coefficient of
0.719.

Marcuson, Ballard, and Cooper (1979) developed a site-specific
correlation between V_s and N for natural and fill materials at
Fort Peck Dam located near Glasgow, Montana. Simple linear relations
were selected and found to predict V_s within 25 percent of the measured value most of the time. These relations are considered applicable only for the specific conditions used.

Seed, Idriss, and Arango (1983) suggested using the following
equations for sands and silty sands:

$$G_{max} = 65 \ N \ \text{(tsf)} \tag{3}$$

and

$$V_s = 185 \ N^{0.5} \ \text{(fps)} \tag{4}$$

These equations were developed primarily for use in liquefaction analysis of sand deposits. It may be inferred from this study that these
relations apply to SPT N-values measured using a rope-and-cathead
system. No distinction of sampler inner diameter was made.

Imai has been involved in V_s correlations since at least 1970
when he published the results of his initial study (Imai and Yoshimura,
1970). Since then, he has coauthored three other papers (Imai and
Yoshimura, 1975; Imai, Fumoto, and Yokota, 1975; and Imai and Tonouchi,

1982) that address V_s correlations involving SPT N-value using a progressively larger data base of measurements. All data were collected using measurements made with a downhole borehole receiver at sites throughout Japan. It is presumed that later studies incorporated all data from previous studies.

In the first three studies, Imai and others found it difficult to distinguish the effect of soil type or geologic age on N versus V_s correlations. Nevertheless, differentiation among these data groups indicated that values of V_s tended to fall in specific ranges.

General relations were presented representing all soil types in each study.

In the fourth study, Imai and Tonouchi (1982) determined that correlations among different soil type and geologic groups were worthy of examination contrary to previous conclusions. Best-fit relations proposed indicate that division of data among both soil type and geologic age groups has a significant effect on the relation representative of the data and the corresponding correlation coefficient. Therefore, it appears as though soil type and geologic age should be used to estimate V_s. However, correlation coefficients for these subdivisions are lower than those of all data combined.

Correlations performed by Imai and Tonouchi (1982) using G_{max}, in general, are more accurate than V_s correlative equations for granular soil categories and less accurate for cohesive soil categories (based strictly on correlation coefficients). There is no apparent explanation for a dichotomy in accuracies between granular and cohesive soil groups.

The correlation coefficient for all data used by Imai and Tonouchi (1982) in N versus G_{max} correlations is essentially equal to that for N versus V_s correlations. This occurrence may seem trivial but, in actuality, it could be very significant. Shear modulus is usually the required end product. If V_s is used to calculate G_{max}, the value of V_s is squared. Any inherent error in the value of V_s consequently is squared resulting in a less-accurate value of G_{max}. If correlations between N and G_{max} are of the same accuracy, or even slightly less accurate, the G_{max} correlations should be used. This reasoning does not imply necessarily that correlations incorporating G_{max} are always to be preferred. Correlations in which G_{max} was calculated directly from V_s and only an estimated value of ρ are no better than using G_{max} estimated from correlations involving V_s. In other words, values of ρ must be measured also to justify using G_{max} correlations.

The more prominent soil categories presented by Imai and Tonouchi (1982) were used to quantify ranges in data and corresponding error between best-fit relations and the upper and lower bounds (in velocity). Three values of N were selected: 10, 30, and 100 blows/ft. The errors estimated appear to be consistent (independent of N) and average about +50 percent (best-fit V_s to upper bound) and -40 percent (best-fit V_s to lower bound).

Sykora and Stokoe (1983) examined the influence of: relative location of the phreatic surface, geologic age, soil type, previous seismic history, range of N-values, and site specificity on N versus V_s correlations. Analyses of data among groups indicated that only divisions among soil-type groups produced substantially different relationships with improved accuracy. Unfortunately, small quantities of data in a couple of geologic age and soil type data groups precluded making conclusive comments and using these different relationships with any degree of confidence. Site-specific correlations produced significantly different best-fit relations with varying magnitudes of correlation coefficient (0.45 to 0.86).

The general relation determined by Sykora and Stokoe (1983) using crosshole measurements of V_s made only in granular soils is:

$$V_s = 330 \ N^{0.29} \ (\text{fps}) \tag{5}$$

This best-fit relation corresponds to a correlation coefficient of 0.84.

Sykora and Stokoe (1983) compared values of V_s estimated using Equation 5 with the actual data collected to determine the range in maximum deviation for V_s per N-value. The deviation decreased (as a percentage) with increased magnitude of N-value. At an N-value of 20 blows/ft, the error was about plus or minus 55 percent. At an N-value of 125 blows/ft, the error was about plus or minus 32 percent.

The results of all aforementioned recent studies are summarized in Table 2. Studies superseded by work performed by the same investigator

Table 2. Comparison of Recent N-Value Versus V_s Field Correlations

Equation No.	Author(s)	Data Information	Soil Types Used	Reported Equation Shear Modulus, G, tsf	Reported Equation Shear Velocity, V_s, fps
1	Ohsaki and Iwasaki (1973)	200 sites in Japan; 220 sets of data	All	$G = 125 \ N^{0.78}$ (0.886)**	$V_s = 268 \ N^{0.39}$†
2	Ohsaki and Iwasaki (1973)	200 sites in Japan; 220 sets of data	Cohesionless	$G = 66.5 \ N^{0.94}$ (0.852)**	$V_s = 195 \ N^{0.47}$†
3	Ohta and Goto (1978a)	289 sets of data; Japanese soils	All	N.R.††	$V_s = 280 \ N^{0.341}$ (0.719)**
4	Ohta and Goto (1978b)	289 sets of data; Japanese soils	Sands	N.R.	$V_s = 290 \ N^{0.340}$
5	Ohta and Goto (1978b)	289 sets of data; Japanese soils	Gravels	N.R.	$V_s = 309 \ N^{0.340}$
6	Imai and Tonouchi (1982)	1,654 sets of data; Japanese soils	All	$G = 147 \ N^{0.68}$ (0.867)**	$V_s = 318 \ N^{0.314}$ (0.868)**
7	Seed, Idriss, and Arango (1983)	Unknown	Sands	$G = 65 \ N^{1.0}$	$V_s = 185 \ N^{0.5}$
8	Sykora and Stokoe (1983)	229 sets of crosshole data; throughout United States	Granular	N.R.	$V_s = 350 \ N^{0.27}$ (0.84)**

Note: * N = Standard Penetration Resistance N-value (blows/ft); not adjusted to a normalize energy efficiency.
       ** Regression correlation coefficient.
        † Assumed, not reported; $\gamma = 112.4$ pcf, typical for Japanese sands (Ohsaki 1962).
       †† N.R. = Not reported.

or results of site-specific studies are not tabulated. Correlative relations listed represent equations proposed by the respective authors and have not been normalized to a uniform SPT energy ratio.

Effective-Vertical-Stress-Corrected N-value

Few studies exist that examine correlations between effective-vertical-stress-corrected N-value N_1 (Marcuson and Bieganousky, 1977) and V_s. This may be due to the relatively recent acceptance of such correction factors. Consistent with the previous section, all N-values used in this section correspond to reported relations (i.e., not adjusted to normalize to a common SPT energy efficiency).

Seed, Idriss, and Arango (1983) proposed an equation to determine V_s from N_1. The equation to determine V_s from N_1 was reported to be conservative and applicable for sands and silty sands up to a maximum depth of 50 ft. This equation is:

$$V_s = 200 \ N_1^{0.5} \ \text{(fps)} \tag{6}$$

Seed and Idriss (1970), and later Seed et al. (1984) attempted to simplify the equation proposed by Hardin and Drnevich (1972). Seed and Idriss (1970) developed the equation:

$$G = 1000 \ K_2 \left(\bar{\sigma}_m\right)^{1/2} \ \text{(psf)} \tag{7}$$

where K_2 is a shear modulus coefficient. At low shear strain (less than 10^{-4} percent), K_2 is referred to as $\left(K_2\right)_{max}$ corresponding to G_{max}. Parametric studies indicated that $\left(K_2\right)_{max}$ was a function only of void ratio and typically ranged from 30 (loose sands: $e \approx 0.95$) to 75 (dense sands: $e \approx 0.35$). Select data measured in situ at six sites in the United States were used to substantiate this range (although values of $\left(K_2\right)_{max}$ of 166 and 119 for slightly cemented and clayey sands, respectively, from two sites were ignored).

Seed et al. (1984) used the results of laboratory tests on gravels to determine a range in $\left(K_2\right)_{max}$ of 80 to 180 for relatively dense, well-graded gravels. The results were in good agreement with in situ measurements made at four sites, two of which in Caracas, Venezuela.

Seed et al. (1984) proposed a relationship between N_1 and $\left(K_2\right)_{max}$ to calculate G_{max}. This relationship was initiated using a correlative equation involving N, depth, soil type, and geologic age proposed by Ohta and Goto (1976). The relation is:

$$G_{max} = 20,000 \ \left(N_1\right)^{1/3} \left(\bar{\sigma}_m\right)^{1/2} \ \text{psf} \tag{8}$$

and was substantiated by results of laboratory tests and a few field data.

Sykora and Stokoe (1983) used 229 sets of data measured using the crosshole method to correlate N_1 with V_s. They concluded that the use of N_1 to correlate with V_s proved to be considerably less accurate and more inconsistent than N versus V_s correlations. The correlation coefficient for the overall best-fit relation was 0.67 as compared to 0.84 for N versus V_s correlations. Using the various data groups, correlation coefficient for N versus V_s correlations averaged 32 percent less than N versus V_s correlations for the same data group. Sykora and Stokoe (1983) concluded that N_1 is not an appropriate correlative variable to use in estimating V_s. This conclusion can be rationalized since effective stress is known to influence both V_s and N. The normalization of N to $\overline{\sigma}_v$ eliminates an independent variable $(\overline{\sigma}_v)$ from one dependent variable (N) and not the other (V_s).

COMPARISON OF STUDIES

Many correlative relations to estimate V_s or G_{max} exist corresponding to different sets of conditions. These relations have been presented in Tables 1 and 2. A graphical comparison of these relations is desirable to more easily identify variability between studies. Prior to one-to-one comparisons, each study should be considered on as much of an equal basis as possible.

Careful scrutiny of SPT techniques and procedures have been made recently. A study by Seed et al. (1985) that compared energy efficiencies and techniques of typical Japanese SPT equipment and procedures with U.S. equipment and procedures indicates that a one-to-one correspondence of N-values between countries is imprecise. Given that the accuracy of geophysical techniques for measurement of dynamic properties (i.e., V_s) should be roughly equivalent between countries, comparisons between N versus V_s correlations from Japan and the U.S. must be put on an equivalent basis by adjusting N-values to account for differences. Relations reported to this point have not been adjusted to account for differences in energy. Not one study mentioned the SPT procedures and equipment used to collect data.

For graphical comparisons made hereafter, SPT N-values measured in Japan were assumed to correspond to an efficiency of 67 percent of free-fall energy (N_{67}) and were adjusted to standard efficiency of 60 percent (N_{60}) proposed by Seed et al. (1985). This assumes that all data were collected using the rope-and-cathead procedure. Measurements of N made in the U.S. were assumed to be N_{60} which is applicable to a safety hammer operated with a rope and cathead used on many different drill rigs (Seed et al. 1985). Further correction of SPT N-values to account for the variation of inner diameter of the split-spoon sample may be appropriate. Again, however, no mention of the size of samples used was reported in any study. For purposes of graphical comparisons made herein, no corrections were made.

Best-fit relations proposed for all soils are plotted in Figure 1. An appreciable amount of deviation is evident among relations, especially at large N-values. The relation proposed by Kanai (1966) is the most incongruous; therefore, it is highly suspect. The other four relations are grouped together with those proposed by Imai and Tonouchi (1982) and Ohta and Goto (1978a) representing approximate median relations.

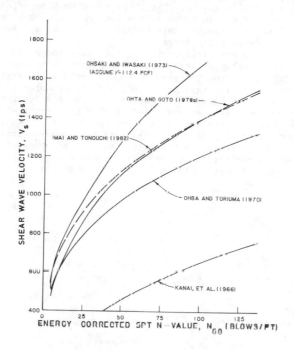

Figure 1. Comparison of results for N versus V_s
correlations (proposed by various studies for all
soils and geologic conditions)

The relation indicated by Ohsaki and Iwasaki (1973) between V_s and N has been inferred based on typical values of unit weight for sand in Japan as stated previously. The assumed value of unit weight could account for the large deviation of this relation among other relations. If a larger value of unit weight had been chosen, the relation would deviate less. In order for the relation to coincide with that of Ohta and Goto (1978a) at an N-value of 50 blows/ft, a unit weight of 152 pcf is required, which is unreasonable. Had a smaller unit weight been used, representative of typical Japanese silt or clay, the relation would deviate even more from the other relations. Therefore, the relation proposed by Ohsaki and Iwasaki (1973) represents an upper bound for reasonable values of unit weight.

Three of the studies examining N versus V_s correlations for
all soils are prominent for different reasons. Imai and Tonouchi
(1982) used a very large data base (1,654 sets of data). Ohta and Goto
(1978b) performed detailed statistical sensitivity analyses of various
factors thought to affect N versus V_s correlations. Ohsaki and
Iwasaki (1973) also paid close attention to various parameters and used
(limited) statistical analysis. The three relationships selected are
very similar for a range in N-value from 5 up to about 30 blows/ft;
beyond that value, the relationship proposed by Ohsaki and Iwasaki
(1973) begins to deviate considerably.
      A number of the correlative studies available and discussed in
this report were developed for granular soils only. The applicability
of best-fit relations from the studies to all soils is uncertain.
Because of this, best-fit relations developed for granular soils are
compared separately in Figure 2. This comparison includes two studies
from the U.S.: Seed, Idriss, and Arango (1983) and Sykora and Stokoe
(1983), both of which were assumed to correspond to N-values measured
with 60 percent energy efficiency. Best-fit relations presented in
Figure 2 are very different at N-values greater than about 25 blows/ft.
Note that the best-fit relation for gravel soils proposed by Ohta and
Goto (1978b) was developed using only eight data points.

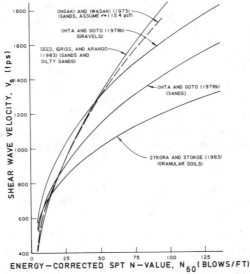

Figure 2. Comparison of results for N versus V_s correlations
        in granular soils (proposed by select studies)

Comparisons of N versus G_{max} correlations were also made since
G_{max} typically is the desired quantity for engineering analyses.
Selected studies which presented equations to estimate G_{max} were by

Seed, Idriss, and Arango (1983), Ohsaki and Iwasaki (1973), and Imai
and Tonouchi (1982). These relations are plotted in Figure 3

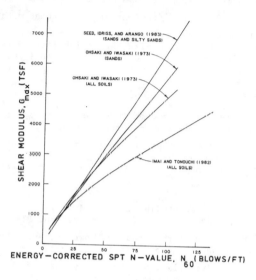

Figure 3. Comparison of results for N versus G_{max} correlations
(proposed by select studies)

corresponding to noted applicable soil types. The four best-fit relations are quite similar below N-values of 25 blows/ft. At N-values
greater than 25 blows/ft, the best-fit relation by Imai and Tonouchi
(1982) begins to diverge; whereas, the other three relations do not
diverge until N-values greater than 50 blows/ft. The relationship proposed by Imai and Tonouchi (1982) for all soils may deviate because of
the inclusion of a wide range of soil types used in the analyses. However, comparisons of equations, developed by Imai and Tonouchi (1982),
summarized in Table 2 for clays and sands of equivalent age do not substantiate this premise. In fact, the equation for alluvial sands plots
well below that for all soils.

Few studies specified an applicable range of N-values for proposed
relationship. Relations drawn in this report were only extended to an
N-value corresponding to the maximum of the data, if available, or any
specified limit.

Interesting similarities exist between the correlative relation
developed by Ohsaki and Iwasaki (1973) and plotted in Figure 3 (and
Figure 2) and a comparative study reported by Anderson, Espana, and
McLamore (1978). The relation by Ohsaki and Iwasaki was determined to
overpredict G_{max} determined at various depths at four different sites
by up to 25 percent. Another investigator (U.S. Army Engineer, 1986)
used correlations proposed by Ohsaki and Iwasaki (1973) and Seed,

Idriss, and Arango (1983) and measured values of V_s to estimate values of N_1 for use in liquefaction potential charts. Calculated values of N_1 were somewhat greater than measured values (i.e., overprediction of V_s). These findings are consistent with the relative location of the best-fit relations among the other proposed relationships. These findings jeopardize the reliability of relations proposed by Seed, Idriss, and Arango (1983) and Ohsaki and Iwasaki (1973).

Influence of Other Parameters

A few parameters have been used to enhance correlations between N and V_s. The two most widely used parameters are geologic age and soil type. The influence of these two parameters as reported in the various studies is discussed herein.

Geologic age. The magnitude of V_s may be very dependent on the geologic age of soil deposits. In all of the field studies examined, the general trend was that older soils exhibited higher velocities than younger soils. Most of the studies indicated well-defined, though not mutually exclusive, ranges in V_s for Holocene- and Pleistocene-age divisions exist. In many of the studies, however, even though there was a distinction between V_s of certain aged soils, there was little effect on the correlative relation between N-value and V_s.

Soil type. Similarly, but to a lesser extent, soil type influences the magnitude of V_s. Soils with wide ranges of grain sizes tend to have smaller average void ratios, and, therefore, exhibit larger values of V_s. Ohta and Goto (1978b) suggested that the use of soil type in correlations involving V_s improves the accuracy because a certain range in void ratio is represented. Hardin and Richart (1963) and Hardin and Drnevich (1972) have proven strong dependence of V_s on void ratio in laboratory tests. Ohta and Goto indicate a systematic change of V_s for soil types where: $\left(V_s\right)_{gravel} \geq \left(V_s\right)_{sand} \geq \left(V_s\right)_{clay}$, mostly due to $e_{gravel} \leq e_{sand} \leq e_{clay}$. The use of soil type as a means to group data, then, seems to reflect the average void ratio of the soils. However, since wide ranges in void ratio are associated with specifying soil type, the influence of soil type is diminished. Of the four variables used in their analyses, the inclusion of soil type in the analyses provided the least improvement to correlation coefficients.

RECOMMENDATIONS

This report was compiled to familiarize practitioners with the evolution and juxtaposition of various shear wave velocity and shear modulus correlations involving SPT N-value so that the applicability of correlations to geotechnical engineering practice can be ascertained for each individual project. General recommendations are provided to assist in solving the question of applicability. These recommendations are:

a. Be cognizant of the methodologies used to conduct the correlative studies which will be used. In particular, the type of geophysical measurements, the details of SPT equipment and measurements, the method of data reduction, the distribution of correlative variables, and the representativeness of the data should be considered. Few studies stipulate in any way the SPT equipment or procedure used.

b. The use of Japanese relationships should be contingent on adjusting N-values for differences in equipment and techniques, in particular, for differences in energy efficiency.

c. Correlative equations proposed by Kanai (1966) differ substantially from nearly all other relationships, producing very low values of V_s or G. Therefore, these equations should not be used.

d. Exercise caution when using relationships proposed by Ohsaki and Iwasaki (1973) and Seed, Idriss, and Arango, (1983). The equations may produce high values of V_s and G_{max} at larger values of N (>25). These relations may only be appropriate for SPT equipment and procedures corresponding to very high energy efficiency and high resistance (e.g., automatic trip hammer and constant diameter sampler).

e. Expect a substantial range of potential error associated with each "best-fit" relation. For SPT N-value versus V_s correlations, this error may range from 150 percent to -40 percent of the calculated value.

f. Future correlative studies should be conducted to account for all significant variables affecting the measurement of SPT N-value and shear wave velocity. In particular, energy efficiency and hammer type of SPT equipment are considered to be very important factors.

g. Reliable V_s correlations should be incorporated into engineering studies to capitalize on the abundant data available and experience of others. Ideally, V_s correlations would be used in all phases of an overall engineering study, including:

(1) Optimizing surface and subsurface (especially seismic geophysical) exploration.

    (2)   Delineating zones with poor soil conditions for
more detailed subsurface investigation.

    (3)   Assigning values of shear modulus to various soil
strata.

    (4)   Design analyses, especially sensitivity analyses.

h.   Correlations should not replace in situ measurements but
rather complement an overall exploration program.

ACKNOWLEDGEMENTS

The study presented in this paper was conducted from January 1986
to September 1987 at the U.S. Army Engineer Waterways Experiment Station in Vicksburg, Mississippi. The authors wish to thank
Messrs. Umehara, Yamamoto, and Inove formerly of the University of
Texas at Austin for translation of technical articles written in
Japanese, and Mr. William Hanks of the Soil Mechanics Division, Geotechnical Laboratory, for drafting assistance.
    Permission was granted by the Chief of Engineers, U.S. Army Corps
of Engineers to publish this report. The views of the authors do not
purport to reflect the position of the Department of the Army or the
Department of Defense.

APPENDIX I.   CONVERSION FACTORS, NON-SI TO SI (METRIC) UNITS OF
              MEASUREMENT

Non-SI units of measurement used in this paper can be converted to
SI (metric) units as follows:

Multiply	By	To Obtain
feet	0.3048	metres
feet per second	0.3048	metres per second
inches	2.54	centimetres
pounds (force) per square foot	47.88026	pascals
pounds (mass) per cubic foot	16.01846	kilograms per cubic metre
tons per square foot	95.76052	kilopascals

APPENDIX II.   REFERENCES

1. Anderson, D., Espana, C., and McLamore, V. 1978. "Estimating In
Situ Shear Moduli at Competent Sites," Proceedings of the Specialty
Conference on Earthquake Engineering and Soil Dynamics, American Soci-
ety of Civil Engineers, Pasadena, Calif, Vol I, pp 181-197.

2. Hardin, B., and Drnevich, V. 1972b. "Shear Modulus and Damping in
Soils: Design Equations and Curves," Journal of the Geotechnical Engineering Division, American Society of Civil Engineers, Vol 98, No. 7,
pp 667-691.

3. Imai, T., Fumoto, H., and Yokota, K. 1975. "The Relation of
Mechanical Properties of Soils to P- and S-Wave Velocities in Japan,"

Proceedings of the Fourth Japanese Earthquake Engineering Symposium (in Japanese; translated by H. Umehara), pp 86-96.

4. Imai, T., and Tonouchi, K. 1982. "Correlation of N-Value with S-wave Velocity and Shear Modulus," Proceedings of the Second European Symposium on Penetration Testing, Amsterdam, The Netherlands, pp 67-72.

5. Imai, T., and Yoshimura, M. 1970. "Elastic Wave Velocities and Characteristics of Soft Soil Deposits," Soil Mechanics and Foundation Engineering, (in Japanese), The Japanese Society of Soil Mechanics and Foundation Engineering, Vol 18, No. 1.

6. _____. 1975. "The Relation of Mechanical Properties of Soils to P- and S-Wave Velocities for Soil Ground in Japan," OYO Corporation Technical Note TN-07.

7. Kanai, K. 1966. "Observation of Microtremors, XI: Matsushiro Earthquake Swarm Areas," Bulletin of Earthquake Research Institute (in Japanese), Vol XLIV, Part 3, University of Tokyo, Tokyo, Japan.

8. Marcuson, W., III, Ballard, R., and Cooper, S. 1979. "Comparison of Penetration Resistance Values to In Situ Shear Wave Velocities," Proceeding of the Second International Conference on Microzonation for Safer Construction, Research & Application, Vol III, San Francisco, Calif.

9. Marcuson, W., III, and Bieganousky, W. 1977. "Laboratory Standard Penetration Tests on Fine Sands," Journal of the Geotechnical Engineering Division, American Society of Civil Engineers, Vol 103, No. 6, pp 565-588.

10. Ohba, S., and Toriuma, T. 1970. "Research on Vibrational Characteristics of Soil Deposits in Osaka, Part 2, On Velocities of Wave Propagation and Predominant Periods of Soil Deposits," Abstracts of Technical Meeting of Architectural Institute of Japan (in Japanese).

11. Ohsaki, Y. 1962. "Geotechnical Properties of Tokyo Subsoils," Soil and Foundations, Vol II, No. 2, pp 17-34.

12. Ohsaki, Y., and Iwasaki, R. 1973. "On Dynamic Shear Moduli and Poisson's Ratio of Soil Deposits," Soil and Foundations, Vol 13, No. 4, pp 61-73.

13. Ohta, Y., et al. 1970. "Elastic Moduli of Soil Deposits Estimated by N-values," Proceedings of the Seventh Annual Conference, (in Japanese), The Japanese Society of Soil Mechanics and Foundation Engineering.

14. Ohta, Y., and Goto, N. 1976. "Estimation of S-Wave Velocity in Terms of Characteristic Indices of Soil," Butsuri-Tanko (Geophysical Exploration) (in Japanese), Vol 29, No. 4, pp 34-41.

15. Ohta, Y., and Goto, N. 1978a. "Empirical Shear Wave Velocity
Equations in Terms of Characteristic Soil Indexes," Earthquake Engineering and Structural Dynamics, Vol 6, pp 167-187.

16. _____ . 1978b. "Physical Background of the Statistically
Obtained S-Wave Velocity Equation in Terms of Soil Indexes,"
Butsuri-Tanko (Geophysical Exploration) (in Japanese; translated by
Y. Yamamoto), Vol 31, No. 1, pp 8-17.

17. Sakai, Y. 1968. "A Study on the Determination of S-Wave Velocity
by the Soil Penetrometer Test" (in Japanese; translated by J. Inove).

18. Seed, H. B., and Idriss, I. M. 1970. "Soil Moduli and Damping
Factors for Dynamic Response Analyses," Report No. UCB/EERC-70/10,
Earthquake Engineering Research Center, University of California,
Berkeley, Berkeley, Calif.

19. Seed, H. B., Idriss, I. M., and Arango, I. 1983. "Evaluation of
Liquefaction Potential Using Field Performance Data," Journal of the
Geotechnical Engineering Division, American Society of Civil Engineers,
Vol 109, No. 3, pp 458-482.

20. Seed, H. B., Tokimatsu, K., Harder, L. F., and Chung, R. M. 1985.
"Influence of SPT Procedures in Soil Liquefaction Resistance Evalua-
tions," Journal of Geotechnical Engineering, American Society of Civil
Engineers, Vol 111, No. 12, pp 1425-1445.

21. Seed, H. B., Wong, R. T., Idriss, I. M., and Tokimatsu, K. 1984.
"Moduli and Damping Factors for Dynamic Analyses of Cohesionless
Soils," Report No. UCB/EERC-84/14, Earthquake Engineering Research Cen-
ter, University of California, Berkeley, Berkeley, Calif.

22. Shibata, T. 1970. "The Relationship Between the N-value and
S-Wave Velocity in the Soil Layer" (in Japanese; as translated by
Y. Yamamoto), Disaster Prevention Research Laboratory, Kyoto Univer-
sity, Kyoto, Japan.

23. Sykora, D. W. 1987. "Examination of Existing Shear Wave Velocity
and Shear Modulus Correlations in Soils," Miscellaneous Paper GL-87-22,
U.S. Army Engineer Waterways Experiment Station, Vicksburg, Miss.

24. Sykora, D. W., and Stokoe, K. H., II. 1983. "Correlations of In
Situ Measurements in Sands With Shear Wave Velocity," Geotechnical
Engineering Report GR83-33, The University of Texas at Austin, Austin,
Tex.

25. U.S. Army Engineers. 1986. Kansas City District, " Dynamic
Analysis of New Milford Creek Dam."

26. Yoshikawa (date unknown), in Japanese, as reported by Sakai, 1968.

DEVELOPING DESIGN GROUND MOTIONS IN PRACTICE

John T. Christian,[1] F. ASCE

ABSTRACT

Describing ground motion for purposes of design involves the interaction of many disciplines and requires that attention be paid to the type of analysis in which the ground motion will be used. Methodologies developed for critical facilities such as nuclear power plants and dams have many similarities. They start from geological and seismological information and work through an orderly procedure to establish a design ground response spectrum, usually based on a standard shape anchored to the value of the peak ground acceleration. Site dependent spectra can also be developed based on suites of motions for similar conditions and earthquakes. Since different approaches will yield different estimates, a method of combining the statistics into an overall site dependent motion is proposed. An alternative to the response spectrum is the power spectral density function combined with other results from the theory of random vibrations. The response spectrum and the power spectral density function can be related, and the power spectral density function offers several advantages, particularly in view of modern developments in dynamic analysis using complex calculus. For non-linear analyses a time-history representation seems to be most appropriate; methods based on linear elasticity cannot recover some of the significant aspects of non-linear response. Many questions remain to be answered, and most of them will require data from field measurements of actual earthquakes.

Introduction

The present conference provides an unusual opportunity for professionals in many disciplines to address a set of related issues that lie near the heart of their common interests - the issues concerned with the proper identification and description of ground motion. These questions are important both because they bear on the scientific understanding of earthquakes but also because they must be answered by the practicing engineer whenever he deals with facilities located in areas subject to seismic hazards. It is appropriate that the conference has been organized under the auspices of the Geotechnical Engineering Division of ASCE, for the geotechnical

[1]Senior Consulting Engineer, Stone & Webster Engineering Corporation, P. O. Box 2325, Boston, Massachusetts 02107.

engineer is often the person in the middle when questions regarding ground motion arise between the various disciplines.

This paper describes some of the methodologies and issues involved in establishing ground motion parameters for use in engineering analysis and design. Therefore, the focus of the discussion is not so much on the description of individual earthquakes as on the development of generalized or enveloping design representations of ground motion. Other speakers will be concerned with local and regional descriptions of seismic events. It is hoped that the interaction among the speakers, panelists, and other participants will yield greater understanding of the perspectives of different disciplines and of how each contributes to the description of ground motion for engineering purposes.

One difficulty that arises at the start of any discussion of this subject is that different engineering applications require different descriptions of the ground motion. Thus, a dynamic analysis using linear techniques in the frequency domain, an evaluation of liquefaction potential, and a non-linear analysis in the time domain may require ground motion parameters. In some cases the description appropriate to one of these analyses may actually be unhelpful for another. Consequently, ground motions for engineering design should be developed with some sensitivity to the analytical techniques in which they will be employed.

Current Procedures for Defining Design Ground Motions

Nuclear Power Plants

The most detailed procedures for developing design ground motions are probably those associated with the siting and design of nuclear power plants. A major reason for this is the existence of a formal and detailed procedure for regulatory review, combined with a very high degree of public visibility. The basic documents, 10CFR100 and 10CFR100 Appendix A, are Federal Regulations and therefore have the force of law. They have been expanded, interpreted, and implemented in a series of Regulatory Guides, Standard Review Plans, and NUREG reports issued by the U. S. Nuclear Regulatory Commission. Although no new plants have been sited in several years, the licensing of plants under construction and the ongoing review of operating plants have led to a continual evolution of the regulatory philosophy. If only as a backdrop for other approaches, it is worth reviewing these procedures briefly.

The first step in developing the design ground motion is a comprehensive evaluation of the geology and tectonic structure of the site and its surrounding region. As it is beyond the scope of the present paper to elaborate on this part of the process, little will be said about it except to note that this essential effort is often time consuming, requires careful attention by competent people, and can employ any reasonable geological or geophysical tools to arrive at an answer.

Next, the seismicity of the site and region must be described. This may involve identifying faults capable of generating earthquakes and describing their seismogenic properties. It may also involve identifying and describing regions of generally uniform seismicity (called seismo-tectonic provinces).

The largest earthquake ground motion for which the plant is to be designed (called the Safe Shutdown Earthquake or SSE) is determined to be the maximum of three possible conditions. The first is found by moving to the site the epicenter of the largest anticipated event in the surrounding province. The second is found by moving the epicenters of the largest events in the other provinces to the nearest points on the province boundaries and attenuating their motions to the site. The third is found by moving the foci of the largest events on any capable faults to the points on the faults nearest the site and attenuating their motions to the site. The source events are described by an appropriate magnitude scale, if the data exist in that form, but in many cases only historical intensities are available. The resulting motion at the site is expressed in terms of peak particle acceleration.

A second ground motion, called the Operating Basis Earthquake or OBE, is usually taken to have half the peak particle acceleration of the SSE, but probabilistic arguments have been advanced successfully for using other values (e.g., Christian et al., 1978). In any case, both earthquakes are usually described by the peak particle acceleration.

The standard response spectra from U. S. NRC Regulatory Guide 1.60 (1973) are multiplied by the values of the peak particle accelerations to obtain the spectra of the ground motion corresponding to the SSE and OBE. Different spectral shapes apply for vertical and horizontal motion.

For purposes of dynamic analysis a set of artificial earthquakes is developed. These must have response spectra that fall above the prescribed spectra at all except a limited number of points, and the motions in different directions must be statistically independent to a prescribed degree.

The above description does not include a number of developments that have taken place in recent years and does not present the details of the procedures. However, it does summarize the idealized version of the process that leads from the initial examination of the site to providing the designer with a ground motion he can use.

Large Dams

The Committee on Earthquakes of the United Sates Committee on Large Dams (USCOLD) (1985) has published guidelines for selecting seismic parameters for dam projects. Obviously, these do not have the regulatory force of the NRC procedures. They do reflect more recent experience and are intended for a much broader range of situations. The committee starts by identifying primary factors to be considered. These include the regional geological setting, the seismic history, and

the local geological setting. They then address the selection of the
design earthquakes, distinguishing among five earthquakes:

The Maximum Credible Earthquake (MCE), which is "the largest
reasonably conceivable earthquake that appears possible" for
a province or structure without regard to probability or
recurrence interval.

The Controlling Maximum Credible Earthquake (CMCE), which is
"the most critical of all the MCEs capable of affecting a
dam."

The Maximum Design Earthquake (MDE), which "will produce the
maximum level of ground motion for which the dam should be
designed or analyzed" and may be less than the CMCE in some
circumstances.

The Operating Basis Earthquake (OBE), which is "the level of
ground motion at the site with a 50% probability of not being
exceeded in 100 years."

The Reservoir-Induced Earthquake (RIE), which is the maximum
ground motion associated with the "filling, drawdown, or
presence of the reservoir."

The choice of which of these earthquakes is to be used for design
and analysis depends on several additional factors, such as the
consequences of failure, the hazard rating of the site, the type of
dam, its functional requirements, the risk rating of the dam and
reservoir, and the consequences of errors in estimating the risk. For
each of the earthquakes chosen, several sets of parameters must then be
identified. First is the set of parameters describing the
characteristics of the source, the site, and the intervening region
through which the motion is propagated. The resulting ground motion is
characterized by one or more peak motion parameters. The committee
recognizes that this usually means the peak ground acceleration, but it
indicates that the duration of the strong motion is important as well.
The input to the analytical process is then usually expressed in the
form of response spectra, artificial earthquakes, or both. Finally,
probabilistic analyses of the seismic hazard can be used to guide the
choice of many of the parameters.

Summary

This brief review shows some differences between two approaches to
developing ground motions for design but it reveals a great many more
similarities. In both cases the procedure can be divided into five
steps:

Describe the geology of the site and region.

Describe the seismicity and tectonic framework.

Determine what design earthquake is required.

From the above results, determine the values of parameters governing the earthquake.

Express the detailed design earthquake by appropriate spectra or time histories of motion.

However, the processes of design and analyses do not stop when the design earthquake has been described; often they have barely begun. At least three further steps are involved:

Use the design earthquake to analyze the structures, soils, or components.

Compare the results of analyses with codes and criteria.

Build or modify the facility to conform to the analyses and the codes and criteria.

The process of design is often iterative, and it should have as few seams as possible between the different disciplines. Therefore, it is worth emphasizing again that the choice of a ground motion to be used in design should be influenced by the way it will be employed in analysis. This requires some communication between the user and the developer of the design ground motion.

Comments on Current Methods

Regional and Local Effects in General

Other papers in this conference deal with the present state of the art for geological input, characterization of strong ground motion on a regional scale, local variations in ground motion, and techniques for analyzing local effects. Therefore, these topics are not covered here in any detail, but a few remarks are in order about some of the engineering problems that can arise.

Ideally the description of the region would be couched in terms of tectonic behavior and would be related to the behavior of specific faults or other geologic structures. In many cases the available tectonic and structural information is incomplete or virtually non-existent. It is well known that in regions of generally sparse seismicity, like eastern North America, it is very difficult to relate observed historical seismicity to geologic structures and that some of the inferences are questionable at best. The choice of an appropriate level of ground motion depends heavily on the historical record, much of which is not in the form of instrumental observations. Many magnitudes of past events have been estimated from intensities or sizes of felt areas, which are often themselves derived from historical observations of variable accuracy. Estimating ground motion in such a region is fraught with much uncertainty.

It might be argued that the difficulty of describing rationally the seismicity in a region of sparse earthquake activity is mitigated by the fact that the overall risk to the public is commensurately small as well and that attention should be concentrated on zones in which large and frequent earthquakes can be expected. Unfortunately, some of the same problems exist even in areas where major tectonic events are under way and are fairly well understood. Medieval and ancient records form the basis for much of the understanding of the seismicity of areas where civilizations have existed for a long time. In more recently settled areas, such as California, many of the recent events have occurred at locations where they were not expected, or beyond the believed end of the causative fault, or on structures thought to be inactive. In areas of major tectonic development, such as Alaska, there is significant recent faulting that is clearly evident on the ground and in satellite imagery but with which no historical earthquake or current micro-earthquake activity can be associated. One must conclude that, regardless of the region, strong ground motion cannot be estimated for engineering purposes on the basis of known active faults alone but must be based on all relevant information, including historical reports in the form of intensities and felt areas.

Campbell (1985) has given a useful and thorough review of the present state of knowledge about attenuation relations. Most of his material is based on recent instrumental observations, but many of the attenuation relations in use in areas of sparse seismicity or sparse instrumentation are still based on data in the form of intensities or felt areas. When the data are instrumental, they are often from small events. Since these may be the only relations available, they are often used in calculations of site effects, including probabilistic hazard evaluations. In the absence of better field data little else can be done, but this is an area that requires further attention from the research community.

Whether there are observable effects of amplification and deamplification caused by the presence of soil layers and how those effects should be calculated have been the subject of much debate over the years. At this time the usual engineering practice is to assume linear visco-elastic models and horizontally polarized and vertically propagating shear waves in performing the calculations. In order to simulate the effects of material non-linearity, equivalent linear properties are chosen on the basis of the level of strain at each elevation in the system. SHAKE (Schnabel, et al., 1972) is the best known of the computer programs for such calculations. The equivalent linear model is clearly and explicitly an approximation for a non-linear material, and one should not expect exact results from it. In particular, it is implicit in such models that the average strain is equally effective for all frequencies. In fact, as Constantopoulos, et al., (1973) demonstrated, small, high frequency hysteresis loops have much lower effective strain than do the larger, low frequency loops. Thus, the effective strain and damping are overestimated for the high frequencies, and this is most evident when such programs are used to deconvolve observed surface motions through relatively soft soil.

It is assumed in most current analyses of site effects that the seismic waves are coherent across the site. Site amplification studies usually assume vertically propagating shear waves, and dynamic analyses of most structures - including large ones like dams - usually have the same input motion across the base. Even when horizontal propagation is specifically prescribed, as in the analysis of stresses in buried pipes, the motion is assumed to have a constant wave form displaced in time. These seem to be reasonable and conservative ways to proceed, but the profession will not be certain until coordinated field measurements of strong ground motion are made across several sites and interpreted.

Standard Response Spectra

The results of the studies of regional and local seismicity, combined with attenuation relations, are usually expressed in the form of peak particle accelerations. It is then customary to multiply the values of peak particle acceleration by the ordinates of standard response spectra to obtain response spectra to be used in design. The best known of these are the shapes proposed by Newmark, et al., (1973), which are the basis of the prescriptions in U. S. Nuclear Regulatory Commission Regulatory Guide 1.60. These were derived from two studies that first assembled a suite of representative strong motion accelerograms and scaled the accelerations in each record proportionately so that all the maximum accelerations or maximum velocities were the same. Then, for each value of damping for which a design spectrum was desired, the response spectra were calculated across a wide range of frequencies. For each frequency the mean and the standard deviation of the logarithms of the ordinates of the response spectra of the different earthquakes were calculated. From these values it was possible to compute estimates of the mean and the 84th percentile response spectra for the suite of earthquakes. The proposed design spectra are sets of straight lines that approximate the 84th percentile values. Since the ordinates of the response spectra at any frequency are found experimentally to be well represented by a log normal distribution, this is a reasonable way to find the shape of a design spectrum.

The suite of records used by Newmark, et al., included records of several earthquakes on a range of foundation materials, and the number of records was limited by the scarcity of strong motion accelerograms available at the time the studies were carried out. Several others have expanded and modified their results. Seed, et al., (1976) used more records and separated them into four categories, based on the type of foundation material. Both Newmark, et al., and Seed, et al., presented their recommendations in the form of standard shapes normalized to a peak particle acceleration, or Zero Period Acceleration (ZPA), of 1.0g. This is the form that has been adopted for most building code applications as well. An interesting alternative is the Japanese proposal (Ohsaki, 1979, and Hisada, et al., 1978) that scales the response spectra according to the peak particle velocity. Since Japanese nuclear power plants are founded on firm ground or rock, the Japanese response spectra do not include effects of records made on softer sites and are, therefore, much less energetic than the American spectra at frequencies below about 2 Hz. Figure 1 (after Christian, 1980) shows a comparison between one of the Japanese spectra and one generated for comparable input using the US NRC standard spectra.

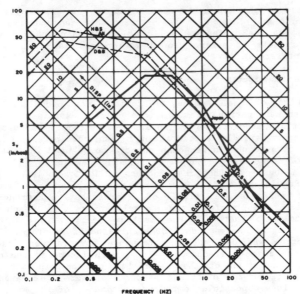

Figure 1. U. S. and Japanese 5% Damped Spectra for M = 7 at 10 Km.

Standard response spectra provide a very effective tool for describing ground motion in a way that incorporates a great deal of information about historically observed earthquake motions and local variation in seismic effects. However, several problems remain. First, the standard spectra are usually scaled from a value of the peak particle acceleration, which must be established before the spectra can be drawn. Because of the statistical techniques used to create the standard spectra, there tends to be much larger variance in the low frequency end of the spectra than at the high frequency end. Indeed, in some applications, once the peak particle acceleration has been fixed, there is no variance for frequencies above 33 Hz. Consequently, such standard spectra anchored to fixed values of accleration will not reproduce observed statistical variability at the high frequencies of the response spectra.

Second, the response spectrum is really intended for the analysis of single-degree-of-freedom systems. When there are many degrees of freedom, and particularly when different levels of damping exist for different portions of the system, it is not always clear how the analysis should proceed. One popular approach is to combine modal responses by the square root of the sum of the squares of the modal responses, but arguments have been raised for other techniques when the modal frequencies are closely spaced or when the range of frequencies represents the relatively rigid response of the higher modes. Artificial time histories whose response spectra correspond to the design response spectra have been used to account for different damping throughout the system. These techniques are approaches to problems that arise in the use of design ground motions rather than in the development of the motion itself, but the geotechnical engineer or seismologist involved in describing the ground motion ought to be aware of the problems that arise in trying to use the particular form of ground motion prescribed.

Artificial Time Histories

Artificial time histories were originally developed as band-pass filtered random vibrations for theoretical studies of structural response. At that time most structural analysis was done in the time domain either by modal superposition or by direct marching. An artificial time history, whose response spectrum agreed with the design response spectrum, became a convenient way to use the existing analytical methods to obtain the response corresponding to the desired input spectrum. Such an approach has much to recommend it, but it also has the disadvantages that the analysis is still using one record with its own phasing of the different components and that it does not include those aspects of ground motion not describable by random vibration theory. Perhaps more important are two other problems, one deriving from the artificial earthquake and the other inherent in the standard response spectrum itself.

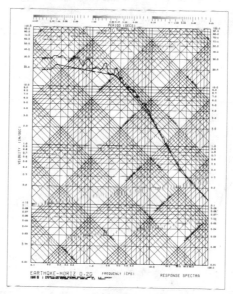

Figure 2. Response Spectra Prescribed and from Artificial Record

Figure 2 shows a standard response spectrum and the response spectrum for an artificial earthquake that is intended to match the standard spectrum. Because regulatory bodies, in particular the US NRC, have mandated that a very small number of points in the jagged response spectrum for the artificial earthquake may fall below the standard spectrum, the mean trend of the artificial spectrum lies well

above the target smooth spectrum, and the response at most frequencies
is correspondingly higher than for the target spectrum. At those
frequencies for which the artificial spectrum has local peaks, the
response can be very much higher than required by the standard
spectrum. This all means that the artificial earthquake introduces a
very substantial additional conservatism in the calculated response of
the system. The alternative of matching the mean trend to the target
shape would result in lower responses than desired for those
frequencies that have low points in the artificial spectrum. Given the
present methods of generating artificial earthquakes using a finite
number of frequencies, it is inevitable that their response spectra
will be jagged and that the problems of added conservatism or
undesirably low response will exist. This is a problem that arises
from the artificial earthquake itself.

Figure 2 also shows that the mean shape of the response spectrum for
the artificial earthquake matches the shape of the target spectrum only
approximately. This happens because the shape of the standard response
spectrum is an artificial simplification of the statistically computed
response spectrum and may not be reproducible from any single actual
record. A further effect of the artificial shapes is that a single
time history does not produce repsonse spectra that match the design
response spectra for all values of damping equally well. These
problems arise from the shapes of the design spectra themselves and not
from the artificial earthquake.

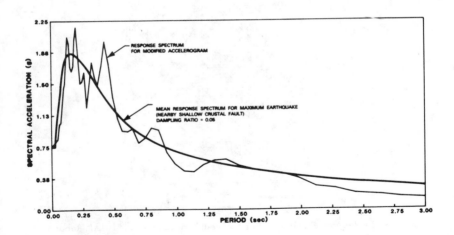

Figure 3. Response Spectrum from Artificial Record Composed from
          Actual Accelerograms

It should be noted that another way to generate an artificial
earthquake is to combine the scaled accelerograms from several actual

earthquakes into a single record. This procedure, which is used most commonly in dam design, does capture many of the properties of actual recordings that are lost when random vibration theory is used, but it is much harder to obtain a good match with the target spectrum. In most practical applications the artificial time history is found to generate a response spectrum that matches the target spectrum quite roughly, and no attempt is made to have a spectrum that envelopes the target spectrum. Figure 3 shows a typical example of the response spectrum for an artificial time history generated by this method. In this case a system whose natural frequencies lie at the peaks of the artificial earthquake would have a very conservative computed response, but one with resonant frequencies at the valleys of the spectrum would have unconservative results.

Artificial earthquakes generated by the methods of random vibration theory are not suited for calculations in which such aspects of real records as the changes of frequency content over time are important. Examples are calculations of non-linear response and of cumulative sliding displacement. However, the time history created by combining portions of actual records does preserve much of the relevant content of actual accelerograms and is much better suited to calculating non-linear response. As shown in Figure 3, the response spectrum in that case may have very large changes over rather modest ranges of frequency, and a change of input material properties can result in large changes in computed displacements due to a shift in the frequency at which the response occurs.

Site Dependent Analysis

Many have argued that, instead of proceeding by first estimating a single parameter such as the peak particle acceleration or the site intensity and then using a standard response spectrum scaled to that value, it would make more sense to develop the design ground motion specifically for the site, incorporating as many parameters as possible in one analysis. An analysis of this sort might involve first establishing the seismic and tectonic environment that governs the design motion and then assembling a suite of accelerograms made under similar circumstances. The response spectra (or other descriptions of ground motion) derived from this suite would form the basis for the design motion.

Hunt, et al., (1986) describe one analysis of site dependent spectra. Although the analysis was done in an area of low seismicity and was motivated by a need to justify an existing design earthquake rather to develop a new one, the analytical approaches and problems are typical of those encountered in other similar analyses.

Seismological considerations indicated that the design earthquake should be equivalent to an event near the site with m_b = 4.75. The site has about 110 feet of sand and gravels lying horizontally over very competent bedrock. There are few recorded acclerograms that satisfy both these conditions - a situation that occurs quite frequently when one attempts to find historical records that match the site conditions. Therefore, two approaches were used. One used recordings made at the surface of sites with soil overlying rock during nearby events with magnitudes between about 4.5 and 6.5. The other used recordings made on rock outcroppings during similar earthquakes and used one-dimensional amplification theory to develop corresponding surface motions. In both cases the accelerograms were scaled to an equivalent magnitude of 4.75 using the relations proposed by Nuttli (1984) and Nuttli and Herrmann (1984).

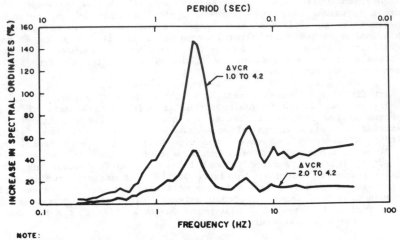

NOTE:
EACH CURVE REPRESENTS THE 50TH PERCENTILE
PERCENT CHANGE IN SPECTRAL ORDINATES FOR THE
NOTED CHANGE IN VELOCITY CONTRAST RATIO (VCR)

Figure 4. Change in Spectra Due to Velocity Contrast Ratio

The sites at which all the records in the first group were made were underlain by rock that had substantially less competence than the rock at the site in question, and the records had to be adjusted accordingly. This was done on the basis of one-dimensional amplification studies using several values of the ratio between the shear wave velocity of the soil and the bedrock. At the site that ratio was 4.2, but the average value at the sites of the recordings was 2.0. Figure 4 shows the increase in calculated ordinates of the 5% damped response spectrum when the velocity contrast ratio is changed from 2.0 to 4.2. The statistics of the suite of records were corrected by these amounts. Figure 4 also shows the effect of moving from a homogeneous case (velocity contrast ratio of 1.0) to a ratio of 4.2. It is clear that the effects of a stiffer underlying layer at the site than at the recording station are significant but not nearly as drastic as those that would occur in moving from a homogeneous case to the present one. It should also be clear that similar approaches can be employed to correct for other differences between the recording sites and the site being studied.

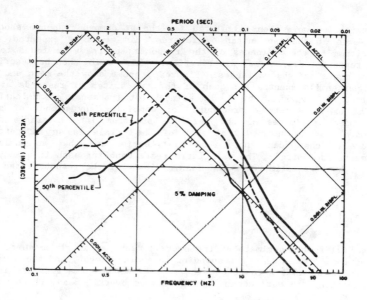

Figure 5. Response Spectra from Similar Sites - 5% Damping

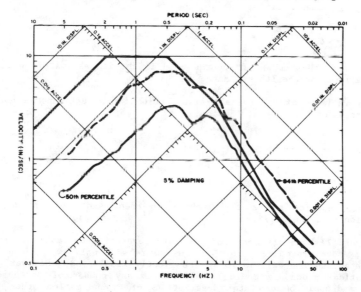

Figure 6. Response Spectra from Records on Rock Sites Propagated
through Soil - 5% Damping

Statistics were then calculated for the 5% damped response spectra from each of the two suites of records. As would be expected, the logarithms of the ordinates of the response spectra were found to be well represented by normal distributions, which could then be described by the means and the variances. Figures 5 and 6 show the mean and the 84th percentile spectra for each of the two suites of rescords at 5% damping. A suggestion by K. F. Reinschmidt (personal communication) can be expanded into a procedure for combining any number of such independent estimates of the response spectrum into a single best estimate. It is assumed that each of the trials is an independent measure of the underlying normal distribution and that an unbiased estimator is desired. The calculations are done for each frequency in turn. The best estimate of the mean ordinate of the response spectrum is

$$E(\mu) = \sum_{i=1}^{m} w_i \, \bar{x}_i \tag{1}$$

where $E(\mu)$ is the estimate of the overall mean, $\bar{x}_i$ is the mean of the ith estimate, w_i is the corresponding weighting factor, and there are m independent estimates to be combined. The weighting factors are found by minimizing the variance of the estimated overall mean, which leads to

$$w_i = \frac{n_i/s_i^2}{\sum_{i=1}^{m} n_i/s_i^2} \tag{2}$$

where n_i is the number of records used in the ith estimate and s_i^2 is the variance of the ith estimate.

The variance of the overall distribution can be estimated from the properties of the χ^2 distribution:

$$s^2 = \frac{1}{\sum_{i=1}^{m} (n_i-1)} \sum_{i=1}^{m} (n_1-1) \, s_i^2 \tag{3}$$

In this equation s^2 is the desired variance, and the rest of the notation is as before.

Figure 7 shows the mean and the 84th percentile estimates computed by the above procedure. In this case there were only two independent estimates, so the calculations were greatly simplified. However, the statistical procedure could be applied to any number of independent cases and could be used for descriptions of ground motion other than the response spectrum so long as the ordinates are normally or lognormally distributed. In this case the lognormal distribution

applies, and the means and variances must be understood to be the means
and variances of the logarithms of the ordinates.

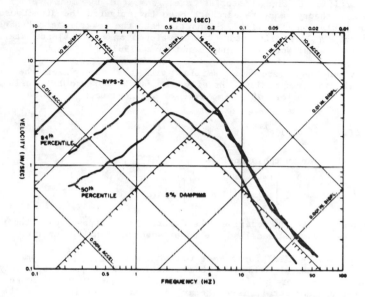

Figure 7. Response Spectra Combining Results of Both Analyses

Power Spectral Density Functions

The majui impetus tor the use of artificial time histories of
acceleration was the desire to avoid uncertainties associated with the
combination of modal responses calculated from response spectra. If
the responses were to be combined by the square root of the sum of
their squares, or some other statistically justified method, there were
questions of when in the analytical process the combination would occur
and how modes closely spaced in frequency would be combined. There
were also the problems that values of modal damping may be diffiuclt or
impossible to calculate for many realistic structures and that, even
when they can be calculated, they may not satify the modal
orthogonality relations. The way out was to use the artificially
generated time histories of motion mentioned earlier.

A factor in the acceptance of articial earthquakes was the fact that
almost all dynamic structural analysis was performed by direct modal
integration in the time domain. Since the mechanisms for dynamic
analysis with prescribed input of time histories of motion were already
in place, it was easy to apply them without modification to the new
artificial motions. Today many of the computer programs for dynamic
analysis, and almost all of the modern programs for analyzing dynamic
soil-structure interaction, operate in the frequency domain. In fact,
one of the first tasks performed is to transform the input time history
into the frequency domain by a Fast Fourier Transform. Since the use

of artificial time histories originated when there were no other commonly known techniques for dynamic structural analysis and since most artificial time histories are generated by means of random vibration theory, many researchers have investigated the direct use of random vibration theory and, in particular, the power spectral density function. A large literature has developed on the subject, starting with the seminal work of Crandall and Mark (1963).

Many engineers are not familiar with the concept of the power spectral density function. Briefly, it can be derived by considering the Fourier transform of a time history of acceleration:

$$A(\omega) = \int_{-\infty}^{+\infty} a(t) \, e^{i\omega t} \, dt \tag{4}$$

The total power in the signal is

$$\text{Total power} \equiv \int_{-\infty}^{+\infty} |a(t)|^2 \, dt = \int_{-\infty}^{+\infty} |A(\omega)|^2 \, d\omega \tag{5}$$

The quantity $|A(\omega)|^2$ describes the distribution of power over the range of frequencies and is called the power spectral density function.

Since $A(\omega) = A(-\omega)$ for a real function $a(t)$, $|A(\omega)|^2$ is symmetrical about the vertical axis, and the function could just as easily be defined only for positive values of frequency. To make the integrations correct it is customary to make the ordinates of the so-called single-sided function twice those of the double-sided function defined over both positive and negative frequencies. Some authors work with frequency (f) in Hertz instead of frequency (ω) in radians per second, and a factor of 2π must be introduced. This means that the user must be careful to establish the conventions that are being used in the particular paper or computer program that he is using.

The power spectral density function is obviously closely related to the Fourier spectrum. When dealing with a particular signal, one must be careful to consider the effects of the finite duration of the record and of smapling at discrete points. Frequency domain techniques usually assume that the record is repeated endlessly, so corrections must be made at the low frequency end or trailing zeros appended to the record before it is transformed. The power spectral density function can also be used to represent the power content of an infinite set of band-pass filtered random vibrations, and it is in this form that it has found its greatest use.

If $H(\omega)$ is the complex transfer or amplification function, then the output power spectral density function $\psi(\omega)$ is calculated from the input power spectral density function $\phi(\omega)$ from

$$\psi(\omega) = |H(\omega)|^2 \, \phi(\omega) \tag{6}$$

Since $H(\omega)$ is directly calculated in many modern computer programs for analysis of dynamic soil-structure interaction, it is relatively easy to find the prower spectral density at any point once the input is given.

It is then necessary to convert the output power spectral density function into some measure of the local response, such as a response spectrum. A large body of literature has developed on this subject, and the reader is referred to Crandall and Mark (1963), Vanmarcke (1976), Kaul (1978), and Igusa and Der Kiureghian (1985), among others, for elaboration of the details involved. In all these approaches the response is associated with a probability of exceedance; that is, the computed finite response has a certain probability of being exceeded. An exceedance probability of zero implies an infinite response.

A further complication in these analyses has to do with the duration of the record. It will be remembered that the theory underlying power spectral density functions involves signals that are essentially infinitely long compared to the periods of interest, but this is not the case for earthquake records. The duration of the record enters into the calculations in two ways. First, the random vibrations are being sampled over a finite time, and the probability of exceedance is correspondingly reduced. Second, for the lower frequencies the duration of the record may not be enough for full response to develop. The reduced response is often simulated by employing an increased effective damping.

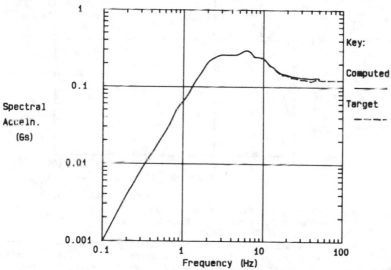

Damping Ratio = 0.05

Figure 8. Response Spectra from Hunt, et al., (1986) and from Matching Power Spectral Density Function (84th %ile)

Kaul (1978), Sundararajan (1980), Unruh and Kana (1980), and others have described how a power spectral density function can be computed from a given response spectrum. The usual technique starts from a

method of computing the repsonse spectrum from the power spectral
density function and then does the reverse calculation by repeated
iterations. Christian, et al. (1988) have implemented these approaches
on a micro-computer and combined it with a method of generating the
amplification function from modal data. Figure 8 shows the 84th
percentile response spectrum for 5% damping computed by Hunt. et al.
(1986) and the back-calculated response spectrum computed from the
corresponding power spectral density function. Figure 9 shows the same
comparison for the 5% damped response spectrum of U. S. NRC Regulatory
Guide 1.60 (1973), normalized to 0.25g. It is clear that the results
for the higher frequencies do not match as well in this case. It is
difficult to match idealized response spectra at higher frequencies
when using either artificial time histories or power spectral density
functions because the response is strongly influenced by the energy at
lower frequencies. In effect, the idealized spectral shape requires so
much energy at the lower frequencies that the response at the higher
frequencies cannot be reduced to the level of the idealized shape.

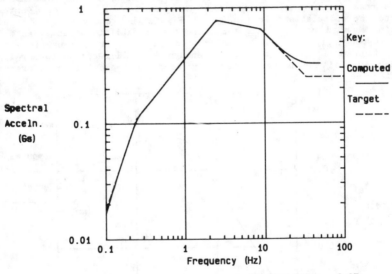

Figure 9. Response Spectra from NRC R. G. 1.60 and from
Matching Power Spectral Density Function

Instead of computing power spectral density functions to match
idealized response spectra, it would be more rational to prescribe
appropriate input power spectral density functions directly from the
ground motion parameters. One way to do this would be to evaluate
power spectral density functions from available records in much the
same way as site dependent design response spectra have been developed.
Another is to use an idealized shape for the power spectral density
function and modify it according to seismological models. Safak (1988)
provides an example of this approach, starting from the modified
Tajimi-Kanai idealized power spectral density function:

$$S(\omega) = S_0 \frac{\left[1 + \left(2\zeta_g \frac{\omega}{\omega_g}\right)^2\right]}{\left[\left[1 - \left(\frac{\omega}{\omega_g}\right)^2\right]^2 + \left(2\zeta_g \frac{\omega}{\omega_g}\right)^2\right]} \frac{\left[\left(\frac{\omega}{\omega_k}\right)^4\right]}{\left[\left[1 - \left(\frac{\omega}{\omega_k}\right)^2\right]^2 + \left(2\zeta_k \frac{\omega}{\omega_k}\right)^2\right]} \quad (7)$$

The terms in the first set of large brackets on the right hand side comes from the orgiginal work of Tajimi (1960) and Kanai (1957); the frequency ω_g and damping ζ_g were originally intended to account for the properties of the geological stata through which the random vibrations propagate. The terms in the second set of large brackets represent a correction proposed by Clough and Penzien (1975) to reduce the power spectral density function at very small frequencies and avoid infinite calculated responses. Safak observes that for firm sites the recommended values are $\zeta_g = \zeta_k = 0.62$, $\omega_g = 0.39$ rad/sec, and $\omega_k = 0.041$ rad/sec and points out that "almost all the stochastic seismic response studies of the past have used this model for ground motion description."

It is clear that power spectral density functions can be derived by using a general model and determining the values of the parameters from seismological data. Safak describes such an approach and provides references to much of the supporting literature. Alternatively, one could evaluate power spectral density functions for a suite of recordings from events corresponding to the circustances of interest and compute the statistics of the desired design power spectral density function. The latter approach is similar to that used for site dependent response spectra.

Shinozuka, et al., (1985) have taken a somewhat different approach. They determined values of the parameters in the original Tajimi-Kanai power spectral density function, not including the Clough-Penzien terms, so that the time histories generated by the resulting power spectral density function would envelop the US NRC Regulatory Guide 1.60 response spectra in the manner prescribed by the Nuclear Regulatory Commission. They propose $\zeta_g = 0.98$ and $\omega_g = 1.57$ rad/sec. Since the NRC rules require that the calculated response spectra not fall below the target spectra except at a very small number of points, the average response spectra are substantially above the target spectra. Consequently, the power spectral density function proposed by Shinozuka, et al., leads to calculated response spectra that are much larger than the targets. Figure 10 shows this result. Random vibration methods, whether they use power spectral density functions directly or by implication (e. g., Igusa and Der Kiureghian, 1985), are attractive in part because they avoid the excessive conservatism of the enveloping artificial time history. Power spectral density functions that reproduce the artificial earthquake instead of the response spectrum would eliminate this advantage and do not seem to be a desirable development.

There has been a great deal of work done to develop an understanding of the power spectral density functions that might describe actual earthquake motions. Somewhat less effort has gone into educating the engineering community in the use of random vibration techniques.

Nevertheless, these methods offer substantial advantages over direct integration using artificial time histories for the analysis of linear systems, and it is to be hoped that descriptions of ground motion in terms of power spectral desnity methods will become more widely reported.

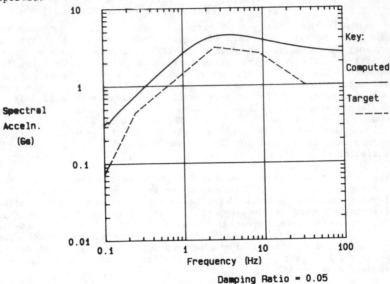

Damping Ratio = 0.05

Figure 10. Response Spectrum Calculated from Power Spectral Density Developed to Match Artificial Time History

Probabilistic Seismic Hazard Calculations

The calculation of seismic hazard has developed considerable currency since the early developments by Cornell (1968) and his fellow workers. A major aid to this process has been the general availability of McGuire's (1976) computer program. Of course, there has been a great deal of further research and elaboration beyond these initial efforts, but the essence of seismic hazard calculations is to be found in the work done in the late 1960's and early 1970's. It is worth noting that a very detailed examination of the methodology has been carried out by the Electric Power Research Institute. Although reports on the methodology, the seismicity evaluations, the attenuation relations, and other details of the study are becoming available, detailed results for specific locations are not, and the computer programs needed to extend the results are available only under limited and expensive circumstances.

Most seismic hazard studies report the results in terms of the annual probability of equalling or exceeding some single parameter describing the ground motion, usually peak particle acceleration. Even though, for example, techniques have long been available for computing hazard as a function of the frequency of motion, the difficulty of determining the proper values of parameters for even the simplified analyses and the inertia of current practice have combined to restrict most analyses to single parameters. It is clear that all the problems that arise from the use of a single parameter in deterministic descriptions exist for probabilistic descriptions as well.

Perhaps the major difficulties in carrying out a seismic hazard analysis lie in establishing the values of the input. It is difficult to establish a consensus not only about the geometric description of the regional and local seismicity but also about the minimum and maximum levels of activity at the sources and about the attenuation relations. The EPRI study involves one of the most comprehensive investigations of this aspect of hazard evaluation and includes the use of peer groups and other mechanisms for assigning levels of confidence to what are inevitably imperfectly understood parameters.

Probabilistic analyses are most effective when they are used to compare alternative risks or courses of action. That is, they are helpful in establishing which source of hazard is most important, which hazard is more important than another, and so on. They are much less effective in establishing a level of loading that corresponds to some pre-determined level of hazard, particularly when that level is so low as to be outside the normal range of human experience. It is often argued that probabilistic analyses provide results that can be used by the public to determine policy and that the public should determine what it considers an acceptable level of risk. Unfortunately, the public, including most of the engineering public, has little or no grasp of what is meant by, say, an annual hazard of 10^{-5} occurrences or a mean return period of 100,000 years. It is only when risks and hazards are compared that one can grasp what is important and can make rational choices.

It is worth noting that regulatory bodies, the public, and the engineering community have a great deal of difficulty in establishing an acceptable level of risk for very unlikely events. Such decisions always involve working at the tails of the probability distributions, where the results are not very robust. When the analyses involve more common events, it is much easier for the engineering community to make confident recommendations about acceptable risks. For example, the U. S. Committee on Large Dams (1985) recommends that the Operating Basis Earthquake should have a 50% probability of not being exceeded in 100 years. The U. S. Nuclear Regulatory Commission at one time considered that the OBE should have a 30% probability of exceedance in the life of the plant, which was assumed to be 40 years. The first of these corresponds to a mean return period of 144 years, and the second to a mean return period of 112 years. These numbers are relatively close and reflect a general agreement about the level of hazard appropriate for a loading that might realistically occur during the life of the structure.

One useful feature of probabilistic hazard calculations is that, because the total hazard is usually computed by combining the contributions of all sources, the individual contributions are usually available to the analyst. This provides quantitative information on the relative importance of different sources and can be quite useful in establishing where further efforts should be directed. It can also be used to check the validity of the probabilistic analysis. For example, it is not unusual to establish two principal sources of seismicity: a large distant one with relatively infrequent occurrence of earthquakes, and a smaller nearer one with relatively frequent events. A check of the detailed output from a seismic hazard calculation is a valuable check on the consistency of the hazard calculation and the assumed model.

Non-Linear Analyses

Non-linear models comprise everything that is not linear, so the term covers a very broad range of behavior. Nevertheless, regardless of the type of non-linear model used or the severity of the non-linearity, the fact of performing a non-linear analysis presumes that there is some aspect of the behavior of the structure or the material properties that cannot be described adequately by a linear model. In this context, a non-linear model is one that attempts to reproduce the details of the incremental behavior and not one that replaces the non-linear behavior with an approximately equivalent linear one.

One very clear result from computational experience with non-linear models is that their reponse is sensitive to many details of the modeling and of the input. Usually the sensitivity must be determined experimentally, and often the results are counterintuitive. These facts imply that ground motion for use in such analyses should be as close as possible to actual seismic ground motion as possible. Simply reproducing the overall frequency content and acceleration level is not enough. The change in frequency content with time, the variation of level of motion with time, maximum accelerations, maximum displacement levels, and certainly duration of strong motion are among the factors that can affect the results.

Therefore, unless and until there is much better understanding of the factors contributing to strong ground motion, non-linear analyses should be performed with actual recordings as input. This implied that there will be a suite of appropriate records in order to ensure that significant aspects of the response are not missed due to some artifact of the input record. In practice it is usually found that the particularities of the material properties and geometry combine to make one record give the largest displacement, stress, or other significant output. Artificial records generated by band-pass filtered random vibration techniques should not be used for non-linear analysis.

Conclusions

Establishing ground motion for design requires the interaction of many disciplines. Attention must be paid not only to the seismological aspects of the problem but also to the type of analysis in which the ground motion will be used. Methodologies developed for critical facilities such as nuclear power plants and dams start from geological and seismological information and work through an orderly procedure to establish a design ground response spectrum, usually based on a standard shape anchored to the value of the peak ground acceleration.

Although there are differences in regulatory climate and philosophy, the approaches are in many respects very similar.

Site dependent spectra can also be developed based on suites of motions for similar conditions and earthquakes. Since different approaches will yield different estimates, a method of combining the statistics into an overall site dependent motion has been developed. It involves establishing the mean by minimizing the variation of the estimated mean and establishing the variation by combining the distributions of the variances of the different estimates.

An alternative to the response spectrum is the power spectral density function combined with other results from the theory of random vibrations. The response spectrum and the power spectral density function can be related, and the power spectral density function offers several advantages, particularly in view of modern developments in dynamic analysis using complex calculus. More descriptions of ground motion in terms of power spectral density functions and greater education of engineers in its use would greatly assist in the application of these powerful techniques. Power spectral density functions should be selected to represent the underlying process of ground motions and not to reproduce specific artificial motions that provide envelopes to other representatons of ground motion.

For non-linear analyses a time-history representation seems to be most appropriate; methods based on linear elasticity cannot recover some of the significant aspects of non-linear response. Usually it is necessary to use a suite of accelerograms, but it is also usually the case that one of them produces the dominant result.

Much progress has been made in quantitative description of earthquake ground motions. This includes work on measures of real engineering significance, such as the power spectral density function. It is possible that the sophistication with which the seismological comminuty can describe ground motion may exceed that with which the engineering community uses it. Of course, many questions remain to be answered, and most of them will require data from field measurements of actual earthquakes. They will also require cooperation between the seismological community and the engineering community so that each may understand what the other is trying to do and the problems it faces.

REFERENCES

Campbell, Kenneth W., (1985) "Strong Motion Attenuation Relations: A Ten Year Perspective," Earthquake Spectra, Vol. 1, No. 4, pp. 759-804.

Christian, John T., (1980) "Comparison of Japanese Design Earthquake Response Spectra with Those Prescribed by U. S. NRC Regualtory Guide 1.60," Nuclear Engineering and Design, Vol. 61, pp. 369-382.

Christian, John T.; Borjeson, Ralph W.; and Tringale, Philip T., (1978) "Probabilistic Evaluation of OBE for Nuclear Plant," Journal of the Geotechnical Engineering Division, ASCE, Vol. 104, No. GT7, pp. 907-919.

Christian, John T.; Harizi, Philip D.; and Hall, John R., Jr., (1988) "Seismic Excitation and Measured Modal Behavior Combined by Power Psectral Density Methods," Res Mechanica, to be published.

Clough, R. W., and Penzien, J., (1975) Dynamics of Structures, McGraw-Hill, New York.

Constantopoulos, Ioannis V.; Roesset, Jose M.; and Christian, John T., (1973) "A Comparison of Linear and Exact Nonlinear Analysis of Soil Amplification," Proceedings, Fifth World Conference on Earthquake Engineering, Rome, Italy, paper 225.

Cornell, C. Allin, (1968) " Engineering Seismic Risk Analysis," Bulletin of the Seismological Society of America, Vol. 59, No. 5, pp. 1583-1606.

Crandall, Stephen H., and Mark, William D., (1963) Random Vibration in Mechanical Systems, Academic Press, New York and London.

Hisada, T.; Ohsaki, Y.; Watabe, M.; and Ohta, T., (1978) "Design Spectra for Stiff Structures on Rock," Second International Conference on Microzonation, San Francisco, California, Vol. III, pp. 1187-1198.

Hunt, Donald D.; Christian, John T.; Chang, Thomas Y. H.; and Cadena, Pedro A., (1986) "Determining Design Response Spectra for a Site of Low Seismicity on the Basis of Historical Records," Proceedings, Third U. S. National Conference on Earthquake Engineering, Charleston, South Carolina, Vol.II, pp. 1141-1152.

Igusa, Takeru, and Der Kiureghian, Armen, (1985) "Generation of Floor Response Spectra Including Oscillator-Structure Interaction," Earthquake Engineering and Structural Dynamics, Vol. 13, pp. 661-676.

Kanai, K., (1957) "Semi-Empirical Formula for the Seimic Characteristics of the Ground," Bulletin of the Earthquake Research Institute, University of Tokyo, Japan, Vol. 35, pp. 308-325.

Kaul, Maharaj K., (1978) "Stochastic Characterization of Earthquakes through their Response Spectrum," Eathquake Engineering and Structural Dynamics, Vol. 6, pp. 497-509.

McGuire, Robin K., (1976) "FORTRAN Computer Program for Seismic Risk Analysis," U. S. Geological Survey Open File Report 76-67.

Newmark, Nathan M.; Blume, John A.; and Kapur, Kanwar K., (1973) "Seismic Design Spectra for Nuclear Power Plants," Journal of the Power Division, ASCE, Vol. 99, No. PO2, pp. 287-303.

Nuttli, Otto W., (1984) "Instrumental Data," included in Nuttli, O. W.; Rodriguez, R.; and Herrmann, R. B., "Strong Ground Motion Studies for South Carolina Earthquakes," prepared for U. S. Nuclear Regulatory Commission, NUREG/CR-3755.

Nuttli, Otto W., and Herrmann, Robert B., (1984) " Ground Motion of Mississippi Valley Earthquakes," Journal of Technical Topics in Civil Engineering, ASCE, Vol. 110, No. 1, pp. 54-69.

Ohsaki, Y., (1979) "Guideline for evaluation of basic design earthquake ground motions," adopted as appendix to Regulatory Guide for Aseismic Design of Nuclear Power Reactor Facilities, Japan.

Safak, Erdal, (1988) "Analytical Approach to Calculation of Response Spectra from Seismological Models of Ground Motion," Earthquake Engineering and Structural Dynamics, Vol. 16, No. 1, pp. 121-134.

Schnabel, Per B.; Lysmer, John; and Seed, H. Bolton, (1972) "SHAKE: A Computer Program for Earthquake Response Analysis of Horizontally Layered Sites," Report EERC-72-12, University of California, Berkeley.

Seed, H. Bolton; Ugas, Celso; and Lysmer, John, (1976) "Site-Dependent Spectra for Earthquake Resistant Design," Bulletin of the Seismological Society of America, Vol. 66, No. 1, pp.221-243.

Shinozuka, M.; Mochio, T.; Jeng, D.; and Chokshi, N. C., (1985) "Power Spectral Density Functions Compatible with Design Response Spectra," Proceedings, Eighth International Conference on Structural Mechanics in Reactor Technology, Brussels, Vol. K2, pp. 69-74.

Sundararajan, C., (1980) "An Iterative Method for the Generation of Seismic Power Spectral Density Functions," Proceedings, Second ASCE Conference on Civil Engineering and Nuclear Power, Knoxville, Tennessee.

Tajimi, H., (1960) "A Statistical Method of Determining the Maximum Response of a Building Structure During an Earthquake," Proceedings of the Second World Conference on Earthquake Engineering, Tokyo, Japan, Vol. II, pp. 781-797.

U. S. Committee on Large Dams (1985) "Guidelines for Selecting Seismic Parameters for Dam Projects," prepared by the Committee on Earthquakes.

U. S. Nuclear Regulatory Commission (1973) Regulatory Guide 1.60, Design Response Spectra for Seismic Design of Nuclear Power Plants, Washington, D. C.

Unruh, James F., and Kana, Daniel D., (1981) "An Iterative Procedure for the Generation of Consistent Power/Response Spectrum," Nuclear Engineering and Structural Design, Vol. 66, pp. 427-435.

Vanmarcke, Erik H., (1976) "Structural Response to Earthquakes," Seismic Risk and Engineering Decisions, C. Lomnitz and E. Rosenblueth, eds., Elsevier Scientific Publishing Co., Amsterdam, pp. 287-337.

SIMULATION OF ACCELERATION TIME HISTORIES CLOSE TO LARGE EARTHQUAKES

David J. Wald, L. J. Burdick and Paul G. Somerville[1]

ABSTRACT

A procedure for the simulation of near-field ground motions due to extended fault ruptures has been developed and validated using strong motion recordings of several events. Accelerograms are generated by summing the contributions of subfault elements in such a way as to simulate a large slip over an extended rupture surface. While gross aspects of the source rupture process are treated deterministically using a kinematic model, stochastic aspects are included to simulate the irregularity in rupture velocity and slip velocity. Gross aspects of wave propagation are modeled by theoretical Green's functions calculated using generalized rays. Detailed aspects of the source radiation at high frequencies, as well as unmodeled propagational aspects such as scattering, are included empirically by using multiple recordings of a small earthquake. The radiation from the fault is represented in a manner that empirically includes the observed breakdown in coherence of the radiation pattern at high frequencies. Large scale asperities are introduced via a variable slip distribution over the fault surface. We have confirmed that a large region of slip on the 1979 Imperial Valley fault plane, identified by several authors using low frequency local and teleseismic waves, also accounts for the largest phase on most accelerograms of the event. Simulated time histories and response spectra at stations close to the fault agree well with the recorded motions.

INTRODUCTION

At present, the strong ground motion frequencies that have been successfully modeled (less than 1 Hz) lie below the frequency range of most interest in earthquake engineering. Part of the problem stems from the fact that there are, at present, inadequacies in both theoretical and quasi-empirical methods of simulating ground motions at higher frequencies. Theoretical calculations of strong ground motion are generally successful at predicting longer period (displacement and velocity) motions (Hartzell and Helmberger, 1982; Olson and Apsel, 1982; Hartzell and Heaton, 1983 and Liu and Helmberger, 1985). These synthesis methods usually involve the representation of a fault plane by a distribution of computed point source responses that are lagged and summed to simulate the propagation of rupture over the fault surface. Although the computation of theoretical Green's functions has become highly accurate in recent years (for example, Helmberger and Harkrider, 1978 and Yao and Harkrider, 1983), there is an upper frequency limit of

[1]all at Woodward-Clyde Consultants, 566 El Dorado Street, Pasadena, CA 91101.

about 1 Hz in waveform modeling approaches because of the lack of detailed knowledge of the local three-dimensional velocity structure and an insufficient understanding of faulting dynamics at higher frequencies.

An alternative to theoretical methods of simulating high frequency ground motion from large earthquakes is to use observed seismograms from small earthquakes as empirical Green's functions in place of theoretical responses. The advantage of using empirical Green's functions is that propagation and source complexities are implicitly included. This approach was first proposed by Hartzell (1978) and Wu (1978). Many investigators have expanded this technique and have taken advantage of most well instrumented moderate earthquakes to model strong ground motions of larger events (see Hartzell (1985) for a partial summary). While there has been success in estimating the strong ground motion from large earthquakes, users of this method generally rely on extensive data sets from small foreshocks or aftershocks. As discussed by Hadley et al. (1982), the empirical Green's function approach has severe limitations imposed by the requirement that small earthquakes be well recorded over a large range of distances and depths and at uncomplicated sites to properly define the Green's functions required to simulate a large event.

SEMI-EMPIRICAL SIMULATION PROCEDURE

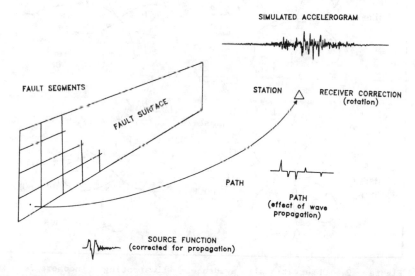

Figure 1. The semi-empirical ground motion simulation procedure. For each subfault, an empirical source function with the appropriate radiation pattern is convolved with the Green's function and corrected for the receiver function. The ground motion contributions of each subfault are then lagged and summed in accordance with the geometry of rupture of the Hartzell and Heaton (1983) model.

In order to avoid these limitations, we have used the semi-empirical approach (Hadley *et al.*, 1982) in order to gain the benefits of both theoretical and empirical techniques. Our simulation procedure is summarized in Figure 1. Accelerograms are generated by summing the contributions of subfault elements in such a way as to simulate a large slip over an extended rupture surface. Large scale asperities are introduced by allowing the slip on each of the subfault elements to vary. While gross aspects of the source rupture process are treated deterministically using a kinematic model, stochastic aspects are included to simulate the irregularity in rupture velocity and slip velocity. Gross aspects of wave propagation are modeled by theoretical Green's functions calculated using generalized rays. This allows us to accurately represent the contributions from all source depths and distances that are otherwise limited by the depth and ranges of the aftershock recordings using the empirical summation approach. Detailed aspects of the source radiation at high

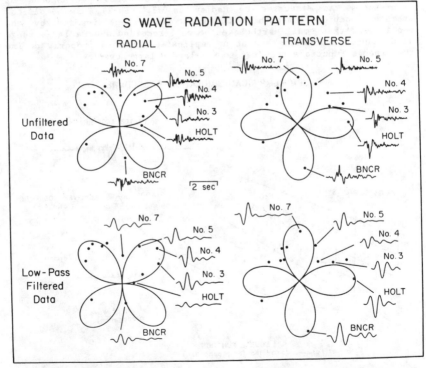

Figure 2. Observational evidence illustrating incoherence in radiation pattern. The top diagrams show the normalized amplitudes of the radial and transverse components of acceleration for the 23:19, 1979 Imperial Valley aftershock plotted as a function of azimuth on the theoretical SV and SH radiation patterns. Bottom diagrams show the amplitudes for the same data after low pass filtering (see text for details).

frequencies, as well as unmodeled propagational aspects such as scattering, are included empirically by using multiple recordings of a small earthquake.

In an extension of the Hadley et al. (1982) approach, we allow the radiation from each subfault to be represented in a manner that empirically includes the observed breakdown in coherence of the radiation pattern at high frequencies, illustrated in Figure 2, observed by Liu and Helmberger (1985). In the top portion of Figure 2, the radial and transverse amplitude ratios (normalized to total S amplitude) for the Imperial Valley aftershock accelerograms depart significantly from the theoretical radiation patterns indicating little coherence in radiation at high frequencies. After lowpass filtering the data to remove energy above 2 Hz, (shown in the lower part of the figure) the amplitude ratios agree quite well with the theoretical radiation patterns and distinct radiation nodes and lobes begin to appear. Hence, while theoretical radiation pattern values are valid at low frequencies, Figure 2 illustrates that the amplitude predicted theoretically may be in error by a factor of three or more at higher frequencies. Rather than theoretically correcting source functions for the desired radiation value and misrepresenting the high-frequency radiation pattern, our approach is to incorporate frequency-dependent radiation pattern effects empirically. Fifteen recordings of the Imperial Valley aftershock are used as a set of empirical source functions having a distribution of locations over the focal sphere. The empirical source function selected for use with a given subfault element is the one whose low-frequency, theoretical radiation pattern value is closest in absolute value to the theoretical radiation pattern value required for the given subfault element. Since no radiation pattern correction is applied to the empirical source functions, the actual frequency-dependent radiation pattern is included implicitly. The use of multiple empirical source functions has the added benefit of providing additional samples of the radiated source spectrum, while preventing a single recording from dominating the resulting ground motions.

VALIDATION OF GROUND MOTION SIMULATION PROCEDURE

We have selected the 1979 Imperial Valley earthquake for the validation of the ground motion simulation procedure. This earthquake was widely recorded on strong motion instruments and has been widely studied. As a result, all of the information that we need to perform simulations of the mainshock recordings is available. First, the local velocity structure is well known from detailed refraction surveys (Fuis et al., 1982). Second, a magnitude 5 aftershock provided a large set of three component accelerograms suitable for use as empirical source functions, and the aftershock's source parameters have been studied in detail by Liu and Helmberger (1985). Third, faulting models of the mainshock have been derived using waveform modeling methods by several authors (Hartzell and Helmberger, 1982; Olson and Apsel, 1982; Hartzell and Heaton, 1983 and Archuleta, 1984).

An important aspect of all of these fault models is the heterogeneous distribution of slip that is inferred to have occurred on the fault plane. These slip distribution models, which we shall refer to as asperity models, were derived from observations of relatively low frequency waves, such as strong motion velocity or displacement records and teleseismic short-period and long-period records. Such heterogeneity in slip distribution is observed quite generally in earthquakes in a wide variety of tectonic environments. However, it has not been known whether the regions of the fault that radiate strongly at lower frequencies also radiate strongly at the higher frequencies of engineering interest.

Thus an important adjunct to our goal of validating our ground motion simulation procedure is to test the hypothesis that regions that have been identified as asperities on the Imperial Fault on the basis of their strong low frequency radiation also radiated strong accelerations. Fortunately, the application of moveout analysis to the El Centro array recordings allows us to unambiguously identify the largest accelerations as originating from a large slip-distribution asperity that is common to all of the previously derived faulting models. A similar conclusion was obtained by Spudich and Cranswick (1984) from analysis of the Differential Array recordings of the earthquake. The definition of a satisfactory model of high frequency radiation from the fault enables us to obtain good agreement between observed and simulated ground motions of the Imperial Valley earthquake.

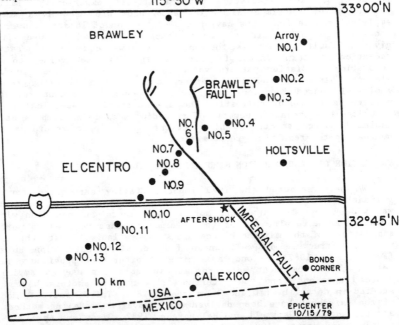

Figure 3. Map of the Imperial Valley showing the surface trace of the Imperial Fault, the El Centro strong motion array and other stations of interest to this study (after Hartzell and Heaton, 1983). The epicenters of the 15 October, 1979 mainshock 23:19 aftershock are indicated by stars.

The rupture segment of the 1979 Imperial Valley earthquake and the location of strong motion recording stations is shown in Figure 3. All of the fault model parameters that we have used are derived from Model 31 of Hartzell and Heaton (1983). The purely strike-slip, vertical fault is 40 km long, 12 km wide, and has a seismic moment of 5.1×10^{25} dyne-cm. The slip distribution model of Hartzell and Heaton (1983) is shown in the top panel of Figure 4, and our discretization of this model is shown in the lower panel. We have allowed the slip to extend to somewhat greater depth than Hartzell and Heaton (1983).

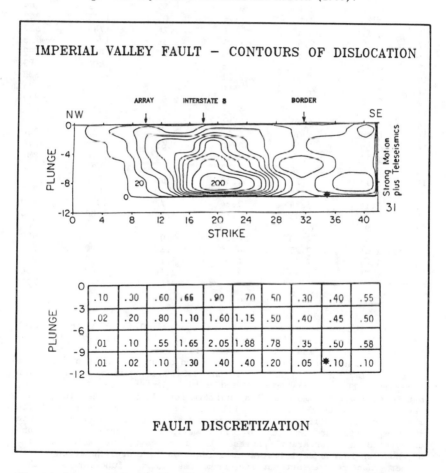

IMPERIAL VALLEY FAULT – CONTOURS OF DISLOCATION

FAULT DISCRETIZATION

Figure 4. Fault rupture model 31 of Hartzell and Heaton (1983) showing contours of dislocation in centimeters on the Imperial Fault plane (top). The bottom panel shows our discretization of model 31 with assigned grid element slip weighting factors. The location of the hypocenter is indicated by a star.

TABLE 1   VELOCITY STRUCTURE

Thickness	α (km/sec)	β (km/sec)	ρ (gm/cm³)	Q_α	Q_β
1	0.21	1.69	0.35	1.52	212	6
2	0.21	1.72	0.50	1.56	212	6
3	0.21	1.93	0.70	1.74	212	15
4	0.21	2.10	0.90	1.89	1271	164
5	0.21	2.25	1.15	2.03	1765	342
6	0.34	2.50	1.50	2.26	1765	512
7	0.48	2.67	1.64	2.36	1765	738
8	0.32	2.85	1.74	2.39	1800	799
9	0.32	3.45	2.08	2.48	1800	799
10	0.80	3.69	2.21	2.51	1886	833
11	0.16	4.20	2.50	2.60	1898	834
12	0.16	4.55	2.71	2.63	1913	837
13	0.40	4.75	2.75	2.65	1934	842
14	0.40	4.92	2.84	2.65	1949	855
15	0.50	5.09	2.94	2.65	1968	866
16	0.50	5.37	3.10	2.65	1978	876
17	1.13	5.65	3.26	2.65	1990	887
18	1.14	5.68	3.28	2.66	2000	900
19	1.14	5.72	3.30	2.68	2000	900
20	1.15	5.75	3.32	2.70	2000	900
21	1.16	5.79	3.34	2.72	2000	900
22	0.75	5.83	3.36	2.74	2000	900
23	0.97	5.85	3.38	2.76	2000	900
24	1.44	7.20	4.17	3.07	2000	900
25	1.45	7.27	4.20	3.10	2000	900
26	1.47	7.34	4.24	3.12	2000	900
27	0.75	7.42	4.28	3.14	2000	900

The velocity structure model for the calculation of theoretical Green's functions, modified from Fuis et al. (1982), is given in Table 1. The Q structure model is from Liu (1983). It can be seen from this Q model that nearly all of the attenuation occurs in the top kilometer of the valley. This suggests that the total attenuation along a given ray path will not vary significantly with distance from the source, but instead will be more or less uniform for all ray paths. The strong near-surface velocity gradient will also tend to produce similarity in the raypaths at different ranges. Accordingly, it is not necessary to include different amounts of attenuation for Green's functions calculated at different ranges. In that case, we can use the attenuation in the empirical source function to represent attenuation, rather than remove attenuation from the source function and then include it in the calculated Green's function. In this way, we can avoid specifying poorly known aspects of attenuation at high frequency, and instead include these implicitly within the empirical source functions.

The El Centro array intersects the surface trace of the Imperial Fault at right angles near the northern end of the 1979 rupture segment, as shown in Figure 3. Observed and simulated accelerograms are plotted as a function of distance to the fault trace in Figures 5 through 8. The profiles are centered on the fault trace, with the stations to the northeast above and the stations to the southwest below. The profiles contain lines showing the moveout across the array of arrivals from two locations on the fault. The curved, dashed line is a smooth curve drawn through the approximate arrival time of the energy contributed from the center of the asperity. A large acceleration pulse at this time is evident in both the observed and simulated accelerograms and on both horizontal components in Figures 5

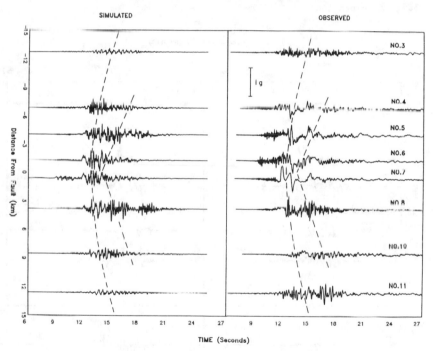

ACCELERATIONS (230 COMPONENT)

Figure 5. Simulated accelerograms using the Hartzell and Heaton (1983) faulting model (left) and observed data (right) for the fault normal components of the El Centro Array stations No.3 through No.11. Accelerations are plotted as a function of perpendicular distance to the surface fault trace. Travel time in seconds is displayed along the bottom of the profile. The amplitude scale shown is the same for all accelerations. The smooth curve represents the approximate arrival time of energy from the asperity. The straight line represents the arrival time of energy from the closest element on the fault plane.

and 6. We know from the simulations that this pulse originated from the region of large slip (the asperity) that lies between the epicenter and the array. The moveout of this pulse across the array uniquely determines its origin as being the asperity.

The straight, dashed line is a curve through the approximate arrival time of energy from the subfault element nearest the array. Since the asperity is centered below Interstate 8 (Figures 3 and 4) and the rupture propagated north-westward from the hypocenter, energy from the asperity always arrives before energy from the closest part of the fault at any of the array stations. Energy from the nearest portion of the fault is apparent in both the observed and simulated accelerograms. However, the amplitude of this motion is less than that originating from the asperity, despite the fact that its source is much closer. This indicates that the asperity did in fact radiate large accelerations.

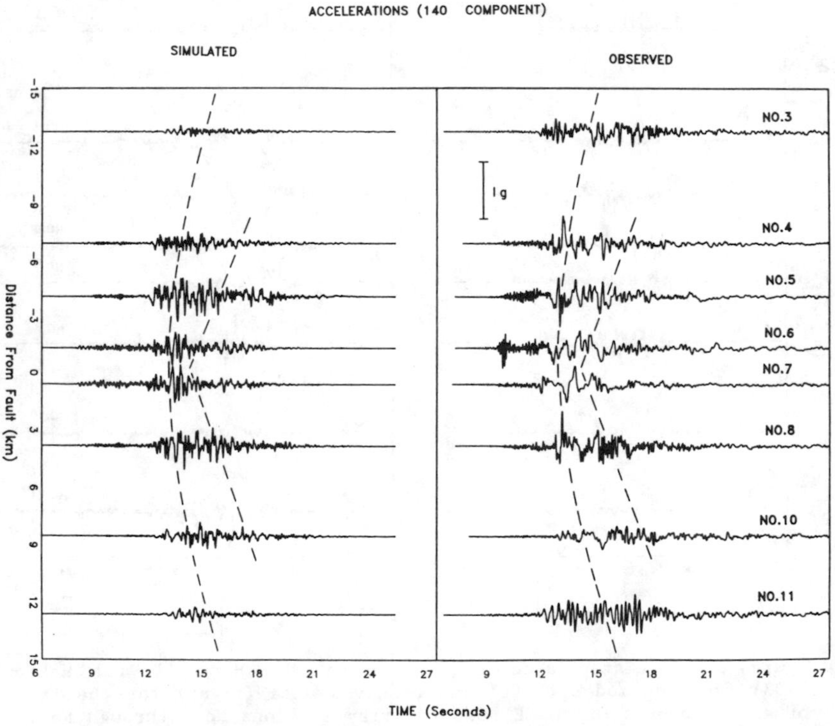

Figure 6. Simulated accelerograms using the Hartzell and Heaton (1983) faulting model (left) and observed data (right) for the fault parallel components of the El Centro Array stations No.3 through No.11. (Same format as in Figure 5)

Imposing a uniform slip distribution on the fault produces simulated accelerograms shown in Figures 7 and 8 which lack arrivals having the moveout of the asperity and whose duration characteristics have less resemblance to the observed accelerograms. We conclude that the asperity model is required in order to explain the observed accelerations. Although this asperity model was determined by longer period near-field velocity and teleseismic records, it is clear that the same asperity model plays an important role in determining the character and content of higher frequency accelerations.

The peak ground accelerations of the horizontal components of motion for the El Centro Array stations profiled in the previous section are plotted as a function of distance to the fault in Figure 9. The simulations agree well with the observations out to distances of about 10 kilometers. Beyond 10 km, the simulations underestimate the observed data. Hartzell and Heaton (1983) obtained a similar result in their velocity simulations using the same faulting model.

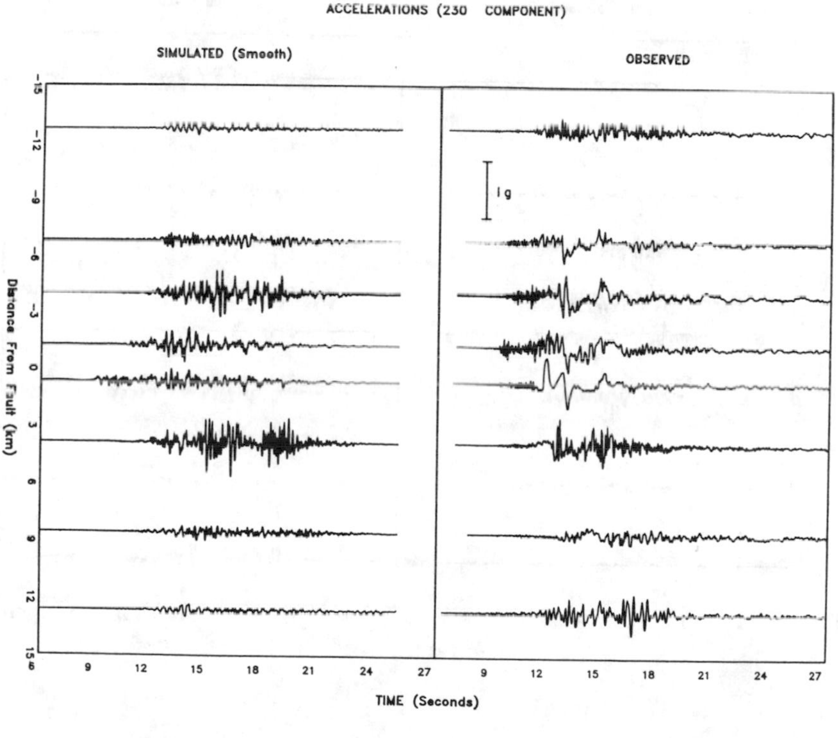

ACCELERATIONS (230 COMPONENT)

SIMULATED (Smooth)   OBSERVED

Figure 7. Simulated accelerograms using a smooth rupture model (left) and observed data (right) for the fault normal components of the El Centro Array stations No.3 through No.11. (Same format as in Figure 5)

The discrepancy is presumably due to inaccuracies in the representation of complex wave propagation effects in the Imperial Valley.

The simulated horizontal component accelerations and response spectra at El Centro array stations 5 and 8, which lie on opposite sides of the fault at distances of about 4 km, are compared with the recorded data in Figures 10 and 11. The acceleration response spectra show good agreement in shape and absolute amplitude at frequencies greater than about 1 Hz, and underpredict the data at lower frequencies. This may be partially due to the simplicity of the Green's functions that we use: they contain neither near-field terms nor surface waves. The empirical source functions derived from the aftershock recordings are also deficient in surface waves due to the depth of that event.

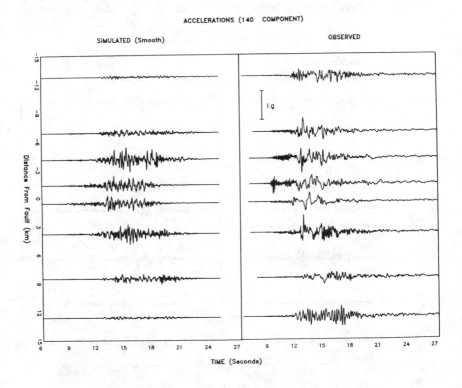

Figure 8. Simulated accelerograms using a smooth rupture model (left) and observed data (right) for the fault parallel components of the El Centro Array stations No.3 through No.11. (Same format as in Figure 5)

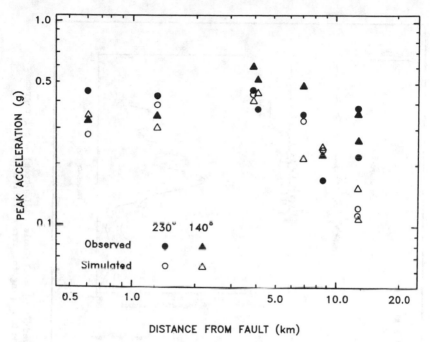

Figure 9. Attenuation of peak acceleration with distance from the fault of simulations (open symbols) compared with the data (solid symbols).

CONCLUSIONS

We have validated a procedure for the simulation of strong ground motions close to large earthquakes by comparing the observed and simulated accelerograms of the Imperial Valley earthquake. In this process, we have demonstrated that a region of large slip on the fault plane that was identified from lower frequency waves was also a strong source of high frequency waves. We have obtained good agreement between observed and simulated ground motions, as represented in time histories, response spectra, and peak acceleration attenuation with distance. The simulation procedure has been further validated against the strong motion recordings of the 1987 Whittier Narrows earthquake (Wald et al., 1988). These successful tests indicate that it is appropriate to use the procedure for the site-specific estimation of ground motions close to large earthquakes.

ACKNOWLEDGMENTS

This work was sponsored by Pacific Gas and Electric Company under the direction of Dr. Y.B. Tsai.

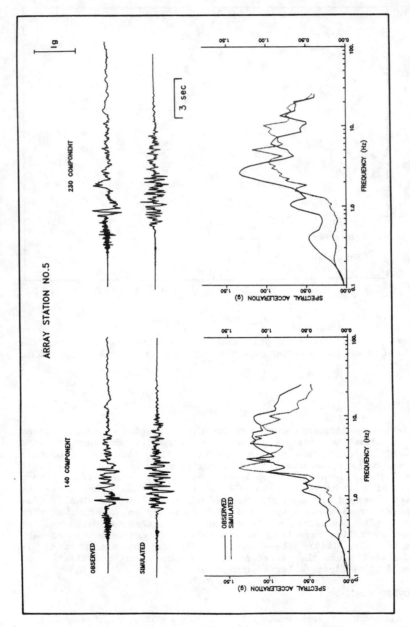

Figure 10. Comparison of observed (top) and simulated (bottom) accelerograms and response spectra with 5 percent damping for array station No.5.

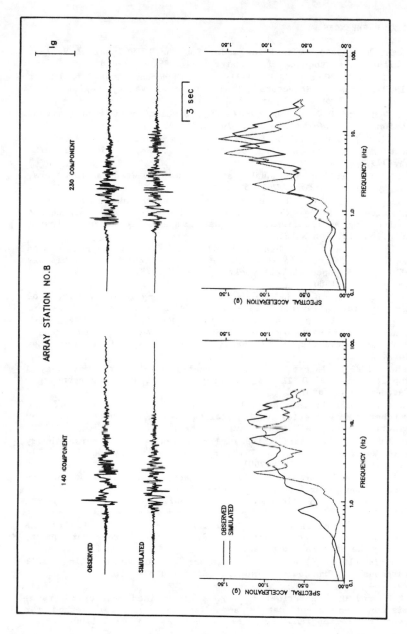

Figure 11. Comparison of observed (top) and simulated (bottom) accelerograms and response spectra with 5 percent damping for array station No. 8.

APPENDIX.-REFERENCES

Archuleta, R. (1984). A faulting model for the 1979 Imperial Valley, California, earthquake, *J. Geophys. Res.*, **89**, 4559-4585.

Fuis, G.S., W.D. Mooney, J.H. Healy, G.A. McMechan, and W.J. Lutter (1982). Crustal Structure of the Imperial Valley region, *U.S. Geol. Surv. Profess. Paper 1254*, 25-49.

Hadley, D.M. and D.V. Helmberger (1980). Simulation of strong ground motions, *Bull. Seism. Soc. Am.* **70**, 617-630.

Hadley, D.M., D.V. Helmberger, and J. A. Orcutt (1982). Peak accelerations and scaling studies, *Bull. Seism. Soc. Am.* **72**, 959-979.

Hartzell, S. (1978). Earthquake aftershocks as Green's functions, *Geophys. Res. Letters*, Vol. 5, 1-4.

Hartzell, S. (1985). The use of small earthquakes as Green's functions, Strong Ground Motion Simulation and Earthquake Engineering Applications, Earthquake Engineering Research Institute, publication No.85-02.

Hartzell, S.H. and T.H. Heaton (1983). Inversion of strong-ground motion and teleseismic waveform data for the fault rupture history of the 1979 Imperial Valley earthquake of 1979, *Bull. Seism. Soc. Am.* **73**, 1553-1584.

Hartzell, S.H. and D.V. Helmberger (1982). Strong-motion modeling of the Imperial Valley earthquake, *Bull. Seism. Soc. Am.* **72**, 571-596.

Helmberger, D.V. and D.G. Harkrider (1978). Modeling earthquakes with generalized ray theory, Proceedings of IUTAM Symposium: Modern Problems in Elastic Wave Propagation, John Wiley & Sons, Inc., New York, 499-518.

Liu, H. (1983). Interpretation of near-source ground motion and implications, *Ph.D Thesis*, California Institute of Technology, Pasadena, California.

Liu, H. and D.V. Helmberger (1985). The 23:19 Aftershock of the October 1979 Imperial Valley earthquake: more evidence for an asperity, *Bull. Seism. Soc. Am.* **75**, 689-708.

Olson, A.H. and R.J. Aspel (1982). Finite faults and inverse theory with applications to the 1979 Imperial Valley earthquake, *Bull. Seism. Soc. Am.* **72**, 1969-2001.

Spudich, P. and E. Cranswick (1984). Direct observations of rupture propagation during the 1979 Imperial Valley earthquake using a short baseline accelerometer array, *Bull. Seism. Soc. Am.* **74**, 2083-2114.

Wald, D.J., L.J. Burdick, P.G. Somerville (1988). Simulation of the recorded accelerograms from the 1987 Whittier Narrows Earthquake, *Earthquake Spectra*, in press.

Wu, Francis (1978). Prediction of strong ground motion using small earthquakes, *Proceedings 2nd Int. Conf. on Microzonation*, vol.II, 701-704.

Yao, Z.X. and D.G. Harkrider (1983). A generalized reflection-transmission coefficient matrix and discrete wavenumber method for synthetic seismograms, *Bull. Seism. Soc. Am.* **73**, 1685-1700.

NEAR FIELD GROUND MOTIONS ON ROCK FOR LARGE SUBDUCTION EARTHQUAKES

R.R. Youngs[1] M.ASCE, S.M. Day[2], and J.L. Stevens[3]

ABSTRACT

Attenuation relationships are presented for estimating peak acceleration and spectral velocities on rock sites in the near field of large subduction zone earthquakes. The attenuation relationships were developed from regression analysis of recorded ground motions and numerical simulations of ground motions for large earthquakes. The empirical data consists of the available recordings obtained on rock from 60 earthquakes including the 1985 events in Chile and Mexico. Attenuation relationships developed from the recorded data provide estimates of near field ground motions for events up to moment magnitude M_w 8. However, the empirical data does not constrain the form of near field magnitude scaling needed to estimate the ground motions for events larger than M_w 8.

Near field ground motions for events of magnitude $M_w \geq 8$ were simulated by superposition of a large number of subevents. The source models for the subevents were derived from finite difference simulations of faulting and wave propagation was modeled using ray theory. The numerical simulations were used to extrapolate the empirical attenuation relationships to larger magnitude events. The resulting attenuation relationships indicate that near field (20 to 40 km source-to-site distances) high frequency ground motions from great subduction zone thrust earthquakes are not expected to have greatly different amplitudes than may result from large shallow crustal earthquakes at similar distances.

INTRODUCTION

The evaluation of seismic hazards along actively subducting margins may require the estimation of near field (<50 km) ground motions from large interface thrust earthquakes. The existing published attenuation relationships for ground motions from subduction zone earthquakes (Iwasaki and others, 1978; Sadigh, 1979; NOAA, 1982; Mori and others, 1984; Vyas and others, 1984; Kawashima and others, 1984) typically indicate that at distances greater than 50 km from the earthquake rupture ground motions from subduction zone earthquakes are substantially larger than those from shallow crustal earthquakes. Use of the published relationships for estimation of near field motions that might result from large interface thrust earthquakes requires extrapolation beyond the empirical data base, which consists of recordings at distances greater than 50 km primarily from events of magnitude $\leq M_w$ 7.5. These extrapolations require assessment of the appropriate form of near field distance and magnitude scaling, concepts of some controversy even for shallow crustal earthquakes. In addition, the published relationships have been derived largely on the basis of soil site recordings.

[1] Geomatrix Consultants, 1 Market Plaza, Spear St Tower, Suite 717
    San Francisco, CA 94105
[2] San Diego State University, Geological Sciences, San Diego, CA 92182
[3] S-Cubed, P.O. Box 1620, La Jolla, CA 92038-1620

The occurrence of magnitude M_w 8 events in Chile and Mexico during 1985 provided a significant extension of the existing strong motion data base. Figure 1 provides a scattergram of the available recordings on rock or rock-like material (shear wave velocity ≥ 750 m/sec) and on soil sites from subduction zone earthquakes. As can be seen, the 1985 recordings have significantly expanded the data base for large magnitude, near field strong motion recordings on rock. These data now provide a reasonable basis for estimating near field motions for events of magnitude ≤ M_w 8.

In this study attenuation relationships for estimating peak horizontal accelerations and 5-percent damped horizontal spectral velocities on rock from subduction zone earthquakes in the magnitude range 5 ≤ M_w ≤ 8 were developed from regression analyses of the strong motion data shown in Figure 1. The soil site data were used to investigate the variance structure of the data and to aid in testing various hypotheses. The results of numerical simulations of ground motions from large plate interface thrust events provided the bases for extrapolation of the attenuation relationships to earthquakes of magnitude > M_w 8.

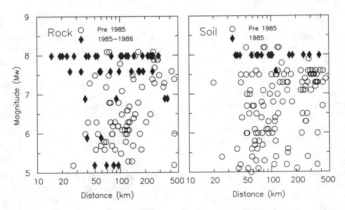

Figure 1 - Distribution of strong motion data from subduction zone earthquakes used in this study.

ANALYSIS OF RECORDED GROUND MOTIONS

Strong Motion Data - The data set collected for this study consists of peak acceleration values for 197 rock recordings and 389 soil recordings and response spectra from 20 rock recordings. The primary sources of the data were: for Alaska, Beavan and Jacob (1984); for Chile and Peru, Saragoni and others (1982, 1985); for Japan, Mori and Crouse (1981); for Mexico, Bufaliza (1984), Anderson and others (1986, in press a and b); and for the Solomons, Crouse and others (1980). Event magnitudes were converted to moment magnitude, M_w, (Hanks and Kanamori, 1979) following the procedure used by Beavan and Jacob (1984) in assembling a strong motion data catalog for Alaska. The Hanks and Kanamori (1979) relationship was used to obtain M_w for events with published seismic moments. If no seismic moment was reported for an event, then the surface wave magnitude was used, provided that it fell in the appropriate range of M_s 5 to 7.5, consistent with the definition of the moment magnitude scale. If only body wave magnitude was reported, then m_b values in the range of 5 to 6 were converted to M_s using the relationship $M_s = 1.8 \cdot m_b - 4.3$ proposed by Wyss and Habermann (1982) and the resulting value was taken to be equal to moment magnitude. The distance measure used was closest distance to the rupture surface. If no rupture area has been published for an event (typically the case for events smaller than M_w 7.5), then hypocentral distance was used.

The distinction between soil and rock sites was made on the basis of the site conditions listed in the various data sources. The recording station at the Geophysical Institute in Lima, Peru was classified as a rock-like site on the basis of the reported subsurface shear wave velocities and evaluations of the site response during past earthquakes (Repetto and others, 1980). The recording station at the School of Engineering in Santiago, Chile was also classified as a rock-like site as it is located on deposits similar in nature to those underlying the Lima site. It should be noted that several of the recording stations for the 1985 Chile earthquake listed as located on rock in Wyllie and others (1986) are actually located on soil deposits, notably the stations at Llolleo (Algermissen, 1985) and Melipilla (Algermissen, personal communication 1987). The recordings obtained on very soft lake deposits such as those in the Mexico City area were not included in the soil data sets as they may represent soil sites with special amplification characteristics.

The ground motion parameters (peak horizontal acceleration, 5% damped horizontal spectral velocity) were characterized in terms of the geometric average of the two horizontal components of motion. This approach was used to remove the effect of component-to-component correlations that affect the validity of statistical tests assuming individual components of motion represent independent measurements of ground motion (Campbell, 1987). The results of the regression analyses indicate no significant difference between the estimates of variance about the median relationships obtained using the geometric mean of two components and obtained using both components as independent data points.

Attenuation Relationship for Peak Horizontal Acceleration on Rock - Figure 2 shows the peak horizontal acceleration data used in the analysis. As indicated, the peak accelerations recorded on soil are significantly higher on average than those recorded on rock over the full range of magnitudes and distances. Although exhibiting a large degree of scatter, the data show a trend toward near field distance saturation of ground motion levels that is currently accepted by most investigators for ground motions from shallow crustal earthquakes. Consequently, the general form of attenuation model used for shallow crustal ground motions was employed in the analysis of the data. The specific form of the relationship is:

$$\ln(a_{max}) = C_1 + C_2 M_w - C_3 \ln[R + C_4 \exp(C_5 M_w)] - R + \epsilon \tag{1}$$

where a_{max} is peak acceleration, R is closest distance to the zone of rupture in kilometers, represents an anelastic attenuation coefficient, C_i are coefficients determined from the data, and ϵ represents a normally distributed random error with zero mean. Initial analyses of the data indicated that had to be constrained to ≥ 0 to produce physically reasonable positive values for both the rock and soil data sets. Therefore, the term R was dropped from the attenuation relationship.

The parameters of Equation 1 were obtained from the data sets shown

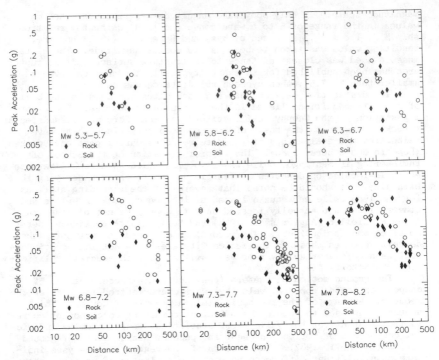

Figure 2 - Peak horizontal acceleration data for subduction zone earthquakes recorded on rock and soil.

in Figure 2 using nonlinear multiple regression techniques. The resulting residuals are plotted against magnitude and distance in Figure 3.
Inspection of the residuals indicated no trend with regards to distance
and a reduction in variance with increasing magnitude. Similar dependence of the variance in ground motion on earthquake magnitude has been
reported by Sadigh (1983) and Abrahamson (1988). Both Sadigh (1983) and
Abrahamson (1988) suggest that the variability in the variance for peak
ground motions can be modeled by a linear relationship between magnitude
and standard deviation. The coefficients of such a relationship can be
obtained by minimizing the expression (Gallant, 1987):

$$\Sigma(|e_i| - \sqrt{\pi}\sigma^*/\sqrt{2})^2 \tag{2}$$

where e_i is the error in predicting the ith data point and σ^* represents
the functional form for the standard error, assumed in this analysis to
be linear in terms of magnitude. The differences in the variance estimates for the soil and rock data sets were not statistically significant
and the errors from the two data sets were combined to estimate σ^*,
yielding:

$$\sigma^* = 1.55 - 0.125M_w \tag{3}$$

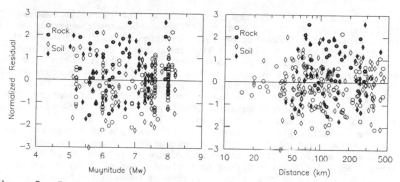

Figure 3 - Residuals from fitting Equation 1 to data shown in Figure 2. Open symbols denote interface events and solid symbols denote intraslab events.

A weighted least squares approach (Draper and Smith, 1981; Gallant, 1987) was used for all subsequent analyses with weights inversely proportional to the variance as defined by Equation 3.

Subduction zone earthquakes can be grouped into two basic types of events: low angle thrust earthquakes occurring on plate interfaces and high angle, predominately normal faulting earthquakes occurring within the downgoing plate. As it has been suggested that the type of fault rupture may have an effect on median ground motion levels (McGarr 1984; Campbell, 1987) possible differences between ground motions from interface and intraslab events were investigated. The differentiation between interface and intraslab events was done on the basis of focal mechanisms, when reported, or on the basis of focal depth, with events below a depth of 50 km considered to be intraslab events. While it is unlikely that interface events would occur at depths greater than 50 km, intraslab events do occur at depths less than 50 km, and it is possible that some intraslab events have been misclassified as interface events.

The residuals for interface and intraslab events are shown in Figure 3 by the open and solid symbols, respectively. As can be seen, the residuals for intraslab events are on average higher than those for interface events although the trend is not as obvious for the soil data as it is for the rock data. Application of the likelihood ratio test for nonlinear regression models suggested by Gallant (1975a, b) indicates that the hypothesis that the coefficients of Equation 1 are the same for intraslab and interface events can be rejected at the 0.05 percentile level for both the rock and soil data sets. As the intraslab events tend to be both deeper and to produce higher ground motions, the possibility of including a term proportional to depth of rupture in Equation 1 was explored. The results indicate that no significant reduction in the standard error is achieved beyond separation of the data into the two subsets.

To test the significance of the observed differences in the residuals shown in Figure 3, Equation 1 was modified to include a "dummy" variable (Draper and Smith, 1981) to identify data from interface and intraslab events, yielding the relationship:

$$\ln(a_{max}) = C_1 + C_2 M_w - C_3 \ln[R + C_4 \exp(C_5 M_w)] + BZ_t + \epsilon \qquad (4)$$

where Z_t is zero for interface events and one for intraslab events. The coefficient B measures the average difference between the ground motions from interface and intraslab events. Equation 4 was fit to the data (using weights based on Equation 3), resulting in an average value of B = 0.54 for the soil and rock data sets. Application of the likelihood ratio test indicates that the hypothesis that B = 0 can be rejected at the 0.05 percentile level for both rock and soil data sets. Further extensions of the model to include a modifying effect of rupture type on the other coefficients of Equation 1 produced no further decrease in the estimated variance and were rejected.

The validity of the systematic difference between ground motions from interface and intraslab events was further investigated using only the data from sites with multiple recordings of both types of earthquakes. Equation 4 was modified to include a set of dummy variables, one for each site, to remove the differences between the median ground motions at a site and the overall median over all sites — in essence removing the effects of systematic site effects. Coefficients C_1 through C_5 were held fixed at the values obtained from the full data set and a regression was performed to obtain the individual site terms and B, the rupture type term. The resulting value of 0.55 agrees very well with the value obtained from the unconstrained regression using the full data set.

The residuals were also examined for evidence of systematic differences between different subduction zones. Systematic differences have been reported previously from examination of teleseismic records (Hartzell and Heaton, 1985; Houston and Kanamori, 1986) and strong motion response spectral ordinates for soil site recordings (Crouse and others, 1988). Both Hartzell and Heaton (1985) and Crouse and others (1988) were unable to identify any correlation between differences in ground motions and physical characteristics of the various subduction zones they studied.

Preliminary analysis of the data set developed for this study (Youngs and others, 1987) suggested that systematic differences may also exist in peak accelerations recorded on rock from different subduction zones. However, subsequent reclassification of site conditions at some of the recording stations and the inclusion of additional data has yielded rock site and soil site data sets that do not exhibit statistically significant differences in the peak accelerations among the different subduction zones.

The resulting median attenuation relationship for rock is:

$$\ln(a_{max}) = 19.16 + 1.045M_w - 4.738\ln[R+205.5\exp(0.0968M_w)] + 0.54Z_t \quad (5)$$

This relationship is compared with the recorded data in Figure 4.

Attenuation Relationships for Spectral Velocity - Attenuation relationships for spectral velocity (S_v) on rock were developed using the procedures employed by Sadigh (1983, 1984). This involves developing relationships for the ratio S_v/a_{max} as a function of magnitude and distance and then applying these to attenuation relationships for peak ground acceleration. The advantages of this approach are that there is a much larger data base of peak acceleration data (Figure 1) than spectral response data (only 20 recordings) for establishing magnitude and distance scaling of absolute levels of ground motion and the use of spectral shapes results in attenuation relationships for various periods

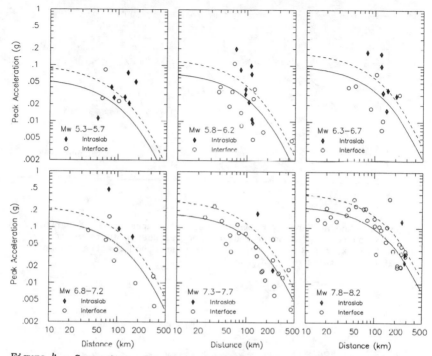

Figure 4 - Comparison of medium attenuation relationships for interface (solid curves) and intraslab (dashed curves) earthquakes with recorded data for rock sites.

that are consistent over the full range of magnitudes and distances to which the relationships apply.

The procedure involves three steps: first, developing a spectral shape for a reference size event for which there is abundant data; second, developing relationships to scale the shape to other magnitudes; and third, computing the standard error of the absolute spectral values about the attenuation relationship. For this analysis, a reference magnitude of M_w 8 was used as there is a relatively large number of rock site recordings for which response spectra are available.

Figure 5a presents median (mean of $\ln[S_v/a_{max}]$) spectral shapes for 5 percent damping developed from magnitude M_w 7.8 to 8.1 data from distances less than 150 km and greater than 150 km. As can be seen, there is a significant difference in spectral shape for the two distance ranges. Because the interest in this study was on near field ground motions, the spectral shape for the < 150 km distance data was used to develop spectral velocity attenuation relationships. Figure 5b shows the smoothed spectral shape for the < 150 km data in terms of spectral acceleration. The maximum spectral acceleration amplification is 2.25 at a period of 0.15 seconds.

The second step is the specification of the variation in spectral shape with earthquake magnitude. Figure 6 presents plots of the ratio

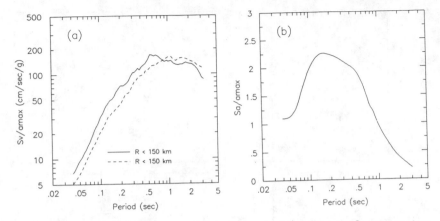

Figure 5 - Median 5-percent damped spectral shapes for M_w 8 earthquakes recorded on rock. (a) shows median spectral shapes in terms of S_v/a_{max} for R < 150 km and R > 150 km. (b) shows smoothed spectral shape for R < 150 km in terms of S_a/a_{max}.

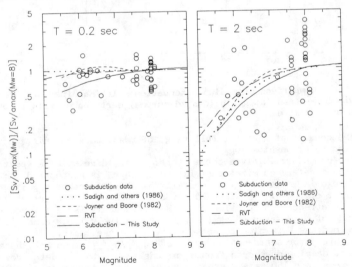

Figure 6 - Relative spectral amplification (5 percent damping) as a function of magnitude.

$[S_v/a_{max}(M_w)]/[S_v/a_{max}(M_w=8)]$ derived from the available response spectra data for recordings on rock sites for periods of vibration of 0.2 and 2.0 seconds. The ratios were obtained by dividing the spectral amplifications for individual events by the amplitude for a magnitude M_w event defined by the smoothed spectral shape shown in Figure 5. Shown also are three curves representing the relative spectral amplification for shallow crustal events derived on the basis of empirical attenuation relationships (Joyner and Boore, 1982; Sadigh and others, 1986) and on the basis of theoretical models employing random vibration theory to estimate ground motions (Hanks and McGuire, 1981; Boore, 1983, 1986). As can be seen, the data for subduction zone earthquakes follows the general trend defined by the relationships for shallow crustal earthquakes. Accordingly, the form of the relationship for spectral amplification employed by Sadigh (1983) was used. Specifically:

$$\ln(S_v/a_{max}) = C_6 + C_7(C_8 - M_w)C_9 \qquad (6)$$

Equation 6 was fit to the data for periods between 0.1 and 3 seconds. In conducting the regression C_8 was fixed at 10 to provide for complete saturation at this magnitude level and C_9 was fixed at 3, representing an average of the values obtained for periods greater than 1 second which exhibit significant magnitude effect on spectral shape. Applying these constraints, parameter C_7 could be fit by the relationship:

$$C_7 = -0.0145 - 0.0063\ln(T) \qquad (7)$$

where T is period of vibration of the oscillator. Parameter C_6 was then fixed to yield spectral amplifications specified for magnitude M_w 8 events by the smoothed spectral shape shown in Figure 5. The resulting relationships are shown in Figure 6 for periods of 0.2 and 2 seconds. The spectral amplifications at these periods are given by:

$$\begin{aligned}
T &= 0.2 \text{ sec} & \ln(S_v/a_{max}) &= 4.278 - 0.0044(10 - M_w)^3 \\
T &= 2.0 \text{ sec} & \ln(S_v/a_{max}) &= 4.960 - 0.0189(10 - M_w)^3
\end{aligned} \qquad (8)$$

in units of cm/sec/g.

The third step is specification of the standard error in $\ln(S_v)$. The standard error was estimated by computing the residuals of the response spectral values about the median estimates from Equations 5, 6, and 7 and Figure 5. The resulting values were less than or equal to the standard deviation for peak acceleration and Equation 3 is considered appropriate for estimating the standard error at all periods.

The attenuation relationships for rock sites defined by Equations 5 through 7 and Figure 5 are considered valid in the magnitude range 5 ≤ M_w ≤ 8 and for distances of 15 to 150 km. Extrapolation to larger magnitude events than have been recorded requires specification of the appropriate near field magnitude scaling law for ground motions. Past applications of the general form of the attenuation relationship defined by Equation 1 have typically followed two limiting cases. One approach has been to assume that the scaling of ground motions with magnitude is independent of distance, implying parameter $C_5 = 0$. Examples of this approach are the attenuation relationships developed by Joyner and Boore (1981, 1982) for western U.S. strong motion data. Attenuation relationships based on self-similar scaling of earthquake source spectra and random vibration theory (Hanks and McGuire, 1981; Boore, 1983, 1986) also imply distance independent magnitude scaling, except for the modifying effect of anelastic attenuation at large distances. The second approach has been to assume ground motions are independent of magnitude at zero distance, im-plying parameter $C_5 = -C_2/C_3$. Examples of this approach are the attenuation relationships developed by Campbell (1981,

1987).
        It is likely that the true form of near field magnitude scaling law
is intermediate between the above limiting cases. The attenuation relationships developed by Seed and Schnable (1980), Sadigh (1983, 1984) and
Sadigh and others (1986) are examples of intermediate magnitude scaling
laws. These relationships exhibit distance independent magnitude scaling for events below magnitude 6.5 and impose the constraint of magnitude independent peak accelerations at zero distance for events of magnitude greater than 6.5. Joyner (1984) has proposed that there is a
critical earthquake above which the self-similar scaling of earthquake
source spectra no longer applies. He suggests that the high frequency
corner of the source spectrum becomes fixed for events that rupture the
entire width of the seismogenic zone, resulting in a reduction by a
factor of about 2 in the increase in ground motion amplitude per unit
increase in magnitude for events above the critical size. Joyner estimates the critical magnitude to be approximately 6.5 for crustal events
in the western U.S.

GROUND MOTION SIMULATIONS

        The appropriate form of the near field magnitude scaling was evalu-
ated on the basis of a series of simulated ground motions for large subduction zone thrust earthquakes. The simulated motions were obtained by
the superposition of the motions from a large number of subevents propagated to the recording site using ray theory (Day and Stevens, 1987).
In the earthquake source model the rupture surface is represented by an
assemblage of subregions of average dimension a. The radiation from
each subregion slip episode is obtained numerically from a dynamic
simulation of faulting based on three-dimensional finite difference
solutions to propagating crack problems (Day, 1982a, b,; Stevens and
Day, 1985). The radiated seismic pulses are scaled to the prescribed
values of subregion dimension and local subregion stress drop. A large
earthquake rupture is simulated by a kinematically prescribed superposition of subregion radiations with a stochastic element incorporated.
        Figure 7 shows schematically how subregion contributions are com-
bined. Each frame in Figure 7 is a snapshot of rupture at a given time.
A global rupture front sweeps the fault with a prescribed rupture velocity of 90 percent of the shear wave velocity. When a subregion is
subsumed by the global rupture front, a subevent is triggered in that
subregion. Shading in Figure 7 indicates subregions that are actively
slipping and an arrow denotes subregions in which a slip episode has
been completed. The stress drop and source dimension of the subevents
are selected randomly from a specified distribution. The mean subevent
stress drop and source dimension were prescribed so as to optimize
agreement between recorded and simulated motions for the 1985 Mexico and
Chile M_w 8 events.
        Expansion of the rupture front and the consequent triggering of
adjacent subregions will reload a subregion, and the model permits
repeated failure of previously slipped regions. This is illustrated in
Figure 7 where, for example, subregion A is triggered at time t_2, is
locked at time t_3, but is then reloaded by expansion of the rupture
front and triggers again at t_4. This secondary event and other secondary subevents additionally load and trigger adjacent regions. For an
overall fault dimension that is large compared to the subregion size,
this retriggering may have to occur repeatedly to build up sufficient
slip to accord with the prescribed seismic moment of the earthquake

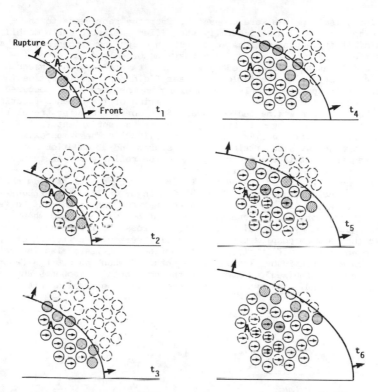

Figure 7 – Schematic representation of rupture propagation, illustrating subevent initiation, reloading, and subsequent secondary failure of previously slipped subregions. Each frame is a snapshot of rupture at a given time. Shading indicates subregions which are actively slipping: an arrow is drawn in a given subregion for each completed slip episode.

being simulated. The number of slip episodes required, N, is just the average fault slip associated with the prescribed seismic moment of the simulation divided by the average subevent slip. If we express the subevent slip in terms of the subevent dimension, a, and subevent stress drop, $\Delta\tau_L$, then N is given by

$$N = W\Delta\tau_G/a\Delta\tau_L \qquad (9)$$

where $\Delta\tau_G$ is the apparent global stress drop and W is fault width. These slip events are distributed randomly over the rise time, denoted T_R, which is obtained from the relationship

$$T_R = K\,W/v_r \qquad 1/2 < K < 1 \qquad (10)$$

where v_r is the rupture velocity.

The seismic pulses were propagated from the subregions to the recording site using a horizontally stratified anelastic earth model. Two simplifying approximations were made in computing ground motions at the surface. First, it was assumed that the Green's function of the earth model can be approximated by geometric ray theory. A consequence of this assumption is the neglect of near field contributions, leading to accurate computations only for wavelengths that are small compared to the source-to-site distance. Second, variation of the Green's function over the subregion was neglected, apart from corrections for travel time variations due to changes in the ray path length (equivalent to the Fraunhofer approximation, Aki and Richards, 1980). These approximations, together with uncertainty of the source and attenuation models at high frequency, limit the frequency range for reliable computation of ground motions to 0.2 to 10 Hz.

The model was first tested by simulating ground motions from the main shock and aftershocks of the 1983 Coalinga, California earthquake sequence (Stevens and Day, 1987). These analyses indicated that better agreement between recorded and simulated motions was obtained when a simple square subevent model with uniform stress distribution was used rather than a more complex model with a slip-weakening criterion of failure and a non-uniform stress distribution. The tests also indicated that much more consistent results were obtained when the Green's functions were "homogenized" by averaging the horizontal components and averaging the radiation pattern coefficient over the focal sphere.

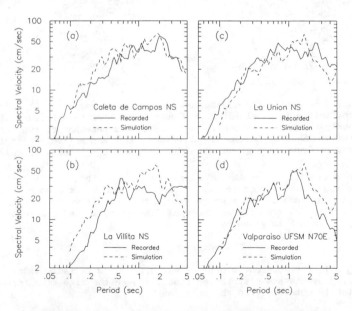

Figure 8 - Comparison of 5-percent damped response spectra for recorded and simulated motions from M_w 8 earthquakes: (a), (b), and (c) from September 19, 1985 Michoacan earthquake, and (d) from March 3, 1985 Chile earthquake.

The ability of the model to generate near field ground motions from
large subduction zone thrust earthquakes was tested by simulating the
near field recordings from the M_w 8.0 September 19, 1985 Michoacan,
Mexico and March 3, 1985 Valparaiso, Chile earthquakes (Day and Stevens,
1987). The simulations were performed using a subregion size of 2.5 km,
an average local stress drop of 38 bars and wave propagation characteristics estimated from the P-wave velocity model given by Havskov and
others (1983). Figure 8 compares the response spectra for simulations
of the ground motions at the three rock sites located above the rupture
of the Michoacan earthquake and one rock site located above the rupture
of the Valparaiso earthquake (the only rock site response spectra presently available to us) with the response spectra for the recorded
motions. As can be seen, the response spectra for the simulations are
in good agreement with those for the recorded motions.

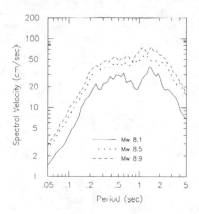

Figure 9 - Response spectra (5 percent damping) for simulated motions on
rock at a distance of 30 km from magnitude M_w 8.1 to 8.9 events.

The appropriate form of the near field ground motion scaling with
magnitude was evaluated by examining the response spectra for a series
of simulated ground motions from events in the magnitude range of M_w 7.7
to 8.9 at sites located 30 to 40 km above the rupture surface. Figure 9
shows typical response spectra for the simulated accelerograms. Figure
10 compares the relative amplitudes of smoothed velocity spectra for
various magnitude events obtained from the simulations with the empirical near field magnitude scaling defined by Equations 5 and 8. Also
shown in Figure 10 are the scaling relationships obtained by imposing
the two limiting conditions of distance independent magnitude scaling
(C_5 = 0) and magnitude independence at zero distance ($C_5 = -C_2/C_3$) on
the empirical data. There is considerable scatter in the simulation
results but a linear fit to the data (dotted line) suggests that the
magnitude scaling law for events above magnitude M_w 8 should have a
flatter slope than indicated by the empirical data from earthquakes of
magnitude < M_w 8.

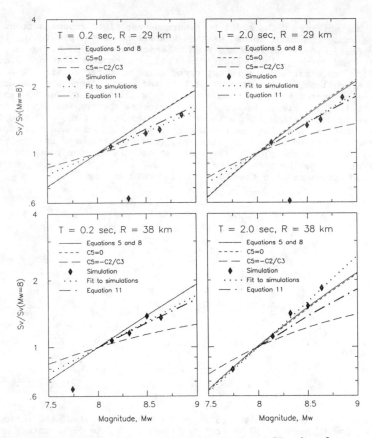

Figure 10 - Comparison of near field magnitude scaling based on simulated motions with scaling developed from empirical attenuation relationships.

DERIVED ATTENUATION RELATIONSHIPS

An attenuation relationship for peak acceleration applicable to earthquakes larger than M_w 8 was developed by constraining parameter C_5 of Equation 1 such that the resulting near field magnitude scaling approximated that indicated by the results of the simulations shown in Figure 10 for events of magnitude $> M_w$ 8 and matched the median values given by Equation 5 for M_w = 8. The resulting relationship is:

$$\ln(a_{max}) = 19.16 + 1.045M_w - 4.738\ln[R+154.7\exp(0.1323M_w)] \qquad (11)$$

The response spectral shape scaling relationships were not altered from those given by Equations 6 and 7 as the magnitude scaling of the spectral shapes for the simulated motions was found to be similar to that obtained for the empirical data.

Figure 11 compares estimates of median response spectra for ground motions on a rock site 30 km above the rupture surface of earthquakes in the magnitude range of M_w 7.25 to 9.5 with response spectra for similar conditions published by Heaton and Hartzell (1986, 1987). Heaton and Hartzell developed their estimates of average response spectra on the basis of ground motions simulated using the empirical Green's function technique. They used empirical Green's functions developed from strong motion recordings of events in the magnitude range of M_w 7 to 7.5. As indicated in Figure 11, the estimates developed from the relationships presented in this paper compare well with Heaton and Hartzell's spectra except at periods longer than about 0.8 seconds. It is likely that the differences at longer periods are primarily due to Heaton and Hartzell's use of recordings obtained on soil sites for their Green's functions. Empirical observations of shallow crustal earthquakes show that long period motions on soil sites are typically a factor of 2 greater than those on rock sites.

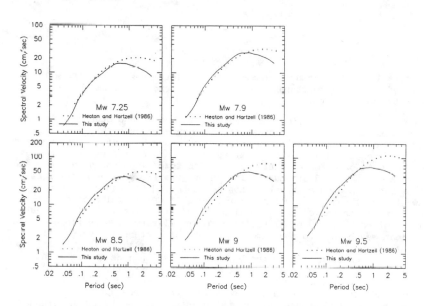

Figure 11 - Comparison of median response spectra (5 percent damping) for various magnitude events at a distance of 30 km from rupture surface predicted using the relationships developed in this study with estimates presented by Heaton and Hartzell (1986).

DISCUSSION

The attenuation relationships developed in this study confirm the results of other investigators in that ground motions from subduction zone earthquakes are expected to be significantly larger than for shallow crustal earthquakes at source-to-site distances greater than 50 km. Figure 12 compares the attenuation relationship developed in this study for peak accelerations on rock from interface thrust earthquakes (Equa-

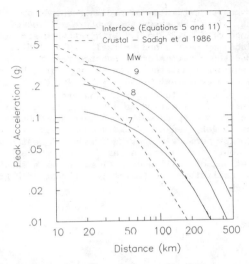

Figure 12 - Comparison of median peak accelerations on rock from large interface thrust earthquakes with peak accelerations from shallow crustal earthquakes.

tion 5) with the relationship developed by Sadigh and others (1986) for rock motions from shallow crustal earthquakes. The relationships for crustal earthquakes are shown only for M_w 7 and 8 events. The curve for M_w 8 represents a significant extrapolation beyond the existing strong motion data base for crustal events. The comparison shown in Figure 12 indicates that at distances of 100 km or greater, peak accelerations from subduction zone thrust events are expected to be about a factor of two larger than those from similar size shallow crustal events. However, at distances of 20 to 40 km both relationships give similar estimates of peak ground acceleration. The similarity of near field ground motions may be due to the extended nature of rupture for very large events.

Differences at the source between subduction zone thrust events and shallow crustal events may also account for the relatively low near field amplitudes for very large events. Zhuo and Kanamori (1987) report that subduction zone thrust events have lower source spectra at frequencies around 1 Hz compared to crustal events like the 1971 San Fernando, California and 1983 Coalinga, California, events. Data from these two events played a significant role in the development of the shallow crustal relationships shown in Figure 12.

ACKNOWLEDGEMENTS

The work reported in this paper was sponsored by the Washington Public Power Supply System.

APPENDIX - REFERENCES

Abrahamson, N.A., 1988, Statistical properties of peak ground accelerations recorded by the SMART 1 array: Bulletin of the Seismological Society of America, v. 78, n. 1, pp. 26-41.
Aki, K., and P.G. Richards, 1980, Quantitative Seismology, Volume I, W.H. Freeman and Co., San Francisco, 512 p.

Algermissen, S.T. (ed.), 1985, Preliminary report of investigations of the central Chile earthquake of March 3, 1985: U.S. Geological Survey Open File Report 85-542.

Anderson, J.G., J.N. Brune, J. Prince, E. Mena, P. Bodin, P. Onate, R. Quaas, and S.K. Singh, 1986, Aspects of strong motion from the Michoacan, Mexico earthquake of September 19, 1985: Proceedings of the 18th Joint Meeting on Wind and Seismic Effects, Washington D.C., May 12-16.

Anderson, J.G., R. Quass, D. Almora M., J.M. Velasco, E. Gueuara O., L.E. de Pavia G., A. Gatierrez R., and R. Vazquez L., (in press a), Guerro, Mexico accelerogram array - Summary of data collected in the year 1985: report prepared jointly by the Instituto de Ingenieria-UNAM and the Institute of Geophysics and Planetary Physics-UCSD.

Anderson, J.G., R. Quass, D. Almora M., J.M. Velasco, E. Gueuara O., L.E. de Pavia G., A. Gatierrez R., and R. Vazquez L., (in press b), Guerro, Mexico accelerogram array - Summary of data collected in the year 1986: report prepared jointly by the Instituto de Ingenieria-UNAM and the Institute of Geophysics and Planetary Physics-UCSD.

Beavan, J., and K.H. Jacob, 1984, Processed strong-motion data from subduction zones: Alaska: Report prepared by Lamont-Doherty Geological Observatory of Columbia University, 250 p.

Boore, D.M., 1983, Stochastic simulation of high-frequency ground motionsbased on seismological models of the radiated spectra: Bulletin of the Seismological Society of America, v. 73, n. 6A, pp. 1865-1894.

Boore, D.M., 1986, Short period P- and S-wave radiation from large earthquakes: Implications for spectral scaling relations: Bulletin of the Seismological Society of America, v. 76, n. 1, pp. 43-64.

Bufaliza, M.A., 1984, Atenuacion de Intensidades sismicas con la distancia en sismos Mexicanos: Tesis, Maestro en Ingenieria (Estructuras), UNAM, Facultad De Ingenieria, 94 p.

Campbell, K.W., 1981, Near-source attenuation of peak horizontal acceleration: Bulletin of the Seismological Society of America, v. 71, n. 6, pp. 2039-2070.

Campbell, K.W., 1987, Predicting strong motion in Utah: Assessment of Regional Earthquake Hazard and Risk Along the Wasatch Front, Utah, U.S. Geological Survey Open File Report 87-585, v. II, pp. L-1-90.

Crouse, C.B, Hileman, J.A., Turner, B.E., and G.R. Martin, 1980, Compilation, assessment and Expansion of the strong motion data base: U.S. Nuclear Regulatory Commission NUREG/CR-1660, 125 p.

Crouse, C.B., Y.K. Vyas, and B.A. Schell, 1988, Ground motions from subduction-zone earthquakes: Bulletin of the Seismological Society of America, v. 78, n. 1, pp. 1-25.

Day, S.M., 1982a, Three-dimensional simulation of fault dynamics: rectangular faults with fixed rupture velocity: Bulletin of the Seismological Society of America, v. 72, n. 3, pp. 705-728.

Day, S.M., 1982b, Three-dimensional simulation of spontaneous rupture propagation: Bulletin of the Seismological Society of America, v. 72, n. 6, pp. 1881-1902.

Day, S.M., and J.L. Stevens, 1987, Simulation of ground motion from the 1985 Michoacan, Mexico earthquake (abs.): Eos, v. 68, n. 44, p. 1354.

Draper, N., and H. Smith, 1981, Applied Regression Analysis, Second Edition, John Wiley and Sons, New York, 709 p.

Gallant, A.R., 1975a, Nonlinear regression: The American Statistician, v. 29, n. 2, pp. 73-81.

Gallant, A.R., 1975b, Testing a subset of the parameters of a nonlinear regression model: Journal of the American Statistical Association, v. 70, n. 352, pp. 927-932.

Gallant, A.R., 1987, Nonlinear Statistical Models, John Wiley and Sons, New York, 610 p.

Hanks, T.C., and H. Kanamori, 1979, A moment-magnitude scale: Journal of Geophysical Research, v. 84, no. B5, pp. 2348-2350.

Hanks, T.C., and R.K. McGuire, 1981, The character of high frequency strong ground motion: Bulletin of the Seismological Society of America, v. 71, n. 6, pp. 2071-2095.

Hartzell, S.H., and Heaton, T.H., 1985, Strong ground motion for large subduction zone earthquakes: Earthquake Notes, v. 55, n. 1, p. 6.

Havskov, J., S.K. Singh, E. Nava, T. Dominguez, and M. Rodriguez, 1983, Playa Azul, Michoacan, Mexico earthquake of 25 October, 1981 (M_s = 7.3): Bulletin of the Seismological Society of America, v. 73, no. 2, pp. 449-457.

Heaton, T.H., and S.H. Hartzell, 1986, Estimation of strong ground motions from hypothetical earthquakes on the Cascadia subduction zone, Pacific Northwest: U.S. Geological Survey Open-File Report 86-328, 69 p.

Heaton T.H., and S.H. Hartzell, 1987, : Science, v. 236, pp. 162-168.

Houston, H., and H. Kanamori, 1986, Source characteristics of the 1985 Michoacan, Mexico earthquake at periods of 1 to 30 seconds: Geophysical Research Letters, v. 13, n. 6, pp. 596-600.

Iwasaki, T., T. Katayama, K. Kawashima, and M. Saeki, 1978, Statistical analysis of strong-motion acceleration records obtained in Japan: Proceedings of the Second International Conference on Microzonation for Safer Construction, Research and Application, v. II, pp. 705-716.

Joyner, W.B., 1984, A scaling law for the spectra of large earthquakes: Bulletin of the Seismological Society of America, v. 74, n. 4, pp. 1167-1188.

Joyner, W.B., and D.M. Boore, 1981, Peak horizontal acceleration and velocity from strong-motion records including records from the 1979 Imperial Valley, California earthquake: Bulletin of the Seismological Society of America, v. 71, n. 6, pp. 2011-2038.

Joyner, W.B., and D.M. Boore, 1982, Prediction of earthquake response spectra: U.S. Geological Survey Open File Report 82-977.

Kawashima, K., Aizawa, K., and K. Takahashi, 1984, Attenuation of peak ground motion and absolute acceleration response spectra: Proceedings of the Eighth World Conference on Earthquake Engineering, San Francisco, v. II, pp. 257-264.

McGarr, A., 1984, Scaling of ground motion parameters, state of stress, and focal depth: Journal of Geophysical Research, v. 89, n. B8, pp. 6969-6979.

Mori, A.W., and C.B. Crouse, 1981, Strong motion data from Japanese earthquakes: World Data Center A for Solid Earth Geophysics, Report SE-29, U.S. Department of Commerce, National Oceanic and Atmospheric Administration, Bolder Colorado.

Mori, J., Jacob, K.H., and J. Beavan, 1984, Characteristics of strong ground motions from subduction zone earthquakes in Alaska and Japan: Earthquake Notes, v. 55, no. 1, p. 16.

NOAA, 1982, Development and initial application of software for seismic exposure evaluation, report by Woodward-Clyde Consultants, prepared for National Oceanic and Atmospheric Administration, Two volumes, May.

Repetto, P., I. Arango, and H.B. Seed, 1980, Influence of site characteristics on building damage during the October 3, 1974 Lima earthquake: University of California, Berkeley Earthquake Engineering Research Center Report EERC-80/41, 63 p.

Sadigh, K., 1979, Ground motion characteristics for earthquakes originating in subduction zones and in the western United States: Sixth Pan-American Conference, Lima, Peru.

Sadigh, K., 1983, Considerations in the development of site-specific spectra: Proceedings of Conference XXIII, Site-Specific Effects of Soil and Rock on Ground Motion and the Implications for Earthquake Resistant Design, U.S. Geological Survey Open-File Report 83-845, pp. 423-458.

Sadigh, K., 1984, Characteristics of strong motion records and their implications for earthquake-resistant design, EERI Seminar No. 12, Stanford, California, July 19, 1984, EERI Publication No. 84-06, v. 12, pp. 31-45.

Sadigh, K., J.A. Egan, and R.R. Youngs, 1986, Specification of ground motion for seismic design of long period structures: Earthquake Notes, v. 57, n. 1, p. 13.

Saragoni, R., Crempien, J., and R. Araya, 1982, Caracteristicas experimentales de los movimientos sismicos Sudamericanos: Universidad de Chile, Revista Del Idiem, v. 21, n. 2, pp. 67-88.

Saragoni, R., Fresard, M., and P. Gonzalez, 1985, Analisis de los acelerogramas del terremoto del 3 de Marzo de 1985: Universidad de Chile, Departmento de Ingenieria Civil, Publicacion SES 1 4/1985(199).

Seed, H.B. and P. Schnabel, 1980, presented in Seed, H.B. and I.M. Idriss, 1980, Ground Motions and Soil Liquefaction During Earthquakes: Earthquake Engineering Research Institute, p. 35.

Stevens, J.L., and S.M. Day, 1985, The physical basis of m_b, M_s and variable frequency magnitude methods for earthquake/explosions discrimination: Journal of Geophysical Research, v. 90, n. B4, pp. 3009-3020.

Vyas, Y.K., Crouse, C.B., and Schell, B.A., 1984, Ground motion attenuation equations for Benioff Zone earthquakes offshore Alaska: Earthquake Notes, v. 55, no. 1, p. 17.

Wyllie, L.A., B. Bolt, M.E. Durkin, J.H. Gates, D. McCormick, P.D. Smith, N. Abrahamson, G. Castro, L. Escalante, R. Luft, R.S. Olsen, and J. Vallenas, 1986, The Chile earthquake of March 3, 1985: Earthquake Spectra, v. 2, n. 2, pp. 249-512.

Wyss, M. and R.E. Habermann, 1982, Conversion of m_b to M_s for estimating the recurrence time of large earthquakes: Bulletin of the Seismological Society of America, v. 72, n. , pp. 1651-1662.

Youngs, R.R., C.-Y. Chang, and J.A. Egan, 1987, Empirical estimates of near field ground motions for large subduction zone earthquakes (abs.): Seismological Research Letters: v. 58, n. 1, p. 22.

Zhuo, Y., and H. Kanamori, 1987, Regional variation of the short-period (1 to 10 second) source spectrum: Bulletin of the Seismological Society of America, v. 77, no. 2, pp. 514-529.

Earthquake Ground Motions
for
Design and Analysis of Dams

J. Lawrence Von Thun[1], Member, ASCE
Louis H. Roehm[2]
Gregg A. Scott[3], Member, ASCE
John A. Wilson[4], Member, ASCE

Abstract

Historical bedrock earthquake records are examined relative to index
parameters corresponding to areas under response spectra, between specified periods related to the natural periods of most dams. It is
judged that this better accounts for the frequency content of the
earthquake records than other parameters such as peak ground acceleration. Attenuation relationships are developed from the historical
data for use in establishing or comparing ground motions used in
analyses. These relationships are then compared to current practice.

I. Introduction

The earthquake loading for which a structure is designed or evaluated
is specified in a general way in terms of magnitude and distance, and
more specifically in terms of ground motion parameters. As more
seismotectonic and geologic information has been gathered and used in
the last 20 years, a tendency toward larger earthquakes for use in the
design and analysis of major structures such as dams has occurred.
Selection of the design earthquake and associated ground motion parameters is in some cases critical to the ability of a dam to satisfy
safety criteria. The selection of ground-motion parameters associated
with a design earthquake both within the USBR (Bureau of Reclamation)
and within the engineering profession, has been influenced by the per-
ceived need for conservatism on a project and the recommendations of
consultants. For design earthquakes of similar magnitude and
distance, the ground motions have varied greatly. A more rational,
consistent selection process should be used, assuming that uncer-
tainities in the seismotectonic environment and the importance of the
proposed facility are accounted for in the selection of the earthquake
magnitude and distance.

[1]Senior Technical Specialist, USBR, Denver, Colorado 80225
[2]Technical Specialist, USBR, Denver, Colorado 80225
[3]Technical Specialist, USBR, Denver, Colorado 80225
[4]Geotechnical Engineer, USBR, Denver, Colorado 80225

A review of current USBR practice in specifying ground motions from the viewpoint of earthquake engineering identified significant variations between the available data base and the ground motion selections being made by the profession, as well as by the USBR. To address this inconsistency, a procedure was developed which can serve as a check of the ground motions specified for a site to verify con-sistency with the available data base or to estimate the degree of conservatism. Alternatively, the procedure may be used as a guide for the development of motions for a site which would therefore be directly related to the existing data base. It is hoped that this paper will form the initial basis for more consistent selection of ground motions for the seismic design and analysis of dams within the USBR. Improvements to the proposed guidelines are anticipated from additional data, suggestions from outside reviewers, and experience gained in their application to current work.

A number of parameters have been used to classify earthquake ground motions, including peak acceleration, peak velocity, peak displace-ment, and variations of these values, such as effective peak accelera-tion. However, the final evaluation of the effect of an earthquake on a dam should be measured in terms of its structural response. This must take into consideration the frequency content of the earthquake record. Response spectra provide a good representation of this fre-quency content; whereas, commonly used scaling parameters such as peak acceleration do not. The authors consider spectrum intensity to be a good representation of response spectra for comparison purposes, and it was selected as the parameter to be evaluated in these studies. Velocity spectrum intensity, as used in this paper, is defined as the area under the velocity response spectrum between periods of 0.1 and 2.5 seconds (Housner, 1959). The parameter "spectrum intensity" as defined above was intended to represent the potential structural response of a wide variety of buildings.

Earthfill and rockfill dams have been shown to have natural periods which, depending on their size, shape, or composition, may range from 0.6 to 2.0 seconds (Makdisi and Seed, 1977.) Because design earth-quakes with significant amounts of energy concentrated in the periods of greatest interest are difficult to locate, design earthquakes are generally selected for embankment dams without regard to the shape of the response spectrum. An appropriate scaling parameter for embank-ment dams should cover a broad range of periods; therefore, velocity spectrum intensity covering periods from 0.1 to 2.5 seconds has been selected.

The use of spectrum intensity to define ground motions for the analy-sis of concrete dams is not a new concept (Tarbox et al., 1979). However, computations and field tests have demonstrated that the natural periods of concrete dams are generally less than 0.5 second (i.e., Rouse and Bouwkamp, 1967, and Rea et al., 1972). Therefore, "acceleration spectrum intensity," which is introduced and defined in this study as the area under the acceleration response spectrum bet-ween periods of 0.1 to 0.5 seconds, is an appropriate indicator of the potential response of concrete dams to a given earthquake record.

In recent years data on ground-motion records have been collected which indicate the possibility that earthquake events resulting from thrust faults produce a greater energy release (total elastic strain energy at the source and/or radiated strain energy at the site) than do normal or strike-slip fault events, for a comparable earthquake magnitude value (McGarr, 1982, Campbell, 1984, 1985, Weichert et al., 1986). Campbell's empirical analysis (1984, 1985), found that reverse and reverse-oblique mechanisms are associated with ground motions approximately 30-40 percent larger than strike-slip mechanisms. McGarr (1982), using theoretical considerations to relate limits for peak accelerations to the state of stress in the earth's crust, computed much higher bounds on peak acceleration for a compressional state of stress (reverse and thrust faults) than for an extensional stress state (normal faults). Since the magnitude scales are a measure of the peak motion of the earthquake in a specific frequency range rather than a measure of the integral effect of the motion, such a consistent bias by source mechanism is possible. It could be important to take this into consideration when selecting ground motions for design and analysis of dams. Verification of a general trend showing thrust fault events to be stronger than strike-slip or normal fault events is difficult because of the inherent variation in earthquake records, the scarcity of large magnitude events with local ground-motion records, the effects of local site conditions, and the inconsistency in magnitude and distance reporting. However, the historical spectrum intensity data were also examined to determine if such a trend exists.

II. Earthquake Record and Parameter Selection

Table 1 lists data for the historical earthquake records that were selected in this study. Conflicting and ambiguous information concerning existing earthquake records is found throughout the literature. Therefore, the sources of information are listed and discussed in this section. The accelerograms of the historical records selected for evaluation were obtained from NOAA (Coffman and Godeaux, 1985), where possible. The spectrum intensity values were calculated from the accelerogram records or from published response spectra using pseudo-spectral values at 5 percent damping. In establishing the spectrum intensity for an event, typically a value is available for each component of ground motion. The larger of the two horizontal components was used to describe the earthquake. The other parameters listed in table 1 are from the following published sources:

1. Coyote Lake - Brady, et al. (November, 1980)
2. San Francisco - Seed, et al. (1974)
3. Imperial Valley - Joyner and Boore (1981), and Brady, et al. (April, 1980)
4. Lytle Creek - Seed, et al. (1974)
5. Mammoth Lakes - Turpen (1980)
6. Friuli, Italy - TVA (1978), and Sabetta and Pugliese (1987)
7. Irpinia, Italy - Sabetta and Pugliese (1987)
8. San Fernando - Woodward-Clyde Consultants (1977)
9. Parkfield - Seed et al. (1974)

Table 1. - Data for Earthquake Rock Records

Earthquake	Richter magnitude*	Fault mechanism	Recording station	Distance (km)	Velocity spectrum intensity (0.1-2.5) (cm)	Acceleration spectrum intensity (0.1-0.5) (cm/s)
San Fernando, 2-9-71	6.4	Thrust	A Lake Hughes No. 4	25	30	112
			B Lake Hughes No. 9	24	17	93
			C Lake Hughes No. 1	30	72	92
			D Santa Felicia Dam	28	33	113
			E Fairmont Reservoir	30	24	66
			F Santa Anita Dam	28	16	115
			G Griffith Park	18	58	157
			H CIT Seismology Lab	20	41	177
Friuli, Italy, 9-15-76	6.0	Thrust	I S. Rocco	5	58	188
Friuli, Italy, 9-15-76	6.1	Thrust	J S. Rocco	12	31	108
Friuli, Italy, 9-11-76	5.9	Thrust	K S. Rocco	12	15	86
Friuli, Italy, 9-11-76	5.5	Thrust	L S. Rocco	19	7	38
Parkfield, 6-27-66	5.6-6.4	Strike-slip	M Temblor	7	52	285
Koyna, India, 12-11-67	6.5	Strike-slip	N Koyna Dam	3	87	364
San Francisco, 3-22-59	5.3	Strike-slip	O Golden Gate Park	11	11	81
Coyote Lake, 8-6-79	5.7	Strike-slip	P San Martin	0	69	230
			Q Gilroy No. 1	8	31	89
			R Gilroy No. 6	1	160	210
Imperial Valley 10-15-79	6.6	Strike-slip	S Superstition Mt. (286)	26	21	135
Helena, 10-31-35	6.0	Normal	T Carrol College	5	45	126
Oroville, 8-1-75	5.7	Normal	U Oroville Dam	10	13	61
Irpinia, Italy, 11-23-80	6.9	Normal	V Sturno	19	169	262
			W Bagnoli Irpino	8	118	102
Irpinia, Italy, 11-23-80	6.3	Normal	X Sturno	19	20	63
			Y Bagnoli Irpino	8	15	31
Lytle Creek, 9-12-70	5.4	Uncertain**	Z Devil's Canyon	19	11	109
			AA Allen Ranch	21	10	60
Mammoth Lakes, 5-25-80	6.5	Uncertain**	BB Long Valley Dam d/s	13	34	91
Mammoth Lakes, 5-25-80	6.7	Uncertain**	CC Long Valley Dam d/s	12	20	90
Mammoth Lakes, 5-27-80	6.3	Uncertain**	DD Long Valley Dam d/s	14	49	148

* Use of the word magnitude throughout this paper refers to the Richter scale.
** Local complications - assumed to be normal or strike-slip events for plotting purposes.

10. Koyna, India - Woodward-Clyde Consultants (1977)
11. Helena - Woodward-Clyde Consultants (1977)
12. Oroville - Woodward-Clyde Consultants (1977)

An attempt was made to utilize distances to the causative fault zone of energy release, rather than epicentral distances. However, only epicentral distances could be found for the Lake Hughes No. 1 (Chang, 1978), Devil's Canyon (TVA, 1978), and Long Valley Dam (Turpen, 1980) recording stations. The reporting of causative fault distance, epicentral distance, or distance from zone of energy release in the literature is inconsistent and unclear. In many publications, the exact meaning of the reported distance is not provided, and attempts to determine the meaning have been unsuccessful. It is not uncommon to find the same earthquake record given various distances in different publications. Seed, et al. (1974), list significantly larger distances for three of the San Fernando recording stations. Conflicting distances can also be found for the Carroll College and Temblor recording stations. A review of other reports indicates that the smallest reported distances are probably correct.

An attempt was also made to utilize only earthquake records from rock sites. The Taft recording station (Kern County earthquake of 1952) represents about the only earthquake greater than M7 recorded, but was not recorded on rock (Chang, 1978). Conflicting information can be found concerning the geologic conditions at the Castaic Old Ridge Route recording station (San Fernando earthquake of 1971). However, in general, this appears to be a stiff soil site (Seed, et al., 1974, and Chang, 1978). There is also some question as to whether the Tolmezzo recording station (Friuli, Italy earthquakes of 1976) should be treated as a rock site (TVA, 1978 and Sabetta and Pugliese, 1987). Westaway and Jackson (1987) indicate that the Calitri recording station (Irpinia, Italy earthquake of 1980) may be on rock. However, the spectrum shape is not typical of a rock record and Sabetta and Pugliese (1987) indicate that it is a soil site. Other information concerning the Irpinia, Italy earthquakes is similar in both publications. Although again conflicting information can be found, it appears that the Lake Hughes No. 12 (San Fernando earthquake of 1971), and Wrightwood recording station (Lytle Creek earthquake of 1970) are founded on shallow alluvium (Chang, 1978). It was suggested that the Logan, Utah, earthquake of 1962, and the Central Greece earthquakes of 1981, might provide additional rock records for normal fault-related events. However, in both cases, the strong motion instruments were located on several hundred feet of soil (for example, Shannon and Wilson, Inc., and Agbabian Associates, 1980). The data corresponding to these nonrock sites are not included in this analysis.

The Borah Peak, Idaho, earthquake of 1983 was also suggested as a possible source of earthquake records. However, no recording stations were within 90 km of the epicenter (Jackson and Boatwright, 1985). The Pacoima Dam record (San Fernando earthquake of 1971) is generally considered to be influenced by topography and local cracking of the instrument house foundation (Woodward-Clyde Consultants, 1977). Thus, the large recorded ground motion is not representative, and was not

included in this analysis. There is some question as to whether the
Koyna record represents free field motions since the instrument was
located in a concrete gravity dam. However, the response spectrum
appears to be representative, and the record was included in the data
base since it represents about the only close-in record for a large
magnitude event.

III. Development of Attenuation Curves from Historical Data

Using acceleration spectrum intensity, the attenuation relationships
for historical earthquake rock records were sketched as shown on
figure 1. This figure includes the data from table 1 for all earthquake mechanisms. Using the velocity spectrum intensity data from
table 1, attenuation relations for historical events, independent of
source mechanism, were sketched as shown on figure 2.

These curves are intended to portray a "reasonable design basis" estimate of spectrum intensity that could be experienced for a given
magnitude event at a given distance from the causative fault. In
establishing the "reasonable design basis" estimate, observations of
events that create a smaller spectrum intensity than that indicated by
the curves would be anticipated, but there would be few observations
at higher levels. However, due to the lack of significant amounts of
data for various earthquake magnitudes, the curves sketched for this
study were drawn on a strictly judgmental basis. Because conservative
criteria are used by the USBR for the specification of MCE's (maximum
credible earthquakes), selection of a ground motion to represent the
MCE by establishing a relationship which envelops all of the ground-
motion data would compound conservatism and may create an unrealistic
loading.

Further examination of the data on figures 1 and 2 shows that while a
generalized relationship between earthquake magnitude and spectrum
intensity may exist, there are clearly inadequate data to draw conclusions with regard to comparisons of thrust events, and normal and
strike-slip events.

It should be noted that most of the data used in developing these
curves are from areas of the western United States and Italy, where
surface faulting generally accompanies large earthquakes. Therefore,
the recommendations of this paper are considered to be applicable for
areas with a similar seismotectonic setting. Areas such as the
central United States, the Pacific Northwest subduction zones, and
many foreign sites may not show a similar response.

IV. Comparisons With Current Practice

    A. Bureau of Reclamation

    In general, concrete dams have been evaluated for earthquake
    loading using linear elastic finite element techniques. For such
    analyses, the USBR has used synthetic accelerograms generated to
    match the frequency characteristics of a response spectrum selected
    for the site.

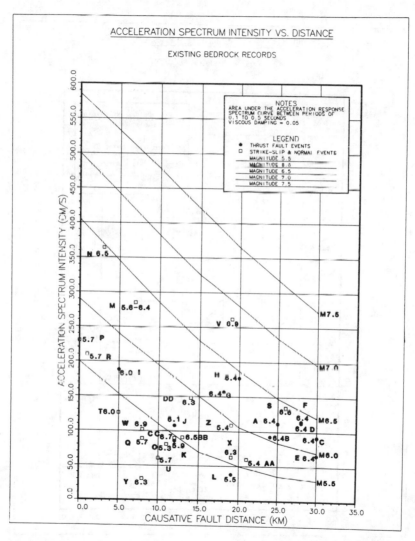

Figure 1

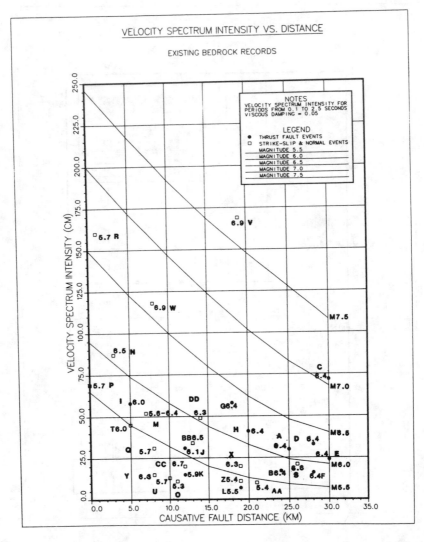

VELOCITY SPECTRUM INTENSITY VS. DISTANCE

EXISTING BEDROCK RECORDS

Figure 2

To date, synthetic accelerograms have been generated for seven response spectra representing earthquakes of Richter magnitudes from 5.5 to 7.5. Each spectrum was developed independently of the others, and in some cases included input from outside consultants. Consequently, consistency was not maintained in their development. Three of the seven independent spectra were subsequently scaled to estimate ground motions at other distances.

In order to compare current seismic loading practice in the USBR to historical records, the acceleration spectrum intensity data for the synthetic accelerograms are compared on figure 3 to the curves developed from the historical data. This comparison shows that the synthetic records are conservative (especially for lower magnitude earthquakes) with respect to the sketched "reasonable design basis curves" except for one M7.5 record and one M6.5 record.

Design earthquake ground motions for analysis of embankment dams have usually been obtained in the USBR by scaling historical earthquake accelerograms according to published procedures (Seed and Idriss, 1982, Krinitsky and Chang, 1977). Scaling parameters for historical records may be determined by evaluating peak acceleration, peak velocity, and/or velocity spectrum intensity. These parameters are used to produce an earthquake ground motion on bedrock representing a particular magnitude event at a specified distance, with an appropriate duration of strong shaking.

The first step in this procedure is to select existing natural earthquake records which are as close as possible to the given MCE's in magnitude, distance, and peak acceleration to minimize the amount of scaling. If the dam is founded on deep alluvium, an accelerogram recorded on alluvium (such as the 1940 El Centro record) may be selected. Once one or two records are selected, their characteristics (peak acceleration, peak velocity, velocity spectrum intensity) are obtained and compared with the desired values (Seed and Idriss, 1982, or Krinitsky and Chang, 1977).

Scaling factors are then computed for peak acceleration, peak velocity, and velocity spectrum intensity by dividing the desired value by the recorded value. Frequently, though, if it is evident that the peak velocities and the velocity spectrum intensities will be reasonably close to or greater than the desired values after scaling, accelerograms are scaled by peak acceleration only. This is often done for earthquake records which are accepted as containing relatively large amounts of total energy, such as the HBS-M7.2-2K-81 accelerogram developed by H. B. Seed by combining and modifying the Taft and Pacoima accelerograms (Seed, personal communication, 1983). This earthquake accelerogram has been referred to as the Pacoima-Taft Modification II record during the last 4 years of use in the USBR. In some cases, the scaling factors may cover a wide range indicating that, for example, peak acceleration may have to be unreasonably large in order to get

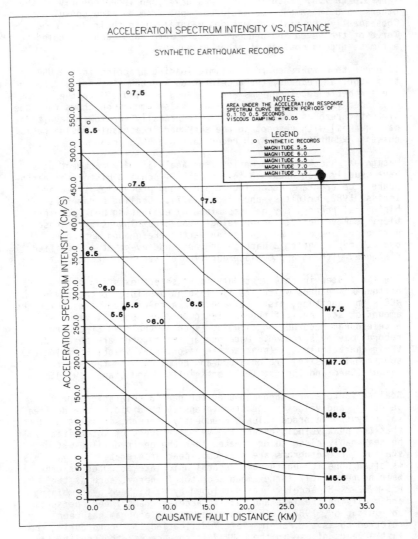

ACCELERATION SPECTRUM INTENSITY VS. DISTANCE

SYNTHETIC EARTHQUAKE RECORDS

Figure 3

velocity spectrum intensities or peak velocities into the desired range of values. It may be appropriate in such cases to try to locate a more compatible accelerogram rather than to perform excessive scaling. Sometimes, however, this is not possible and it becomes necessary to separate the accelerogram record into several segments and scale each part by different amounts.

If the duration of the record needs to be changed, it is done by deleting part of it to shorten it, or by combining records to lengthen it. Another alternative to lengthen a record is to repeat selected portions within a single accelerogram record.

Figure 4 shows the comparison of the magnitude-distance attenuation curves from figure 2 with representative records currently used in design and analysis of embankment dams. With the exception of one magnitude 6.0 event, the velocity spectrum intensity of records currently in use are conservative when compared to the curves developed from the historical data.

In the case of near-field magnitude 6.5-7.5 events, the design earthquakes used for analysis of embankment dams are conservative. For example, the velocity spectrum intensity of the HBS-M7.2-2K-81 accelerogram record which is extensively used for large, near-field earthquakes is far greater than what has occurred in almost all historical cases. It is known to produce a large response in the 1- to 1.5-second period range. Use of conservative ground motions has not created a hardship in design and analysis due to the fact that well constructed embankment dams have an acceptable calculated response to even very strong seismic shaking.

The velocity spectrum intensity of a specific design earthquake is dependent on the velocity spectrum intensity of the recorded motion selected for use. This is shown on figure 4 in which design earthquakes for Richter magnitude 6.5, 7.0, and 7.5 were developed by scaling the HBS-M7.2-2K-81 record, and accelerograms for Richter magnitude 5.5 and 6.0 were developed from the Koyna earthquake record. The difference in velocity spectrum intensity of these two records is the cause of the large gap between the Richter magnitude 6.0 and 6.5 velocity spectrum intensities. The Richter magnitude 7.5 earthquake at 30 km distance was generated by combining and scaling the Taft and Koyna records.

Although the current USBR practice for specifying ground motions for concrete and embankment dams has generally resulted in reasonable or conservative earthquake spectrum contents, this has occurred more as a result of coincidence than by plan. Certainly current practice does not require conformance of design response spectra content with the historical data. Acceleration spectrum intensity appears to provide a valuable indicator of the range and level of earthquake motions important for the analysis and design of concrete dams, and velocity spectrum intensity provides the same for embankment dams. It therefore appears appropriate that for

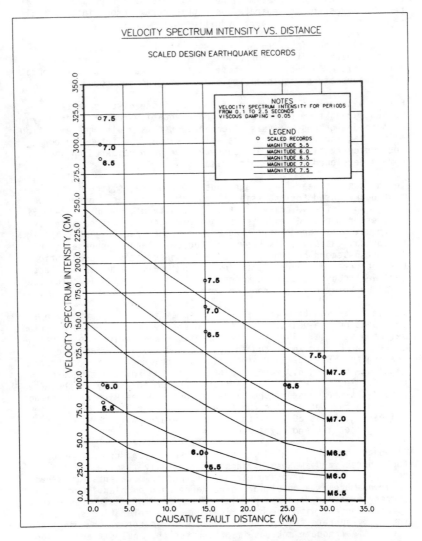

Figure 4

consistency, design earthquake records should be selected so as to reasonably match the attenuation curves developed from these parameters.

B. Seed and Idriss Method

The spectrum scaling approach provided by Seed and Idriss (1982) is fairly widely used in the profession, and provides another basis for comparing the spectral content of actual earthquakes with a design approach. Thus, this approach was compared to the spectrum intensity attenuation curves developed from historical events for five earthquake Richter magnitudes at three distances. The Seed and Idriss spectra, for purposes of the comparison, were obtained as follows:

1. The average peak acceleration was determined from figure 17 of Seed and Idriss (1982). Values of peak acceleration were obtained for magnitudes 5.5, 6.0, 6.5, 7.0, and 7.5 at distances of 5 km, 15 km, and 25 km.

2. Normalized spectral acceleration values for rock were obtained from figure 30 of Seed and Idriss (1982). Values for the mean and the mean plus one standard deviation spectra were obtained at 0.1 second intervals.

3. The normalized velocity response spectra values were computed from the normalized acceleration response spectra values using the relationship $S_v = S_a (T/2\pi)$ were S_v is the spectral velocity, S_a is the spectral acceleration and T is the period.

4. The spectrum intensities were computed as the area under the normalized spectral velocity and normalized spectral acceleration curves from 0.1 to 2.5 seconds and 0.1 to 0.5 second, respectively. These spectrum intensities are for a zero period spectral acceleration value of 1.0.

5. These spectrum intensities were then multiplied by the respective peak acceleration values for magnitude and distance. The results are spectrum intensities that represent a magnitude and distance for both the mean and mean plus one standard deviation spectra. The values are plotted on figures 5 and 6.

The Seed and Idriss data, when compared to the idealized velocity spectrum intensity curves from existing records (figure 5), show that:

1. The velocity spectrum content of the mean spectra proposed by the Seed and Idriss method matches the "reasonable design basis" cuves drawn on the basis of actual records well at the magnitude 6.0 level, but is significantly less for larger magnitude events.

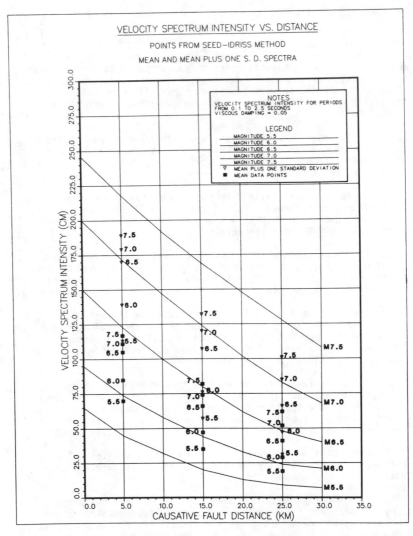

Figure 6

2. The velocity spectrum content of the mean plus one standard deviation spectra proposed by the Seed and Idriss method matches the "reasonable design basis" curves well at the magnitude 7.0 level, but overestimates the lower magnitude events and underestimates the magnitude 7.5 events.

3. The current records being used by the USBR for embankment dam analysis generally have velocity spectrum intensity values well in excess of the Seed and Idriss spectra for earthquakes equal to or greater than magnitude 6.5 (figure 4 compared to figure 5).

4. The Seed and Idriss data are grouped together because of scaling by peak acceleration, which current attenuation relationships show has a small variation as a function of magnitude. It follows then that scaling by peak acceleration would likewise not show significant distinction in spectral content between earthquake magnitudes. However, Idriss (1985) indicates that a spectrum shape that varies with magnitude is being considered. This approach may result in greater variation of spectrum intensity with earthquake magnitude.

The same type of comparison was made for the acceleration spectrum intensity (figure 6). There is more separation for the Seed and Idriss data points at various magnitudes for the acceleration spectrum intensity plot than for the velocity spectrum intensity plot. However, the curves are still grouped together such that they overestimate the acceleration spectrum intensity for some magnitudes, and underestimate it for others using the mean spectra. The Seed and Idriss mean plus one standard deviation spectra overestimate the "reasonable design basis" curves for acceleration spectrum intensity in all cases.

V. Conclusions

The evaluation of the existing, usable ground-motion data characterized according to spectrum intensity concepts indicated that:

A. There is considerable variation between what may be described as a reasonable design basis spectrum content for a given earthquake magnitude as established from the data base of historical ground motions and the spectrum content of ground motions currently being recommended for analysis and design.

B. There was insufficient evidence to conclude, for design basis purposes, that thrust faults produce predictably greater ground motion than strike slip and normal faults based on spectrum intensity as a comparison parameter.

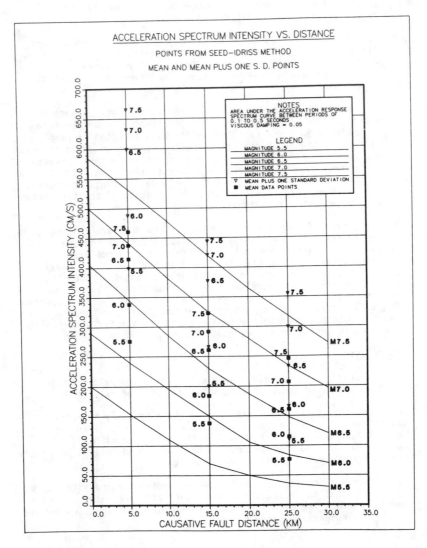

Figure 6

Scaling design earthquakes by acceleration or velocity spectrum intensity should better account for the frequency content of earthquake
records. Therefore, it is recommended that ground motions for analysis of concrete dams be scaled such that the acceleration spectrum
intensity approximates that given by the relation shown on figure 1,
and that the bedrock ground motions for embankment dam analysis be
scaled such that the velocity spectrum intensity approximates that
given by figure 2. If historical records are selected for analysis
based on figures 1 or 2, at least two records should be used in case
one of the records contains low amplitude vibrations corresponding to
the natural period(s) of the dam. This procedure is expected to provide a better representation of the structural response of dams than
scaling by peak acceleration or other less descriptive parameters. It
should also be useful in determining how reasonable ground motions
developed by other methods are. As discussed previously, the procedure is considered to be appropriate since conservative criteria are
introduced in other parts of the seismic analysis procedure, and it is
undesirable to compound conservatism.

Acknowledgements

The assistance of Mr. Kiran Adhya and Mr. Richard Munoz in the preparation of this paper is gratefully acknowledged.

Appendix. References

Brady, A. G., V. Perez, and P. N. Mork, "The Imperial Valley
Earthquake, October 15, 1979, Digitization and Processing of
Accelerograph Records," Open-File Report 80-703, U.S. Geological
Survey, Menlo Park, California, April, 1980.

Brady, A. G., P. N. Mork, V. Perez, and L. D. Porter, "Processed Data
from the Gilroy Array and Coyote Creek Records, Coyote Lake,
California, 6 August 1979," Open-File Report 81-42, U.S. Geological
Survey, Menlo Park, California, November 1980.

Campbell, K. W., "Near-Source Attenuation of Strong Ground Motion for
Moderate to Large Earthquakes - An Update and Suggested
Application to the Wasatch Fault Zone of North-Central Utah,"
Proceedings, Workshop on Evaluation of Regional and Urban
Earthquake Hazards and Risk in Utah, Open-file Report 84-763, U.S.
Geological Survey, 1984.

Campbell, K. W., "Strong Motion Attenuation Relations: A 10-year
Perspective," Earthquake Spectra, Vol. 1, No. 4, pp. 1825-1839,
1985.

Chang, F. K., "Catalogue of Strong Motion Earthquake Records - Vol. 1,
Western United States, 1933-1971," State-of-the-Art for Assessing
Earthquake Hazards in the United States, Report 9, Miscellaneous
Paper S-73-1, U.S. Army Corp of Engineers, Waterways Experiment
Station, Vicksburg, Mississippi, 1978.

Coffman, J. L., and S. Godeaux, "Catalog of Strong-Motion Accelerograph Records," Report SE-38, World Data Center A for Solid Earth Geophysics, National Geophysical Data Center, National Oceanic and Atmospheric Administration, Boulder, Colorado, 1985.

Housner, G. W., "Behavior of Structures During Earthquakes," Journal of the Engineering Mechanics Division, American Society of Civil Engineers, Vol. 85, No. EM4, pp. 109-129, 1959.

Idriss, I. M., "Evaluating Seismic Risk in Engineering Practice," Proceedings, 11th International Conference on Soil Mechanics and Foundation Engineering, Vol. 1, pp. 255-320, San Francisco, California, 1985.

Jackson, S. M., and J. Boatwright, "The Borah Peak, Idaho Earthquake of October 28, 1983 - Strong Ground Motion," Earthquake Spectra, Vol. 2, No. 1, pp. 51-69, 1985.

Joyner, W. B., and D. M. Boore, "Peak Horizontal Accleration and Velocity from Strong-Motion Records Including Records from the 1979 Imperial Valley, California, Earthquake," Bulletin of the Seismological Society of America, Vol. 71, No. 6, pp. 2011-2038, 1981.

Krinitsky, E. L., and F. K. Chang, "Specifying Peak Motions for Design Earthquakes," State-of-the-Art for Assessing Earthquake Hazards in the United States, Report 7, Miscellaneous Paper S-73-1, U.S. Army Corps of Engineers, Waterways Experiment Station, Vicksburg, Mississippi, 1977.

Makdisi, F. I., and H. B. Seed, "A Simplified Procedure for Estimating Earthquake-Induced Deformations in Dams and Embankments," Report No. EERC 77-19, University of California, Berkeley, California, 1977.

McGarr, A., "Upper Bounds on Near-Source Peak Ground Motion Based on a Model of Inhomogeneous Faulting," Bulletin of the Seismological Society of America, Vol. 72, No. 6, pp. 1825-1841, 1982.

Rea, D., C. Liaw, and A. Chopra, "Dynamic Properties of Pine Flat Dam," Report No. EERC 72-7, University of California, Berkeley, California, 1972.

Rouse, G., and J. Bouwkamp, "Vibration Studies of Monticello Dam," Research Report 9, U.S. Bureau of Reclamation, Denver, Colorado, 1967.

Sabetta, F., and A. Pugliese, "Attenuation of Peak Horizontal Acceleration and Velocity from Italian Strong-Motion Records," Bulletin of the Seismological Society of America, Vol. 77, No. 5, pp. 1491-1513, 1987.

Seed, H. B., C. Ugas, and J. Lysmer, "Site-Dependent Spectra for Earthquake-Resistant Design," Report No. EERC 74-12, University of California, Berkeley, California, 1974.

Seed, H. B., and I. M. Idriss, "Ground Motions and Soil Liquefaction During Earthquakes," Earthquake Engineering Research Institute, Berkeley, California, 1982.

Shannon and Wilson, Inc., and Agbabian Associates, "Geotechnical and Strong Motion Data from U.S. Accelerograph Stations," for Nuclear Regulatory Commission, NUREG/CR-0985, Vol. 3, 1980.

Tarbox, G. S., K. J. Dreher, and L. R. Carpenter, "Seismic Analysis of Concrete Dams," 13th International Congress on Large Dams, New Delhi, Vol. 2, pp. 963-994, 1979.

Tennessee Valley Authority (TVA), "Justification of the Seismic Design Criteria used for the Sequoyah, Watts Bar, and Bellefonte Nuclear Power Plants," Civil Engineering Branch, Division of Engineering Design, 1978.

Turpen, C. D., "Strong-Motion Instrumentation Program Results from the May, 1980, Mammoth Lakes, California Earthquake Sequence," California Division of Mines and Geology, Special Report 150, 1980.

Weichert, D.H., R. J. Wetmiller, and P. Munro, "Vertical Earthquake Acceleration Exceeding 2 g? The Case of the Missing Peak," Bulletin of the Seismological Society of America, Vol. 76, No. 5, pp. 1473-1478, 1986.

Westaway, R., and J. Jackson, "The Earthquake of 23 November 1980 in Campania-Basilicata (Southern Italy)," Geophysical Journal of the Royal Astronomical Society, Vol. 90, pp. 375-443, 1987.

Woodward Clyde Consultants, "Earthquake Evaluation Studies of the Auburn Dam Area," Vol. 8, Earthquake Ground Motions, San Francisco, California, 1977.

On Seismically Induced
Pore Pressure and Settlement

Albert T. F. Chen*, M. ASCE

Two different approaches are used to estimate pore pressures and settlement in a 50-ft (15.2-m) sand deposit subjected to a variety of earthquake loadings. Although the two approaches seem consistent in predicting the occurrence of liquefaction, the results show that they are quite divergent in estimating pore-pressure build-ups and magnitude of ground settlement.

INTRODUCTION

Recent advances in geotechnical engineering provide engineers with many alternatives for making estimates of earthquake-induced pore-pressure build-ups and settlements of saturated cohesionless deposits. To an engineer the major decision would be to choose between an empirical approach or an analytical one. Empirical approaches usually are based on field performance whereas analytical ones generally incorporate soil behaviors that have been observed in the laboratory.

Despite the availability of both types of approach, the relative effectiveness of either type has not been established. Centrifuge tests, capable of verifying the validity of a particular soil model, cannot sufficiently reflect the complexity of actual field conditions. Data obtained through instrumentation of potentially liquefiable sites offer some insights but are not abundant enough to clarify all the uncertainties. Consequently, further understanding of the effectiveness (and limitations) of each type of approach is needed.

To better understand the suitability of both the empirical and the analytical approach for engineering practice, a comparison of these two approaches was attempted. The empirical method chosen for the comparison is one developed by Seed and his coworkers (Seed et al., 1983; Seed et al., 1985; Tokimatsu and Seed, 1987) This method, hereby denoted as the simplified procedure, is chiefly based on the field performance of various sites and has provisions for pore-pressure and settlement estimates, in addition to evaluation of the liquefaction potential of a given site. The analytical method chosen for the comparison is the effective stress analysis developed by Finn

---

*Research Civil Engineer, U.S. Geological Survey, Menlo Park, CA

and his coworkers (Finn et al., 1977; Martin et al., 1975). This method, together with the computer program DESRA2 (Lee and Finn, 1978), has been demonstrated fequently in recent literature on geotechnical engineering (for example: National Research Council, 1985).

The comparison focuses on the estimated values of pore pressure and settlement of a given site under a variety of earthquake loadings. Findings from the comparison are presented in this paper. It is noted that results and comments derived therefrom are not necessarily restricted to the applications of these two particular methods but should be valid for other empirical and analytical methods as well.

A COMPARISON

The site considered for the comparison is a hypothetical but nevertheless well-known one. It consists of a 50-ft (15.2-m) uniform deposit of saturated sand over an elastic half space. The sand is identified as an uniform angular quartz sand (Crystal Silica No. 20) which has a relative density of 45% and permeability of 0.003 ft/sec (0.09 cm/sec). The water table is 5 ft (1.5 m) below the ground surface. The same site has been used frequently by Finn and his coworkers for the development of their effective stress model and for the computer program DESRA2. The soil properties required for input to DESRA2 are thus readily available as given in Fig. 1.

The SPT (standard penetration test) blowcounts at the site for input to the simplified procedure are not available but can be obtained from the relative density of the Crystal Silica sand. According to Tokimatsu and Seed (1987), the $(N_1)_{60}$-value of a sand is directly related to its relative density where $(N_1)_{60}$ is the modified SPT blowcount corrected for hammer energy and overburden pressure (Seed et al., 1985). For a 45% relative density, the corresponding $(N_1)_{60}$ is 9.3. This $(N_1)_{60}$-value together with the unit weight of the sand and the ground water condition is sufficient to characterize the site for the simplified procedure.

The seismic inputs used in the comparison are based on two strong-motion records. One is the N65E component at Cholame-Shandon Station 2 during the 1966 Parkfield earthquake (M_L=5.8). The other is the N21E component at the Lincoln School tunnel in Taft during the 1952 Kern County earthquake (M_L=7.2). These records are then scaled to different peak accelerations to provide different shaking intensities. Designations for the input earthquakes considered for the comparison are listed in Table 1.

The 1978 version of DESRA2 used in this study (Lee and Finn, 1978) has no provision for settlement computation. The program, however, does compute the volumetric strain due to shear straining for each soil sublayer and the settlement of the soil deposit can be obtained by integrating these volumetric strains with respect to depth. In addition, DESRA2 assumes that a liquefied soil sublayer

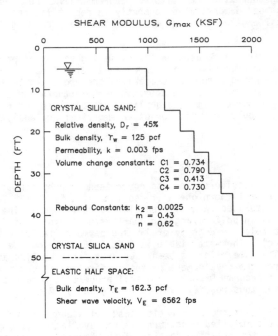

SHEAR MODULUS, G_{max} (KSF)

CRYSTAL SILICA SAND:

Relative density, D_r = 45%
Bulk density, γ_w = 125 pcf
Permeability, k = 0.003 fps
Volume change constants: C1 = 0.734
C2 = 0.790
C3 = 0.413
C4 = 0.730

Rebound Constants: k_2 = 0.0025
m = 0.43
n = 0.62

CRYSTAL SILICA SAND

ELASTIC HALF SPACE:

Bulk density, γ_E = 162.3 pcf
Shear wave velocity, V_E = 6562 fps

Figure 1.-- Site parameters for running program DESRA2
(1 ft=0.305 m; 1 pcf=16 kg/m³; 1 ksf=47.9 kPa)

Table 1. -- List of Input Earthquakes

Designation of Earthquake Input	Strong Motion Record Used	Peak Acceleration (g)	
PK2	1966 Parkfield	1/2	
PK4	1966 Parkfield		1/4
PK8	1966 Parkfield	1/8	
PK12	1966 Parkfield		1/12
KC2	1952 Kern County	1/2	
KC4	1952 Kern County		1/4
KC8	1952 Kern County	1/8	
KC12	1952 Kern County		1/12

maintains an effective stress equal to 5% of the overburden stress and the execution of the program automatically terminates as soon as more than half of the soil sublayers become liquefied. No attempt was made to modify these criteria for this study.

To evaluate the liquefaction potential, the simplified procedure first estimates the average (shear) stress ratio, τ_{av}/σ'_0, developed in a soil sublayer. This stress ratio at a given depth is given by:

$$\frac{\tau_{av}}{\sigma'_0} = 0.65 \cdot \frac{a_{max}}{g} \cdot \frac{\sigma_0}{\sigma'_0} \cdot r_d \tag{1}$$

in which a_{max} = maximum acceleration at the ground surface; g = gravitational acceleration; σ_0 and σ'_0 are the total and the initial effective overburden pressure respectively at the depth considered; and r_d = a stress reduction factor. The stress reduction factor generally varies from a value of 1 at the ground surface to a value of 0.9 at a depth of about 30 ft (9.6 m). In this study, r_d is obtained from Fig. 2 which represents an average of the range of values of r_d's established for a wide variety earthquake motions and soil conditions

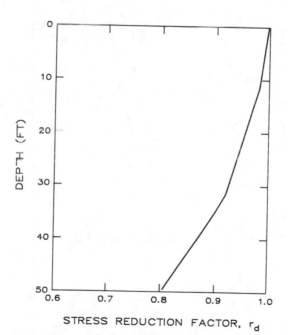

Figure 2. -- Stress reduction factor as a function of depth
(1 ft = 0.305 m)

(Seed and Idriss, 1982; Chen, 1986). The stress ratio is then compared with the liquefaction resistance determined from a chart for a given earthquake magnitude similar to that shown in Fig. 3 (Seed et al., 1985). The factor of safety against liquefaction is defined as the ratio of the liquefaction resistance to τ_{av}/σ'_0. The factor of safety against liquefaction is then used to determine the number of equivalent cycles induced by the earthquake, N_e. In turn, N_e helps to determine the pore pressure build-up according to the relation shown in Fig. 4 in which N_1 is the number of equivalent cycles required to cause liquefaction (DeAlba et al., 1976; Seed et al., 1983). Finally, the settlement of the sand layer is estimated on the basis of the volumetric strain developed in each sublayer. The volumetric strain is assumed to be dependent on the average stress ratio, τ_{av}/σ'_0, and the modified penetration blowcount, $(N_1)_{60}$, as proposed in Fig. 5 by Tokimatsu and Seed (1987).

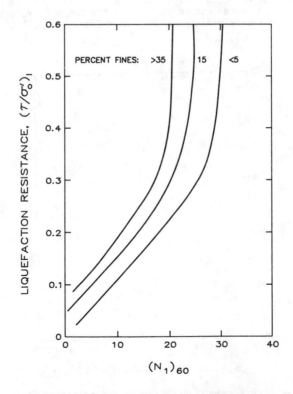

Figure 3. -- Variation of liquefaction resistance witn $(N_1)_{60}$ in silty sands for M=7.5 earthquakes (Seed et al., 1985).

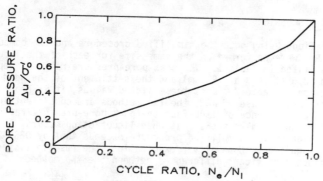

Figure 4. -- Average rate of pore pressure build-up in cyclic
simple shear tests (Seed et al., 1983).

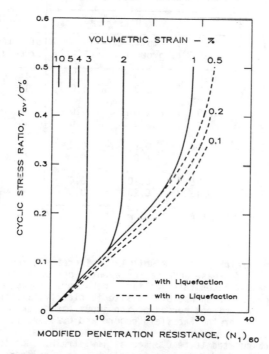

Figure 5. -- Proposed chart for evaluation of volumetric strain for
saturated clean sands during 7.5-magnitude earthquakes
(Tokimatsu and Seed, 1987).

RESULTS AND DISCUSSION

Computations using both the simplified procedure and the computer program DESRA2 were performed on the same site for each of the input earthquakes listed in Table 1. The peak pore-pressure ratio at depths of 22.5 and 32.5 ft (6.9 and 9.9 m) and the settlement of the sand layer are given in Table 2. From these listed values, it is apparent that, except for the case of KC8, the two methods are consistent in predicting the occurrence of liquefaction (peak pore-pressure ratio equals 1.0). Table 2 also shows that when liquefaction occurs, the settlement value by the simplified procedure exceeds that by DESRA2 by an order of magnitude. Upon closer examination, the results listed further reflect significant differences between the two methods considered.

Table 2. -- Comparison of Results from Different Procedures

EVENT	PEAK PORE-PRESSURE RATIO, $\Delta u/\sigma'_0$				SETTLEMENT (in)	
	at 22.5 ft		at 32.5 ft			
	simplified procedure	DESRA2	simplified procedure	DESRA2	simplified procedure	DESRA2
PK2	1.0	1.0	1.0	1.0	14.60	0.92
KC2	1.0	1.0	1.0	1.0	14.60	1.15
PK4	1.0	1.0	1.0	1.0	14.30	1.39
KC4	1.0	1.0	1.0	1.0	14.60	1.45
PK8	0.51	0.77	0.52	0.89	0.51	0.97
KC8	1.0	0.63	1.0	0.64	11.30	0.81
PK12	0.08	0.38	0.10	0.41	0.04	0.30
KC12	0.21	0.46	0.25	0.46	0.48	0.49

Note: 1 in = 2.54 cm; 1 ft = 0.305 m.

When the strain level in a soil layer increases, the shear modulus tends to decrease and the fundamental period of the soil layer lengthens. DESRA2 takes this nonlinear behavior into account and the frequency-dependent response of the site is clearly illustrated in Table 2. The predominant period of the input earthquakes is estimated to be at 0.6 sec for the Parkfield record and 0.25 sec for the Kern County record. The fundamental period of the 50-ft sand deposit at the site is calculated to be 0.3 sec at low strain. Due to increasing strain level, the fundamental period of the site is lengthened to match the predominant period of the Parkfield input. Consequently, according to results from DESRA2, the pore-pressure response due to the Parkfield input starts to exceed that from the Kern County input as the input peak acceleration increases from g/12 to g/8 even though the magnitude is considerably larger for the Kern County input. On the other hand, unless additional measures are taken, the simplified procedure always results in higher pore-pressure response for larger input earthquake magnitudes at a given peak acceleration. The additional measures may include to choose a different r_d from that given in Fig. 2 for different cases or to estimate the state of stress in the soil deposit by other means. Either option, however, will require a great deal of first-hand experience and considerable engineering judgment. Otherwise, the simplified procedure cannot be used to evaluate the frequency-dependent aspects of the response of a site.

It has been shown that settlements predicted by the simplified procedure are consistent with recorded values from several sites in Japan. Consequently, settlement values given in Table 2 by the simplified procedure are probably in the right order of magnitude. By comparison, the values implied by DESRA2 seem to be on the low side for cases in which liquefaction was involved, regardless of whether appropriate actions are taken to correct for effects of multi-directional shaking. Recent centrifuge studies at the California Institute of Technology have shown that the settlement predicted using DESRA2 was about half the value measured (Hushmand et al., 1987). This difference is attributed to inadequate analytical modeling of post-liquefaction behavior and will be discussed below.

It is also noted from Table 2 that settlement values for those cases where liquefaction took place are quite similar in magnitude according to the simplified procedure. This is because volumetric strain seems to depend less on shear stress ratio for certain $(N_1)_{60}$ values when liquefaction occurs. As shown in Fig. 5, the volumetric strain stays at slightly less than 3% for $(N_1)_{60}$=9.3 as long as the shear stress ratio exceeds 0.1.

OTHER CONSIDERATIONS

Post-liquefaction behavior of sands is not yet well understood. This in turns makes analytical predictions of settlement associated with liquefaction difficult. Modeling studies using centrifuges show that the physical process of post-liquefaction phenomenon is not governed by consolidation alone and suggest that the sedimentation-solidification-consolidation theory would be a more appropriate alternative (Hushmand et al., 1987). However, even consideration of sedimentation and solidification does not necessarily explain the post-liquefaction behavior of sand deposits as observed in the field.

On the basis of the recurrence of liquefaction at many sites, some investigators suggest that "....earthquakes are incapable of densifying originally loose deposits to a stable mass...." (Ambraseys and Sarma, 1969), and that "Such action can leave the layers in a looser, more susceptible condition than they were before the earthquake" (Youd, 1984). More recently, the NRC Committee on Earthquake Engineering has endorsed the possibility that "Local volume

change occurs, but the sand as a whole remains at a constant volume
and is 'globally undrained'" (National Research Council, 1985). These
local volume changes associated with water-content redistribution
complicate the post-liquefaction behavior and are no doubt a critical
factor that governs the magnitude of deposit settlement after
liquefaction.

Obtaining the soil properties needed for an analytical procedure
also has interesting implications. Properties derived specifically as
input to DESRA2 have been scarce and the understanding of the role of
each parameter is not common knowledge either. The values of 0.572,
1.629, 1.311, and 1.413 were reported recently for the shear-volume-
change constants C1 to C4 respectively for a fine Nevada sand at a
relative density of 44% (Hushmand et al., 1987). By comparing this
set of constants with the set listed in Fig. 1 for a different sand at
a relative density of 45%, the differences in numerical values are
rather obvious despite the similarity in relative density. Pore
pressure responses from DESRA2 on the same site but with the new set
of C1-C4 parameters are even more surprising. The corresponding
effects indicate a substantial reduction in the peak pore-pressure
ratio. At both 22.5- and 32.5-ft (6.9- and 9.9-m) depths, the peak
pore-pressure ratio was reduced from 0.46 to 0.20 in the case of KC12,
whereas in the case of KC4, the sublayers at these depths register a
peak pore-pressure ratio of 0.60 instead of being liquefied. The
simplified procedure assumes that pore-pressure build-up depends only
on the cycle ratio, N_e/N_l, (DeAlba et al., 1976) and therefore has no
provisions to account for this type of dependency. However, if the
response of a deposit is so sensitive to the values of this type of
material constants and if these constants can have as wide a range of
values at a given relative density as indicated, one wonders whether
the state-of-the-art laboratory and field procedures are adequate and
practical enough to provide good quality input to analytical
procedures such as DESRA2.

CONCLUSIONS

A comparison of analytical and empirical methods to estimate the
pore-pressure build-up and the settlement of a 50-ft (15.2 m) sand
layer from a variety of earthquake loadings was presented. Although
specific methods were used in this comparison, the conclusions drawn
should also apply to other analytical or empirical methods as well.

As far as build-ups of seismically induced pore-pressure are
concerned, empirical procedures need additional consideration to
include frequency-dependent response owing to different input
earthquakes. The empirical procedures, however, are probably
incapable of taking into account any markedly different volume-change
behaviors of cohesionless soils at similar relative densities. At the
present time the major roadblock in confidently predicting the
settlement of liquefied deposits is the inability to model post-
liquefaction behavior of sands. It appears that any advance in the
state of the art on this subject will be dependent on the ability to
predict local water-content redistribution within the deposits during

earthquake shaking. Meanwhile, empirical procedures, developed on the
basis of the field performance of many sites, are expected to provide
an estimate of settlement that may be viewed as an upper bound value
for design considerations.

For practical applications, it must be emphasized that, although
the analytical method is more versatile and can take many more factors
into consideration, the benefit of using an analytical method is
always restricted by the difficulty and the effort required to obtain
appropriate input. Improved field and laboratory procedures will play
an important role in improving this situation.

REFERENCES CITED

Ambraseys, N., and Sarma, S. (1969). "Liquefaction of Soils Induced by
    Earthquakes." Bulletin of the Seismological Society of America,
    v. 59, no. 2, pp. 651-664.

Chen, A. T. F. (1986). "PETAL2--PEnetration Testing And Liquefaction:
    An Interactive Computer Program." U.S. Geological Survey Open-
    File Report No. 86-178, Menlo Park, CA 94025.

DeAlba, P., Sood, H. B., and Chan, C. K. (1976) "Sand Liquefaction in
    Large-Scale Simple Shear Tests." Journal of the Geotechnical
    Engineering Division, ASCE, v. 102, no. GT9, pp. 909-927.

Finn, W. D. L., Lee, K. W., and Martin, G. R. (1977). "An Effective
    Stress Model for Liquefaction." Journal of the Geotechnical
    Engineering Division, ASCE, v. 103, no. GT6, pp. 517-533.

Hushmand, B., Crouse, C. B., Martin, G., and Scott, R. F. (1987).
    "Site Response and Liquefaction Studies Involving the
    Centrifudge." Soil Dynamics and Liquefaction, Elsevier, pp. 3-24.

Lee, M. K. W., and Finn, W. D. L. (1970). "DESRA-2 Dynamic Effective
    Stress Response Analysis of Soil Deposits with Energy
    Transmitting Boundary Including Assessment of Liquefaction
    Potential." Soil Mechanics Series No. 38, Department of Civil
    Engineering, University of British Columbia, Vancouver, B.C.,
    Canada.

Martin G. R., Finn, W. D. L., and Seed, H. B. (1975). "Fundamentals of
    Liquefaction under Cyclic Loading." Journal of the Geotechnical
    Engineering Division, ASCE, v. 101, no. GT5, pp. 423-438.

National Research Council (1985). "Liquefaction of Soils during
    Earthquakes." Report No. CETS-EE-001, National Technical
    Information Service, Springfield, VA 22161.

Seed, H. B., and Idriss, I. M. (1982). Ground Motions and Soil
    Liquefaction during Earthquakes, EERI Monograph Series,
    Earthquake Engineering Research Institute.

Seed, H. B., Idriss, I. M., and Arango, I. (1983). "Evauation of Liquefaction Potential Using Field Performance Data." Journal of Geotechnical Engineering, ASCE, v. 109, no. 3, pp.458-482.

Seed, H. B., Tokimatsu, K., Harder, L. F., and Chung, R. M. (1985). "Influence of SPT Procedures in Soil Liquefaction Resistance Evaluations." Journal of Geotechnical Engineering, ASCE, v. 111, no. 12, pp.1425-1445.

Tokimatsu, K., and Seed, H. B. (1987). "Evaluation of Settlements in Sands due to Earthquake Shaking." Journal of Geotechnical Engineering, ASCE, v. 113, no. 8, pp. 861-878.

Youd, T. L. (1984). "Recurrence of Liquefaction at the Same Site." Proceedings of the Eighth World Conference on Earthquake Engineering, Prentice-Hall, Inc., Englewood Cliffs, NJ, v.3, pp.231-238.

THE LIQUEFACTION OF SAND LENSES DURING AN EARTHQUAKE

by

Luis E. Vallejo[*], A.M. ASCE

ABSTRACT

One of the most dramatic events of the Alaska earthquake of 1964 was the enormous landslide along the shoreline in the Turnagain Heights of Anchorage. The slide extended about 2593 meters along the shoreline and retrogressed inland a distance of about 366 meters. The liquefaction of sand lenses present in the clay that formed most of the area that failed was found to play a major role in the development of the landslide at the Turnagain Heights. In this paper, the effects of the liquefaction of sand lenses in the ground that contains them are presented. These effects are analyzed using the principles of Linear Elastic Fracture mechanics (LEFM) theory.

INTRODUCTION

One of the most dramatic events of the Alaska earthquake of 1964 was the enormous landslide along the shoreline in the Turnagain Heights of Anchorage. The shoreline in this area was formed by bluffs some 21 meters high and having slopes of about 1½:1 down the bay. The slide extended about 2593 meters along the shoreline and retrogressed inland a distance of about 366 meters. The liquefaction of sand lenses present in the clay that formed most of the area that failed was found to play a major role in the development of the landslide at the Turnagain Heights (4) (Fig. 1).

Very little is known about the effects that the liquefaction of sand lenses have on the ground surrounding them as well as on any structure sitting on the surface of this ground. In this paper, an analysis of these effects is presented using the principles of Linear Elastic Fracture Mechanics (LEFM) theory (1-5).

THE EFFECT OF ONE SAND LENSE

Stress Analysis

When a sand lense is embedded in a clay layer (Fig. 1),

---

[*] Assistant Professor, Department of Civil Engineering, University of Pittsburgh, Pittsburgh, PA 15261

493

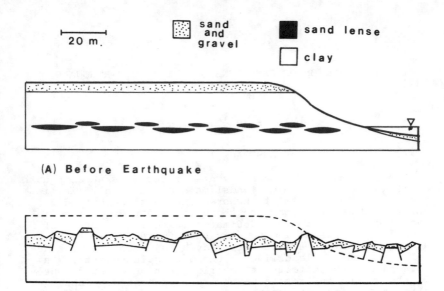

Fig. 1   Sand lenses and the development of the Turnagain
         Landslide in Anchorage, Alaska (4).

the sand lense serves to stiffen the clay and increase its
resistance with respect to static shear and compressive
loads (4). However, during an earthquake the sand lense
may liquefy if the sand in the lense is saturated and loose.
If it liquefies, the sand lense takes the form of a fluid
that fills a cavity that is contained within a clay layer
(Fig. 1). The walls of the cavity that contains the sand
lense are subjected to a pressure u that develops in the
liquefied sand.

Figure 2 depicts a clay element that contains a crack
filled by a liquefied sand. The liquefied sand exerts a
pressure u on the walls of the crack. This pressure u is
the result of the forces induced by an earthquake and the
overburden pressure at the location of the liquefied sand
lense (Fig. 1). Fig. 2 also indicates the type of stresses
to which the clay element is subjected to in the field.
These stresses are comprised of an earthquake induced shear
stress, τ_h , the overburden pressure, σ_v , and the horizontal earth pressure, σ_h .

Figure 2 also shows the stresses, σ_x , σ_y , and τ_{xy} that
the shear stress, τ_h , the overburden pressure, σ_v, the horizontal earth pressure, σ_h , and the fluid pressure u

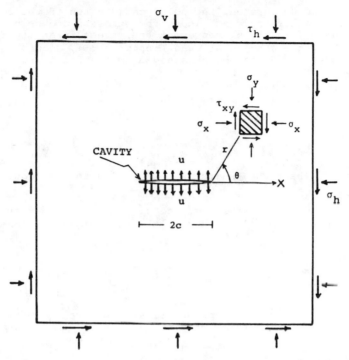

Fig. 2   System of stresses acting on a clay element
          containing a liquefied sand lense.

induce in the clay surrounding the liquefied sand lense.
The stresses σ_x , σ_y , and τ_{xy} can be calculated using the
principles of Linear Elastic Fracture Mechanics Theory as
follows (1-5)

$$\sigma_x = \frac{k_1}{(2r)^{1/2}} \cos\frac{\theta}{2}\left(1 - \sin\frac{\theta}{2}\sin\frac{3\theta}{2}\right) - \frac{k_2}{(2r)^{1/2}}\sin\frac{\theta}{2}\left(2 + \cos\frac{\theta}{2}\cos\frac{3\theta}{2}\right) \quad [1]$$

$$\sigma_y = \frac{k_1}{(2r)^{1/2}} \cos\frac{\theta}{2}\left(1 + \sin\frac{\theta}{2}\sin\frac{3\theta}{2}\right) + \frac{k_2}{(2r)^{1/2}}\sin\frac{\theta}{2}\cos\frac{\theta}{2}\cos\frac{3\theta}{2} \quad [2]$$

$$\tau_{xy} = \frac{k_1}{(2r)^{1/2}} \cos\frac{\theta}{2}\sin\frac{\theta}{2}\cos\frac{3\theta}{2} + \frac{k_2}{(2r)^{1/2}}\cos\frac{\theta}{2}\left(1 - \sin\frac{\theta}{2}\sin\frac{3\theta}{2}\right) \quad [3]$$

where r and θ represent the polar coordinates of the point
near the tip of the crack where the stresses are being calculated, and k_1 and k_2 represent the stress intensity factors that according to Sih and Liebowitz (5), Gdoutos (2),
and Kobayashi (3) can be obtained from the following relationships (Fig. 2)

$$k_1 = (\sigma_v - u) (c)^{\frac{1}{2}} \qquad [4]$$

and

$$k_2 = (\tau_h) (c)^{\frac{1}{2}} \qquad [5]$$

where c is half the length of the crack with the liquefied
sand.

The principal stresses σ_1 and σ_3 at a point near the
sand lense can be obtained from the following relationship
that uses Eqs. [1] to [5]

$$\sigma_{1,3} = \frac{\sigma_x + \sigma_y}{2} \pm \left[(\frac{\sigma_x - \sigma_y}{2})^2 + (\tau_{xy})^2 \right]^{\frac{1}{2}} \qquad [6]$$

The direction of the principal stresses can be obtained
from the following relationships

$$\psi = (1/2) \tan^{-1} \left[\frac{2 \tau_{xy}}{(\sigma_x - \sigma_y)} \right] \qquad [7]$$

and

$$\lambda = (\psi + \pi/2) \qquad [8]$$

where ψ is the angle of inclination with respect to the X
axis (Fig. 2) of the line of action of σ_3 . λ represents
the inclination with respect to the X axis of the line
of action of σ_1 .

A computer program that uses Eqs. [1] to [8] was written
in order to calculate and plot the magnitude and direction
of the principal stresses in regions near one liquefied
sand lense.

Principal Stresses in Clay Surrounding a Liquefied Sand Lense

When plotting the principal stresses in clay regions
surrounding a liquefied sand lense, the case considered
involves a clay element with a liquefied sand lense that
develops a pressure u that is equal to the value of the

overburden pressure σ_v (Fig. 3).

An analysis of Fig. 3 shows that the clay surrounding
the liquefied sand lense developed tensile as well as
compressive stresses. These stresses were concentrated in
regions close to the tips of the cavity containing the
liquefied sand lense. For the case of a left-lateral shear
stress, τ_h , induced by an earthquake, and a pressure in
the liquefied sand u equal to the overburden pressure, σ_v,
the upper right and lower left sections of the clay surrounding the cavity developed tensile stresses. The
upper left sections and lower right sections of the cavity
developed compressive stresses. Since clay is generally
weaker in tension than in compression, the clay surrounding
the cavity that is subjected to tensile stresses will fail.
This will cause the extension of the cavity filled with
liquefied sand.

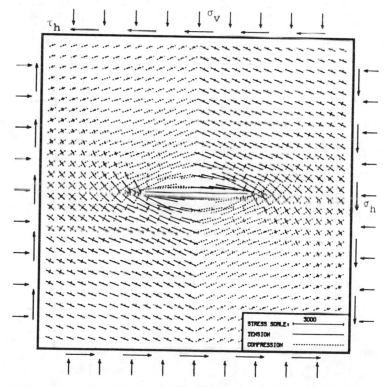

Fig. 3 Principal stresses around a liquefied sand lense.
σ_v = 2000 , σ_h = 1000 , τ_h = 1000, u= 2000 units.

Angle of Propagation of the Cavity Containing the Liquefied Sand Lense
     In order to calculate the angle of propagation of the
cavity that contains the liquefied sand subjected to a
system of stresses as shown in Fig. 2, Fig. 4 will be used.
Fig. 4 is similar to Fig. 2, except that the stresses at
a point close to the tip of the cavity containing the liquefied sand lense are given in a polar mode. The tangential stress, σ_θ , at a point at a distance r from the
tip of the cavity can be obtained from the following relationship (1,2,5)

$$\sigma_\theta = \frac{1}{(2r)^{\frac{1}{2}}}\ \cos\frac{\theta}{2}\ (\ k_1\ \cos^2\frac{\theta}{2} - \frac{3}{2}\ k_2\ \sin\theta\ ) \qquad [9]$$

     According to Erdogan and Sih (1), the cavity containing
the liquefied sand will propagate when σ_θ (Fig. 4) reaches
a maximum value. That is, crack propagation will take
place in a radial direction from its crack tip and the

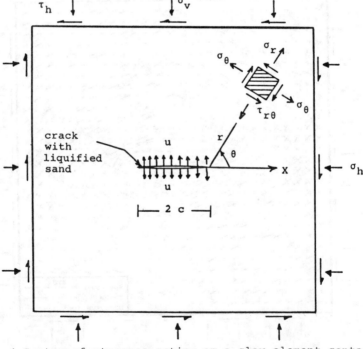

Fig. 4 System of stresses acting on a clay element contain-
       ing a liquefied sand lense. Polar mode of stresses.

direction of crack propagation is normal to the maximum
tangential stress σ_θ. Hence the direction of crack propagation takes place when θ (Figs. 2 and 4) reaches a value
equal to α that can be obtained from

$$\frac{d\sigma_\theta}{d\theta} = 0 \qquad\qquad [10]$$

and

$$\frac{d^2\sigma_\theta}{d\theta^2} < 0 \quad \text{at} \quad \theta = \alpha \qquad\qquad [11]$$

The direction of crack propagation, $\theta = \alpha$ can be
obtained by the use of Eqs. [9] , [10], [11] , [4] and [5].
Applying the conditions expressed by Eqs. [10] and [11]
together with Eq. [9] after some mathematical manipula-
tions one arrives at the following equation from which
$\theta = \alpha$ can be obtained

$$k_1 \sin \alpha + k_2 (3 \cos \alpha - 1) = 0 \qquad\qquad [12]$$

For the case in which σ_v is equal to u (Fig. 3),
according to Eq. [4] , k_1 becomes equal to zero and Eq.
[12] can be written as

$$(3 \cos \alpha - 1) = 0 \qquad\qquad [13]$$

The solution to Eq. [13] gives the angle of propaga
tion of the cavity containing the liquefied sand lense.
This solution gives an angle α equal to 70.5 degrees as
shown in Fig. 5.

Thus, when a horizontal sand layer liquefies in the
field under an earthquake shear stress, τ_h , an overburden
pressure, σ_v , and a fluid pressure , u , the cavity that
contains the liquefied sand lense propagates toward the
ground surface following a direction that is equal to 70.5
degrees with respect to the plane of the cavity (Fig. 5).
The presence of semi-vertical cracks containing liquefied sand have been observed in the field after the occur-
ence of earthquakes (4).

The Effect of the Liquefaction of One Sand Lense on the
Ground Surface
      The effects of the liquefaction of one sand lense on
the ground surrounding it have been depicted in

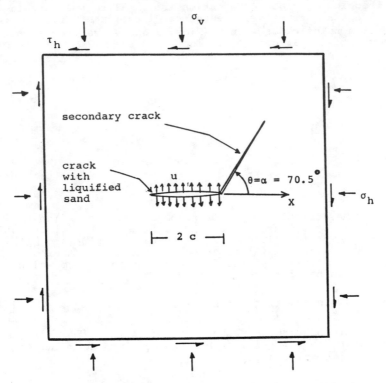

Fig. 5   Direction of propagation of the cavity that
         contains a liquefied sand lense.

Figure 6. As a result of earthquake induced shear stresses, a saturated sand lense located at a depth h below
the ground surface liquefies (Fig. 6). The combined
effect of the earthquake induced shear stress, the overburden pressure, and the pressure developed by the liquefied sand causes the cavity that contain the liquefied
sand to propagate itself in the form of a secondary tensile crack that is inclined at 70.5 degrees with respect
to the plane of the cavity.

     The secondary tensile crack serves as a path through
which the liquefied sand moves to the ground surface and
forms there a sand crater. The empty cavity that contained the sand lense closes as a result of the existing overburden pressure. The closing of the empty cavity causes
not only the collapse of the clay above it, but the generation of a large basin of depression on the ground surface above the cavity (Fig. 6)

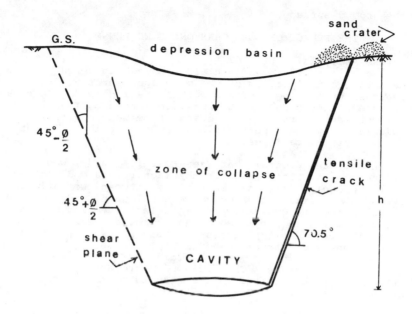

Fig. 6   Effects of the liquefaction of a sand lense on
          the ground surrounding it.

The collapsed zone and the basin of depression is delimited on one side by the tensile crack that originates
from the right tip of the cavity and is inclined at 70.5
degrees with respect to the plane of the cavity, and on
the other side by a shear plane that is inclined at
45 - φ/2 with respect to the vertical and originates from
the left tip of the cavity (Fig. 6). The shear plane
forms when the clay above the cavity collapses and moves
downwards toward the cavity.

If a structure such as a building, bridge, railroad,
or a pipeline is located on the depression basin (Fig. 6),
it could experience damage from the differential settlements that are the result of the slope changes in the
ground surface (7). These slope changes are caused by the
downward movement of the material above the cavity that
contained the sand lense. The size of the basin of depression depends upon the dimensions of the closing cavity as well as on the location of it with respect to the
ground surface. According to Voight and Pariseau (6), the
larger the dimensions of a closing cavity and the shallower its location with respect to the ground surface, the
larger will be the basin of depression that develops at

the ground surface.

THE EFFECT OF MORE THAN ONE SAND LENSE

Stress Analysis
Equations [1] to [8] can also be used to calculate the
magnitude and direction of the principal stresses in clay
regions surrounding cavities with liquefied sand. This can
be done by applying the superposition method. That is,
the stresses in regions along and in between cavities are
calculated by the superposition of the elastic fields of
the individual cavities.

A computer program that uses Eqs. [1] to [8] was developed to calculate and plot the magnitude and direction
of the principal stresses in regions between two cavities.
Two of the cavities considered were alligned, and two were
not alligned and were arranged in a left-stepping manner.
The cavities were subjected to an overburden stress, σ_v ,
equal to 2000 units of stress; a shear stress, τ_h , equal
to 1000 units of stress; a horizontal stress, σ_h , equal
to 1000 units of stress; and a fluid pressure u^h equal to
2000 units of stress. The fluid pressure acted normal to
the walls of the cavities. The results of the stress
analysis are shown in Figures 7 and 8.

An analysis of Figs. 7 and 8 shows that large zones of
tensile stresses developed in the regions between the two
cavities. Clay is assumed to enclose the cavities. Since
it is known that clay is very weak in tension, tensile
cracks will then develop in the regions between the cavities. These tensile cracks will form in a direction normal to the direction of the tensile stresses in the regions between the two cavities as shown in Figs. 7 and 8.
These figures show that the two alligned and left-stepping
cavities interacted with each other by developing tensile
cracks that joined them.

The Effect of the Liquefaction of More Than One Sand Lense on the Ground Surface.
If a clay layer contains an array of saturated sand
lenses as shown in Fig. 1, the effect of an earthquake on
these sand lenses is: (a) to cause their liquefaction,
and (b) to cause the development of secondary tensile
cracks that connect the cavities (Figs. 7 and 8).

The interaction of an array of cavities as a result of
an earthquake, causes the formation of a continuous failure surface which is formed by the cavities and the secondary tensile cracks that connect them. The continuous
failure surface is filled by a liquified sand that develops a fluid pressure, u , equal to the overburden
pressure. Any material above this continuous failure surface will be very unstable and will glide, fail, and break

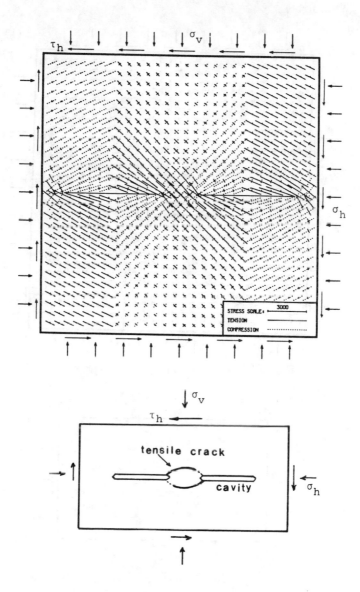

Fig. 7 Principal stresses and cavity interaction for the case of two alligned cavities with σ_v = 2000, τ_h = σ_h = 1000, and u = 2000 units of stress.

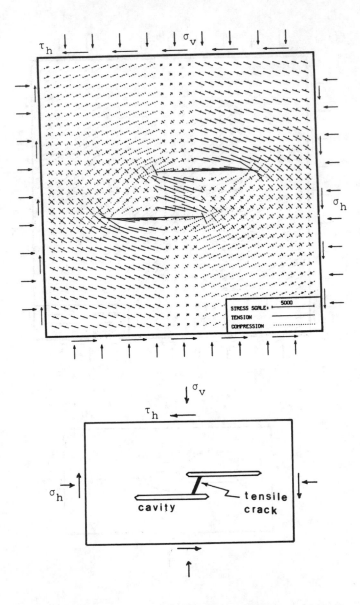

Fig. 8   Principal stresses and cavity interaction for the case of two overlapping left-stepping cavities with $\sigma_v = u = 2000$, $\sigma_h = \tau_h = 1000$ units of stress.

into pieces. This type of failure, resulting from the
interaction of cavities that contained liquefied sand,
occurred during the Alaska earthquake of 1964 and formed
the Turnagain Heights landslide. This landslide was
recorded by Seed (4) and is shown in Fig. 1

CONCLUSIONS

Using the principles of Linear Elastic Fracture Mechanics theory, the effects of the liquefaction of sand
lenses on the ground that contains them have been analyzed.
It was determined that:

(1) When one saturated sand lense liquefies, the liquefied sand exterts pressure on the walls of the
cavity that contains it. This pressure acting together with an earthquake shear stress and the overburden pressure cause large concentrated stresses in
the clay regions surrounding the liquefied sand lense.
These concentrated stresses are both tensile as well
as compressive in nature.

(2) The tensile stresses in the clay surrounding the
cavity filled by a liquefied sand, causes the extension of the cavity. This cavity extension takes the
form of a secondary tensile crack that develops at
one of the tips of the cavity. The tensile crack
propagates at an angle of 70.5 degrees with respect
to the plane of the cavity. The tensile crack propagates toward the ground surface with the help of
the pressures developed by the liquefied sand. The
tensile crack serves as drainage path for the liquefied sand.

(3) The overburden stress closes the cavity that
contained the sand lense and causes the collapse of
the ground above it as well as the formation of a
basin of depression at the ground surface level. The
collapse zone above the cavity is delimited by a
tensile crack that extends from one of the tips of
the cavity at 70.5 degrees with respect to the plane
of the cavity, and a shear plane that extends from
the other tip of the cavity. The shear plane is
inclined at $45 - \phi/2$ with respect to the vertical.
ϕ is the anfle of internal friction of the material
above the cavity. Any structure located on the depression basin could experience damage as a result of
differential settlements.

(4) If more than one sand lense exist in a clay layer,
the liquefaction of the sand lenses causes the clay
between the sand lenses to develop large zones of
tensile stresses. The tensile stresses cause the clay
to develop tensile cracks that connect the cavities

that contain the liquefied sand. The joining of the
cavities and the secondary tensile cracks, produces a
continuous failure surface. This failure surface is
filled by a liquefied sand under pressure. Any
material above this failure surface is very unstable
and fails when it glides over the failure surface and
breaks itself into pieces.

ACNOWLEDGEMENTS

The work described here was supported by Grant No.
ECE-8414931 to the University of Pittsburgh from the National Science Foundation, Washington, D.C. This support
is gratefully acknowledged.

The Author also gives special thanks to Mr. Mahiru
Shettima for his help with the computer work.

APPENDIX I - REFERENCES

1. Erdogan, R., and Sih, G.C., "On the Crack Extension in
   Plates Under Plain Loading and Transverse Shear,"
   Journal of Basic Engineering, ASME, Vol. 85, 1963,
   pp. 519-527.

2. Gdoutos, E.E. Problems of Mixed Mode Crack Propagation.
   The Hague: Nijhoff Press, 1984.

3. Kobayashi, S., "Fracture Criteria for Anisotropic Rocks",
   Memoirs of the Faculty of Engineering of Kyoto
   University, Vol. 32, No. 3, 1970, pp. 307-333.

4. Seed, H.B., "Landslides During Earthquakes Due to Soil
   Liquefaction," Journal of the Soils Mechanics and
   Foundations Division of ASCE, Vol. 94, No. SM5, 1968.

5. Sih, G.C., and Liebowitz, H., "Mathematical Theories of
   Brittle Fracture. In: Fracture, Edited by H. Liebowitz, Vol. II, 1968, Academic Press, pp. 67-190.

6. Voight, B., and Pariseau, W., "State of Predictive Art
   in Subsidence Engineering," Journal of the Soil Mechanics and Foundation Division of ASCE, Vol. 96,
   No. SM2, 1970, pp. 721-750.

7. Wahls , H.E., "Tolerable Settlement of Buildings,"
   Journal of Geotechnical Engineering Division, ASCE,
   Vol. 107, No. GT11, 1981, pp.1489-1504.

APPENDIX II - NOTATION

The following symbols are used in this paper:

c = half length of a cavity

r = radius measured from tip of cavity

k_1 = stress intensity factor, Mode 1

k_2 = stress intensity factor, Mode 2

α = angle of cavity extension

θ = angle measured from tip of cavity and from the plane of the cavity

σ_x, σ_y, τ_{xy} = normal and shear stresses at a point in the vicinity of the tip of cavity.

σ_r = radial stress

σ_θ = tangential stress

$\tau_{r\theta}$ = shear stress

$\sigma_{1,3}$ = principal stresses

σ_v = overburden stress

σ_h = horizontal stress

τ_h = earthquake induced shear stress

u = pressure developed by liquefied sand.

ψ = direction angle for σ_3

λ = direction angle for σ_1

Performance of Foundations
Resting on Saturated Sands

Raj Siddharthan* A.M. ASCE and Gary M. Norris** M. ASCE

ABSTRACT

A method to evaluate deformations of foundations resting on saturated sand is described. Nonlinear hysteretic soil behavior and the stiffness and strength degradation are accounted for. The computed responses agree with those recorded in centrifuge tests on surface foundations. The importance of such factors as the location of water table, the contact pressure, depth of embedment and strength excitation has been investigated.

INTRODUCTION

Almost all major earthquake damage reports contain accounts of movements or complete failures of foundations such as bridge abutments. Ross et al (1969) who assessed damage to highway bridge foundations after the 1964 Alaska earthquake, reported that as much as 50% of the foundations surveyed suffered severe or moderate damage. Even where damage was moderate or minor, there was evidence of bridge joints closing indicating lateral displacement of the abutments. The most common reason for the poor performance of foundations resting on saturated cohesionless soils has been the loss of strength and stiffness of the foundation soil caused by the generation of residual porewater pressure.

One of the major factors that needs to be considered in the performance evaluation of foundations is the proper modeling of the surrounding foundation soils. This is because the foundation displacements are, in most cases, the result of deformation of the foundation soil. The deformations in loose saturated cohesionless soils can be divided into three basic components (Chang, 1984; Siddharthan, 1984): (1) a cyclic component associated with cyclic load, (2) a residual component due to a softening of the soil, and (3) additional deformation due to grain slip. Since the duration of the seismic excitation is short, the deformation during the excitation occurs under the condition of no volume change. Therefore only the first two components of deformation occur during shaking. The third component which reflects the deformation due to interparticle slip occurs afterward with the dissipation of the induced porewater pressure.

---
*Assistant Professor, Civil Engineering Dept., University of Nevada-Reno.
**Associate Professor, Civil Engineering Dept., University of Nevada-Reno.

Several laboratory investigations on the settlement of foundations have been reported in the literature.   De Alba et al. (1976), Heller (1977), and Yoshimi and Tokimatsu (1977) have reported on the behavior of small foundations subjected to base excitation on large shake tables. Recently, Whitman and Lambe (1982) and Finn et al. (1984a, 1987) have reported the results of a series of centrifuge tests carried out at Cambridge University. The associated study of the data has revealed these interesting features: a) a substantial part of the footing settlement is caused by a softening of the deposit; b) the porewater pressure ratios in the soil directly beneath the foundation are substantially less than the ratios in the soil at a distance from the foundation; c) during excitation, the foundation moves steadily downward with little or no intermittent upward movement; and d) at larger excitations, substantially higher settlements occur even though the porewater pressure ratios directly under the structure are small.

Since foundation settlement during excitation is primarily a function of the softening of deposit, a method of analysis carried out in the time domain that accounts for the generation of residual porewater pressure is necessary. Furthermore, nonlinear hysteretic soil behavior should be accounted for. Chang (1984), Siddharthan (1984), and Bouckovalas et al. (1986) have presented methods to compute the seismic displacement of foundations based on effective stresses. The methods proposed by Chang (1984) and Bouckovalas et al.   (1986) do not account for the structure's inertia effect upon foundation response nor for other dynamic soil-structure interaction effects. While these factors may not be important under wave loading conditions, the effects of these factors for seismic loading conditions are important. The applicability of the   method proposed by Siddharthan (1984) that accounts for these factors has been verified to a limited extent by a comparison of computed residual porewater pressures and acceleration histories with centrifuge test results after Finn et al. (1984a,b).

In this paper, Siddharthan's two - dimensional effective stress method of analysis is used to evaluate the deformation response of surface and embedded foundations.

First, the computed deformation response of a surface foundation is compared with available records of observed response from centrifuge tests (Siddharthan and Norris, 1987 a,b). Thereafter, the influence of such factors as the location of the water table, the contact pressure, and the strength of excitation upon the response of surface and embedded foundations is reported.

PROPOSED METHOD OF ANALYSIS

In the dynamic analyses of soil-structure systems, it is often assumed that the loading imposed by seismic excitation can be superimposed on the long-term (static) equilibrium conditions.   Since the dynamic soil properties, such as strength and stiffness, depend upon insitu stresses, the static stress condition should be evaluated first. The incremental elastic (tangent) approach which simulates layer construction (Duncan and Chang, 1970) can be used for this purpose. Thereafter, the incremental displacements between two time steps during the base excitation can be computed by solving the following set of dynamic incremental equilibrium equations.

$$[M] \{\Delta \ddot{x}\} + [C] \{\Delta \dot{x}\} + [K]_t \{\Delta x\} = -[M] \{I\} \Delta \ddot{U}_g(t) \tag{1}$$

in which $[M]$ is the diagonal mass matrix, $[C]$ is the viscous damping matrix, $[K]_t$ is the tangent stiffness matrix, $\{I\}$ is the unit vector, $\Delta \ddot{U}_g(t)$ is the increment in either the horizontal or the vertical base acceleration, and $\{\Delta x\}$, $\{\Delta \dot{x}\}$, and $\{\Delta \ddot{x}\}$ are the incremental displacements, velocities, and accelerations of the nodes relative to the base. Depending on whether the equation is written for the horizontal (x) or the vertical (y) direction, $\Delta U_g(t)$ is the increment in the corresponding component of base acceleration.

The stress-strain relationships and the poreweater pressure generation model used in the analysis are an extension of the procedures developed by Finn et al. (1977) for level ground conditions. Since a detailed description of these procedures is available in the literature they will not be discussed here. In the proposed two-dimensional analysis, soil response is modeled by combining the effects of shear and normal stresses. In shear, the soil is treated as in the case of level ground: a nonlinear hysteretic material exhibiting Masing behavior during unloading and reloading. The initial or skeleton shear stress-strain relationship is assumed to be hyperbolic. To model a condition of no volume change during excitation, the tangent bulk modulus is assumed to be a large value. The tangent stiffness matrix needs to be modified at every time step to reflect the current stress-strain state. Hysteretic damping is accounted for by the above mentioned modification of the tangent stiffness matrix. A small amount of viscous damping of a Rayleigh type may be included in the analysis to account for damping from any nonhysteretic source. An equivalent critical damping of approximately 2% was used in the study. The cyclic component of the displacement at any time is simply the algebraic sum of the incremental values.

In an effective stress response analysis, the seismically induced porewater pressures need to be assessed. The Martin-Finn-Seed (1975) porewater pressure model has been used for this purpose. However, this model has been modified to account for the presence of initial static shear stress. The initial static shear stress influences the liquefaction potential curve and the rate of generation of porewater pressure (Vaid et al., 1979; Chang, 1982). In this model, the incremental residual porewater pressures are evaluated by first computing the incremental plastic volumetric strain that occurs under drained conditions and then multiplying this by the rebound modulus. When using the Martin-Finn-Seed equation to compute the incremenal plastic strain, the shear strain γ_{xy} is employed. Other researchers have proposed similar approaches (Arulanandan et al., 1985; Bouckovalas et al., 1984). The porewater pressure model constants are selected to match the rate of porewater pressure generation and the liquefaction potential curves.

There is a limit to the amount of porewater pressure that can develop in a soil element. This limit is easy to estimate in the case of triaxial or simple shear tests in the laboratory. This is because the ultimate state of stress has to be on the Mohr failure envelope and given the boundary conditions of the sample the effective stress path followed by the sample is predictable (Chang, 1982). However, in a two-dimensional analysis, the effective stress path followed by an element cannot be predicted beforehand; therefore, the maximum residual porewater pressure that develops cannot be assessed independent of the dynamic loading. However, one way of limiting the porewater pressure that might develop would be to only allow generation until the stress path touches the Mohr envelope. It will be seen later that this treatment leads to much lower porewater pressures at locations under the foundation as compared to the values observed in the centrifuge tests. Consequently, it was decided to permit porewater pressure generation to continue until the effective minor principal stress reaches a certain value. Poulos et al. (1985) suggest that, for a given soil, this value is only a function of the void ratio. After initially touching the failure envelope, the stress path should travel down the failure envelope until the above criterion is satisfied. With movement down the Mohr envelope, the Mohr circle of stress decreases in size. Accordingly, correction forces are applied in the finite element program to cause such a reduction in the affected elements.

To determine the stress path followed by an element, one needs to compute the state of the effective stresses during the dynamic loading. The method employed to do this is discussed below.

The incremental nodal equilibrium equations, with consideration of the porewater pressure in the elements, can be expressed (Christian et al., 1970) as

$$\{\Delta P\} = [K]_t \{\Delta \delta\} + [K^*]\{\Delta U\} \qquad (2)$$

in which $\{\Delta P\}$ is the incremental load vector, $\{\Delta \delta\}$ represents the incremental nodal displacements, $[K^*]$ is the porewater pressure stiffness matrix, and $\{\Delta U\}$ represents the element porewater pressures. This equation is used to compute the current effective stresses and deformations of the deposit due to the increase in the porewater pressures. In other words, $\{\Delta \delta\}$ represents the displacements due to the softening of the deposit. Note that, in solving Eq. 2, $\{\Delta P\} = 0$ while the values of $\{\Delta U\}$ are computed based on the porewater pressure generation model. This part of the analysis is considered to be an extension of the initial static analysis; and, thus, stress dependent tangent shear and bulk modulus values are used in the computations. The effective stresses so calculated are then used to update soil property values that depend upon the effective stress.

When a dry sand sample is subjected to cyclic loading, plastic volume change, E_{vp}, occurs due to interparticle slip. The occurrence of E_{vp} is the reason for the porewater pressure generation in saturated sands under undrained cyclic loading conditions. When this porewater pressure is allowed to drain, a volumetric strain equal to E_{vp} occurs in the sample resulting in settlement. A number of researchers have studied the problem of settlement due to E_{vp} in both dry and saturated sandy deposits (Silver et al., 1971; Lee et al., 1974; Pyke et al., 1975). They concluded that the plastic volume change depends upon the amplitude of cyclic shear strain and the previous loading history, while it is independent of vertical stress.

During seismic excitation, the effective stresses decrease due to the generated porewater pressures; and then, after the excitation, the effective stresses increase as porewater pressures dissipate. In principle, accounting for E_{vp} is possible by an appropriate selection of soil property values during this unloading-reloading process. However, the selection of appropriate soil properties leads to more uncertainties. Accordingly, the following approximate procedure is employed: 1) nonlinear stress dependent elastic bulk (tangent) modulus values are used during the loading and unloading process and 2) then independently computed values of E_{vp} are used to assess deformations due to grain slip after porewater pressure dissipation is complete. This procedure is similar to that of Marr et al (1981) and Bouckovalores et al., (1986) who studied the ware induced permanent deformations of offshore structures.

The Goodman type of slip or contact elements (Goodman et al 1967) have been incorporated into the analysis to represent the interface characteristics between the soil and the structural elements. The properties of the slip elements were assumed to be elastic and perfectly plastic with failure at the interface given by a Mohr-Coulomb failure criterion. The structure was modeled with elastic plane strain elements having no damping and a very high stiffness values to represent a rigid structure.

APPLICATION OF THE METHOD OF ANALYSIS

Centrifuge Tests

Whitman and Lambe (1982) reported a series of seismic tests on centrifuged models in which the centrifugal acceleration was about 80 g. Fig. 1 shows the model used in these tests. A uniform deposit of Leighton-Buzzard #120/200 sand of an average relative density of 56% supported a circular solid brass foundation (with the foundation representing a structure). The thickness of the deposit was 151 mm, and the diameter of the foundation was 113mm. The corresponding prototype deposit was 12.1 m in height, and the foundation was 9.0 m in diameter. The average bearing stress beneath the foundation under prototype conditions was 130 kN/m^2. The deposit was instrumented to measure the settlement of the foundation and the deposit and, also, the porewater pressure at a number of locations in the deposit. In the tests, the input excitation was intended to be (prototype scale) ten sinusoidal pulses with a constant period of 1.0 sec. However, the actual input motion was somewhat different due to resonances and problems with mechanical linkage clearances especially during the initiation of the base motion. In the first series of tests, the water table was located very close to the surface at 0.5 m; and, in the other, it was located at 1.3 m below the foundation.

A number of interesting observations were reported by Whitman and Lambe. For the case where the peak acceleration was 0.17 g, they found that, at points away from the foundation, the porewater pressure rose quickly to become equal to the vertical overburden pressure. Furthermore, during the excitation, the foundation settled

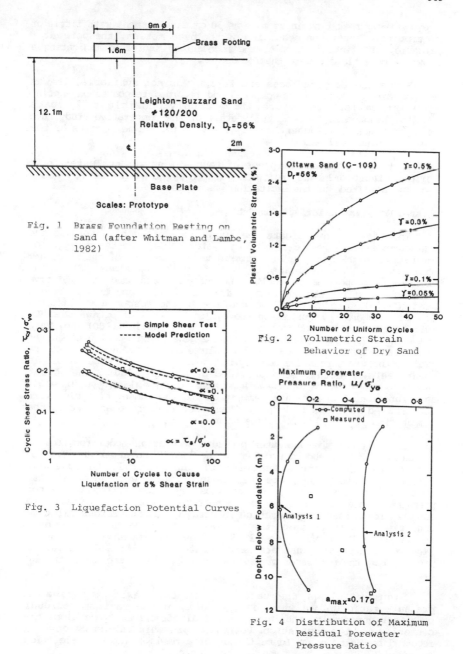

Fig. 1 Brass Foundation Resting on
Sand (after Whitman and Lambe,
1982)

Fig. 2 Volumetric Strain
Behavior of Dry Sand

Fig. 3 Liquefaction Potential Curves

Fig. 4 Distribution of Maximum
Residual Porewater
Pressure Ratio

progressively resulting in as much as 30 cm displacement with little or
no tipping. After the cessation of the base motion, the porewater
pressure, at points just under the foundation, increased indicating a
flow of pore fluid toward the foundation.

Since the recorded base excitation was not available, it was
not possible to compare the computed and recorded responses on a point
by point basis. Therefore, it was only possible to compare
results phenomenologically. Full detail of the comparative study have
been presented by Siddharthan et al (1987a,b). Some results of the
study are presented below.

In this study, the response of foundations was evaluated for a
uniform sinusoidal base excitation of ten cyles with a period of 1.0
sec. The amplitude of the excitation was varied.

Seismic Response of Surface Foundations

In order to investigate the importance of such factors as the
contact pressure, the location of the water table, and the strength of
excitation on the behavior of foundations the response of the structure
shown in Fig. 1 was evaluated for a variety of conditions. Since the
engineering properties of #120/200 Leighton Buzzard sand at a relative
density, D_r, of 56% are not readily available, the properties of Ottawa
sand (C-109) were used in their place in the response evaluation.
Extensive laboratory studies on the cyclic behavior of Ottawa sand (C-
109) have been reported in the literature by Bhatia (1980) and Vaid
and Finn (1979). The porewater pressure generation model used in the
analysis requires that the plastic volume change behavior of the dry
sand at the given relative density be known. From the data on simple
shear tests on Ottawa sand provided by Bhatia (1980) , the volume
change characteristics of Ottawa sand at a relative density of $D_r = 56\%$
as a function of shear strain amplitude and the number of cycles were
constructed as shown in Fig. 2. These curves were then used in the
computation of the plastic volume change.

In addition, the porewater pressure generation model requires an
evaluation of the rebound modulus of the sand (Finn et al. 1977). The
parameters (which are a function of the initial static shear
stress, τ_s) used to evaluate the rebound modulus were selected   so
that the model was able to reproduce the rate of the porewater
pressure generation and also the liquefaction potential curves. Data
on porewater pressure generation in Ottawa sand at various relative
densities as a function of the initial static shear stress has been
reported by Vaid and Finn (1979). The interpolated laboratory
liquefaction curves that are relevant for $D_r = 56\%$ and the curves
predicted by the porewater pressure generation model are shown in Fig.
3.

Figure 4 shows the variation in the maximum porewater
pressure ratio (u/σ_{yo}) which is the ratio of the maximum residual
porewater pressure to the initial vertical effective stress along the
centerline of the foundation. Porewater pressure ratios which were
reported by Whitman and Lambe (1982) for a maximum input acceleration
(a_{max}) of 0.17g are also given in the figure.

Two analyses were performed as part of the present study. In the first, the porewater pressure generation was stopped as soon as the effective stress path touched the Mohr-Coulomb envelope (Analysis 1). In the other, the porewater pressure generation was allowed to continue until the minor effective stress reached a critical value, σ_{3crit} (Analysis 2). As mentioned previously, correction forces were applied so that the stress path would then follow along the failure envelope. Poulos et al. (1985) report that σ_{3crit} depends upon the particle size distribution, particle characteristics such as angularity, and the void ratio. For undisturbed fine grained sandy soils, σ_{3crit} may be less than 1 kPa and a high as 100 kPa. Since no data is available on Leighton-Buzzard sand, a lower bound value of 2kPa, that is appropriate for subrounded fine grain particles was assigned for σ_{3crit}. It should be noted here that while the selection of this value affects the maximum porewater pressure in the elements just below the structure, it does not affect the deformation response of the structure.

From Fig. 4, it appears that, while the first analysis underpredicts the porewater pressure, the second overpredicts it. There are a number of reasons for the differences between the computed and predicted responses. Apart from the fact that the base excitation between the two is different, plane strain conditions were assumed in the computations while an axisymmetric condition exists in the model and prototype. Secondly, since σ_{3crit} is a function of either overburden pressure or void ratio, it should vary from point to point. It was decided not to adjust σ_{3crit} because of uncertainties in its selection. Furthermore, it should be noted that selection of a lower bound value for σ_{3crit} will also result in higher computed porewater pressures in Analysis 2.

It should be noted that, while the porewater pressures predicted under the foundation by the aforementioned two methods differed from the observed values, the predictions at points away from the foundation (free field) matched the recorded values. This can be better explained by considering the stress paths of a number of points. Figure 5 shows the effective stress paths for points A, B, and C, which are on a horizontal line located at 8.4m below the surface (in the prototype). This plot is a (q,p) plot where $q = (\sigma_1 - \sigma_3)/2$ and $p = (\sigma_1 + \sigma_3)/2$. It should be noted that, before the earthquake, the stresses corresponding to point A (beneath the centerline of the foundation) are quite close to the failure line. Then, as porewater pressure is generated, the stress path touches the failure envelope soon after the initiation of excitation. Note by comparison, that the stress paths of points B and C which are further away from the structure meet the failure line at a much later time. Porewater pressure generation at these points was stopped only after the minor principal effective stress reached σ_{3crit}. A closer examination of the rate of porewater pressure generation at these points reveals that, at A, the porewater pressure increased very quickly and it stabilized within two seconds after the initiation of excitation. However, at C (free field) the porewater pressure increased at a slower rate and stabilized when it approached the initial overburden pressure. This phenomenon was also observed in the results reported by Whitman and Lambe (1982).

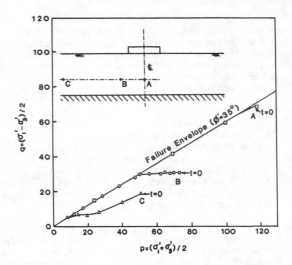

Fig. 5   Effective Stress Paths of Elements
(Units: kN/m^2)

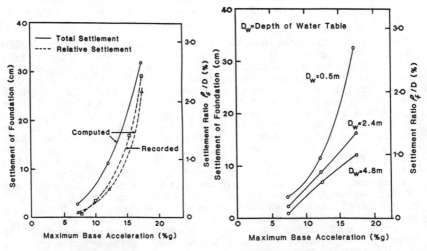

Fig. 6   Final Settlement
of Foundation

Fig. 7   Influence of the
Location of Water Table

Computed and recorded porewater pressure response at various points showed that while the porewater pressure ratio in the soil directly below the foundation varied between 0.1 and 0.5, the ratios in the free field varied between 0.95 and 1.0. The lower porewater pressure ratio values evaluated in the elements directly under the foundation should be interpreted with extreme care. This is because even though the ratios are rather small, Whitman and Lambe report that the final settlement of the foundation relative to the surrounding soil was as much as 30 cm, indicating substantially lower stiffnesses for the elements beneath the foundation. This is reinforced by the stress path plots, which indicate that low porewater ratios are enough to force failure in the elements directly below the foundation thus causing low stiffness values.

Figure 6 presents the computed final settlement (ρ_f) of the foundation as a function of base excitation for Analysis 2. The final settlement is computed by adding the settlement due to grain slip to the residual displacements evaluated at the end of the base excitation. One will note that the settlement is also presented as a percentage of the thickness of the deposit (ρ_f/D). The settlements given by Whitman and Lambe (1982) are relative settlements evaluated with respect to the surrounding soil. Therefore, computed and recorded relative settlements are also shown in the figure. Even though Whitman and Lambe have presented a substantial amount of centrifuge test data, only three test results are shown in Fig. 6. This is because the centrifuge tests at Cambridge University were carried out in consecutive runs on the same soil going from a lower base acceleration to a higher one. In this type of test procedure, the sand progressively densifies and, thus, the properties differ from those of the original soil. Therefore, only the test results that correspond to a first excitation were plotted.

One will note from Fig. 6 that the foundation settlement is small when base acceleration is low, but it increases rapidly with an increase in the strength of the base excitation. The computed and recorded settlements are in very good agreement. The shape of these settlement curves is very similar to that obtained on shaking table tests by Yoshimi and Tokimatsu (1977).

The influence of the location of the water table on the final settlement of the foundation is shown in Fig. 7. Three cases are presented corresponding to the water table at 0.5, 2.4, and 4.8m form the surface. One will note that the settlement is substantially reduced when the water table is lowered. The settlement was reduced by as much as 50% when the water table was lowered from 0.5 to 2.4m from the surface. A lower water table leads to higher effective stresses in the deposit and, thus, to higher strengths and stiffnesses. Furthermore, it reduces the volume of soil that is likely to lose strength since only saturated soils are so affected. A substantial reduction in settlement was also observed by Whitman and Lambe (1982) who report that settlement was reduced by as much as 25% when the water table was lowered from 0.5m to 1.3m.

The influence of contact pressure on the final settlement of the foundation is depicted in Fig. 8. The contact pressure is increased from 0.5q_o to 1.5q_o by appropriately increasing the unit weight of the foundation. Here q_o = 130 kN/m^2. As expected, an increase in contact pressure can substantially affect the settlement of the foundation but only at higher accelerations. At low base excitation, there is little effect. This is because the settlement caused by softening of the deposit is minimal at low base excitation. However, at higher base excitation, the higher contact pressure leads to a substantial increase in the component of settlement due to a softening of the deposit. A similar variation in foundation settlement was observed in the shaking table tests performed by Yoshimi and Tokimatsu (1977).

Seismic Response of Buried Foundations

In order to investigate the influence of the depth of burial, D_b, on the seismic response of foundations, the case where D_b = 3m was analyzed. In this case, the water table was located 0.5m below the base of the foundation. The thickness of the soil deposit below the base of the foundation and the contact pressure were 12.1m and 130 kN/m^2 respectively. The height of the structure was 9.0m (see the insert in Fig. 9).

Fig. 9 shows the final settlement of the foundation as a function of the strength of the base excitation. The settlement reduces substantially as the depth of embedment increases. For example, in the case of a base excitation of amplitude 0.17g, the settlement of the foundation where D_b = 3m is only 27% of the settlement of the surface foundation.

Table 1   Response of Embedded and Surface Foundations

	a_{max}=0.17g D_b=0 D_b=3m		a_{max}=0.12g D_b=0 D_b=3m		a_{max}=0.07g D_b=0 D_b=3m	
Final Settlement of the Foundation (cm)	54.9	15.2	16.7	10.0	3.6	1.4
Final Horizontal Disp. of the Foundation (cm)	27.7	51.8	10.7	6.7	0.10	0.0
Maximum Horizontal Acceleration (%g)	22.4	22.4	24.9	25.1	12.0	18.1
Maximum Vertical Acceleration (%g)	6.5	8.4	6.8	10.4	2.9	4.9

The final settlement of the foundation is mainly due to softening of the foundation soil and the grain slip. The component of settlement caused by the softening of the deposit is small in the case of an embedded foundation. This is because lower porewater pressures develop in the foundation soil. The amount of settlement caused by grain slip seems to be somewhat unpredictable by comparison. The cyclic shear strains are the ones that cause grain slip (or volumetric

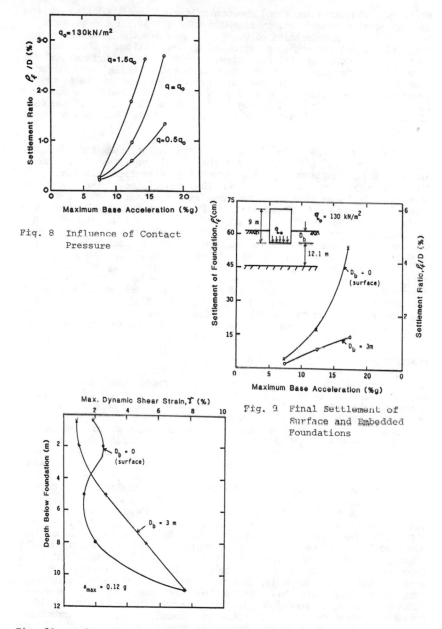

Fig. 8 Influence of Contact Pressure

Fig. 9 Final Settlement of Surface and Embedded Foundations

Fig. 10 Maximum Dynamic Shear Strain Distribution

strain). The variation along the centerline of the foundation of the maximum cyclic shear strain is shown in Fig. 10 for the case of a maximum input acceleration of 0.12g. These are the maximum values computed during the entire duration of the excitation. They do not necessarily occur at the same time. Results for a surface and an embedded foundation (D_b = 3m) are reported. While the embedded foundation results in lower shear strains in the elements closer to the foundation, the shear strains at depths are higher than those computed for surface foundations. This may be due to the fact that, in the case of an embedded foundation, the foundation and the surrounding soil attempt to move as a unit.

Table 1 shows the response computed at the top of the structure by the program. While the final vertical (settlement) and horizontal displacement and horizontal accelerations are the values that correspond to the point on the centerline, the vertical acceleration values correspond to those at the corner. Table 1 clearly shows the influence of embedment. Except for the final predicted horizontal displacement in the case of a_{max} = 0.17g, the embedded foundations give substantially lower vertical and horizontal displacement values than the corresponding surface foundations. This may be because, during the higher excitation, the structure and adjacent soil have a tendency to slide as a unit on top of the bottom soil layers. The horizontal maximum acceleration of the foundation seems to be unaffected by the depth of the embedment at higher excitation; but, at low excitation (a_{max} = 0.07g), it shows higher magnification in the case of D_b = 3m. The maximum vertical acceleration of the foundation, on the other hand, shows higher values compared to the surface foundation for all cases reported.

CONCLUSION

A method to evaluate seismic deformations that accounts for all the important factors that affect the cyclic behavior of saturated sandy soils is presented. Two very important factors considered in the analysis are the nonlinear hysteretic soil behavior and the soil's stiffness and strength degradation. The deformation evaluation includes components of deformation due to softening of the soil and grain slip that occur as the porewater pressure dissipates. This method has been used to evaluate the response of surface and embedded rigid foundations. The computed responses agree phenomenologically with those reported form available centrifuge and shaking table tests.

A parametric study relative to surface and embedded rigid foundations resting on a fine uniform medium dense sand (D_r = 56%) deposit has been reported. In the investigation, a uniform sinusoidal base excitation of ten cycles with a period of 1.0 sec. was used. The amplitude of the input motion was varied. This study reveals that it is important to keep track of the effective stress paths followed by the soil elements, and secondly, that porewater pressures seem to increase even after the path has intersected the Mohr failure envelope. Furthermore, care should be taken in interpreting values of residual porewater pressure ratio in elements directly under the structure. This is because low values of porewater pressures can still indicate failure (low stiffness) in those elements.

The parametric study confirms that factors such as the location of water table, the contact pressure, the depth of embedment and the strength of base excitation have a significant influence on the settlement response. For example, settlement was reduced by as much as 50% when the water table was lowered from 0.5 to 2.4m below the surface; and at low levels of base excitation, the foundation settlement was low but increased rapidly as the strength of excitation increased. Responses such as the final settlement and the horizontal displacement were always much smaller for embedded foundations than surface foundations. One exception was that at high base excitation, the tendency of the embedded foundation to move as a unit with the surrounding soil resulted in higher horizontal displacements.

ACKNOWLEDGEMENTS

Special thanks are due to Cindy Renslow for her excellent typing. This study was funded by an Engineering Research Initiation Grant from the Engineering Foundation. Such support is gratefully acknowledged.

REFERENCES

Arulanandan, K. and K Mureleetharan (1985). "Soil Liquefaction - A Boundary Value Problem," Research Report, Department of Civil Engineering, University of California, Davis.

Bhatia, S.K. (1980). "The Verification of Relationships for Effective Stress Method to Evaluate Liquefaction Potential of Saturated Sands," Thesis submitted in partial fulfillment of the requirements for a Ph.D., University of British Columbia, Vancouver, Canada.

Bouckovalas, G., A.W. Marr, and J.T. Christian (1986). "Analyzing Permanent Drift Due to Cyclic Loads," Journal of the Geotechnical Engineering Division, ASCE, GT6. Vol. 112:579-593.

Chang, C.S. (1982). "Residual Undrained Deformation From Cyclic Loading," Journal of the Geotechnical Engineering Division, ASCE, GT4. Vol. 108 pp. 637-646.

Chang C.S. (1984). "Analysis of Earthquake Induced Footing Settlement," 8th World Earthquake Engineering Conference, San Francisco. pp. 87-94.

Christian, J.T. and J.W. Boehmer (1970). "Plane Strain Consolidation by Finite Elements," Journal of the Soil Mechanics and Foundation Engineering, ASCE, SM4. Vol. 96 pp. 1435-1455.

DeAlba, P., H.B. Seed, and C.K. Chan (1976). "Sand Liquefaction in Large Scale Simple Shear Tests," Journal of the Geotechnical Engineering Division, ASCE, GT9. pp. 909-928.

Finn, W.D.L., K.W. Lee, and G.R. Martin (1977). "An Effective Stress Model for Liquefaction," Journal of the Geotechnical Engineering Division, ASCE, GT6, Vol. 103 pp. 517-533.

Finn, W.D.L., R. Siddharthan, F. Lee, and A.N. Schofield (1984a). "Seismic Response of Drilling Islands in a Centrifuge Including Soil-Structure Interaction," Proceedings, 16th Annual Offshore Technical Conference, Houston, Texas. pp. 399-406.

Finn, W.D.L. and R. Siddharthan (1984b). "Seismic Response of Caisson-Retained and Tanker Islands," Proceedings, 8th World

Conference on Earthquake Engineering, San Francisco, pp. 857-865.
Finn, W.D.L., R.S. Steedman, M. Yogendrakumar, and R. Ledbetter
    (1985). "Seismic Response of Gravity Structures in a
    Centrifuge," Proceedings, 17th Annual Offshore Technical
    Conference, OTC paper 4885, pp. 389-394.
Finn, W.D.L., and Yogendrakumar, M., (1987). "Centrifugal Modelling
    and Analysis of Soil-Structure Interaction," Fifth Canadian
    Conference on Earthquake Engineering, Ottawa, July, pp. 453-460.

Goodman, R.E., Taylor, R.L. and Breeke, T.L., (1968). "A Model for the
    Mechanics of Jointed Rock," Journal of the Soil Mechanics and
    Foundation Engineering Division, ASCE, Vol. 94, SM3, May, pp.
    637-659.
Heller, L.W. (1977). "Discussion on Sand Liquefaction in Large-Scale
    Single Shear Tests," Journal of the Geotechnical Engineering
    Division, ASCE, GT7. pp. 834-836.
Lee, K.L. and A. Albaisa (1974). "Earthquake Induced Settlements in
    Saturated Sands," Journal of the Geotechnical Engineering
    Division, ASCE, GT4. Vol. 100 pp. 387-486.
Marr, W.A., and Christian, J.T., (1981) "Permanent Displacement due to
    Cyclic Loading," Journal of the Geotechnical Engineering
    Division, ASCE, Vol. 107, No. GT6, Aug. pp. 1129-1149.
Poulos, S.J., G. Castro, and J.W. France (1985). "Liquefaction
    Evaluation Procedure," Journal of the Geotechnical Engineering
    Division, ASCE, #6. Vol. 111 pp. 772-792.
Pyke, R., H.B. Seed, and C.K. Chan (1975). "Settlement of Sands
    Under Multi-directional Shaking," Journal of the Geotechnical
    Engineering Division, ASCE, Vol. 101, GT4, pp. 379-398.
Ross, G.A., Seed, H.B., and Migliaccio, R., (1969). "Bridge Foundation
    Behavior in Alaska Earthquake," Journal of the Soil Mechanics and
    Foundations Engineering Division, ASCE, Vol. 95, SM4, July, pp
    1007-1036.
Siddharthan, R. (1984). "A Two-dimensional Nonlinear Static and
    Dynamic Response Analysis of Soil Structure," Thesis submitted
    in partial fulfillment of the requirements for a Ph.D., Univ. of
    British Columbia, Vancouver, Canada.
Siddharthan, R. and Norris, G.M., (1987a). "Seismic Displacement
    Response of Surface Footings in Sand," Fifth Canadian Conference
    on Earthquake Engineering, Ottawa, July, pp. 407-416.
Siddharthan, R. (1987b). "Dynamic Effective Stress Response of Surface
    and Embedded Footings in Sand," Final Report Submitted to
    Engineering Foundation, Report No. CCEER.87-1, Department of
    Civil Engineering, University of Nevada-Reno.
Silver, M.L. and H.B. Seed (1971). "Volume Change in Sands During
    Cyclic Loading," Journal of the Soil Mechanics and Foundation
    Engineering Division, ASCE, SM9, 97 pp. 1171-1182.
Vaid, Y.P. and W.D.L. Finn (1979). "Effect of Static Shear on
    Liquefaction Potential," Journal of the Geotechnical Engineering
    Division, ASCE, GT10. 105 pp. 1233-1246.
Whitman, R.V. and P.C. Lambe (1982). "Liquefaction: Consequences for a
    Structure," Proceedings of Conference on Soil Dynamics &
    Earthquake Engineering, Southampton. pp. 941-949.
Yoshimi, Y. and K. Tokimatsu (1977). "Settlement of Buildings on
    Saturated Sand During Earthquakes," Soils and Foundations, #1,
    17, pp. 23-38.

DYNAMIC ANALYSIS IN GEOTECHNICAL ENGINEERING

W.D. Liam Finn*

ABSTRACT

Developments in dynamic analysis of soil structures since 1978 are critically reviewed. Attention is limited to methods that are beginning to be used in engineering practice. These methods are based either on elastic-plastic theories or on direct simulation of the nonlinear hysteretic stress-strain curves. The advantages and disadvantages of the methods and the problems of applying them in practice are discussed. The two types of methods are compared by application to two example problems. The newer methods are also contrasted with those used in current engineering practice.

INTRODUCTION

Dynamic response analysis as practised widely today had its origin in the pioneering research of Seed and his co-workers at the University of California at Berkeley to explain ground response phenomena during the major earthquakes in Alaska and Japan in 1964. Seed and Idriss (1967) modelled the nonlinear hysteretic stress-strain properties of soils by using an equivalent linear elastic method of analysis. This model was later incorporated into the computer program SHAKE (Schnabel et al., 1972) which is widely used to predict site specific ground motions. Within the next three years, the model was generalized to two dimensions in the finite element programs QUAD-4 (Idriss et al., 1973) which operates in the time domain and FLUSH (Lysmer et al., 1975) which operates in the frequency domain. QUAD-4 has been used primarily for the analysis of the seismic response of earth and rockfill dams and FLUSH for soil-structure interaction studies. The successful simulation of the large scale deformations that occurred in the Lower San Fernando Dam during the San Fernando earthquake of 1971 marks the greatest achievement of this type of analysis (Seed et al., 1973).

The rapid development of soil dynamics and the achievements of these early years are described in the proceedings of the first major conference devoted to soil dynamics and the geotechnical aspects of earthquake engineering held under the auspices of the American Society of Civil Engineers at Pasadena, California, in 1978 (Youd, 1978). Progress in dynamic response analysis of soil structures and soil-structure interaction systems since then will be reviewed in this paper.

---

*Professor, Department of Civil Engineering, University of British Columbia, 2324 Main Mall, Vancouver, B.C., V6T 1W5, Canada.

The important elements in a dynamic response analysis are

(a) input motions
(b) site-structure model (geometrical, geological, geotechnical)
(c) static and dynamic soil properties
(d) constitutive model of soil response to loading
(e) method of analysis (computer program)
(f) validation (field data, model test data).

This review focusses primarily on the last three items.

The development of constitutive relations for soils has been one of the most active research areas in geotechnics in recent years. The stimulus for active research on constitutive modelling came from the limitations inherent in the equivalent linear method. Equivalent linear analysis is based on total stresses and the effects of seismically induced porewater pressures on strength and stiffness cannot be taken into account continuously during the analysis. In addition, since the analysis is elastic, it cannot be used to compute permanent deformations directly.

The state-of-the-art for analyzing permanent deformations was assessed in a report on earthquake engineering research by the National Research Council of the United States (NRC, 1982) as follows:

"Many problems in soil mechanics, such as safety studies of earth dams, require that the possible permanent deformations that could be produced by earthquake shaking of prescribed intensity and duration be evaluated. Where failure develops along well-defined failure planes, relatively simple elasto-plastic models may suffice to calculate displacements. However, if the permanent deformations are distributed throughout the soil, the problem is much more complex and practical, reliable methods of analysis are not available."

Therefore a major motivation for the development of more general constitutive relations has been the need to model nonlinear behaviour in terms of effective stresses and to provide reliable estimates of porewater pressures and permanent deformations under cyclic loading due to earthquakes and storm waves.

The effective stress methods fall into two main categories, elastic-plastic methods and those based on direct modelling of the nonlinear hysteretic stress-strain response. A few methods are based on endochronic theory which is a radically different concept. The number and complexity of the elastic-plastic models would render a superficial review of very limited utility. So this review will attempt to give the geotechnical engineer an appreciation of the basic concepts of the general types of elastic-plastic models which have already found their way into practice. The direct nonlinear simulations, in contrast, are based on concepts familiar to most geotechnical engineers and therefore are more easily understood. How these two approaches differ from each other will be described and the implications of these differences for practical applications will be illustrated by a comparative computational study of the seismic response of a level saturated sand site and of a structure embedded in a saturated sand foundation.

Typical examples of important cyclic loading problems are the response of nuclear power plants and earth dams to earthquake loading and offshore oil and gas production platforms to both storm wave and earthquake loading. However, despite the importance of these problems and the need for more representative and comprehensive analyses of them, the vast amount of research on constitutive models in recent years has had little impact on geotechnical engineering practice in these areas. Some reasons for this will be discussed later.

In 1980, an international workshop on generalized stress-strain relations in geotechnical engineering was held in Montreal, Canada to test the predictive capability of twelve selected constitutive relations on the same data base (Yong and Ko, 1981). This conference was the first of its kind and resulted in a much clearer understanding of the limitations of existing constitutive models and the substantial inherent difficulties in developing a truly general model of soil response to static and dynamic loading.

A similar workshop was held in Cleveland in 1987 (Saada and Bianchini, 1987). On this occasion more than thirty constitutive models were evaluated, a number indicative of the rapid growth in research on constitutive modelling.

When confronted with this intellectually intimidating smorgasbord, one can appreciate the comments of the reviewer (R.H.G.P., 1985) of an earlier compilation of such methods (Pande and Zienkiewicz, 1982):

"Faced with a specific problem in the field of dynamic loading or pseudo-static loading, the reader would find this book vastly tempting, but bewildering. Rather like a gourmet with a once-in-a-life-time opportunity to visit a great restaurant with all expenses paid. Every dish is a marvellous creation: but which one to choose?"
A major aim of this review is to demystify some of the more exotic items on the menu and give some guidance on appropriate selections for different occasions.

First however, equivalent linear analysis will be revisited to examine how it is used in predicting permanent deformations. This provides the point of departure and justification for consideration of the newer methods.

EQUIVALENT LINEAR ANALYSIS

The fundamental assumption of the equivalent linear method of analysis (EQL) is that the dynamic response of soil, a nonlinear hysteretic material, may be approximated satisfactorily by a damped elastic model if the properties of that model are chosen appropriately. The appropriate properties are obtained by an iterative process.

In QUAD-4 analyses, for example, the stress-strain properties of the soil are defined in each finite element by the Poisson ratio, shear strain dependent shear moduli and equivalent viscous damping ratios. Extensive data on shear strain dependent moduli and damping have been presented by Seed and Idriss (1970) and Seed et al. (1986).

A major application of equivalent linear analysis has been in the estimation of seismically induced displacements in earth and rockfill dams. However since the analysis with strain compatible properties is elastic, permanent deformations in these structures cannot be computed directly. This difficulty is surmounted by one of two approaches. The first approach uses the acceleration output from the analysis, the second the cyclic shear stress output. If the fill does not suffer significant degradation in strength during shaking, the deformations may be derived from the acceleration data. If significant strength loss occurs due to high porewater pressures, then the deformations may be estimated from the shear stress output with the help of laboratory test data. The structure of the procedure for estimation of displacements is shown in Fig. 1.

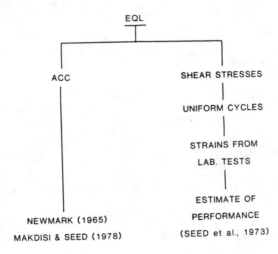

Figure 1.   Procedures for Estimating Dam Deformations.

Deformations from Acceleration Data

The earthquake induced displacements of a potential sliding block of a dam cross-section can be estimated from acceleration data as follows.

A potential sliding surface in the dam is analyzed statically to find the inertia force $F_I = (W/g)a_y$ required to cause failure (Fig. 2) where W is the weight of the sliding block and g the acceleration due

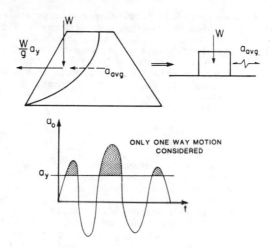

Figure 2.   Newmark's Method for Estimating Permanent Deformations.

to gravity. The yield acceleration a_y is then deduced from this force.
The average acceleration time-history of the sliding block is obtained
from a QUAD-4 or other equivalent linear analysis using the technique
described by Makdisi and Seed (1978). The yield acceleration is deduc-
ted from the average acceleration time history and the net acceleration, shown shaded in Fig. 2, is available to generate permanent
displacements. The analysis is conducted on an equivalent model of a
sliding block on a plane with only one way motions allowed (Fig. 2).

     This type of analysis was pioneered by Newmark (1965).   The
Makdisi and Seed (1978) version takes into account the flexibility of
the dam when calculating the average acceleration acting on the
potentially sliding block.

     In practice the procedure proposed by Makdisi and Seed (1978) for
determining the average acceleration of the potentially sliding block
is sometimes not followed.   Their procedure is based on the method
described by Chopra (1966) and involves calculating the resultant
dynamic force acting on the base of the sliding block and dividing this
force by the mass of the block to get the average acceleration.   Sometimes in practice the average acceleration is computed by averaging the
accelerations at nodal points of the finite element mesh within the
sliding block.   The latter procedure will generally yield a more
conservative estimate of the average acceleration.

     Makdisi and Seed (1978), on the basis of extensive dynamic
analysis studies, have presented charts that allow estimates of the
displacements in well-designed earth and rock fill dams up to 70 m in
height to be made conveniently without resorting to dynamic analysis.

The Makdisi-Seed method is particularly appropriate when there is no significant degradation in strength and stiffness of the fill in the dam during earthquake shaking. In these circumstances deformations tend to be localized along planes of maximum obliguity. If there is significant strength degradation, the deformation pattern will probably be more diffused. In this case the stress route to the estimation of strains and deformations is, in theory at least, more applicable. However, in current practice, the residual strength is frequently used in the Makdisi-Seed type of analysis.

Deformations from Stress Data

When the stress route to deformations is followed (Seed et al., 1973; Seed, 1979a), the shear stress history in each finite element is first converted to an equivalent number of uniform stress cycles. The strain potential of a finite element is considered to be the strain developed in a laboratory specimen consolidated to the same stress conditions as the finite element and subjected to the same number of equivalent uniform cycles of loading. These strains are incompatible because they are determined from isolated element tests but they are an indication of probable performance of the dam during an earthquake. A procedure for harmonizing these strains to give a continuous deformation field has been presented by Serff et al. (1976).

An extreme but very important example of strength degradation is when the earthquake shaking triggers liquefaction in sands and the undrained shear strength drops below the magnitude of the driving shear stresses on a potential slip surface (Poulos, 1981; Poulos et al., 1985; Castro et al., 1985). It would be reasonable to expect that such a phenomenon would be evident in cyclic triaxial tests conducted under cyclic stress and consolidation conditions similar to those in the corresponding finite element in the dam. However Castro et al. (1985) have demonstrated that it is impossible to retrieve samples of loose sands for testing which will be representative of field conditions even using the best conventional sampling methods.

They demonstrated that the density of the samples considered representative of conditions in the liquefied upper shell of the Lower San Fernando Dam in the analysis reported by Seed et al. (1973) were too high. The steady state strengths of these samples were in the range 180 kPa - 1,140 kPa which are significantly higher than those required for stability. When the densities were corrected for the effects of sampling, handling and consolidation prior to testing, the steady state strengths dropped by a factor of 20 to 9 kPa - 57 kPa.

The procedure for correcting steady state strengths for the effects of sampling, handling and consolidation prior to testing is difficult and requires a very high level of skill. In an attempt to avoid this problem, Seed (1987) back-figured (steady state) residual strengths from case histories of flow failures and related the strengths to the corrected standard penetration resistance $(N_1)_{60}$ measured in situ before the earthquake. The procedure for correcting the N values is given by Seed et al. (1985). Recently Seed (1988) has

made minor modifications to his 1987 correlations between residual
strength and $(N_1)_{60}$ as a result of new data from the Chilean earthquake
of 1985 and a re-evaluation of several critical cases in the original
data set.

Both acceleration and stress methods of deformation analysis provide the opportunity to compare the predicted deformations of a dam
under study with the actual responses of other dams which have been
shaken by earthquakes. This comparison provides a calibration of the
analysis and a basis for judging the implications of the computed
deformations for safety and design.

The best type of calibration is possible when the analysis can be
tuned by site specific data from previous events at the same site. An
outstanding example of this type of application is the prediction of
displacements at two sites in Anchorage, Alaska by Idriss (1985).

Calibration of Analysis by Site Data

In 1982, the State of Alaska planned the construction of a major
office complex in downtown Anchorage at one of two sites designated
Site 1B and Site 3 (Fig. 3). The sites were close to the 4th Avenue

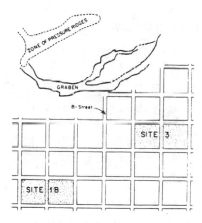

Figure 3.   Locations of Proposed Sites in Relation to Previous Slides
(adapted from Idriss, 1985).

slide that occurred during the 1964 Alaska earthquake and therefore
there was concern about the potential for large deformations at these
sites in future earthquakes.

The deformations in the 4th Avenue slide were caused by large
blocks of soil moving on a thin flat shear zone. This type of
deformation mode is ideally suited to analysis by the Newmark method.

The strategy adopted in assessing the potential for lateral deformation at the proposed sites was to tune the Newmark method of analysis
until the deformations in 1964 could be adequately simulated and then
to use the tuned method to predict the deformations at Site 1B and Site
3.

When it is not possible to calibrate the method of analysis by
site specific data, much more reliance must be placed on the inherent
reliability of the method of analysis itself. Data from case histories
at other locations may be useful in establishing the reliability of the
method. But uncertainties in interpreting site conditions, establishing quantitative estimates of soil parameters, specifying input motions
and defining the delimited region to be analyzed make extensive parametric studies necessary in critical situations when site specific
response data are not available.

The methods of analysis considered so far are a unique blend of
theory and empiricism and are based on an easily understood procedure
for capturing some of the nonlinear aspects of soil response to seismic
loading. They are widely used in practice and are familiar to many
geotechnical engineers. More sophisticated models will now be reviewed
which give a more detailed and coherent picture of dynamic soil
response .

ONE-DIMENSIONAL (1-D) NONLINEAR ANALYSIS

The simplest approach to modelling the nonlinear hysteretic
stress-strain behaviour of soil under cyclic loading is to model the
stress-strain paths analytically. This was first done by Streeter et
al. (1974) who used a Ramberg-Osgood (1943) representation of the
stress-strain curve and solved the equations of motion using the method
of characteristics. Their method of analysis, based on total stresses
and applicable to level ground conditions, was incorporated into the
computer program CHARSOIL.

In 1975 a porewater pressure generation model for cyclic loading
was developed by Martin et al. (1975), applicable to level ground
conditions. This model which allows the porewater pressures during
cyclic loading to be calculated from the cyclic strain history opened
the way to dynamic effective stress analysis.

The first dynamic effective stress method of 1-D analysis was
presented in 1976 by Finn et al. (1976). The most recent version of
this analysis is incorporated in the computer program DESRA-2 (Lee and
Finn, 1978). The stress-strain behaviour of the soil in shear was
represented by a hyperbolic initial loading curve (skeleton curve).
Stress-strain paths during unloading and reloading were defined by the
Masing criterion (1926). In 1986 a two-dimensional (2-D) version of
DESRA called TARA-3 was released (Finn et al., 1986). The constitutive
model in this program and a more general version of the
Martin-Finn-Seed porewater pressure model applicable to 2-D analyses
will be described in detail later. The constitutive model in DESRA is
just the shear component of the TARA-3 model. Therefore the reader is

referred to the description of TARA-3 for details on shear stress-strain behaviour and on the porewater pressure model.

An extensive verification of the DESRA-2 program was conducted by Finn and Bhatia (1980) under cyclic simple shear conditions using a variety of loading patterns; constant stress, constant strain and irregular loading. Experimental and computed response data for an irregular pattern of strain history shown in Fig. 4, agree very closely.

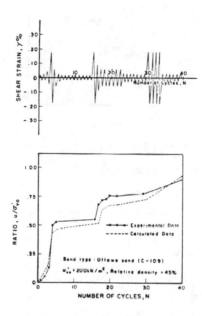

Figure 4. Measured and Computed Porewater Pressures Under Irregular Cyclic Loading (after Finn and Bhatia, 1980).

1-D nonlinear effective stress methods were also developed by Martin and Seed (1978), Ishihara and Towhata (1982) and Dikmen and Ghaboussi (1984). The results of comparative total stress studies between the various methods are shown in Fig. 5. It is clear that all the nonlinear methods give essentially similar results for total stress analyses. The SHAKE analysis gives a greater response than any of the nonlinear methods. This is sometimes the case because the SHAKE analysis is elastic and can build up a higher response due to resonance when there is a match between site period and the period of the input motion. This problem is not so acute in the nonlinear analyses because the site period is constantly changing due to the variable moduli. Porewater pressures predicted by effective stress analyses using DESRA-2 and LASS IV are shown in Fig. 6. They agree very closely.

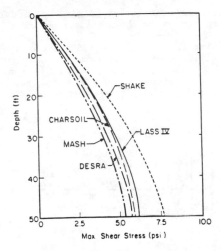

Figure 5. Shear Stresses Computed by Various Nonlinear Methods
(adapted from Finn et al., 1978; Martin and Seed, 1978;
and Dikmen and Ghaboussi, 1984).

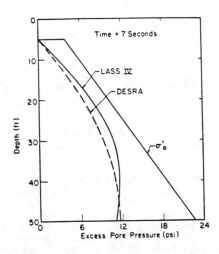

Figure 6. Vertical Distribution of Porewater Pressure Computed by
DESRA-2 and LASS IV (after Dikmen and Ghaboussi, 1984).

The utility and reliability of 1-D direct nonlinear analysis will be demonstrated by the analysis of a seismic test of a centrifuged model and two case histories from the field.

Validation of DESRA-2 by Centrifuge Tests

Hushmand et al. (1987) conducted simulated earthquake loading tests on centrifuge models of saturated sand in the centrifuge at the California Institute of Technology. The test container (Fig. 7) consisted of a stack of rectangular aluminum rings. Bearings in slotted groves between the rings minimized inter-ring friction.

Fine Nevada sand with a mean grain size of 0.1 mm was used to construct the model. The test was carried out at a centrifugal acceleration of 50 g.

The saturated model was subjected to input accelerations with frequency content and duration corresponding to those in the N-S acceleration component of the 1940 El Centro earthquake. The motions were scaled to give a peak base acceleration of 0.55 g.

The measured porewater pressures at two locations are shown in Fig. 8 and indicate that liquefaction occurred throughout the entire depth of the sand after about 5 s. The porewater pressures computed by DESRA-2 are also shown in Fig. 8 and are almost identical to the measured pressures. One notable difference is that the computed pressures do not show any oscillations. This results from the fact that DESRA-2 computes only the residual porewater pressures due to plastic volumetric compactive strains in the sand. These are the porewater pressures that control stability. It does not compute the transient porewater pressures resulting from instantaneous changes in the total mean normal stress. The measured and computed vertical distributions of porewater pressures with depth at t = 4s are shown in Fig. 9 and compare very closely.

Hushmand et al. (1987) also show that DESRA-2 predicts accelerations quite well.

This independent validation of DESRA-2 is a strong indication of its capability to model the effective stress response of saturated sands.

Seismic Response of Owi Island Test Site

Owi Island test site: Owi Island No. 1 is an artificial island located on the west side of Tokyo Bay. A test site at the south end of the island is instrumented to record porewater pressures and ground accelerations during earthquakes. The soil profile at the site, is shown in Fig. 10. The depths from which undisturbed samples, S, were recovered are also shown in Fig. 10.

The Mid-Chiba earthquake, with a magnitude M = 6.1, shook the Tokyo Bay area on September 25, 1980 and generated the porewater pressures and accelerations shown in Fig. 11. The maximum horizontal

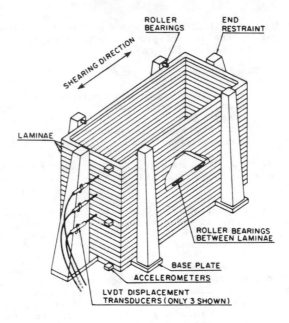

Figure 7.   Stacked Ring Container for Centrifuge Test on Sand Layer
            (adapted from Hushmand et al., 1987).

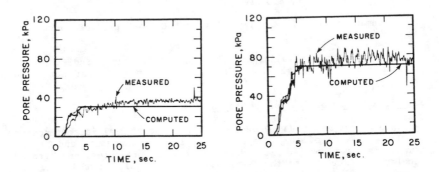

Figure 8.   Porewater Pressures at Two Locations in Centrifuge Model
            (adapted from Hushmand et al., 1987).

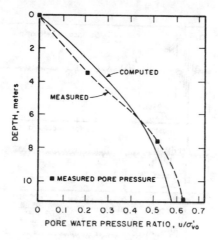

Figure 9. Comparison of Measured and Computed Porewater Pressures
(t = 4s)(adapted from Hushmand et al., 1987).

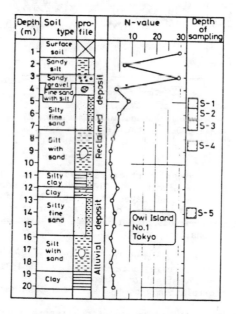

Figure 10.   Soil Profile at Owi Island Test Site (after Ishihara
et al., 1981).

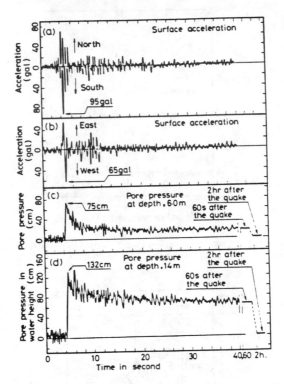

Figure 11.   Surface Accelerations and Porewater Pressures Recorded at
Owi Island During the Mid-Chiba Earthquake of 1980 (after
Ishihara et al., 1981)

accelerations at the ground surface were 95 gals in the N-S direction
and 65 gals in the E-W direction. The rise in porewater pressure was
0.75 m of water in the sand layer at a depth of 6 m and 1.32 m at a
depth of 14 m. Fourier spectra of the acceleration records indicate
that the predominant periods of motion were 0.64 sec and 0.5 sec in the
E-W and N-S directions, respectively.

All the properties required for the analysis of Owi Island No. 1
by DESRA-2 were obtained using data usually available from conventional
site and laboratory investigations. For example, the shear moduli were
obtained from a correlation with the standard penetration resistance N.

Full details of the instrumentation, recorded data, and the site
investigations on Owi Island and the associated laboratory testing have
been described by Ishihara et al. (1981). A more detailed description
of the analysis of the site is by Finn et al. (1982).

Comparison of Field and Computed Responses: The recorded ground motions are shown in Fig. 12a; those computed by DESRA-2 in Fig. 12b. Except for minor differences in frequency and magnitude in the 8-10 s range, the computed motions are very similar to the recorded motions.

The porewater pressures recorded at the 6 m depth are shown to an expanded scale in Fig. 13a. During the low level shaking of the first 4 s, the response was elastic and porewater pressures developed only in response to changes in the total applied mean normal stresses. Such porewater pressures result from the elastic coupling of soil and water. With the onset of more severe shaking, plastic volumetric strains are induced and these result in the development of residual porewater pressures which are independent of the instantaneous states of stress. These pressures accumulate with continued plastic volumetric deformation. Residual porewater pressure is indicated by the steep rise and sustained level in recorded porewater pressure in Fig. 13a. During shaking, the varying applied sresses continue to generate small instantaneous fluctuations in the porewater pressure which are superimposed on the larger residual porewater pressures. The gradual decay in the sustained level of porewater pressure is due to dissipation of porewater pressure by drainage. At this stage in the excitation, the dissipation of porewater pressure by drainage exceeds the generation by low level excitation.

The computed porewater pressures are shown in Fig. 13b and are very similar to the recorded values. The computed records do not show any fluctuations because DESRA-2 computes only the residual porewater pressures which control stiffness and stability and ignores pressures due to transient changes in total mean normal stresses.

Recorded and computed porewater pressures for the sand layer at a depth of 14 m are shown in Figs. 14a and 14b, respectively. DESRA-2 results compare very favourably with the recorded values. Dissipation of porewater pressure is negligible in the lower sand compared to the upper sand because it is capped by a clay layer instead of by pervious fill. The DESRA-2 program can take these different drainage conditions into account during the dynamic analysis.

Liquefaction Studies at Ishinomaki Port, Japan

During the 1978 Miyagi-Ken-Oki earthquake of magnitude M=7.4, some sites liquefied in Ishinomaki Port on the east coast of Honshu near Sendai (Fig. 15). These events were analyzed using the program DESRA-2 (Lee and Finn, 1978) by the Port and Harbour Research Institute, Ministry of Transport, Japan and a major report on the studies has been published (Iai et al., 1985). Investigations were conducted at six sites designated A through F (Fig. 15). Sites A,B,D and F liquefied. Two sites in the port C and E, which did not liquefy, were also studied to see whether DESRA-2 could discriminate between these sites and those which liquefied.

Accelerations were recorded on a SMAC-B2 strong motion accelerograph on a rock outcrop near Kaihoku Bridge (Fig. 15). The motions, shown in Fig. 16, have a peak acceleration of 293 gals in the W42N

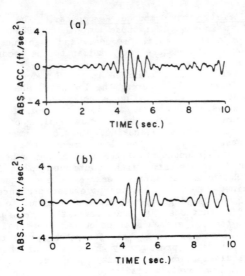

Figure 12. (a) Measured and (b) Computed Surface Accelerations at Owi
Test Site.

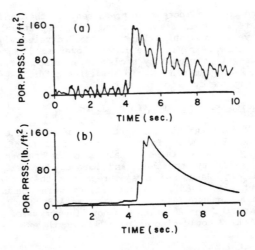

Figure 13. (a) Measured and (b) Computed Porewater Pressures at Depth
of 6 m at Owi Test Site.

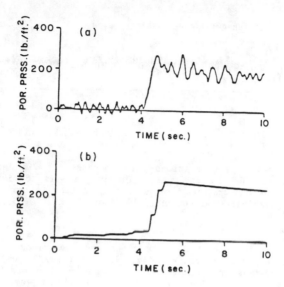

Figure 14. (a) Measured and (b) Computed Porewater Pressures at Depth of 14 m at Owi Test Site.

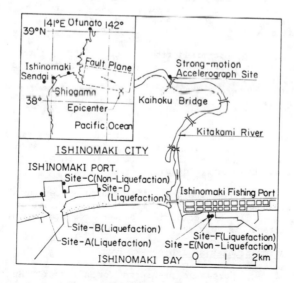

Figure 15. Map Showing Earthquake Epicenter, Accelerograph Site and Study Sites at Ishinomaki Port.

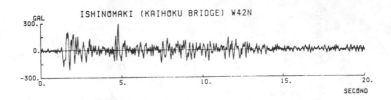

Figure 16.   Ground Accelerations near Kaihoku Bridge.

direction.   The W42N accelerations were increased by 20% and used as input motions for DESRA-2 analyses.   The increase is to simulate the effects of the second component of horizontal motion following Pyke et al. (1974) and Ishihara and Yamazaki (1980).

Soil Conditions

The soil conditions at all six sites are summarized in Fig. 17. The standard penetration resistances, N, (in blows/30 cm) and the unconfined compression strengths, q_u, were measured before the earth-quake except for the N-values at Site B which were measured afterwards.   However, comparative studies of N-values before and after the earthquake at Site D which also liquefied during the earthquake indicate little difference in the N-values.   Therefore the N-values at Site B were considered reasonably representative of site conditions before the earthquake.   The unconfined compression strengths for the clays below 40 m in depth were estimated by extrapolation assuming normal consolidation.

All properties were determined from the conventional site investi-gations conducted before and after the earthquake and from static and cyclic laboratory tests.
The procedure for determining most of the properties for dynamic analysis from the field and laboratory data is illustrated in the flow chart in Fig. 18.

The liquefaction resistance was determined for the sands at a relative density D_r = 65%, using constant volume cyclic simple shear tests.   Resistances at other relative densities less than D_r = 65% were assumed proportional to relative density.

To check the validity of the properties and parameters used in the dynamic analyses, DESRA-2 was used to simulate the liquefaction resist-ance curve at D_r = 65%.   The comparison between computed and measured liquefaction resistance is very good as shown in Fig. 19.

Results of Simulation Studies at Ishinomaki Port

A summary of peak porewater pressures for all sites during the earthquake is given in Fig. 20.   The simulation of liquefaction events

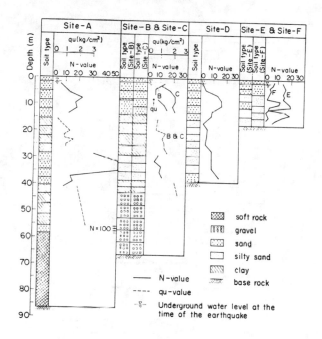

Figure 17. Soil Conditions at Study Sites.

by DESRA-2 agrees with what happened in the field as shown in Table 1. The results suggest that a dynamic effective stress analysis can usefully discriminate between the liquefaction potentials of various sites. In addition the analysis gives data on deformations, accelerations and porewater pressures for the design of post-earthquake rehabilitation works or new construction.

Conclusions from Validation Studies

The DESRA-2 program simulated the acceleration time-histories and porewater pressures at the Owi Island test site and in the centrifuge

TABLE 1

Comparison of Observed Site Performance and Simulation Results

Site	Site-A	Site-B	Site-C	Site-D	Site-E	Site-F
Liquefaction Observed	Yes	Yes	No	Yes	No	Yes
Liquefaction Computed	Yes	Yes	No	Yes	No	Yes

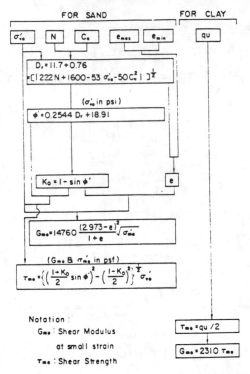

Figure 18.   Determination of Parameters for Dynamic Analysis of
              Ishinomaki Sites.

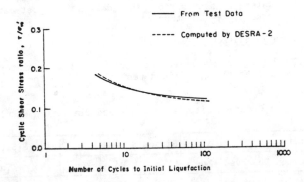

Figure 19.   Comparison of Simulated and Measured Liquefaction
              Resistance.

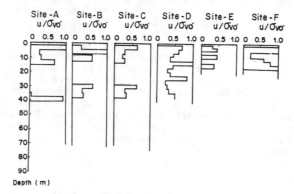

Figure 20. Peak Porewater Pressure Distributions at 6 Ishinomaki Sites.

test quite accurately. The agreement between predicted and computed responses in the centrifuge test is especially compelling since it was a completely independent study. The simulations at the Ishinomaki site showed that DESRA-2 could reliably predict the potential for liquefaction.

In both field studies, the analyses were conducted using soil data obtained in conventional site investigations except for the cyclic triaxial tests in the Owi Island study and cyclic simple shear tests in the case of Ishinomaki. As will be seen later when the 2-D version of DESRA-2 is discussed, technology has advanced so that even these tests would not be necessary today for preliminary or parametric studies. The ability to determine the soil properties for dynamic analysis using conventional procedures is a key factor in determining the acceptability of a method of dynamic analysis. It is one of the more attractive features of both the direct non-linear and equivalent linear methods.

Another attractive feature of nonlinear effective stress dynamic analysis is that it gives a very complete and coherent picture of dynamic response.

Analysis by Elastic-Plastic Methods

Level sites may also be analysed in terms of effective stresses using elastic-plastic methods. Discussion of these more general methods is deferred to a later section where they are examined in a two-dimensional context.

TWO-DIMENSIONAL NONLINEAR ANALYSIS

Dynamic effective stress analysis of soil structures such as embankment dams and soil-structure interaction systems is the major challenge of soil dynamics.

A program TARA-3 developed by Finn et al. (1986) will be described here which can be used to model nonlinear effective stress response of 2-D structures. It is a major extension in capability and efficiency of an earlier program developed by Siddharthan and Finn (1982) and is a 2-D version of the DESRA program described in the previous section. It has been the subject of detailed verification using data from simulated earthquake tests on centrifuged models. Data from one of these verification studies will be presented to show the capability of the program to estimate porewater pressures and accelerations and to model soil-structure interaction.

Method of Analysis

An incremental approach has been adopted to model nonlinear behaviour using tangent shear and bulk moduli, G_t and B_t respectively. The incremental dynamic equilibrium forces $\{\Delta P\}$ are given by

$$[M]\{\Delta\ddot{x}\} + [C]\{\Delta\dot{x}\} + [K]\{\Delta x\} = \{\Delta P\} \qquad (1)$$

where $[M]$, $[C]$ and $[K]$ are the mass, damping and stiffness matrices respectively, and $\{\Delta x\}$, $\{\Delta\dot{x}\}$, $\{\Delta\ddot{x}\}$ are the matrices of incremental relative displacements, velocities and accelerations. The viscous damping is of the Rayleigh type and its use is optional. Very small amounts of viscous damping are used, typically equivalent to less than 1% of critical damping in the dominant response mode. Its primary function is to control any high frequency oscillations that may arise from numerical integration. Damping is primarily hysteretic and is automatically included as the hysteretic stress-strain loops are executed during analysis.

The stiffness matrix is a function of the current tangent moduli during loading or unloading. The use of shear and bulk moduli allows the elasticity matrix $[D]$ to be expressed as

$$[D] = B_t[Q_1] + G_t[Q_2] \qquad (2)$$

where $[Q_1]$ and $[Q_2]$ are constant matrices for the plane strain conditions usually considered in analyses. This formulation reduces the computation time for updating $[D]$ whenever G_t and B_t change in magnitude because of straining or porewater pressure changes.

Stress-Strain Behaviour in Shear

The behaviour of soil in shear is assumed to be nonlinear and hysteretic and exhibits Masing behaviour (1926) during unloading and reloading. Masing behaviour provides hysteretic damping.

The relationship between shear stress τ and shear strain γ for the initial loading phase under either drained or undrained loading conditions is assumed to be hyperbolic and given by

$$\tau = f(\gamma) = \frac{G_{max}\ \gamma}{(1 + (G_{max}/\tau_{max})\ |\gamma|)} \tag{3}$$

in which G_{max} is the maximum shear modulus and τ_{max} is the appropriate shear strength. This initial loading or skeleton curve is shown in Fig. 21(a). The unloading-reloading curves are modelled using the Masing criterion. This implies that the equation for the unloading curve from a point (γ_r, τ_r) at which the loading reverses direction, is given by

$$\frac{\tau - \tau_r}{2} = \frac{G_{max}(\gamma - \gamma_r)/2}{1 + (G_{max}/2\tau_{max})\ |\gamma - \gamma_r|} \tag{4}$$

or

$$\frac{\tau - \tau_r}{2} = f(\frac{\gamma - \gamma_r}{2}) \tag{5}$$

The shape of the unloading-reloading curve is shown in Fig. 21(b).

Finn et al. (1976) proposed rules for extending the Masing concept to irregular loading. They suggested that unloading and reloading curves follow the skeleton loading curve when the magnitude of the previous maximum shear strain is exceeded (Fig. 21c). When the current loading curve intersects the previous loading curve the stress-strain curve follows the previous loading curve (Fig. 21d).

The tangent shear modulus, G_t, for a point on the skeleton curve is given by

$$G_t = \frac{G_{max}}{(1 + \frac{G_{max}|\gamma|}{\tau_{max}})^2} \tag{6}$$

At a stress point on an unloading or reloading curve G_t is given by

$$G_t = \frac{G_{max}}{(1 + \frac{G_{max}}{2\tau_{max}}\ |\gamma - \gamma_r|)^2} \tag{7}$$

Stress-Strain Behaviour in Hydrostatic Compression

The response of the soil to uniform all round pressure is assumed to be nonlinearly elastic and dependent on the mean normal stress.

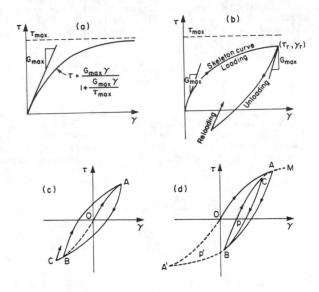

Figure 21.   Nonlinear Hysteretic Loading Paths.

Hysteretic behaviour, if any, is neglected in this mode. The relationship between tangent bulk modulus, B_t, and mean normal effective
stress, σ_m', is assumed to be in the form

$$B_t = K_b P_a \left(\frac{\sigma_m'}{P_a}\right)^n \qquad\qquad (8)$$

in which K_b is the bulk modulus constant, P_a is the atmospheric pressure in units consistent with σ_m', and n is the bulk modulus exponent.

RESIDUAL POREWATER PRESSURE MODEL

During seismic shaking two kinds of porewater pressures are
generated in saturated sands; transient and residual. The transient
pressures are due to changes in the applied mean normal stresses during
seismic excitation. For saturated sands, the transient changes in
porewater pressures are equal to changes in the mean normal stresses.
Since they balance each other, the effective stress regime in the sand
remains largely unchanged and so the stability and deformability of the

sand is not seriously affected. Therefore these pressures are not modelled in TARA-3.

The residual porewater pressures are due to plastic deformations in the sand skeleton. These persist until dissipated by drainage or diffusion and therefore they exert a major influence on the strength and stiffness of the sand skeleton. Both the shear and bulk moduli are dependent on the effective stresses in the soil and therefore, during an analysis, excess porewater pressures must be continually updated and their effects on the moduli taken continuously into account as described in Finn et al. (1976). The residual porewater pressures are modelled in TARA-3 using the model developed by Martin, Finn and Seed (1975). Therefore computed porewater pressure records will show the steady accumulation of pressure with time but will not show the fluctuations in pressure caused by the transient changes in mean normal stresses.

In the Martin-Finn-Seed model the increments in porewater pressure ΔU that develop in a saturated sand under seismic shear strains are related to the volumetric strain increments $\Delta \epsilon_{vd}$ that occur in the same sand under drained conditions with the same shear strain history. This model applies only to level ground so that there are no static shear stresses acting on horizontal planes prior to the earthquake. The M-F-S model is subsequently modified to include the effects of the initial static shear stresses present in 2-D analyses as described later.

The porewater pressure model is described by

$$\Delta U = \bar{E}_r \cdot \Delta \epsilon_{vd} \qquad (9)$$

in which E_r is the one-dimensional rebound modulus of sand at an effective stress σ_v'.

Under drained simple shear conditions, the volumetric strain increment $\Delta \epsilon_{vd}$ is a function of the total accumulated volumetric strain ϵ_{vd} and the amplitude of the current shear strain γ, and is given by

$$\Delta \epsilon_{vd} = C_1 (\gamma - C_2 \epsilon_{vd}) + \frac{C_3 \epsilon_{vd}^2}{\gamma + C_4 \epsilon_{vd}} \qquad (10)$$

in which C_1, C_2, C_3 and C_4 are volume change constants that depend on the sand type and relative density and may be determined directly by means of drained cyclic simple shear tests on dry or saturated samples. In practice simpler procedures to be described later are used.

An analytical expression for the rebound modulus $\bar{E}_r$, at any effective stress level σ_v', is given by Martin et al. (1975) as

$$\bar{E}_r = \frac{d\sigma_v'}{d\epsilon_{vr}} = (\sigma_v')^{1-m}/[m\,K_2(\sigma_{vo}')^{n-m}] \qquad (11)$$

in which σ_{vo}' is the initial value of the effective stress and K_2, m and n are experimental constants derived from rebound tests in a consolidation ring.

Extension of M-F-S Model to 2-D Conditions

In 2-D analysis in practice, the permanent volume changes due to shearing action are related to the cyclic shear stresses on horizontal planes because the seismic input motions are usually assumed to be shear waves propagating vertically. Therefore, in TARA-3, for computation of $\Delta\epsilon_{vd}$ in equation (10), the shear strain on the horizontal plane, γ_{xy} is substituted in place of γ. Also, σ_v' and σ_{vo}' in equation (11) are replaced by σ_y' and σ_{yo}' respectively, where σ_y' and σ_{yo}' are the current and initial vertical effective stresses.

Static shear stresses are usually present on the horizontal planes in 2-D problems. The presence of initial static shear stresses may significantly affect the cyclic behaviour of sands depending on the relative density of the sand and the level of the initial static shear stress (Seed and Lee, 1966; Vaid and Finn, 1978; Vaid and Chern, 1983). In saturated sands, the rate of development of porewater pressure, the level to which it may rise and the liquefaction potential curve are all dependent on the static shear stress level. These effects are taken into account in the porewater pressure model by specifying model constants as described in the next section, such that they produce a reasonable match of the peak porewater pressures and the rates of porewater pressure generation observed in laboratory samples with different initial static shear stress ratios.

Determination of Porewater Pressure Constants in Practice

The direct measurement of the constants in the porewater pressure model requires cyclic simple shear equipment which is not yet in common use. Therefore, to facilitate the use of TARA-3 in practice, techniques have been developed to derive the constants from the liquefaction resistance curve of the soil. The liquefaction curve may be determined from cyclic triaxial tests and then corrected to simple shear conditions as described by Seed (1979b) or derived directly from Standard Penetration Test data (Seed et al., 1983). In the latter case the constants are derived by a regression process to ensure that the predicted liquefaction curve compares satisfactorily with the field liquefaction curve using the program SIMCYC (Yogendrakumar and Finn, 1986a). If the liquefaction curve has been derived by laboratory tests, the rate of porewater pressure increase is known as well. Then a regression analysis is used to select constants that match both the rate of porewater pressure generation and the liquefaction curve using the program C-PRO (Yogendrakumar and Finn, 1986b).

Effect of Porewater Pressures

As porewater pressures increase in saturated soils during seismic shaking, the effective stresses decrease. The effective stress system is continually updated by solving the equilibrium equations while taking the increments in porewater pressure into account. The incremental displacements, strains and stresses given by this procedure constitute the response of the structure to softening of the elements. The incremental strains are accumulated and they contribute to the permanent deformations of the soil structure. Both shear and bulk moduli which depend on the current mean-normal effective stress must be adjusted to be compatible with the updated effective stress system.

If significant volumetric compaction occurs during seismic loading, the moduli should also be modified to reflect this strain hardening, following procedures outlined by Finn et al. (1976).

SOME SPECIAL FEATURES OF ANALYSIS BY TARA-3

Dynamic analyses are conducted in current engineering practice without including the effects of gravity or previous strains. Initial static analyses are usually conducted only to determine the initial stress conditions so that appropriate initial moduli may be selected. However as strength and stiffness degrade during seismic excitation because of increasing porewater pressures, the structure deforms under the gravitational forces. This effect is taken into account in TARA-3.

TARA-3 also has the capability to begin the dynamic analysis from a zero strain condition as in current practice or from the initial state of strain under static loading. The latter procedure leads to the best modelling of plastic deformations.

For analysis involving soil-structure interaction, it is important to model slippage between the structure and soil. Slip may occur during very strong shaking or even under moderate shaking if high porewater pressures are developed under the structure. TARA-3 contains slip elements of the Goodman type (1968) to allow for relative movement between soil and structure in both sliding and rocking modes of response during earthquake excitation.

The three components of permanent deformation in a soil structure system as a result of earthquake loading are computed by TARA-3. The first component is the dynamic residual deformation that occurs as a result of the hysteretic stress-strain response (Fig. 22). The second component is the deformation under gravity loading when increasing porewater pressures during the earthquake reduce the stiffness of the dam. The third component is the deformation of the system that occurs due to consolidation as the seismically induced residual porewater pressures dissipate. The final post earthquake deformation field computed by TARA-3 is the sum of all three deformation components.

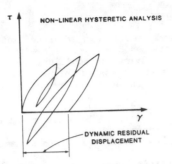

Figure 22.   Residual Displacement due to Cyclic Loading.

VALIDATION OF TARA-3 BY CENTRIFUGE TESTS

The United States Nuclear Regulatory Commission (USNRC) through
the U.S. Army Corps of Engineers sponsored a series of simulated
earthquake loading tests on centrifuged models to verify the nonlinear
dynamic effective stress method of analysis incorporated in TARA-3.
The tests were conducted at Cambridge University in the United Kingdom.
The models included dry and saturated embankments with and without
surface supported and embedded structures.   One of these tests
involving complex soil-structure interaction will be described here
(Steedman, 1986).

A schematic view of a saturated embankment with an embedded structure is shown in Fig. 23.   This configuration with a strong soil-
structure interaction provides a very severe test of the capabilities
of TARA-3 to model dynamic response.   The structure is made from a
solid piece of aluminum alloy and has dimensions 150 mm wide by 108 mm
high in the plane of shaking.   The length perpendicular to the plane of
shaking is 470 mm and spans the width of the model container.   The
structure is embedded a depth of 25 mm in the sand foundation.   Sand
was glued to the base of the structure to prevent slip between
structure and sand.

The foundation was constructed of Leighton Buzzard Sand passing BSS
No. 52 and retained on BSS No. 100.   The mean grain size is therefore
0.225 mm.   The sand was placed as uniformly as possible to a nominal
relative density D_r = 52%.

During the test the model experienced a nominal centrifugal acceleration of 80 g.   The model therefore simulated a structure approximately 8.6 m high by 12 m wide embedded 2 m in the foundation sand.

De-aired silicon oil with a viscosity of 80 centistokes was used
as a pore fluid.   In the gravitational field of 80 g, the structure
underwent consolidation settlement which led to a significant increase
in density under the structure compared to that in the free field.
This change in density was taken into account in the analysis.

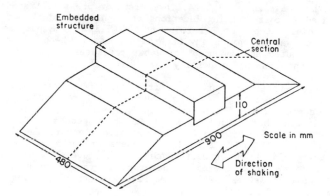

Figure 23. Model Structure Embedded in Saturated Sand Embankment.

The locations of the accelerometers (ACC) and pressure transducers (PPT) are shown in Fig. 24. Analyses of previous centrifuge tests indicated that TARA-3 was capable of modelling acceleration response satisfactorily. Therefore, in the present test, more instrumentation was devoted to obtaining a good data base for checking the ability of TARA-3 to predict residual porewater pressures.

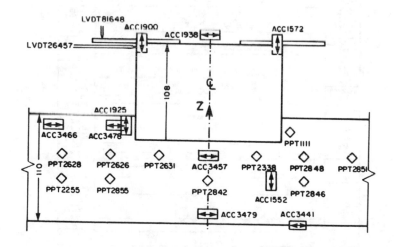

Figure 24. Instrumentation of Model Structure.

As may be seen in Fig. 24, the porewater pressure transducers are duplicated at corresponding locations on both sides of the centre line of the model except for PPT 2255 and PPT 1111. The purpose of this duplication was to remove any uncertainty as to whether a difference between computed and measured porewater pressures might be due simply to local inhomogeneity in density.

The porewater pressure data from all transducers are shown in Fig. 25. These records show the sum of the transient and residual porewater pressures. The peak residual pressure may be observed when the excitation has ceased at about 95 milliseconds. The pressures recorded at corresponding points on opposite sides of the centre line such as PPT 2631 and PPT 2338 are generally quite similar although there are obviously minor differences in the levels of both total and residual porewater pressures. Therefore it can be assumed that the sand foundation is remarkably symmetrical in its properties about the centre line of the model.

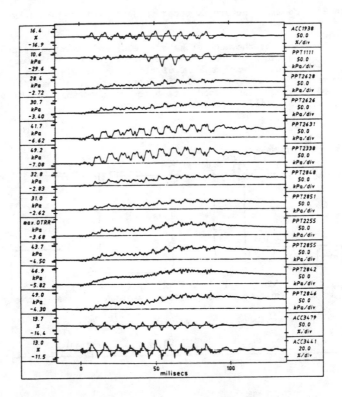

Figure 25.   Porewater Pressure Data from Centrifuge Test.

Computed and Measured Acceleration Responses

The soil-structure interaction model was converted to prototype scale before analysis using TARA-3 and all data are quoted at prototype scale. Soil properties were consistent with relative density.

The computed and measured horizontal accelerations at the top of the structure at the location of ACC 1938 are shown in Fig. 26. They are very similar in frequency content, each corresponding to the frequency of the input motion given by ACC 3441 (Fig. 25). The peak accelerations agree fairly closely.

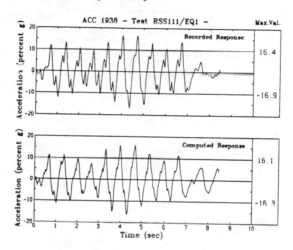

Figure 26. Recorded and Computed Horizontal Accelerations at ACC 1938.

The vertical accelerations due to rocking as recorded by ACC 1900 and those computed by TARA-3 are shown in Fig. 27. Again, the computed accelerations closely match the recorded accelerations in both peak values and frequency content. Note that the frequency content of the vertical accelerations is much higher than that of either the horizontal acceleration at the same level in the structure or that of the input motion. This occurs because the foundation soils are much stiffer under the normal compressive stresses due to rocking than under the shear stresses induced by the horizontal accelerations.

Computed and Measured Porewater Pressures

The porewater pressures in the free field recorded by PPT 2851 are shown in Fig. 28. In this case the changes in the mean normal stresses are not large and the fluctuations of the total porewater pressure about the residual value are relatively small. The peak residual porewater pressure, in the absence of drainage, is given directly by the pressure recorded after the earthquake excitation has

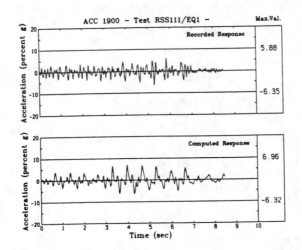

Figure 27.   Recorded and Computed Vertical Accelerations at ACC 1900.

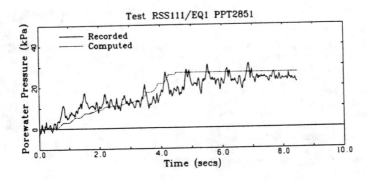

Figure 28.   Recorded and Computed Porewater Pressures at PPT 2851.

ceased.   In the present test, significant shaking ceased after 7
seconds.   A fairly reliable estimate of the peak residual pressure is
given by the record between 7 and 7.5 seconds. The recorded value is
slightly less than the value computed by TARA-3 but the overall
agreement between measured and computed pressures is quite good.

As the structure is approached, the recorded porewater pressures
show the increasing influence of soil-structure interaction. The pressures recorded by PPT 2846 adjacent to the structure (Fig. 29) show
somewhat larger oscillations than those recorded in the free field.
This location is close enough to the structure to be affected by the
cyclic normal stresses caused by rocking. The recorded peak value of
the residual porewater pressure is given by the relatively flat portion
of the record between 7 and 7.5 seconds. The computed and recorded
values agree very closely.

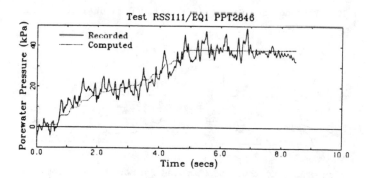

Figure 29. Recorded and Computed Porewater Pressures at PPT 2846.

Transducer PPT 2338 is located directly under the structure near
the edge and was subjected to large cycles of normal stress due to
rocking of the structure. These fluctuations in stress resulted in
similar fluctuations in mean normal stress and hence in porewater
pressure. This is clearly evident in the porewater pressure record
shown in Fig. 30. The higher frequency peaks superimposed on the
larger oscillations are due to dilations caused by shear strains. The
peak residual porewater pressure which controls stability is observed
between 7 and 7.5 seconds just after the strong shaking has ceased and
before significant drainage has time to occur. The computed and
measured residual porewater pressures agree very closely.

Contours of computed porewater pressures are shown in Fig. 31.
They indicate very symmetrical distribution of residual porewater
pressure. Recorded values are also shown in this figure.

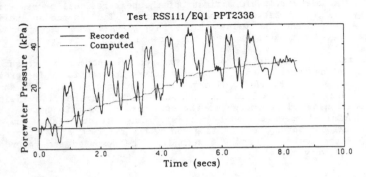

Figure 30.   Recorded and Computed Porewater Pressures at PPT 2338.

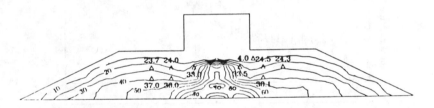

Figure 31.   Contours of Computed Porewater Pressures.

Stress-Strain Response

It is of interest to contrast the stress-strain response of the sand under the structure with that of the sand in the free field.   The stress-strain response at the location of porewater pressure transducer PPT 2338 is shown in Fig. 32.   Hysteretic behaviour is evident but the response for the most part is not strongly nonlinear.   This is not surprising as the initial effective stresses under the structure were high and the porewater pressures reached a level of only about 20% of the initial vertical effective stress.   The response in the free field at the location of PPT 2851 (Fig. 33) is strongly nonlinear with large hysteresis loops indicating considerable softening due to high pore-water pressures and shear strain.   At this location the porewater pressures reached about 80% of the initial effective vertical pressure.

Conclusion from Validation Studies of TARA-3

The comparison between measured and computed responses for the centrifuge model of a structure embedded in a saturated sand foundation demonstrates the wide ranging capability of TARA-3 for performing

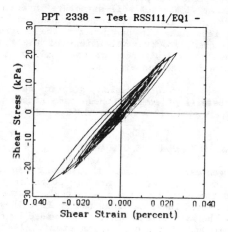

Figure 32.   Stress-Strain Response Under the Structure.

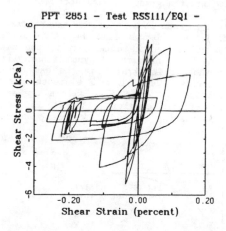

Figure 33.   Stress-Strain Response in the Free Field.

complex effective stress soil-structure interaction analysis with acceptable accuracy for engineering purposes. Seismically induced residual porewater pressures are satisfactorily predicted even when there are significant effects of soil-structure interaction. Computed accelerations agree in magnitude, frequency content and distribution of peaks with those recorded. In particular, the program was able to model the high frequency rocking vibrations of the model structures. This is an especially difficult test of the ability of the program to model soil-structure interaction effects.

Other tests in the verification program were also simulated satisfactorily. Details of some of these simulations may be found in Finn et al. (1984, 1985, 1986) and Finn (1985, 1988).

APPLICATION OF TARA-3 TO ANALYSIS OF DAMS

TARA-3 has been used to estimate the seismic response of a number of dams. In particular, it has been used to determine the peak dynamic displacements and the post-earthquake permanent deformations. Typical results for the proposed Lukwi tailings dam in Papua New Guinea will be presented to show the kind of data that is provided by a true nonlinear effective stress method of analysis (Finn et al., 1987,1988). First, however, the framework of a TARA-3 analysis as applied to dams will be presented.

TARA-3 conducts both static and dynamic analysis. A static analysis is first carried out to determine the stress and strain fields throughout the cross-section of the dam at the end of construction. The program can simulate the gradual construction of the dam.

Dynamic analysis in each element of the dam starts from the static stress-strain condition as shown in Fig. 34. This leads to accumulating permanent deformations in the direction of the smallest residual resistance to deformation. Methods of dynamic analysis commonly used in practice ignore the static strains in the dam and start from the origin of the stress-strain curve in all elements even in those which carry high shear stresses. TARA-3 also allows the analysis to start from the zero stress-strain condition, if it is desired to follow current practice. As shaking proceeds, two phenomena occur; porewater

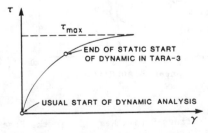

Figure 34. Starting Points for Dynamic Analysis

pressures develop in saturated portions of the embankment and, in the unsaturated regions, volumetric strains and associated settlements develop. The program takes into account the effects of the porewater pressures on moduli and strength during dynamic analysis and estimates the additional deformations due to gravity acting on the softening soil.

At the end of the earthquake, additional settlements occur due to consolidation as the seismically induced residual porewater pressures dissipate. The final deformed shape of the dam results from the sum of permanent deformations due to the hysteretic dynamic stress-strain response, constant volume deformations in saturated portions of the embankment, volumetric strains in unsaturated portions and deformations due to consolidation as the seismic porewater pressures dissipate. The final post-earthquake deformed shape of a saturated embankment of loose sand computed by TARA-3 is shown in Fig. 35. This shows the classical spreading due to high porewater pressures.

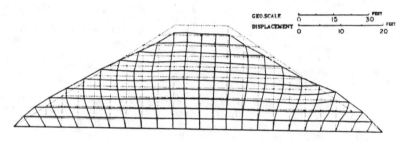

Figure 35.   Deformed Shape of Uniform Saturated Embankment After Earthquake.

The post-earthquake deformed shape of an embankment with a central core is shown in Fig. 36. The water table is about 1.7 m below the crest. Only the upstream segment to the left of the core is saturated and generates high porewater pressure during earthquake shaking. Large deformations occur upstream and the core is strongly deformed towards the upstream side. Although the deformations in this case are contained, they are sufficient to cause severe cracking around the core.

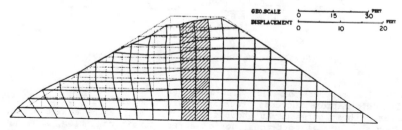

Figure 36.   Deformed Shape of Central Core Embankment After Earthquake.

These examples show the ability of TARA-3 to predict observed deformation modes in embankments during earthquakes.

Lukwi Tailings Dam

The finite element representation of the Lukwi tailings dam is shown in Fig. 37. The sloping line in the foundation is a plane

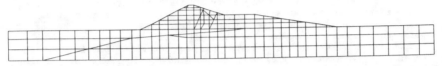

Figure 37. Finite Element Idealization of Lukwi Tailings Dam.

between two foundation materials. Upstream to the left is a limestone with shear modulus G = 6.4 x 10⁶ kPa and a shear strength defined by c' = 700 kPa and φ' = 45°. The material to the right is a siltstone with a low shearing resistance given by c' = 0 and φ' = 12°. The shear modulus is approximately G = 2.7 x 10⁶ kPa. The difference in strength between the foundation soils is reflected in the dam construction. The upstream slope on the limestone is steep whereas the downstream slope on the weaker foundation is much flatter and has a large berm to ensure stability.

The dam was subjected to strong shaking with a peak acceleration of 0.33 g (Fig. 38). The shear stress-shear strain response of the limestone foundation is almost elastic as shown in Fig. 39.

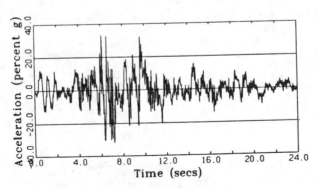

Figure 38. Input Motions for Dynamic Analysis of Dam.

The response of the siltstone foundation is strongly nonlinear. The deformations increase progressively in the direction of the initial static shear stresses as shown in Fig. 40. Since the analysis starts from the initial post-construction stress-strain condition, subsequent large dynamic stress impulses move the response close to the highly

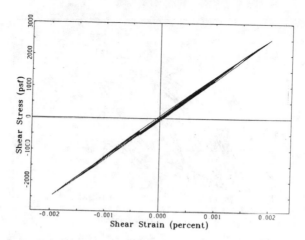

Figure 39. Shear Stress - Shear Strain Response of Limestone
Foundation.

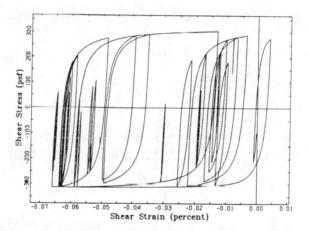

Figure 40. Shear Stress - Shear Strain Response of Siltstone
Foundation.

nonlinear part of the stress-strain curve. It may be noted that the
hysteretic stress-strain loops all reach the very flat part of the
stress-strain curve, thereby ensuring successively large plastic
deformations.

An element in the berm also shows strong nonlinear response with
considerable hysteretic damping (Fig. 41).

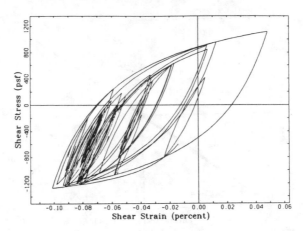

Figure 41.   Shear Stress - Shear Strain Response of Berm.

The acceleration time history of a point near the crest in the steeper upstream slope is shown in Fig. 42.   The displacement time history of the point is shown in Fig. 43.   Note that the permanent deformation is of the order of 25 cm.   Most of this was generated by a large permanent slip which occurred about 8 secs after the start of shaking.   The deformed shape of the central portion of the dam is shown to a larger scale in Fig. 44.

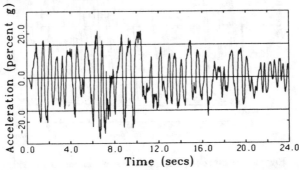

Figure 42.   Computed Accelerations of a Point Near the Crest.

ELASTIC-PLASTIC CONSTITUTIVE MODELS

Plasticity theory has been a very fertile field for the development of constitutive models of soil response to cyclic loading.   Twenty six of the thirty-two constitutive models listed in the preprint volume of the Cleveland workshop on constitutive modelling (Saada and

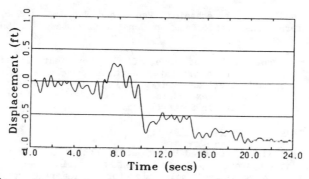

Figure 43. Displacement History of Point Near the Crest.

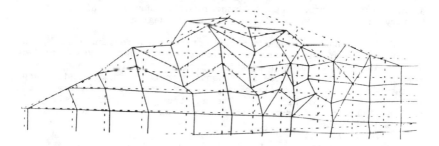

Figure 44. Deformed Shape of the Dam After Earthquake to Enlarged
Scale.

Bianchini, 1987) are based on elastic-plastic theory and these, by no
means, exhaust the number of available models.

Models based on the classical isotropic theory of plasticity such
as the critical state model (Roscoe, Schofield and Wroth, 1958; Roscoe
and Burland, 1968) cannot simulate the porewater pressures and permanent deformations generated by cyclic loading (Carter et al., 1982).

Research over the last ten years has been devoted to the development of more complex elastic plastic models with the potential for
simulating cyclic loading effects while retaining some of the convenient features of classical plasticity theory. Detailed descriptions of
these developments can be found in Pande and Zienkiewicz (1982), in a
report on constitutive laws prepared for the XI International
Conference on Soil Mechanics and Foundation Engineering (Murayama,
1985) and in the proceedings of the following conferences;
International Conference on Numerical Methods in Geomechanics (Kawamoto
and Ichikawa, 1985; Swoboda, 1988), International Symposium on

Numerical Models in Geomechanics (Pande and Van Impe, 1986) and the Second International Conference on Constitutive Laws for Engineering Materials (Desai et al., 1987).

A brief description of some of the key features of general plasticity theory will be presented next to provide a framework for understanding the limitations of classical theory and some of the novel features of the more complex models. A more detailed and very readable description may be found in Scott (1985).

Basic Elements of General Plasticity

On the basis of observed behaviour in cyclic loading tests, soil is a nonlinear, hysteretic, strain-hardening material over the range of strains usually considered tolerable for engineering design. The behaviour in extension and compression may be quite different and the Bauschinger effect (different yield stresses in extension and compression) is usually pronounced. Furthermore, because of their geological history soils are always anisotropic to some degree.

For strain-hardening materials, represented for instance by the piecewise linear stress-strain curve in Fig. 45a, a yield surface exists in stress space separating regions of elastic and plastic response. The yield surface corresponding to the initial elastic range defined by the stress σ_0 is given by the inner curve in the stress space σ_1, σ_2 shown in Fig. 45b. The equation, f=0, defining the yield surface, shapes the mathematical forms of the equations of the computational model.

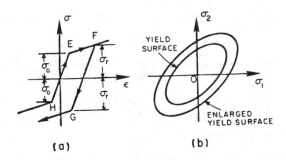

(a)   (b)

Figure 45.   Isotropic Hardening Model

If loading is continued past σ_0 in Fig. 45a to σ_r, what happens to the yield surface? The assumptions made at this point introduce a major distinction between plasticity theories and a radical difference in the predictive capabilities of the resulting plasticity models. It is frequently assumed that unloading from F is elastic and continues until G is reached in the opposite loading sense at a similar stress magnitude. It is clear that the elastic region is now extended from

$2\sigma_0$ to $2\sigma_r$. The new yield surface is given by the outer curve in Fig. 45b. All paths within this enlarged surface are elastic. The yield surface has been expanded uniformly by the loading causing plastic deformation. A material with this type of response is called an isotropically hardening material.

The yield surface is also used to define loading states. For strain-hardening materials, stress-increments directed outward from the yield surface are termed loading states, those directed tangential to it are termed neutral and those directed inwards are termed unloading states. Because of this close connection with states of loading, the yield surface is also called a loading surface. If the yield surface is defined by $f(\sigma_{ij}) = 0$ where σ_{ij} are the stresses, then

for loading $\qquad \dfrac{\partial f}{\partial \sigma_{ij}} \, d\sigma_{ij} > 0$ $\qquad\qquad$ (12)

for neutral loading $\qquad \dfrac{\partial f}{\partial \sigma_{ij}} \, d\sigma_{ij} = 0$ $\qquad\qquad$ (13)

for unloading $\qquad \dfrac{\partial f}{\partial \sigma_{ij}} \, d\sigma_{ij} < 0$ $\qquad\qquad$ (14)

The vector with components $\partial f / \partial \sigma_{ij}$ is in the direction of the normal to the yield surface and the dot product of $\partial f / \partial \sigma_{ij}$ with the stress increment vector $d\sigma_{ij}$ defines the nature of the loading increment and hence the loading state. This precise definition of loading or unloading is one of the advantages of plasticity theory for the analysis of multi-axial stress-states. Generalizations of successful one-dimensional methods not based on plasticity theory to multi-axial stress-states sometimes have difficulty in defining the loading states precisely.

Prager (1955) made a different assumption about the behaviour of the yield surface as loading proceeded from E to F in Fig. 46a. He assumed that the yield surface did not change in size or shape but translated without rotation in stress-space as in Fig. 46b. In effect, he assumed the yield surface to be represented by a frictionless rigid ring which was pushed around the stress-space σ_1, σ_2 (Fig. 46b) by the stress point. As the stress increased from E to F, the stress point pushed the yield surface to the stress-state associated with F. On unloading from this state it is clear that the maximum elastic response range is still $2\sigma_0$ since the yield surface has not changed in size. Yielding will occur again when loading in the opposite sense touches the boundary of the yield surface. From Fig. 46a, it can be seen that the yield stress in extension is now different from that in compression, thus modelling the Bauschinger effect. This type of

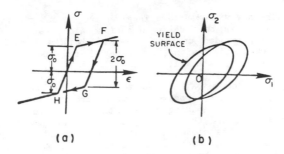

(a)   (b)

Figure 46. Kinematic Hardening Model

hardening is called kinematic hardening as it is specified by the motion of the yield surface. Kinematic hardening is one of the key elements that makes the theory of plasticity a useful model for soils under cyclic loading.

The hardening that occurs beyond initial yield in Figs. 45a and 46a is of the simplest kind - linear. In general, the hardening is nonlinear but can be represented by a piecewise linear approximation. The slope of the line EF (Fig. 46a) is used to define a work-hardening or plastic modulus. In the linear hardening case, a single modulus associated with a single yield surface is sufficient to define the plastic deformations. For the piecewise linear representation a nest of yield surfaces and associated plastic moduli are required, each associated with a particular linear segment of the strain-hardening stress-strain curve. This generalization of the basic model is due to Mroz (1967). When more than one yield surface is used, the translation of the yield surfaces in stress-space must be controlled by some criterion to ensure that the yield surfaces do not intersect. A rule developed by Mroz (1967) is used, The Mroz rule states that if stress point P is on yield surface $f^{(m)}$ then on loading $f^{(m)}$ translates towards the next yield surface $f^{(m+1)}$ along a line PR where R is the point on $f^{(m+1)}$ with outwards normal in the same direction as the normal at P. P and R are called conjugate points.

A crucial element of an incremental theory of plasticity is the flow rule which defines the directions of the plastic strain increments during loading. The strain increments are assumed to be normal to a surface in stress-space called the potential surface, g=0. If the potential surface coincides with the yield surface, the flow rule is called an associated flow rule, otherwise it is called a nonassociated flow rule. Most formulations of plasticity theory for soils assume an associated flow rule for convenience despite strong evidence to the contrary in the case of sands.

The incremental relationship between stress increments and strain-increments is the corner-stone of any computational model. The relationship for elastic response is given by one of the usual formulations

of the theory of elasticity. The relationship between stress increments and plastic strain increments depends on the flow rule. A simple
heuristic development of the relationship is given below for an
associated flow rule.

Only stress increments normal to the yield surface cause plastic
strain increments in strain-hardening materials. The components of the
unit normal, n_{ij}, are defined by

$$n_{ij} = \frac{\partial f/\partial \sigma_{ij}}{\{\partial f/\partial \sigma_{mn} \quad \partial f/\partial \sigma_{mn}\}^{1/2}} \tag{15}$$

where the quantity below the line is the magnitude of the vector with
components $\partial f/\partial \sigma_{ij}$. The normal stress increment components $(d\sigma_{ij})_n$ are
given by

$$(d\sigma_{ij})_n = n_{ij} \ (n_{rs} \quad d\sigma_{rs}) \tag{16}$$

Assuming a homogeneous first order relationship between $d\epsilon_{ij}^p$ and
$(d\sigma_{ij})_n$ the following relationship is obtained

$$d\epsilon_{ij}^p = \frac{1}{H} \ \frac{\partial f}{\partial \sigma_{ij}} \ \frac{\partial f/\partial \sigma_{rs} \quad d\sigma_{rs}}{\partial f/\partial \sigma_{mn} \quad \partial f/\partial \sigma_{mn}} \tag{17}$$

where H is the plastic modulus. This equation looks formidable but in
many cases simplifies considerably. For undrained compression in a
triaxial test and assuming a Von Mises yield surface, Eqn. (17) reduces
to

$$d\epsilon_{ij}^p = \frac{2}{3H} \ d\sigma_{11} \tag{18}$$

where $d\sigma_{11}$ is the increment in axial stress (Scott, 1985).

Classical Cam-Clay Model

The critical state Cam-Clay model familiar to many geotechnical
engineers is based on an isotropic hardening theory of plasticity. The
essential elements of the model are shown in Fig. 47. The yield surface is plotted in q-p space where p is the mean-normal effective
stress and q is the generalized deviatoric stress. Assume that the
elliptical yield surface has been established by isotropic consolidation under p_c'. An associated flow rule is assumed so that plastic
strain increments are normal to the yield surface. The critical state
line, CSL, defines stress states at which deformation occurs at

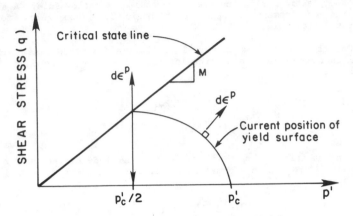

Figure 47.   Elements of Cam-Clay Model

constant volume.   Therefore the yield locus must intersect the CSL so
that plastic strain-increments are vertical and therefore do not give
volumetric strains.

Consider the effects of a sequence of uniform stress cycles starting from a stress point on the yield surface and involving initial
unloading.   The unloading, according to the classical theory, is
elastic and so no volumetric strains or residual porewater pressures
can occur.   On subsequent reloading and unloading, since the loads are
of constant amplitude, only the previously defined elastic region will
be traversed.   Again no volumetric strains and residual porewater
pressures will be generated.   These findings, of course, are directly
contrary to experience.   Carter et al. (1982) resolved this dilemma by
suggesting that during unloading the yield surface retracted with the
stress point to shrink the elastic region and thus ensure plastic
deformations on reloading.   This idea is closely allied to the concept
of kinematic hardening advanced by Prager (1955).

These limitations of classical plasticity have been removed by
abandoning the concept of the single yield surface which expands isotropically.   Instead progressive strain-hardening has been modelled
using multi-yield surfaces with kinematic hardening (Prevost, 1981) or
by using a hardening function with a bounding surface theory (Bardet,
1986; Dafalias and Herrman, 1982; and Pastor and Zienkiewicz, 1986).
These references refer to the latest developments in these approaches.

Multi-Yield Surface Model

The Prevost (1978) model for undrained stress-strain behaviour of
clay will be used to illustrate how multiple yield surfaces can be used
to model stress-strain curves.   In this case Prevost assumed yielding
to be controlled by deviatoric stresses and to occur at constant

plastic volume. This behaviour may be modelled using a Von Mises yield
criterion which is represented by a circle in the deviatoric plane, the
plane normal to the hydrostatic line in stress-space.

A piecewise linear representation of a stress-strain curve
obtained in a triaxial test is shown in Fig. 48a. The behaviour in
extension and compression is different but it is assumed that segments
of similar slope exist in both the extension and compression regions.
The yield surfaces delimiting regions of similar mechanical properties
are shown in Fig. 48b. Note that the centres are displaced with
respect to the origin reflecting the anisotropy induced by previous
stress history and are centered on the y-axis. This implies that the
y-axis is an axis of symmetry. Thus the yield curves are indicative of
a K_0-consolidated sample loaded along the axis in a triaxial test.

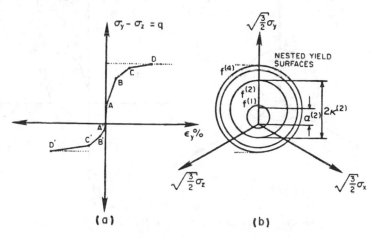

Figure 48. Elements of Multi-yield Surface Model

The stress-strain curve in Fig. 48a was generated by loading from
the origin to D along the y-axis. The region AA' inside the first
yield surface is assumed elastic. During loading when the stress point
reaches A, plastic strains are generated and may be calculated by Eqn.
(17) using the plastic modulus appropriate for the segment AB. On
further loading the stress point reaches B with yield surfaces 1 and 2
touching. The stress point now translates both surfaces giving plastic
strains related to the plastic modulus for the segment BC. Finally
when the point D is reached all surfaces are in contact as shown in
Fig. 49a.

On unloading from D the response is elastic until the diameter of
$f^{(1)}$ has been traversed. During initial compression the elastic range
was $\alpha^{(1)} + K^{(1)}$ where $\alpha^{(1)}$ is the amount the centre of $f^{(1)}$ is
displaced from the origin and $K^{(1)}$ is the radius of the circle. The
range on unloading $2K^{(1)}$. For isotropic material when $\alpha^{(1)} = 0$, the

elastic range in unloading is twice that on loading. In this case, the
multi-yield surface approach to modelling the unloading curve with pure
kinematic hardening is identical to the Masing criterion used in
TARA-3.

On further unloading, plastic deformations again develop with a
plastic modulus appropriate to segment AB being used in segment EF.
This process is repeated for each segment. The unloading curve
obtained from the multi-yield surface approach is given by curve 1 in
Fig. 49b. The corresponding Masing curve is given by curve 2. The

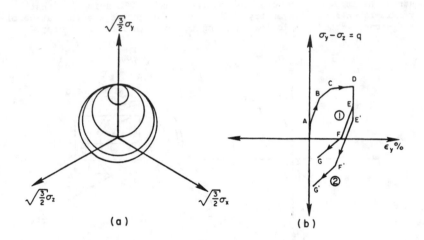

Figure 49. Unloading Paths from Multi-Yield Surfaces

modelling of the unloading curve should be carefully checked by
comparison with the unloading curve in the laboratory test. The
initial loading curve can be matched as closely as desired by taking a
large number of segments with their associated yield surfaces. Close
matching of the unloading curve cannot be guaranteed.

One difficulty with this method in finite element analysis is
keeping track of the locations of all the circles in every finite
element of the domain. This is a major chore which can be avoided by
using the concept of a bounding surface.

Bounding Surface Plasticity

The fundamental concepts of bounding surface plasticity theory
will be explained using the results of a uniaxial compression test
(Bardet 1986).

The stress-plastic strain curve for a cycle of loading and
unloading is shown in Fig. 50. The vertical segment of the curve

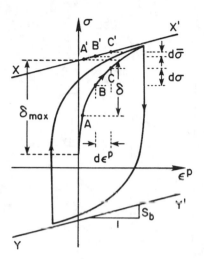

Figure 50. Elements of Bounding Surface Model

represents the elastic region and with infinite slope corresponds to a plastic modulus $H = \infty$. The bounds of admissible stress states for the current level of hardening are given by the lines XX' and YY' with slope S_b. The point representing the limit of elastic response is a distance δ_{max} from the bound XX'.

In the range of plastic strains, the stress-strain curve approaches and eventually merges with the bound XX'. Obviously the slope of the stress-plastic strain curve varies at the distance δ between a given stress-strain state, B, and the corresponding or image point, B', on the bounding line. The function defining the variation in slope with δ is a decreasing function ranging from $H = \infty$ at the distance δ_{max} to $H = S_b$ at the bounding surface.

Consider an incremental stress change, $d\sigma$, at B and a corresponding change $d\bar{\sigma}$ at B' so that the new stress points C and C' are directly in a vertical line. This implies that $d\sigma$ and $d\bar{\sigma}$ cause the same plastic strain increment, $d\epsilon^P$. Therefore $d\bar{\sigma}$ may be obtained from the equation

$$d\sigma/S = d\bar{\sigma}/S_b \qquad (19)$$

Some elements of classical plasticity theory such as the associated flow rule are used with the bounding surface. Therefore the strains at B during application of a stress increment can be conveniently established by making the calculations for the stress increment $d\bar{\sigma}$ on the bounding surface and mapping them to point B.

For general stress conditions the elements of bounding surface plasticity are shown in I-J space in Fig. 51 where I is the first

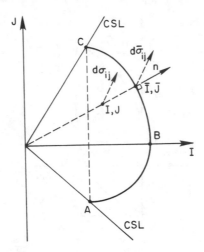

Figure 51. Bounding Surface with Radial Mapping

stress invariant $\sigma_{11}+\sigma_{22}+\sigma_{33}$ and J is the second deviator stress (S_{ij}) invariant given by $J = (1/2S_{ij}S_{ij})^{1/2}$. The elliptical bounding surface may be assumed to have been established by isotropic consolidation. In classical plasticity ABC would be considered the yield surface with all the attendant difficulties for the analysis of cyclic loading. The bounding surface is considered a limiting boundary in stress space beyond which the stress point cannot stray unless the bounding surface can expand as a result of volumetric strain hardening. Plastic strains can occur inside the bounding surface and there may be a nucleus of elastic response. The plastic strains for interior stress points are deduced from what happens at corresponding points on the bounding surface using radial mapping (Bardet, 1986).

Consider a stress increment $d\sigma_{ij}$ applied to an interior point I,J (Fig. 51). The radial line from the origin through I,J cuts the bounding surface at $\bar{I},\bar{J}$ called the image point. It is assumed that the increment of image stress $d\bar{\sigma}_{ij}$ gives the same increment of plastic strain as the stress increment $d\sigma_{ij}$. Assuming an associated flow rule, this may be expressed by the equation

$$\frac{1}{H} n_{ij}d\sigma_{ij} = \frac{1}{H_b} n_{ij}d\bar{\sigma}_{ij}$$ (20)

where

$$n_{ij} = \frac{\partial f/\partial \sigma_{ij}}{(\partial f/\partial \sigma_{mn} \ \partial f/\partial \sigma_{mn})^{1/2}}$$ (21)

H_b is the plastic modulus at the image point and H is the plastic modulus at I,J. H is assumed to vary as a function of the distance δ from the bounding surface from H=∞ at the elastic region to $H_b$on the bounding surface. This function must be established as part of the model calibration. The functional definition of the variation of plastic modulus takes the place of the nested yield surfaces in the multi-yield surface model.

Other rules for radial mapping may be used. For example Dafalias and Hermann (1982) use a projection centre CI_0 on the I-axis instead of the origin (Fig. 52). The location C must be determined as part of the model calibration. For the same bounding surface and critical state line, the use of different origins for radial mapping will define different image points and different directions for the plastic strain increments as shown by points D and B in Fig. 52.

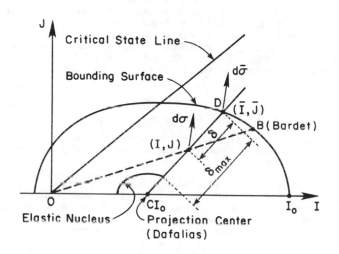

Figure 52. Bounding Surface with Projection Centres

Validation of Constitutive Models

Case Western Reserve University and the Institute de Mecanique of the University of Grenoble have established a data centre for cohesion-less soils. The data bank currently contains data from over 200 tests on 3 sands, obtained from a wide variety of loading paths. One of the sands is the well-known Reid Bedford Sand from near Vicksburg, Mississippi. Half the tests were conducted in a hollow cylinder device at Case; the other half in a cubical testing device at the University of Grenoble.

A workshop was held at Case in July 1987 to publicize the data base to the geotechnical engineering profession and to test the predictive capabilities of various constitutive models (Saada and Bianchini, 1987).

Data from 24 tests were given to each of the predictors to calibrate his model. Tests were divided into four groups. Each group consisted of 3 compression tests, 2 extension tests and an isotropic consolidation test with unloading/reloading. Each group of tests was conducted on each type of sand and in each of the testing devices. Having calibrated his model on the 4 test groups, the predictor was asked to predict the results of the other tests with more complicated stress paths.

Of particular interest for this review are the predictions of response to cyclic torsional shear stresses in the hollow cylinder test. This test was conducted in 3 stages. The specimen was loaded first in compression to a specified stress level and then 5 cycles of sinusoidally varying torsional stress with fixed amplitude were applied to the specimen. Finally failure was induced by increasing the torsional shear stress monotonically. Axial force and deformation, volume change, torque and rotational displacement were measured during the test.

Also of interest is a test in the cubical apparatus with a circular loading path for two cycles.

Predictions of stress-strain curves under monotonic loading were generally quite good but no model did a uniformly good job of predicting volumetric strains.

Typical results for cyclic loading tests are shown in Figs. 53 through 58. Dotted lines indicate the predictions and solid lines the measured responses. Results for a multi-yield surface model are shown in Fig. 53 and for three bounding surface models in Figs. 54 through 56. Although these models are considered among the best of their type, the predictions of the stress-strain paths are fairly crude and the volumetric strains are over-estimated by amounts up to a factor of 2.

Two multi-mechanism models did not fare any better as exemplified by the predictions of the Matsuoka et al. (1986) model in Fig. 57 and those of the Miura-Finn (1987) model in Fig. 58.

Predictions of the stress paths and of volumetric strains in the case of the circular loading paths in the cubical device were also disappointing.

Poor predictions of volumetric strains lead to poor estimates of isotropic hardening and hence of the evolution of bounding surfaces. Also if there are difficulties in predicting reliably the response to 5 cycles of relatively slow loading in a drained test, it is difficult to

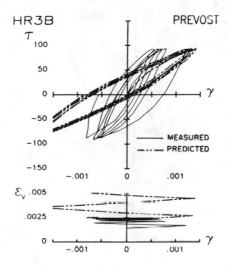

Figure 53. Prediction of Stress-Strain Loops and Volumetric Strains
by Prevost

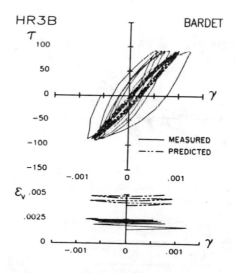

Figure 54. Prediction of Stress-Strain Loops and Volumetric Strains
by Bardet

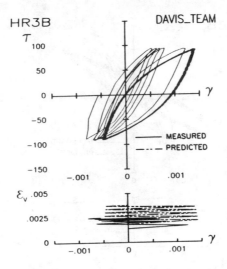

Figure 55.   Prediction of Stress-Strain Loops and Volumetric Strains
             by Davis-Team

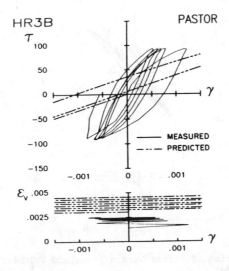

Figure 56.   Prediction of Stress-Strain Loops and Volumetric Strains
             by Pastor

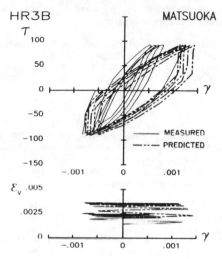

Figure 57. Prediction of Stress-Strain Loops and Volumetric Strains by Matsuoka

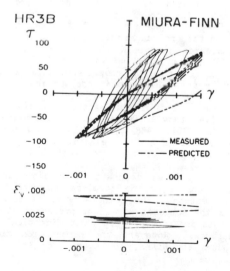

Figure 58. Prediction of Stress-Strain Loops and Volumetric Strains by Miura-Finn

have confidence in predictions of porewater pressures in undrained
tests at high frequency seismic loading.

Full appreciation of the findings of the Cleveland workshop must
await the publication of the proceedings which are expected to contain
the predictors' assessments and explanations of how their models
behaved. But on the basis of the predictions alone given in the preprint volume, it is hard to escape the conclusion that despite their
theoretical generality, the quality of the predictions of the elastic
plastic methods are strongly path dependent. The predictions are good
for loading paths close to those used to calibrate the models. But for
paths far removed from these, such as the cyclic torsional shear and
circular loading paths, the predictions were disappointing. This was
also the writer's experience with the Miura-Finn multi-mechanism model.
Despite obtaining a satisfactory fit with the calibration data, the
predictions for cyclic loading were poor.

Path dependent calibration of model parameters is inconsistent
with the theoretical generality of the elastic-plastic models. If
calibration along stress paths similar to those expected in the field
should be found necessary for satisfactory predictions of response then
much simpler models might prove adequate at much less cost.

Direct Comparison of Nonlinear and Elastic-Plastic Models

A consortium of 18 Japanese construction firms have pooled
resources to develop an industrially applicable computer program DIANA
for static and dynamic analysis of soils and soil-structure interaction
systems (Kawai, 1985). The program can treat fluid saturated porous
media as either 2-phase or 1-phase materials. Among the constitutive
models incorporated in the program are the generalized plasticity model
of Pastor and Zienkiewicz (1986) which was used by Pastor to make
predictions (Fig. 56) at the Cleveland workshop. Also incorporated is
an elastic-plastic multi-mechanism model with some features of the
Matsuoka et al (1986) model, also used in the workshop (Fig. 57).

Sato Kogyo of Japan, one of the partners in the Diana project,
conducted a comparative study of TARA-3 and DIANA using 1-D and 2-D
test problems (Yoshida, 1987, 1988). The DIANA analyses used coupled
fluid-soil skeleton equations and an elastic-plastic multi-mechanism
model. The analyses were conducted on the UNIVAC 1100/61 II computer.

The response of the level saturated site shown in Fig. 59 was
computed by both programs using the acceleration record shown in the
same figure as input motion. The computed surface accelerations are
shown in Fig. 60 and the computed porewater pressures at mid-depth are
shown in Fig. 61. Both programs predict almost identical accelerations
and porewater pressures. However, the computation times are very
different; 35 minutes for DIANA, 38 seconds for TARA-3, for a time
ratio of about 55:1.

The seismic test on the centrifuge model of an embedded structure
described earlier (Fig. 23) was also analyzed on the UNIVAC 1100 by
both programs. The results obtained from the TARA analysis are similar

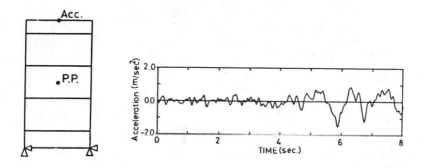

Figure 59. Shake Table Model and Input Accelerations

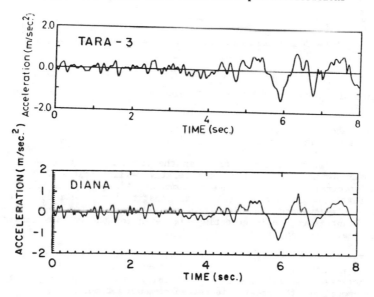

Figure 60. Surface Accelerations of Layer Computed by DIANA and TARA-3

to those described earlier. Analysis for the full 10 seconds of input motion took 2 hours. Selected acceleration results from the DIANA analysis are shown in Fig. 62. Porewater pressure results are shown for a location near the structure (PPT 2846) in Fig. 63 and for a location under the structure (PPT 2338) in Fig. 64.

All DIANA analyses were stopped after 2 seconds of shaking because the time for computation had already exceeded 18 hours. This would

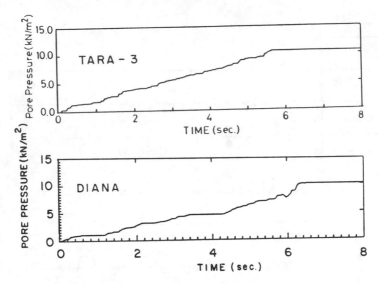

Figure 61.   Measured and Computed Porewater Pressures in Sand Layer.

suggest about 90 hours to process the entire 10 seconds of shaking.
For this example the time ratio was 45:1.

The relative computing times in the comparative study should not
be taken as an indication of the relative efficiencies of DIANA and
TARA-3 programs. The difference in computing times is primarily due to
the relative complexities of the two constitutive models used in the
study. Improvements in DIANA are underway which promise to make a
significant reduction in computing time even for the complex constitutive model (Yoshida, 1988; Shiomi 1988).

For level site responses, the more complex elastic-plastic model
offered no advantages over the simpler model in TARA-3 in return for
the increased complexity, greater difficulty in defining the parameters
of the model and the greatly increased computing time.

Results from the DIANA analysis of the soil-structure interaction
problem are not extensive enough to make final judgements about the
effectiveness of the constitutive model for this type of problem. Over
the 2s of response available, it appears that the accelerations are
simulated with satisfactory accuracy, while the porewater pressures are
low. However the computed porewater pressures compare poorly with the
measured porewater pressures over the 2s of data. Especially surprising is the tendency to negative pressures which indicates that the
model is predicting a significant expansion of the soil structure
during the simulated earthquake. The major benefit expected from using
coupled equations for fluid and soil structure is the ability to model

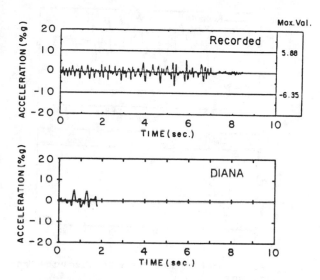

Figure 62. Measured and Computed Vertical Accelerations of the
Structure.

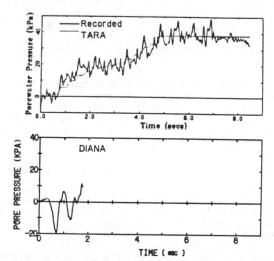

Figure 63. Measured and Computed Porewater Pressures Adjacent to the
Structure.

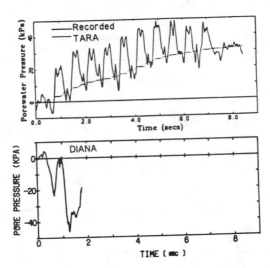

Figure 64.   Measured and Computed Porewater Pressures Under the
Structure.

adequately both the transient and residual porewater pressures.   TARA-3
which does not couple the responses of fluid and soil only predicts the
residual porewater pressures which are shown by the dotted lines in
Figs. 63 and 64.   For this problem the performance of the more
comprehensive model is clearly not as good as the simpler nonlinear
hysteretic model.

The data presented at the Cleveland workshop show how difficult it
is to predict volumetric strains accurately even for relative few
cycles of slow loading.   When coupled equations for soil and fluid are
used, the computed porewater pressures are sensitive to errors in
computed volumetric strains.   Concern about this sensitivity led the
writer to base the DESRA-2 and TARA-3 analyses on uncoupled equations.

CONCLUSIONS

A hierarchy of constitutive models are available for analysis of
the dynamic response of soil structures and soil-structure interaction
systems.   The models range from the relatively simple equivalent linear
model to complex elastic-plastic models.

The equivalent linear model is based on strain dependent moduli
and viscous damping ratios.   These properties are familiar to
practising engineers and standard laboratory procedures are available
for their measurement.   A large body of data is available in the

technical literature on these properties. The model is used in conjunction with iterative elastic analysis, a method which is widely understood. Because of their familiarity with the method of analysis and the properties used in it, practising engineers find the equivalent linear method relatively easy to use.

Since the method is elastic and is based on total stresses, it does not compute porewater pressures or permanent deformations. These are inferred from laboratory tests in which static and dynamic stress conditions in the field are simulated. Since the effects of porewater pressures are not included in the analysis, the computed stresses and accelerations may be overestimated.

The programs based on the equivalent linear method are computationally efficient and make it feasible and convenient to use in parametric studies as an aid to design.

The direct nonlinear method uses low strain moduli and shear strength as model parameters. These may be determined using standard laboratory test procedures, may be measured directly in situ or derived from correlations with in situ index tests such as the SPT or CPT. In analysis, the method tracks the hysteretic stress-strain paths continuously during cyclic loading and unloading and uses tangent moduli to estimate stresses and strains. Permanent deformations including the ratchetting effects typical of cyclic loading can be computed directly. The method includes a porewater pressure generation model which permits effective stress analysis. This model can operate with either laboratory data from cyclic loading tests or in situ estimates of liquefaction potential based on SPT or CPT tests.

The direct nonlinear method of analysis has successfully simulated response to cyclic loading in element tests and the dynamic response of a wide range of centrifuged models including dry and saturated embankments with and without surface and embedded structures. It has also been successful in simulating response in the field. Thus it has been validated not just on element tests with homogeneous stress and strain fields but in the highly inhomogeneous stress fields associated with soil-structure interaction problems.

The effective stress nonlinear method and the total stress equivalent linear method make comparable demands on computational time.

Elastic-plastic models are complex and incorporate parameters not familiar to many practising engineers. The models treat saturated soil as a 2-phase material using coupled equations for the soil and water phases. The coupled equations and the more complex constitutive models make very heavy demands on computing time. Comparative studies involving the elastic-plastic models and the direct nonlinear method raise questions about the cost-effectiveness of the elastic-plastic models at their present state of evolution. The increased cost and complexity do not seem to result in correspondingly better predictions of dynamic response.

Validation studies of the elastic-plastic models suggest that despite their theoretical generality the quality of predictions of response seems to be strongly path dependent. When loading paths are similar to the stress paths used in calibrating the models, the predictions are good. As the loading path deviates from the calibration path, the predictions become less reliable. In particular, the usual method of calibrating these models using data from static compression and extension tests does not seem adequate to ensure reliable estimates of dynamic response for the shear loading paths that are important in many kinds of seismic response studies. It is recommended that calibration studies of the elastic-plastic models for dynamic response should include cyclic torsional shear and/or simple shear stress paths.

All the methods discussed above are path dependent. In the equivalent of linear method, path dependency is introduced through the selection laboratory test conditions to be equivalent to field conditions for estimating porewater pressure and deformations. In the direct nonlinear method, path dependency is also introduced through the manner in which the liquefaction potential or porewater pressure development is determined. In the elastic-plastic models, it arises through the process of calibration. From the simplest model to the most complex, stress path dependency of the quality of response predictions is an obstacle that cannot be avoided at the present stage in the evolution of constitutive models for soils.

No one constitutive model can be recommended for all case. More comparative studies such as that between DIANA and TARA-3 are needed to clarify the relative advantages and disadvantages of the various methods. More validation studies using centrifuge tests are needed to test the robustness and reliability of models in inhomogeneous stress fields. In the meantime, the practising engineer should be well aware of the physical basis and assumptions of any model he uses and the kind of validation testing to which it has been subjected. He should not be beguiled by either the charm of simplicity or the glamour of sophistication in selecting a model for use.

ACKNOWLEDGEMENTS

Critical discussions with P.M. Byrne, J.T. Christian, P.V. Lade, D.J. Naylor, G.N. Pande, M. Pastor, H.B. Seed, T. Shiomi, I.M. Smith, Y.P. Vaid, M. Yogendrakumar, N. Yoshida, T.L. Youd were very helpful to the writer in developing the themes in this paper. Similar discussions with J.P. Bardet, Y.F. Dafalias, and J.H. Prevost at the invited symposium conducted by the Canadian Oil and Gas Lands Administration in Ottawa in July 1987 were extremely helpful in understanding the internal mechanics and operational modes of elastic-plastic modelling. Special thanks are due to M. Yogendrakumar for assistance with all aspects of the production of this review. The text was typed by Ms. Kelly Lamb.

REFERENCES

Bardet, J.P. (1986). Bounding Surface Plasticity Model for Sands,
Journal of the Geotechnical Engineering Division, ASCE, Vol. 112, No.
11, November, pp. 1198-1217.

Biot, M.A. (1941). General Theory for Three Dimensional Consolidation,
Journal of Applied Physics, Vol. 12, pp. 155-164.

Carter, J.P., Booker, R. and Wroth, C.P. (1982). A Critical State Soil
Model for Cyclic Loading, in Soil Mechanics - Transient and Cyclic
Loads. Editors: G.N. Pande and O.C. Zienkiewicz, John Wiley and Sons,
New York, N.Y.

Castro, G., Poulos, S.J. and Leathers, F.D. (1985). Re-examination of
Slide of Lower San Fernando Dam, Journal of Geotechnical Engineering
Division, ASCE, Vol. 111, No. 9, September, pp. 1093-1107.

Chopra, A.K. (1966). Earthquake Effects on Dams, Ph.D. Thesis,
University of California, Berkeley.

Dafalias, Y.F. and Hermann, L.R. (1982). Bounding Surface Formulation
of Soil Plasticity, Soil Mechanics - Transient and Cyclic Loads, G.
Pande and O.C. Zienkiewicz, Eds., John Wiley & Sons, Inc., London,
U.K., pp. 253-282.

Desai, C.S., Krempl, E., Kioussis, P.D. and Kundu, T. (1987). Editors.
Proceedings of the Second International Conference on Constitutive Laws
for Engineering Materials, Theory and Applications, Tucson, Arizona,
January 5-8, 2 Volumes.

Dikmen, S.U. and Ghaboussi, J. (1984). Effective Stress Analysis of
Seismic Response and Liquefaction, Theory, Journal of the Geotech.
Eng. Div., ASCE, Vol. 110, No. 5, Proc. Paper 18790, pp. 628-644.

Finn, W.D. Liam. (1985). Dynamic Effective Stress Response of Soil
Structures; Theory and Centrifugal Model Studies, Proc. 5th Int. Conf.
on Num. Methods in Geomechancis, Nagoya, Japan, Vol. 1, 35-36.

Finn, W.D. Liam. (1988). Nonlinear Dynamic Analysis of Soil-Structure
Interaction Systems, Report to European Research Office of the U.S.
Army, London, England, by Cork Geotechnics Ltd., (in press).

Finn, W.D. Liam and Bhatia, S.K. (1980). Verification of Nonlinear
Effective Stress Model in Simple Shear, Application of Plasticity and
Generalized Stress-Strain in Geotechnical Engineering, ASCE, Editors,
R.N. Yong and E.T. Selig, pp. 241-252.

Finn, W.D. Liam, Iai, S. and Ishihara, K. (1982). Performance of
Artificial Offshore Islands Under Wave and Earthquake Loading, Field
Data Analyses, Proceedings of the 14th Annual Offshore Technology
Conference, Houston, Texas, May 3-6, Vol. 1, OTC Paper 4220, pp. 661-
672.

Finn, W.D. Liam, Lee, W.K. and Martin, G.R. (1976). An Effective Stress Model for Liquefaction, Journal of the Geotechnical Engineering Division, ASCE, Vol. 103, No. GT6, Proc. Paper 13008, 517-533.

Finn, W.D. Liam, Lo, R.C. and Yogendrakumar, M. (1987). Dynamic Nonlinear Analysis of Lukwi Tailings Dam, Soil Dynamics Group, Dept. of Civil Engineering, University of British Columbia, Vancouver, B.C., Canada.

Finn, W.D. Liam, Martin, G.R. and Lee, K.W. (1978). Comparison of Dynamic Analysis of Saturated Sands, Proc. ASCE Geotechnical Engineering Division, Specialty Conference on Earthquake Engineering and Soil Dynamics, Pasadena, California, June 19-21, pp. 472-491.

Finn, W.D. Liam, R. Siddharthan, F. Lee and A.N. Schofield. (1984). Seismic Response of Offshore Drilling Islands in a Centrifuge Including Soil-Structure Interaction. Proc, 16th Annual Offshore Technology Conf, Houston, Texas, OTC Paper 4693.

Finn, W.D. Liam, R.S. Steedman, M. Yogendrakumar, and R.H. Ledbetter. (1985). Seismic Response of Gravity Structures in a Centrifuge, Proc. 17th Annual Offshore Tech. Conf., Houston, Texas, OTC Paper 4885, 389-394.

Finn, W.D. Liam, Yogendrakumar, M., Yoshida, N. and Yoshida, H. (1986). TARA-3: A Program to Compute the Response of 2-D Embankments and Soil-Structure Interaction Systems to Seismic Loadings, Dept. of Civil Engineering, University of British Columbia, Vancouver, Canada.

Finn, W.D. Liam, Yogendrakumar, M., Lo, R.C. and Yoshida, N. (1988). Direct Computation of Permanent Seismic Deformations, Proceedings of the 9th World Conference on Earthquake Engineering, Tokyo and Kyoto, Japan, August (to appear).

Goodman, R.E., Taylor, R.L. and Brekke, T.L. (1968). A Model for the Mechanics of Jointed Rock, Journal of the Soil Mechanics and Foundation Division, ASCE, May, pp. 637-659.

Hushmand, B., Crouse, C.B., Martin, G. and Scott, R.F. (1987). Site Response and Liquefaction Studies Involving the Centrifuge: Structures and Stochastic Methods, Developments in Geotechnical Engineering 45, Editor: A.S. Cakmak, pp. 3-24.

Iai, S., Tsuchida, H. and Finn, W.D. Liam (1985). An Effective Stress Analysis of Liquefaction at Ishinomaki Port During the 1978 Miyagi-Ken-Oki Earthquake, Report of the Port and Harbour Research Institute, Vol. 24, No. 2, pp. 1-84.

Idriss, I.M. (1985). Evaluating Seismic Risk in Engineering Practice, Proceedings of the XI International Conference on Soil Mechanics and Foundation Engineering, San Francisco, Vol. 1, pp. 255-320.

Idriss, I.M., Lysmer, J., Hwang, R. and Seed, H.B. (1983), QUAD-4: A Computer Program for Evaluating the Seismic Response of Soil-Structures by Variable Damping Finite Element Procedures, Report No. EERC 73-16, Earthquake Engineering Research Center, University of California, Berkeley, California, July.

Ishihara, K. and Towhata, I. (1982). Dynamic Response Analysis of Level Ground Based on the Effective Stress Method, in Soil Mechanics - Transient and Cyclic Loads, Edited by G.N. Pande and O.C. Zienkiewicz, John Wiley and Sons, Ltd., new York, pp. 133-172.

Ishihara, K. Shimizu and Yasuda, Y. (1981). Porewater Pressure Measured in Sand Deposits During an Earthquake, Soils and Foundations, Vol. 21, No. 4, December, pp. 85-100.

Ishihara, K. and Yamazaki, F. (1980). Cyclic Simple Shear Tests on Saturated Sand in Multi-Directional Loading, Soils and Foundations, Vol. 20, No. 1, March, pp. 45-60.

Kawai, T. (1985). Summary Report on the Development of the Computer Program "DIANA - Dynamic Interaction Approach and Non-Linear Analysis," Science University of Tokyo.

Kawamoto, T. and Ichikawa, Y. (1985). Editors: Numerical Methods in Geomechanics Nagoya 1985, Proceedings of the Fifth International Conference in Numerical Methods in Geomechanics, Nagoya, April 1-5, 4 Volujes. A.A. Balkema Rotterdam.

Lee, M.K.W. and Finn, W.D.L. (1978). DESRA-2, Dynamic Effective Stress Response Analysis of Soil Deposits with Energy Transmitting Boundary Including Assessment of Liquefaction Potential, Soil Mechanics Series, No. 38, Dept. of Civil Engineering, University of British Columbia, Vancouver, B.C.

Lysmer, J., Udaka, T., Tsai, C.F. and Seed, H.D. (1975). FLUSH: A Computer Program for Approximate 3-D Analysis of Soil-Structure Interaction Problems. Report No. EERC 75-30, Earthquake Engineering Research Center, University of California, Berkeley, California.

Makdisi, F.I. and Seed, H.B. (1978). A Simplified Procedure for Estimating Dam and Embankment Earthquake-Induced Deformations, Journal of Geotechnical Engineering Division, ASCE, Vol. 104, No. GT7, July, pp. 849-867.

Martin, P.P. and Seed, H.B. (1978). MASH - A Computer Program for the Nonlinear Analysis of Vertically Propagating Shear Waves in Horizontally Layered Soil Deposits, EERC Report No. UCB/EERC-78/23, University of California, Berkeley, California, October.

Martin, G.R., Finn, W.D.Liam, and Seed, H.B. 1975). Fundamentals of Liquefaction Under Cyclic Loading, Soil Mech. Series Rpt. No. 23, Dept. of Civil Engineering, University of British Columbia, Vancouver; also Proc. Paper 11284, J. Geotech. Eng. Div. ASCE, Vol. 101, No. GT5, 324-438.

Masing, G. (1926). Eigenspannungen und Verfestigung beim Messing, Proceedings, 2nd Int. Congress of Applied Mechanics, Zurich, Switzerland.

Mroz, Z. (1967). On the Description of Anisotropic Workhardening, Journal of the Mechanics and Physics of Solids, Vol. 15, pp. 163-175.

Murayama, S. (1985). Constitutive Laws of Soils, Japanese Society of Soil Mechanics and Foundation Engineering, Tokyo, 130 pp.

National Research Council of the United States (1982). Earthquake Engineering - 1982, Report by Committee on Earthquake Engineering Research, National Academy Press, Washington, D.C.

Newmark, N.M. (1965). Effects of Earthquake on Dams and Embankments, 5th Rankine Lecture, Geotechnique 15, No. 2, pp. 139-160.

Pande, G.N. and Van Impe, W.F. (1986). Editors. Numerical Models in Geomechanics, Proceedings of the Second International Confrence, Ghent, Belgium, March 31 - April 4. M.Y. Jackson & Son, Redruth, England.

Pande, G.N. and Zienkiewicz, O.C. (1982). Editors: Soil Mechanics - Transient and Cyclic Loads. John Wiley & Sons, Ltd., New York, N.Y.

Pastor, M. and Zienkiewicz, O.C. (1986). A Generalized Plasticity, Hierarchical Model for Sand Under Monotonic and Cyclic Loading, 2nd Int. Symposium on Numerical Models in Geomechanics, Eds. G.N. Pande and W.F. Van Impe, Ghent.

Poulos, S.J. (1981) The Steady State of Deformation, Journal of the Geotechnical Engineering Division, ASCE, Vol. 107, No. GT5, May, pp. 513-562.

Poulos, S.J., Castro, G. and France, J.W. (1985). Liquefaction Evaluation Procedure: Journal of Geotechnical Engineering, ASCE, Vol. 111, No. GT6, June, pp. 772-795.

Prager, W. (1955). The Theory of Plasticity: A Survey of Recent Achievement, Proceedings, Inst. Mech. Eng., London, Vol. 169, pp. 41-57.

Prevost, J.H. (1978). Anisotropic Undrained Stress-Strain Behaviour of Clays, Journal of the Geotechnical Engineering Division, ASCE, Vol. 104, No. GT8, August, pp. 1975-1090.

Prevost, J.H. (1981). DYNAFLOW: A Nonlinear Transient Finite Element Analysis Program, Princeton University, Department of Civil Engineering, Princeton, N.J.

Pyke, R.M., Chan, C.K. and Seed, H.B. (1974). Settlement and Liquefaction of Sands Under Multi-Direction Shaking, Report No. EERC 74-2, Earthquake Engineering Research Center, University of California, Berkeley, February.

Ramberg, W. and Osgood, W.R. (1943). Description of Stress-Strain Curves by Three Parameters, National Advis. Com. Aeronaut., Tech. Note 902, Washington, D.C.

R.H.G.P. (1985). Review of Soil Mechanics - Transient and Cyclic Loads, Editors: G.N. Pande and O.C. Zienkiewicz in Ground Engineering.

Roscoe, K.H. and Burland, J.B. (1968). On the Generalized Stress-Strain Behaviour of Wet Clay, Engineering Plasticity, J. Heyman and F. Leckie, eds., Cambridge University Press, Cambridge, England, pp. 535-609.

Roscoe, K.H., Schofield, A.N. and Wroth, C.P. (1958). On the Yielding of Soils, Geotechnique, Vol. 9, pp. 22-53.

Saada, A. and Bianchini, G.S. (1987). Editors: Proceedings of the International Workshop on Constitutive Equations for Granular Non-Cohesive Soils, Case Western Reserve University, Cleveland, Ohio, July 22-24, to be published by A.A. Balkema.

Schnabel, P.B., Lysmer, J. and Seed, H.B. (1972). SHAKE: A Computer Program for Earthquake Response Analysis of Horizontally Layered Sites, Report No. EERC 72-12, Earthquake Engineering Research Center, University of California, Berkeley, California.

Scott, R.F. (1985). Plasticity and Constitutive Relations in Soil Mechanics, Journal of the Geotechnical Engineering Division, ASCE, Vol. III, No. 5, May, pp. 563-605.

Seed, H.B. (1979a). Considerations in the Earthquake-Resistant Design of Earth and Rockfill Dams, 19th Rankine Lecture, Geotechnique 29, No. 3, pp. 215-263.

Seed, H.B. (1979b). Soil Liquefaction and Cyclic Mobility Evaluation for Level Ground During Earthquakes, Journal of Geotechnical Engineering Division, ASCE, Vol. 105, No. GT2, pp. 201-255.

Seed, H.B. (1987). Design Problems in Soil Liquefaction, Journal of Geotechnical Engineering, ASCE, Vol. 113, No. 7, August, pp. 827-845.

Seed, H.B. (1988). Private Communication.

Seed, H.B. and Idriss, I.M. (1967). Analysis of Soil Liquefaction: Nigatta Earthquake, J. Soil Mechanics and Foundations Division, ASCE, Vol. 93, No. SM3, pp. 83-108.

Seed, H.B. and Idriss, I.M. (1970). Soil Moduli and Damping Factors for Dynamic Response Analyses, Report No. EERC 70-10, Earthquake Engineering Research Center, University of California, Berkeley, California, December.

Seed, H.B., Idriss, I.M. and Arango, I. (1983). Evaluation of Liquefaction Potential Using Field Performance Data, Journal of Geotechnical Engineering Division, Vol. 109, No. 3, pp. 458-482.

Seed, H.B. and Lee, K.L. (1966). Liquefaction of Saturated Sands During Cyclic Loading, Journal of the Soil Mechanics and Foundation Engineering Division, ASCE, Vol. 92, No. SM6, November.

Seed, H.B., Lee, K.L., Idriss, I.M. and Makdisi, F. (1973). Analysis of the Slides in the San Fernando Dams During the Earthquake of Feb. 9, 1971, Report No. EERC 73-2, Earthquake Engineering Research Center, University of California, Berkeley, June.

Seed, H.B., Tokimatsu, K., Harder, L.F. and Chung, R.M. (1985). Influence of SPT Procedures in Soil Liquefaction Resistance Evaluations, Journal of Geotechnical Engineering Division, ASCE, Vol. 111, No. GT12, December, pp. 1425-1445.

Seed, H.B., Wong, R.T., Idriss, I.M. and Tokimatsu, K. (1986). Moduli and damping Factors for Dynamic Analyses of Cohesionless Soils, Journal of the Geotechnical Engineering Division, ASCE, Vol. 112, No. 11, November, pp. 1016-1032.

Serff, N., Seed, H.B., Makdisi, F.I. and Chang, C.Y. (1976). Earthquake Induced Deformations of Earth Dams, Report No. EERC 76-4, Earthquake Engineering Research Center, University of California, Berkeley, September.

Shiomi, T. (1988). Private Communication.

Siddharthan, R. and W.D. Liam Finn. (1982). TARA-2, Two dimensional Nonlinear Static and Dynamic Response Analysis, Soil Dynamics Group, University of British Columbia Vancouver, Canada.

Steedman, R.S. (1986). Embedded Structure on Sand Foundation: Data Report of Centrifuge Model Tests, RSS110 and RSS111. Engineering Department, Cambridge University, Cambridge, England.

Streeter, V.L., Wylie, E.B. and Richart, F.E. (1973). Soil Motion Computations by Characteristics Methods, ASCE National Structural Engineering Meeting, San Francisco, California, Preprint 1952, April.

Vaid, Y.P. and Chern, J.C. (1981). Effect of Static Shear on Resistance to Liquefaction, Soil Mechanics Series No. 51, Department of Civil Engineering, University of British Columbia, Vancouver, Canada.

Vaid, Y.P. and Finn, W.D. Liam (1979). Effect of Static Shear on Liquefaction Potential, Journal of the Geotechnical Engineering Division, ASCE, Vol. 105, GT10, October, pp. 1233-1246.

Yogendrakumar, M. and W.D. Liam Finn. (1986a). SIMCYC2: A Program for Simulating Cyclic Simple Shear Tests on Dry or Saturated Sands, Report, Soil Dynamics Group, Dept. of Civil Engineering, University of British Columbia, Vancouver, Canada.

Yogendrakumar, M. and Finn, W.D. Liam (1986b). C-PRO: A Program for Evaluating the Constants in the Martin-Finn-Seed Porewater Pressure Model", Soil Dynamics Group, Department of Civil Engineering, University of British Columbia, Vancouver, Canada, November.

Yong, R.K. and Ko, H.Y. (1981). Editors: Proceedings of the Workshop on Limit Equilibrium, Plasticity and Generalized Stress-Strain in Geotechnical Engineering, Montreal, Canada.

Yoshida, N. (1987). Oral Presentation at International Symposium on Centrifuge Test Data and Evaluation of Numerical Modelling, Cambridge University, July 1-3.

Yoshida, N. (1988). Private Communication.

Youd, L. (1978). Editor: Proceedings of the ASCE Geotechnical Engineering Division Specialty Conference on Earthquake Engineering and Soil Dynamics, Pasadena, California, June 19-21.

AUTHOR INDEX
Page number refers to first page of paper.